Principles of Food Chemistry
Third Edition

John M. deMan, PhD

Professor Emeritus
Department of Food Science
University of Guelph
Guelph, Ontario

A Chapman & Hall Food Science Book

AN ASPEN PUBLICATION®
Aspen Publishers, Inc.
Gaithersburg, Maryland
1999

The author has made every effort to ensure the accuracy of the information herein. However, appropriate information sources should be consulted, especially for new or unfamiliar procedures. It is the responsibility of every practitioner to evaluate the appropriateness of a particular opinion in the context of actual clinical situations and with due considerations to new developments. The author, editors, and the publisher cannot be held responsible for any typographical or other errors found in this book.

Aspen Publishers, Inc., is not affiliated with the American Society of Parenteral and Enteral Nutrition.

Library of Congress Cataloging-in-Publication Data

deMan, John M.
Principles of food chemistry/
John M. deMan.—3rd ed.
p. cm.
Includes bibliographical references and index.
ISBN 0-8342-1234-X
1. Food—Composition. I. Title.
TX531.D43 1999
664—dc21
98-31467
CIP

About Aspen Publishers • For more than 35 years, Aspen has been a leading professional publisher in a variety of disciplines. Aspen's vast information resources are available in both print and electronic formats. We are committed to providing the highest quality information available in the most appropriate format for our customers. Visit Aspen's Internet site for more information resources, directories, articles, and a searchable version of Aspen's full catalog, including the most recent publications: **http://www.aspenpublishers.com**
Aspen Publishers, Inc. • The hallmark of quality in publishing
Member of the worldwide Wolters Kluwer group.

Editorial Services: Jane Colilla
Library of Congress Catalog Card Number: 98-31467
ISBN: 0-8342-1234-X
Printed in the United States of America

1 2 3 4 5

Table of Contents

Preface

This book was designed to serve as a text for courses in food chemistry in food science programs following the Institute of Food Technologists minimum standards. The original idea in the preparation of this book was to present basic information on the composition of foods and the chemical and physical characteristics they undergo during processing, storage, and handling. The basic principles of food chemistry remain the same, but much additional research carried out in recent years has extended and deepened our knowledge. This required inclusion of new material in all chapters. The last chapter in the second edition, Food Additives, has been replaced by the chapter Additives and Contaminants, and an additional chapter, Regulatory Control of Food Composition, Quality, and Safety, has been included. This last chapter is an attempt to give students some understanding of the scientific basis of the formulation of laws and regulations on food and on the increasing trend toward international harmonization of these laws. A number of important food safety issues have arisen recently, and these have emphasized the need for comprehensive and effective legal controls.

In the area of water as a food component, the issue of the glass transition has received much attention. This demonstrates the important role of water in food properties. Lipids have received much attention lately mainly because of publicity related to nutritional problems. Sutructured lipids, low caloric fats, and biotechnology have received a good deal of attention. Our understanding of the functionality of proteins expands with increasing knowledge about their composition and structure. Carbohydrates serve many functions in foods, and the noncaloric dietary fiber has assumed an important role.

Color, flavor, and texture are important attributes of food quality, and in these areas, especially those of flavor and texture, great advances have been made in recent years. Enzymes are playing an ever increasing part in the production and transformation of foods. Modern methods of biotechnology have produced a gamut of enzymes with new and improved properties.

In the literature, information is found using different systems of units: metric, SI, and the English system. Quotations from the literature are presented in their original form. It would be difficult to change all these units in the book to one system. To assist the reader in converting these units, an appendix is provided with conversion factors for all units found in the text.

It is hoped that this new edition will continue to fulfill the need for a concise and relevant text for the teaching of food chemistry. I express gratitude to those who have provided comments and suggestions for improvement, and especially to my wife, Leny, who has provided a great deal of support and encouragement during the preparation of the third edition.

CHAPTER 1

Water

Water is an essential constituent of many foods. It may occur as an intracellular or extracellular component in vegetable and animal products, as a dispersing medium or solvent in a variety of products, as the dispersed phase in some emulsified products such as butter and margarine, and as a minor constituent in other foods. Table 1–1 indicates the wide range of water content in foods.

Because of the importance of water as a food constituent, an understanding of its properties and behavior is necessary. The presence of water influences the chemical and microbiological deterioration of foods. Also, removal (drying) or freezing of water is essential to some methods of food preservation. Fundamental changes in the product may take place in both instances.

PHYSICAL PROPERTIES OF WATER AND ICE

Some of the physical properties of water and ice are exceptional, and a list of these is presented in Table 1–2. Much of this information was obtained from Perry (1963) and Landolt-Boernstein (1923). The exceptionally high values of the caloric properties of water are of importance for food processing

Table 1–1 Typical Water Contents of Some Selected Foods

Product	Water (%)
Tomato	95
Lettuce	95
Cabbage	92
Beer	90
Orange	87
Apple juice	87
Milk	87
Potato	78
Banana	75
Chicken	70
Salmon, canned	67
Meat	65
Cheese	37
Bread, white	35
Jam	28
Honey	20
Butter and margarine	16
Wheat flour	12
Rice	12
Coffee beans, roasted	5
Milk powder	4
Shortening	0

operations such as freezing and drying. The considerable difference in density of water

Table 1–2 Some Physical Properties of Water and Ice

Water	Temperature (°C)					
	0	20	40	60	80	100
Vapor pressure (mm Hg)	4.58	17.53	55.32	149.4	355.2	760.0
Density (g/cm^3)	0.9998	0.9982	0.9922	0.9832	0.9718	0.9583
Specific heat (cal/g°C)	1.0074	0.9988	0.9980	0.9994	1.0023	1.0070
Heat of vaporization (cal/g)	597.2	586.0	574.7	563.3	551.3	538.9
Thermal conductivity (kcal/m^2 h°C)	0.486	0.515	0.540	0.561	0.576	0.585
Surface tension (dynes/cm)	75.62	72.75	69.55	66.17	62.60	58.84
Viscosity (centipoises)	1.792	1.002	0.653	0.466	0.355	0.282
Refractive index	1.3338	1.3330	1.3306	1.3272	1.3230	1.3180
Dielectric constant	88.0	80.4	73.3	66.7	60.8	55.3
Coefficient of thermal expansion × 10^{-4}	—	2.07	3.87	5.38	6.57	—

Ice	Temperature (°C)						
	0	−5	−10	−15	−20	−25	−30
Vapor pressure (mm Hg)	4.58	3.01	1.95	1.24	0.77	0.47	0.28
Heat of fusion (cal/g)	79.8	—	—	—	—	—	—
Heat of sublimation (cal/g)	677.8	—	672.3	—	666.7	—	662.3
Density (g/cm^3)	0.9168	0.9171	0.9175	0.9178	0.9182	0.9185	0.9188
Specific heat (cal/g°C)	0.4873	—	0.4770	—	0.4647	—	0.4504
Coefficient of thermal expansion × 10^{-5}	9.2	7.1	5.5	4.4	3.9	3.6	3.5
Heat capacity (joule/g)	2.06	—	—	—	1.94	—	—

and ice may result in structural damage to foods when they are frozen. The density of ice changes with changes in temperature, resulting in stresses in frozen foods. Since solids are much less elastic than semisolids, structural damage may result from fluctuating temperatures, even if the fluctuations remain below the freezing point.

STRUCTURE OF THE WATER MOLECULE

The reason for the unusual behavior of water lies in the structure of the water molecule (Figure 1–1) and in the molecule's ability to form hydrogen bonds. In the water molecule the atoms are arranged at an angle

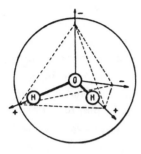

Figure 1–1 Structure of the Water Molecule

of 105 degrees, and the distance between the nuclei of hydrogen and oxygen is 0.0957 nm. The water molecule can be considered a spherical quadrupole with a diameter of 0.276 nm, where the oxygen nucleus forms the center of the quadrupole. The two negative and two positive charges form the angles of a regular tetrahedron. Because of the separation of charges in a water molecule, the attraction between neighboring molecules is higher than is normal with van der Waals' forces.

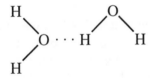

In ice, every H_2O molecule is bound by four such bridges to each neighbor. The binding energy of the hydrogen bond in ice amounts to 5 kcal per mole (Pauling 1960). Similar strong interactions occur between OH and NH and between small, strongly electronegative atoms such as O and N. This is the reason for the strong association in alcohols, fatty acids, and amines and their great affinity to water. A comparison of the properties of water with those of the hydrides of elements near oxygen in the Periodic Table (CH_4, NH_3, HF, DH_3, H_2S, HCl) indicates

that water has unusually high values for certain physical constants, such as melting point, boiling point, heat capacity, latent heat of fusion, latent heat of vaporization, surface tension, and dielectric constant. Some of these values are listed in Table 1–3.

Water may influence the conformation of macromolecules if it has an effect on any of the noncovalent bonds that stabilize the conformation of the large molecule (Klotz 1965). These noncovalent bonds may be one of three kinds: hydrogen bonds, ionic bonds, or apolar bonds. In proteins, competition exists between interamide hydrogen bonds and water-amide hydrogen bonds. According to Klotz (1965), the binding energy of such bonds can be measured by changes in the near-infrared spectra of solutions in *N*-methylacetamide. The greater the hydrogen bonding ability of the solvent, the weaker the C=O···H–N bond. In aqueous solvents the heat of formation or disruption of this bond is zero. This means that a C=O···H–N hydrogen bond cannot provide stabilization in aqueous solutions. The competitive hydrogen bonding by H_2O lessens the thermodynamic tendency toward the formation of interamide hydrogen bonds.

The water molecules around an apolar solute become more ordered, leading to a loss in entropy. As a result, separated apolar groups in an aqueous environment tend to

Table 1–3 Physical Properties of Some Hydrides

Substance	Melting Point (°C)	Boiling Point (°C)	Molar Heat of Vaporization (cal/mole)
CH_4	−184	−161	2,200
NH_3	− 78	− 33	5,550
HF	− 92	+ 19	7,220
H_2O	0	+100	9,750

associate with each other rather than with the water molecules. This concept of a hydrophobic bond has been schematically represented by Klotz (1965), as shown in Figure 1–2. Under appropriate conditions apolar molecules can form crystalline hydrates, in which the compound is enclosed within the space formed by a polyhedron made up of water molecules. Such polyhedrons can form a large lattice, as indicated in Figure 1–3. The polyhedrons may enclose apolar guest molecules to form apolar hydrates (Speedy 1984). These pentagonal polyhedra of water molecules are unstable and normally change to liquid water above 0 °C and to normal hexagonal ice below 0 °C. In some cases, the hydrates melt well above 30 °C. There is a remarkable similarity between the small apolar molecules that form these clathrate-like hydrates and the apolar side chains of proteins. Some of these are shown in Figure 1–4. Because small molecules such as the ones shown in Figure 1–4 can form stable water cages, it may be assumed that some of

the apolar amino acid side chains in a polypeptide can do the same. The concentration of such side chains in proteins is high, and the combined effect of all these groups can be expected to result in the formation of a stabilized and ordered water region around the protein molecule. Klotz (1965) has suggested the term *hydrotactoids* for these structures (Figure 1–5).

SORPTION PHENOMENA

Water activity, which is a property of aqueous solutions, is defined as the ratio of the vapor pressures of pure water and a solution:

$$a_w = \frac{p}{p_o}$$

where

a_w = water activity

p = partial pressure of water in a food

p_o = vapor pressure of water at the same temperature

According to Raoult's law, the lowering of the vapor pressure of a solution is proportional to the mole fraction of the solute: a_w can then be related to the molar concentrations of solute (n_1) and solvent (n_2):

$$a_w = \frac{p}{p_o} = \frac{n_1}{n_1 + n_2}$$

The extent to which a solute reduces a_w is a function of the chemical nature of the solute. The equilibrium relative humidity (ERH) in percentage is ERH/100. ERH is defined as:

$$ERH = \frac{p^{equ}}{p^{sat}}$$

where

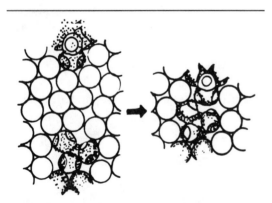

Figure 1–2 Schematic Representation of the Formation of a Hydrophobic Bond by Apolar Group in an Aqueous Environment. Open circles represent water. *Source*: From I.M. Klotz, Role of Water Structure in Macromolecules, *Federation Proceedings*, Vol. 24, Suppl. 15, pp. S24–S33, 1965.

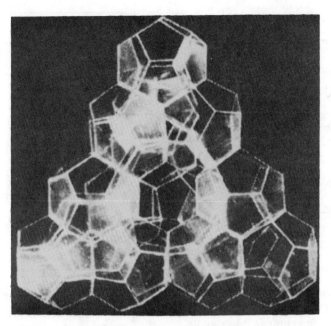

Figure 1–3 Crytalline Apolar Polyhedrons Forming a Large Lattice. The space within the polyhedrons may enclose apolar molecules. *Source*: From I.M. Klotz, Role of Water Structure in Macromolecules, *Federation Proceedings*, Vol. 24, Suppl. 15, pp. S24–S33, 1965.

Crystal Hydrate Formers	Amino Acid Side Chains	
CH_4	$-CH_3$	(Ala)
CH_2 with CH_3, CH_3	$-CH$ with CH_3, CH_3	(Val)
CH_3-CH with CH_3, CH_3	$-CH_2-CH$ with CH_3, CH_3	(Leu)
CH_3-SH	$-CH_2-SH$	(Cys)
CH_3-S-CH_3	$-CH_2-CH_2-S-CH_3$	(Met)
benzene ring	$-CH_2-$ benzene ring	(Phe)

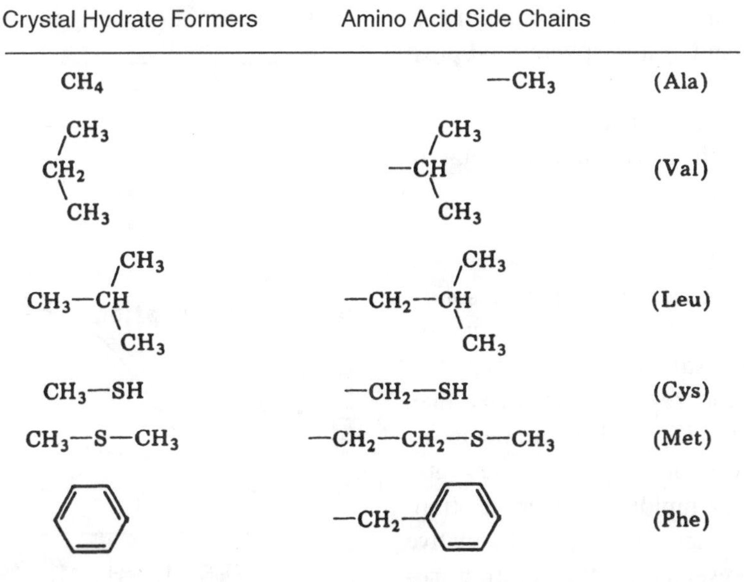

Figure 1–4 Comparison of Hydrate-Forming Molecules and Amino Acid Apolar Side Chains. *Source:* From I.M. Klotz, Role of Water Structure in Macromolecules, *Federation Proceedings*, Vol. 24, Suppl. 15, pp. S24–S33, 1965.

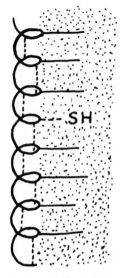

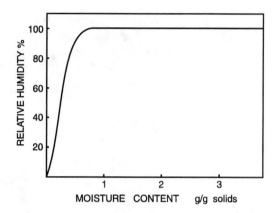

Figure 1–6 Water Activity in Foods at Different Moisture Contents

Figure 1–5 Hydrotactoid Formation Around Apolar Groups of a Protein. *Source*: From I.M. Klotz, Role of Water Structure in Macromolecules, *Federation Proceedings*, Vol. 24, Suppl. 15, pp. S24–S33, 1965.

p^{equ} = partial pressure of water vapor in equilibrium with the food at temperature T and 1 atmosphere total pressure

p^{sat} = the saturation partial pressure of water in air at the same temperature and pressure

At high moisture contents, when the amount of moisture exceeds that of solids, the activity of water is close to or equal to 1.0. When the moisture content is lower than that of solids, water activity is lower than 1.0, as indicated in Figure 1–6. Below moisture content of about 50 percent the water activity decreases rapidly and the relationship between water content and relative humidity is represented by the sorption isotherms. The adsorption and desorption processes are not fully reversible; therefore, a distinction can be made between the adsorption and desorption isotherms by determining whether a dry product's moisture levels are increasing, or whether the product's moisture is gradually lowering to reach equilibrium with its surroundings, implying that the product is being dried (Figure 1–7). Generally, the adsorption isotherms are required for the observation of hygroscopic products,

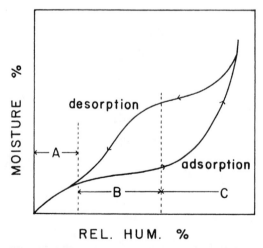

Figure 1–7 Adsorption and Desorption Isotherms

and the desorption isotherms are useful for investigation of the process of drying. A steeply sloping curve indicates that the material is hygroscopic (curve A, Figure 1–8); a flat curve indicates a product that is not very sensitive to moisture (curve B, Figure 1–8). Many foods show the type of curves given in Figure 1–9, where the first part of the curve is quite flat, indicating a low hygroscopicity, and the end of the curve is quite steep, indicating highly hygroscopic conditions. Such curves are typical for foods with high sugar or salt contents and low capillary adsorption. Such foods are hygroscopic. The reverse of this type of curve is rarely encountered. These curves show that a hygroscopic product or hygroscopic conditions can be defined as the case where a small increase in relative humidity causes a large increase in product moisture content.

Sorption isotherms usually have a sigmoid shape and can be divided into three areas that correspond to different conditions of the water present in the food (Figure 1–7). The

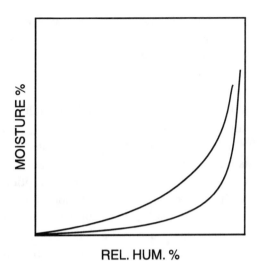

Figure 1–9 Sorption Isotherms for Foods with High Sugar or Salt Content; Low Capillary Adsorption

first part (A) of the isotherm, which is usually steep, corresponds to the adsorption of a monomolecular layer of water; the second, flatter part (B) corresponds to adsorption of additional layers of water; and the third part (C) relates to condensation of water in capillaries and pores of the material. There are no sharp divisions between these three regions, and no definite values of relative humidity exist to delineate these parts. Labuza (1968) has reviewed the various ways in which the isotherms can be explained. The kinetic approach is based on the Langmuir equation, which was initially developed for adsorption of gases and solids. This can be expressed in the following form:

$$\frac{a}{V} = \left[\frac{K}{bV_m}\right] + \frac{a}{V_m}$$

where

a = water activity
b = a constant

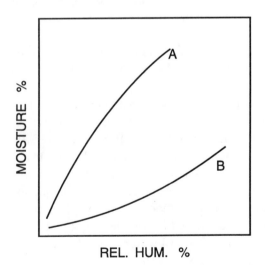

Figure 1–8 Sorption Isotherms of Hygroscopic Product (*A*) and Nonhygroscopic Product (*B*)

K = $1/p_o$ and p_o = vapor pressure of water at T_o

V = volume adsorbed

V_m = monolayer value

When a/V is plotted versus a, the result is a straight line with a slope equal to $1/V_m$ and the monolayer value can be calculated. In this form, the equation has not been satisfactory for foods, because the heat of adsorption that enters into the constant b is not constant over the whole surface, because of interaction between adsorbed molecules, and because maximum adsorption is greater than only a monolayer.

A form of isotherm widely used for foods is the one described by Brunauer et al. (1938) and known as the BET isotherm or equation. A form of the BET equation given by Labuza (1968) is

$$\frac{a}{(1-a)V} = \frac{1}{V_m C} + \left[\frac{a(C-1)}{V_m C}\right]$$

where

C = constant related to the heat of adsorption

A plot of $a/(1-a)V$ versus a gives a straight line, as indicated in Figure 1–10. The monolayer coverage value can be calculated from the slope and the intercept of the line. The BET isotherm is only applicable for values of a from 0.1 to 0.5. In addition to monolayer coverage, the water surface area can be calculated by means of the following equation:

$$S_o = V_m \cdot \frac{1}{M_{H_2O}} \cdot N_o \cdot A_{H_2O}$$

$$= 3.5 \times 10^3 V_m$$

where

S_o = surface area, m^2/g solid

M_{H_2O} = molecular weight of water, 18

N_o = Avogadro's number, 6×10^{23}

A_{H_2O} = area of water molecule, 10.6×10^{20} m^2

The BET equation has been used in many cases to describe the sorption behavior of foods. For example, note the work of Saravacos (1967) on the sorption of dehydrated apple and potato. The form of BET equation used for calculation of the monolayer value was

$$\frac{p}{W(p_o - p)} = \frac{1}{W_1 C} + \frac{C-1}{W_1 C} \cdot \frac{P_o}{P}$$

where

W = water content (in percent)

p = vapor pressure of sample

p_o = vapor pressure of water at same temperature

C = heat of adsorption constant

W_1 = moisture consent corresponding to monolayer

The BET plots obtained by Saravacos for dehydrated potato are presented in Figure 1–11.

Other approaches have been used to analyze the sorption isotherms, and these are described by Labuza (1968). However, the Langmuir isotherm as modified by Brunauer et al. (1938) has been most widely used with food products. Another method to analyze the sorption isotherms is the GAB sorption model described by van den Berg and Bruin (1981) and used by Roos (1993) and Jouppila and Roos (1994).

As is shown in Figure 1–7, the adsorption and desorption curves are not identical. The hysteresis effect is commonly observed; note,

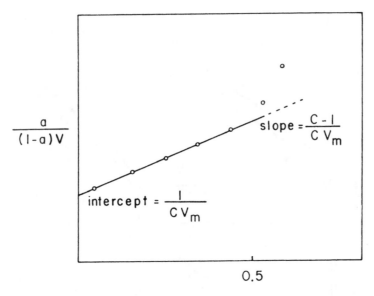

Figure 1–10 BET Monolayer Plot. *Source*: From T.P. Labuza, Sorption Phenomena in Foods, *Food Technol.*, Vol. 22, pp. 263–272, 1968.

for example, the sorption isotherms of wheat flour as determined by Bushuk and Winkler (1957) (Figure 1–12). The hysteresis effect is explained by water condensing in the capil- laries, and the effect occurs not only in region C of Figure 1–7 but also in a large part of region B. The best explanation for this phe- nomenon appears to be the so-called ink bot-

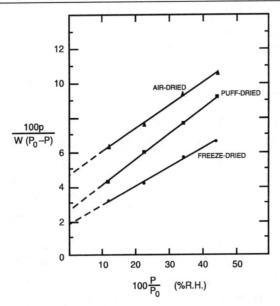

Figure 1–11 BET Plots for Dehydrated Potato. *Source*: From G.D. Saravacos, Effect of the Drying Method on the Water Sorption of Dehydrated Apple and Potato, *J. Food Sci.*, Vol. 32, pp. 81–84, 1967.

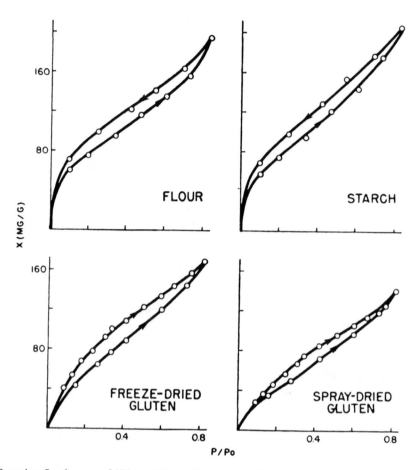

Figure 1–12 Sorption Isotherms of Wheat Flour, Starch, and Gluten. *Source*: From W. Bushuk and C.A. Winkler, Sorption of Water Vapor on Wheat Flour, Starch and Gluten, *Cereal Chem.*, Vol. 34, pp. 73–86, 1957.

tle theory (Labuza 1968). It is assumed that the capillaries have narrow necks and large bodies, as represented schematically in Figure 1–13. During adsorption the capillary does not fill completely until an activity is reached that corresponds to the large radius *R*. During desorption, the unfilling is controlled by the smaller radius *r*, thus lowering the water activity. Several other theories have been advanced to account for the hysteresis in sorption. These have been summarized by Kapsalis (1987).

The position of the sorption isotherms depends on temperature: the higher the temperature, the lower the position on the graph.

This decrease in the amount adsorbed at higher temperatures follows the Clausius Clapeyron relationship,

$$\frac{d(\ln a)}{d(1/T)} = -\frac{Q_s}{R}$$

where

Q_s = heat of adsorption

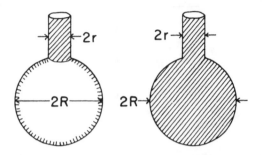

Figure 1–13 Ink Bottle Theory of Hysteresis in Sorption. *Source*: From T.P. Labuza, Sorption Phenomena in Foods, *Food Technol.*, Vol. 22, pp. 263–272, 1968.

R = gas constant
T = absolute temperature

By plotting the natural logarithm of activity versus the reciprocal of absolute temperature at constant moisture values, straight lines are obtained with a slope of $-Q_s/R$ (Figure 1–14). The values of Q_s obtained in this way for foods having less than full monolayer coverage are between about 2,000 and 10,000 cal per mole, demonstrating the strong binding of this water.

According to the principle of BET isotherm, the heat of sorption Q_s should be constant up to monolayer coverage and then should suddenly decrease. Labuza (1968) has pointed out that the latent heat of vaporization ΔH_v, about 10.4 kcal per mole, should be added to obtain the total heat value. The plot representing BET conditions as well as actual findings are given in Figure 1–15. The observed heat of sorption at low moisture contents is higher than theory indicates and falls off gradually, indicating the gradual change from Langmuir to capillary water.

TYPES OF WATER

The sorption isotherm indicates that different forms of water may be present in foods. It is convenient to divide the water into three types: Langmuir or monolayer water, capillary water, and loosely bound water. The bound water can be attracted strongly and held in a rigid and orderly state. In this form

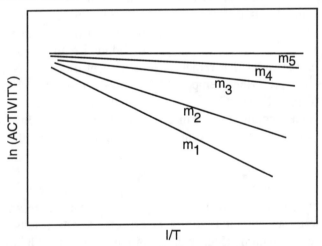

Figure 1–14 Method for Determination of Heat of Adsorption. Moisture content increases from M_1 to M_5. *Source*: From T.P. Labuza, Sorption Phenomena in Foods, *Food Technol.*, Vol. 22, pp. 263–272, 1968.

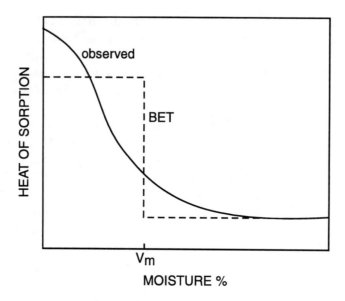

Figure 1–15 Relationship of Heat of Sorption and Moisture Content as Actually Observed and According to BET Theory. *Source*: From T.P. Labuza, Sorption Phenomena in Foods, *Food Technol.*, Vol. 22, pp. 263–272, 1968.

the water is unavailable as a solvent and does not freeze. It is difficult to provide a rigid definition of bound water because much depends on the technique used for its measurement. Two commonly used definitions are as follows:

1. Bound water is the water that remains unfrozen at some prescribed temperature below 0 °C, usually –20 °C.
2. Bound water is the amount of water in a system that is unavailable as a solvent.

The amount of unfreezable water, based on protein content, appears to vary only slightly from one food to another. About 8 to 10 percent of the total water in animal tissue is unavailable for ice formation (Meryman 1966). Egg white, egg yolk, meat, and fish all contain approximately 0.4 g of unfreezable water per g of dry protein. This corre-

sponds to 11.4 percent of total water in lean meat. Most fruits and vegetables contain less than 6 percent unfreezable water; whole grain corn, 34 percent.

The free water is sometimes determined by pressing a food sample between filter paper, by diluting with an added colored substance, or by centrifugation. None of these methods permits a distinct division between free and bound water, and results obtained are not necessarily identical between methods. This is not surprising since the adsorption isotherm indicates that the division between the different forms of water is gradual rather than sharp. A promising new method is the use of nuclear magnetic resonance, which can be expected to give results based on the freedom of movement of the hydrogen nuclei.

The main reason for the increased water content at high values of water activity must be capillary condensation. A liquid with surface tension σ in a capillary with radius *r* is subject to a pressure loss, the capillary pressure $p_o = 2\sigma/r$, as evidenced by the rising of the liquid in the capillary. As a result, there is a reduction in vapor pressure in the capillary, which can be expressed by the Thomson equation,

$$\ln\frac{p}{p_o} = -\frac{2\sigma}{r} \cdot \frac{V}{RT}$$

where

p = vapor pressure of liquid
p_o = capillary vapor pressure
σ = surface tension
V = mole volume of liquid
R = gas constant
T = absolute temperature

This permits the calculation of water activity in capillaries of different radii, as indicated in Table 1–4. In water-rich organic foods, such as meat and potatoes, the water is present in part in capillaries with a radius of 1 μm or more. The pressure necessary to remove this water is small. Calculated values of this pressure are given in Table 1–5 for water contained in capillaries ranging from 0.1 μm to 1 mm radius. It is evident that water from capillaries of 0.1 μm or larger can easily drip out. Structural damage caused, for instance, by freezing can easily result in drip loss in these products. The fact that water serves as a solvent for many solutes such as salts and sugars is an additional factor in reducing the vapor pressure.

The caloric behavior of water has been studied by Riedel (1959), who found that water in bread did not freeze at all when moisture content was below 18 percent (Figure 1–16). With this method it was possible to determine the nonfreezable water. For bread, the value was 0.30 g per g dry matter,

Table 1–4 Capillary Radius and Water Activity

Radius (nm)	Activity (a)
0.5	0.116
1	0.340
2	0.583
5	0.806
10	0.898
20	0.948
50	0.979
100	0.989
1000	0.999

and for fish and meat, 0.40 g per g protein. The nonfreezable and Langmuir water are probably not exactly the same. Wierbicki and Deatherage (1958) used a pressure method to determine free water in meat. The amount of free water in beef, pork, veal, and lamb varies from 30 to 50 percent of total moisture, depending on the kind of meat and the period of aging. A sharp drop in bound water occurs during the first day after slaughter, and is followed by a gradual, slight increase. Hamm and Deatherage (1960b) determined the changes in hydration during the heating of meat. At the normal pH of meat there is a considerable reduction of bound water.

Table 1–5 Pressure Required To Press Water from Tissue at 20°C

Radius	Pressure (kg/cm^2)
0.1 μm	14.84
1 μm	1.484
10 μm	0.148
0.1 mm	0.0148
1 mm	0.0015

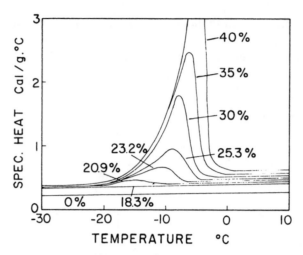

Figure 1–16 Specific Heat of Bread of Different Water Contents (Indicated as %) as a Function of Temperature. *Source*: From L. Riedel, Calorimetric Studies of the Freezing of White Bread and Other Flour Products, *Kältetechn*, Vol. 11, pp. 41–46, 1959.

FREEZING AND ICE STRUCTURE

A water molecule may bind four others in a tetrahedral arrangement. This results in a hexagonal crystal lattice in ice, as shown in Figure 1–17. The lattice is loosely built and has relatively large hollow spaces; this results in a high specific volume. In the hydrogen bonds, the hydrogen atom is 0.1 nm from one oxygen atom and 0.176 nm

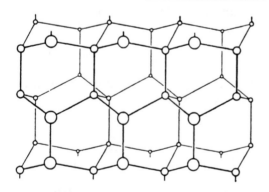

Figure 1–17 Hexagonal Pattern of the Lattice Structure in Ice

from another hydrogen atom. When ice melts, some of the hydrogen bonds are broken and the water molecules pack together more compactly in a liquid state (the average ligancy of a water molecule in water is about 5 and in ice, 4). There is some structural disorder in the ice crystal. For each hydrogen bond, there are two positions for the hydrogen atom: O–H+O and O+H–O. Without restrictions on the disorder, there would be 4^N ways of arranging the hydrogen atoms in an ice crystal containing N water molecules ($2N$ hydrogen atoms). There is one restriction, though: there must be two hydrogen atoms near each oxygen atom. As a result there are only $(3/2)^N$ ways of arranging the hydrogen atoms in the crystal.

The phase diagram (Figure 1–18) indicates the existence of three phases: solid, liquid, and gas. The conditions under which they exist are separated by three equilibrium lines: the vapor pressure line TA, the melting pressure line TC, and the sublimation pressure line BT. The three lines meet at point T,

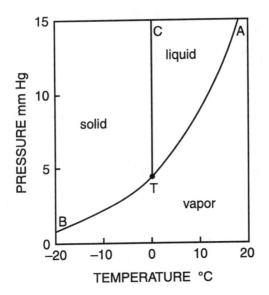

Figure 1–18 Phase Diagram of Water

where all three phases are in equilibrium. Figure 1–18 shows that when ice is heated at pressures below 4.58 mm Hg, it changes directly into the vapor form. This is the basis of freeze drying.

It is possible to supercool water. When a small ice crystal is introduced, the supercooling is immediately terminated and the temperature rises to 0°C. Normally the presence of a nucleus is required. Generally, nuclei form around foreign particles (heterogeneous nucleation). It is difficult to study homogeneous nucleation. This has been studied in the case of fat crystallization, by emulsifying the fat so that it is divided into a large number of small volumes, with the chance of a globule containing a heterogeneous nucleus being very small (Vanden-Tempel 1958). A homogeneous nucleus forms from the chance agglomeration of water molecules in the ice configuration. Usually, such nuclei disintegrate above a critical temperature. The probability of such nuclei forming depends on the volume of water; they are more likely to

form at higher temperature and in larger volumes. In ultrapure water, 1 mL can be supercooled to −32°C; droplets of 0.1 mm diameter to −35°C; and droplets of 1 μm to −41°C before solidification occurs.

The speed of crystallization—that is, the progress of the ice front in centimeters per second—is determined by the removal of the heat of fusion from the area of crystallization. The speed of crystallization is low at a high degree of supercooling (Meryman 1966). This is important because it affects the size of crystals in the ice. When large water masses are cooled slowly, there is sufficient time for heterogeneous nucleation in the area of the ice point. At that point the crystallization speed is very large so that a few nuclei grow to a large size, resulting in a coarse crystalline structure. At greater cooling speed, high supercooling occurs; this results in high nuclei formation and smaller growth rate and, therefore, a fine crystal structure.

Upon freezing, HOH molecules associate in an orderly manner to form a rigid structure that is more open (less dense) than the liquid form. There still remains considerable movement of individual atoms and molecules in ice, particularly just below the freezing point. At 10°C an HOH molecule vibrates with an amplitude of approximately 0.044 nm, nearly one-sixth the distance between adjacent HOH molecules. Hydrogen atoms may wander from one oxygen atom to another.

Each HOH molecule has four tetrahedrally spaced attractive forces and is potentially able to associate by means of hydrogen bonding with four other HOH molecules. In this arrangement each oxygen atom is bonded covalently with two hydrogen atoms, each at a distance of 0.096 nm, and each hydrogen atom is bonded with two other hydrogen atoms, each at a distance of 0.18 nm. This results in an open tetrahedral structure with adjacent oxygen atoms spaced about 0.276 nm apart and separated by single hydrogen atoms. All bond angles are approximately 109 degrees (Figure 1–19).

Extension of the model in Figure 1–19 leads to the hexagonal pattern of ice established when several tetrahedrons are assembled (Figure 1–17).

Upon change of state from ice to water, rigidity is lost, but water still retains a large number of ice-like clusters. The term *ice-like cluster* does not imply an arrangement identical to that of crystallized ice. The HOH bond angle of water is several degrees less than that of ice, and the average distance between oxygen atoms is 0.31 nm in water and 0.276 nm in ice. Research has not yet determined whether the ice-like clusters of water exist in a tetrahedral arrangement, as they do in ice. Since the average intermolecular distance is greater than in ice, it follows that the greater density of water must be achieved by each molecule having some neighbors. A cubic structure with each HOH molecule surrounded by six others has been suggested.

At 0 °C, water contains ice-like clusters averaging 90 molecules per cluster. With increasing temperature, clusters become smaller and more numerous. At 0 °C, approximately half of the hydrogen bonds present at −183 °C remain unbroken, and even at 100 °C approximately one-third are still present. All hydrogen bonds are broken when water changes into vapor at 100 °C. This explains the large heat of vaporization of water.

Crystal Growth and Nucleation

Crystal growth, in contrast to nucleation, occurs readily at temperatures close to the freezing point. It is more difficult to initiate crystallization than to continue it. The rate of ice crystal growth decreases with decreasing temperature. A schematic graphical representation of nucleation and crystal growth rates is given in Figure 1–20. Solutes of many types and in quite small amounts will greatly slow ice crystal growth. The mechanism of this action is not known. Membranes may be impermeable to ice crystal growth and thus limit crystal size. The effect of membranes on ice crystal propagation was studied by Lusena and Cook (1953), who found that membranes freely permeable to liquids may be either permeable, partly permeable, or impermeable to growing ice crystals. In a given material, permeability to ice crystal growth increases with porosity, but is also affected by rate of cooling, membrane composition and properties, and concentration of the solute(s) present in the aqueous phase. When ice crystal growth is retarded by solutes, the ice phase may become dis-

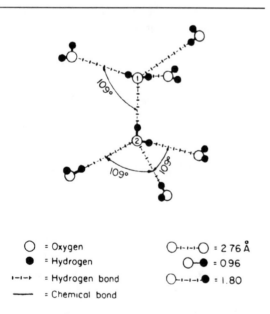

○	= Oxygen	○⊶⊶○ = 2 76 Å
●	= Hydrogen	○—● = 0 96
⊷⊷⊷	= Hydrogen bond	○⊷⊷● = 1.80
——	= Chemical bond	

Figure 1–19 Hydrogen Bonded Arrangement of Water Molecules in Ice

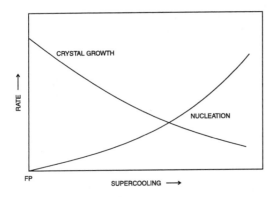

FP = temperature at which crystals start to form.

Figure 1–20 Schematic Representation of the Rate of Nucleation and Crystal Growth

continuous either by the presence of a membrane or spontaneously.

Ice crystal size at the completion of freezing is related directly to the number of nuclei. The greater the number of nuclei, the smaller the size of the crystals. In liquid systems nuclei can be added. This process is known as *seeding*. Practical applications of seeding include adding finely ground lactose to evaporated milk in the evaporator, and recirculating some portion of crystallized fat in a heat exchanger during manufacture of margarine. If the system is maintained at a temperature close to the freezing point (FP), where crystallization starts (Figure 1–20), only a few nuclei form and each crystal grows extensively. The slow removal of heat energy produces an analogous situation, since the heat of crystallization released by the few growing crystals causes the temperature to remain near the melting point, where nucleation is unlikely. In tissue or unagitated fluid systems, slow removal of heat results in a continuous ice phase that slowly moves inward, with little if any nucleation. The effect of

temperature on the linear crystallization velocity of water is given in Table 1–6.

If the temperature is lowered to below the FP (Figure 1–20), crystal growth is the predominant factor at first but, at increasing rate of supercooling, nucleation takes over. Therefore, at low supercooling large crystals are formed; as supercooling increases, many small crystals are formed. Control of crystal size is much more difficult in tissues than in agitated liquids. Agitation may promote nucleation and, therefore, reduced crystal size. Lusena and Cook (1954) suggested that large ice crystals are formed when freezing takes place above the critical nucleation temperature (close to FP in Figure 1–20). When freezing occurs at the critical nucleation temperature, small ice crystals form. The effect of solutes on nucleation and rate of ice crystal growth is a major factor controlling the pattern of propagation of the ice front. Lusena and Cook (1955) also found that solutes depress the nucleation temperature to the same extent that they depress the freezing point. Solutes retard ice growth at 10 °C supercooling, with organic compounds having a greater effect than inorganic ones. At low concentrations,

Table 1–6 Effect of Temperature on Linear Crystallization Velocity of Water

Temperature at Onset of Crystallization (°C)	Linear Crystallization Velocity (mm/min)
−0.9	230
−1.9	520
−2.0	580
−2.2	680
−3.5	1,220
−5.0	1,750
−7.0	2,800

proteins are as effective as alcohols and sugars in retarding crystal growth.

Once formed, crystals do not remain unchanged during frozen storage; they have a tendency to enlarge. Recrystallization is particularly evident when storage temperatures are allowed to fluctuate widely. There is a tendency for large crystals to grow at the expense of small ones.

Slow freezing results in large ice crystals located exclusively in extracellular areas. Rapid freezing results in tiny ice crystals located both extra- and intracellularly. Not too much is known about the relation between ice crystal location and frozen food quality. During the freezing of food, water is transformed to ice with a high degree of purity, and solute concentration in the unfrozen liquid is gradually increased. This is accompanied by changes in pH, ionic strength, viscosity, osmotic pressure, vapor pressure, and other properties.

When water freezes, it expands nearly 9 percent. The volume change of a food that is frozen will be determined by its water content and by solute concentration. Highly concentrated sucrose solutions do not show expansion (Table 1–7). Air spaces may partially accommodate expanding ice crystals. Volume changes in some fruit products upon freezing are shown in Table 1–8. The effect of air space is obvious. The expansion of water on freezing results in local stresses that undoubtedly produce mechanical damage in cellular materials. Freezing may cause changes in frozen foods that make the product unacceptable. Such changes may include destabilization of emulsions, flocculation of proteins, increase in toughness of fish flesh, loss of textural integrity, and increase in drip loss of meat. Ice formation can be influenced by the presence of carbohydrates. The effect of sucrose on the ice formation process

Table 1–7 Volume Change of Water and Sucrose Solutions on Freezing

Sucrose (%)	Volume Increase During Temperature Change from 70°F to 0°F (%)
0	8.6
10	8.7
20	8.2
30	6.2
40	5.1
50	3.9
60	None
70	−1.0 (decrease)

has been described by Roos and Karel (1991a,b,c).

The Glass Transition

In aqueous systems containing polymeric substances or some low molecular weight materials including sugars and other carbohydrates, lowering of the temperature may result in formation of a glass. A glass is an amorphous solid material rather than a crystalline solid. A glass is an undercooled liquid

Table 1–8 Expansion of Fruit Products During Freezing

Product	Volume Increase During Temperature Change from 70°F to 0°F (%)
Apple juice	8.3
Orange juice	8.0
Whole raspberries	4.0
Crushed raspberries	6.3
Whole strawberries	3.0
Crushed strawberries	8.2

of high viscosity that exists in a metastable solid state (Levine and Slade 1992). A glass is formed when a liquid or an aqueous solution is cooled to a temperature that is considerably lower than its melting temperature. This is usually achieved at high cooling rates. The normal process of crystallization involves the conversion of a disordered liquid molecular structure to a highly ordered crystal formation. In a crystal, atoms or ions are arranged in a regular, three-dimensional array. In the formation of a glass, the disordered liquid state is immobilized into a disordered glassy solid, which has the rheological properties of a solid but no ordered crystalline structure.

The relationships among melting point (T_m), glass transition temperature (T_g), and crystallization are schematically represented in Figure 1–21. At low degree of supercooling (just below T_m), nucleation is at a minimum and crystal growth predominates. As the degree of supercooling increases, nucleation becomes the dominating effect. The maximum overall crystallization rate is at a point about halfway between T_m and T_g. At high cooling rates and a degree of supercooling that moves the temperature to below T_g, no crystals are formed and a glassy solid results. During the transition from the molten state to the glassy state, the moisture content plays an important role. This is illustrated by the phase diagram of Figure 1–22. When the temperature is lowered at sufficiently high moisture content, the system goes through a rubbery state before becoming glassy (Chirife and Buera 1996). The glass transition temperature is characterized by very high apparent viscosities of more than 10^5 Ns/m^2 (Aguilera and Stanley 1990). The rate of diffusion limited processes is more rapid in the rubbery state than in the glassy state, and this may be important in the storage stability of certain foods. The effect of water activity on the glass transition temperature of a number of plant products (carrots, strawberries, and potatoes) as well as some biopolymers (gelatin, wheat gluten, and wheat starch) is shown in Figure 1–23 (Chirife and Buera 1996). In the rubbery state the rates of chemical reac-

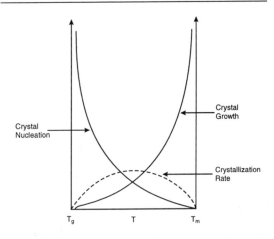

Figure 1–21 Relationships Among Crystal Growth, Nucleation, and Crystallization Rate between Melting Temperature (T_m) and Glass Temperature (T_g)

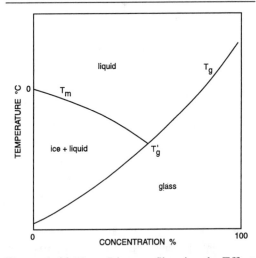

Figure 1–22 Phase Diagram Showing the Effect of Moisture Content on Melting Temperature (T_m) and Glass Transition Temperature (T_g)

tion appear to be higher than in the glassy state (Roos and Karel 1991e).

When water-containing foods are cooled below the freezing point of water, ice may be formed and the remaining water is increasingly high in dissolved solids. When the glass transition temperature is reached, the remaining water is transformed into a glass. Ice formation during freezing may destabilize sensitive products by rupturing cell walls and breaking emulsions. The presence of glass-forming substances may help prevent this from occurring. Such stabilization of frozen products is known as *cryoprotection*, and the agents are known as *cryoprotectants*.

When water is rapidly removed from foods during processes such as extrusion, drying, or freezing, a glassy state may be produced (Roos 1995). The T_g values of high molecu-

lar weight food polymers, proteins, and polysaccharides are high and cannot be determined experimentally, because of thermal decomposition. An example of measured T_g values for low molecular weight carbohydrates is given in Figure 1–24. The value of T_g for starch is obtained by extrapolation.

The water present in foods may act as a plasticizer. Plasticizers increase plasticity and flexibility of food polymers as a result of weakening of the intermolecular forces existing between molecules. Increasing water content decreases T_g. Roos and Karel (1991a) studied the plasticizing effect of water on thermal behavior and crystallization of amorphous food models. They found that dried foods containing sugars behave like amorphous materials, and that small amounts of water decrease T_g to room temperature with

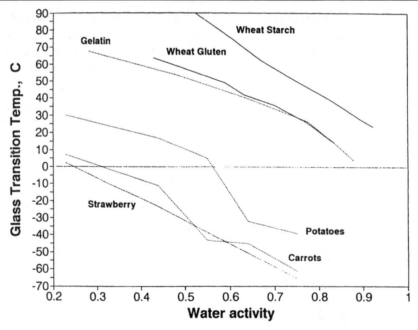

Figure 1–23 Relationship Between Water Activity (a_w) and Glass Transition Temperature (T_g) of Some Plant Materials and Biopolymers. *Source*: Reprinted with permission from J. Cherife and M. del Pinar Buera, Water Activity, Water Glass Dynamics and the Control of Microbiological Growth in Foods, *Critical Review Food Sci. Nutr.*, Vol. 36, No. 5, p. 490, © 1996. Copyright CRC Press, Boca Raton, Florida.

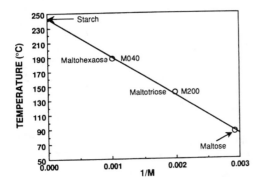

Figure 1–24 Glass Transition Temperature (T_g) for Maltose, Maltose Polymers, and Extrapolated Value for Starch. M indicates molecular weight. *Source*: Reprinted with permission from Y.H. Roos, Glass Transition-Related Physico-Chemical Changes in Foods, *Food Technology*, Vol. 49, No. 10, p. 98, © 1995, Institute of Food Technologists.

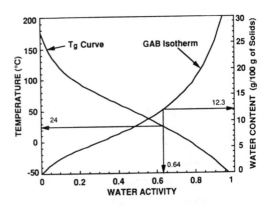

Figure 1–25 Modified State Diagram Showing Relationship Between Glass Transition Temperature (T_g), Water Activity (GAB isotherm), and Water Content for an Extruded Snack Food Model. Crispness is lost as water plasticization depresses T_g to below 24 °C. Plasticization is indicated with critical values for water activity and water content. *Source*: Reprinted with permission from Y.H. Roos, Glass Transition-Related Physico-Chemical Changes in Foods, *Food Technology*, Vol. 49, No. 10, p. 99, © 1995, Institute of Food Technologists.

the result of structural collapse and formation of stickiness. Roos and Karel (1991e) report a linearity between water activity (a_w) and T_g in the a_w range of 0.1 to 0.8. This allows prediction of T_g at the a_w range typical of dehydrated and intermediate moisture foods.

Roos (1995) has used a combined sorption isotherm and state diagram to obtain critical water activity and water content values that result in depressing T_g to below ambient temperature (Figure 1–25). This type of plot can be used to evaluate the stability of low-moisture foods under different storage conditions. When the T_g is decreased to below ambient temperature, molecules are mobilized because of plasticization and reaction rates increase because of increased diffusion, which in turn may lead to deterioration. Roos and Himberg (1994) and Roos et al. (1996) have described how glass transition temperatures influence nonenzymatic browning in model systems. This deteriorative reaction

showed an increased reaction rate as water content increased.

Water Activity and Reaction Rate

Water activity has a profound effect on the rate of many chemical reactions in foods and on the rate of microbial growth (Labuza 1980). This information is summarized in Table 1–9. Enzyme activity is virtually nonexistent in the monolayer water (a_w between 0 and 0.2). Not surprisingly, growth of microorganisms at this level of a_w is also virtually zero. Molds and yeasts start to grow at a_w between 0.7 and 0.8, the upper limit of capillary water. Bacterial growth takes place when a_w reaches 0.8, the limit of loosely

Table 1–9 Reaction Rates in Foods as Determined by Water Activity

Reaction	Monolayer Water	Capillary Water	Loosely Bound Water
Enzyme activity	Zero	Low	High
Mold growth	Zero	Low[*]	High
Yeast growth	Zero	Low[*]	High
Bacterial growth	Zero	Zero	High
Hydrolysis	Zero	Rapid increase	High
Nonenzymic browning	Zero	Rapid increase	High
Lipid oxidation	High	Rapid increase	High

[*]Growth starts at a_w of 0.7 to 0.8.

bound water. Enzyme activity increases gradually between a_w of 0.3 and 0.8, then increases rapidly in the loosely bound water area (a_w 0.8 to 1.0). Hydrolytic reactions and nonenzymic browning do not proceed in the monolayer water range of a_w (0.0 to 0.25). However, lipid oxidation rates are high in this area, passing from a minimum at a_w 0.3 to 0.4, to a maximum at a_w 0.8. The influ-

ence of a_w on chemical reactivity has been reviewed by Leung (1987). The relationship between water activity and rates of several reactions and enzyme activity is presented graphically in Figure 1–26 (Bone 1987).

Water activity has a major effect on the texture of some foods, as Bourne (1986) has shown in the case of apples.

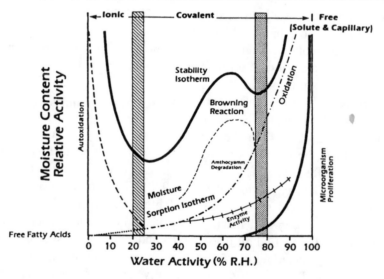

Figure 1–26 Relationship Between Water Activity and a Number of Reaction Rates. *Source*: Reprinted with permission from D.P. Bone, Practical Applications of Water Activity and Moisture Relations in Foods, in *Water Activity: Theory and Application to Food*, L.B. Rockland and L.R. Beuchat, eds., p. 387, 1987, by courtesy of Marcel Dekker, Inc.

WATER ACTIVITY AND FOOD SPOILAGE

The influence of water activity on food quality and spoilage is increasingly being recognized as an important factor (Rockland and Nishi 1980). Moisture content and water activity affect the progress of chemical and microbiological spoilage reactions in foods. Dried or freeze-dried foods, which have great storage stability, usually have water contents in the range of about 5 to 15 percent. The group of intermediate-moisture foods, such as dates and cakes, may have moisture contents in the range of about 20 to 40 percent. The dried foods correspond to the lower part of the sorption isotherms. This includes water in the monolayer and multilayer category. Intermediate-moisture foods have water activities generally above 0.5, including the capillary water. Reduction of water activity can be obtained by drying or by adding water-soluble substances, such as sugar to jams or salt to pickled preserves. Bacterial growth is virtually impossible below a water activity of 0.90. Molds and yeasts are usually inhibited between 0.88 and 0.80, although some osmophile yeast strains grow at water activities down to 0.65.

Most enzymes are inactive when the water activity falls below 0.85. Such enzymes include amylases, phenoloxidases, and peroxidases. However, lipases may remain active at values as low as 0.3 or even 0.1 (Loncin et al. 1968). Acker (1969) provided examples of the effect of water activity on some enzymic reactions. A mixture of ground barley and lecithin was stored at different water activities, and the rates of hydrolysis were greatly influenced by the value of a (Figure 1–27). When the lower a values were changed to 0.70 after 48 days of

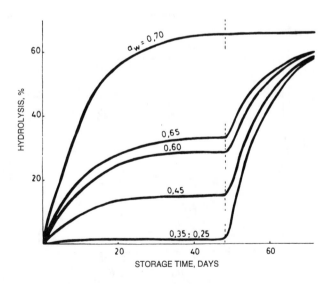

Figure 1–27 Enzymic Splitting of Lecithin in a Mixture of Barley Malt and Lecithin Stored at 30 °C and Different Water Activities. Lower a_w values were changed to 0.70 after 48 days. *Source*: From L. Acker, Water Activity and Enzyme Activity, *Food Technol.*, Vol. 23, pp. 1257–1270, 1969.

storage the rates rapidly went up. In the region of monomolecular adsorption, enzymic reactions either did not proceed at all or proceeded at a greatly reduced rate, whereas in the region of capillary condensation the reaction rates increased greatly. Acker found that for reactions in which lipolytic enzyme activity was measured, the manner in which components of the food system were put into contact significantly influenced the enzyme activity. Separation of substrate and enzyme could greatly retard the reaction. Also, the substrate has to be in liquid form; for example, liquid oil could be hydrolyzed at water activity as low as 0.15, but solid fat was only slightly hydrolyzed. Oxidizing enzymes were affected by water activity in about the same way as hydrolytic enzymes, as was shown by the example of phenoloxidase from potato (Figure 1–28). When the lower a values were increased to 0.70 after 9 days of storage, the final values were lower than with

the sample kept at 0.70 all through the experiment, because the enzyme was partially inactive during storage.

Nonenzymic browning or Maillard reactions are one of the most important factors causing spoilage in foods. These reactions are strongly dependent on water activity and reach a maximum rate at a values of 0.6 to 0.7 (Loncin et al. 1968). This is illustrated by the browning of milk powder kept at 40 °C for 10 days as a function of water activity (Figure 1–29). The loss in lysine resulting from the browning reaction parallels the color change, as is shown in Figure 1–30.

Labuza et al. (1970) have shown that, even at low water activities, sucrose may be hydrolyzed to form reducing sugars that may take part in browning reactions. Browning reactions are usually slow at low humidities and increase to a maximum in the range of intermediate-moisture foods. Beyond this range the rate again decreases. This behavior

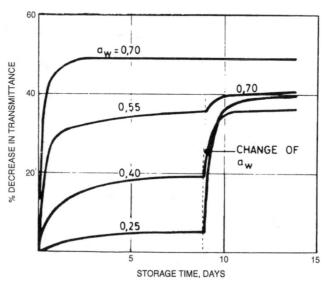

Figure 1–28 Enzymic Browning in the System Polyphenoloxidase-Cellulose-Catechol at 25 °C and Different Water Activities. Lower a_w values were changed to 0.70 after 9 days. *Source*: From L. Acker, Water Activity and Enzyme Activity, *Food Technol.*, Vol. 23, pp. 1257–1270, 1969.

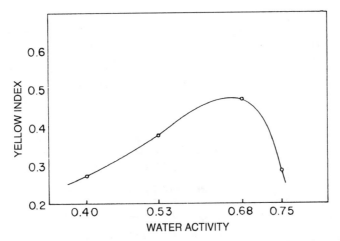

Figure 1–29 Color Change of Milk Powder Kept at 40°C for 10 Days as a Function of Water Activity

can be explained by the fact that, in the intermediate range, the reactants are all dissolved, and that further increase in moisture content leads to dilution of the reactants.

The effect of water activity on oxidation of fats is complex. Storage of freeze-dried and dehydrated foods at moisture levels above those giving monolayer coverage appears to give maximum protection against oxidation. This has been demonstrated by Martinez and Labuza (1968) with the oxidation of lipids in freeze-dried salmon (Figure 1–31). Oxidation of the lipids was reduced as water content increased. Thus, conditions that are

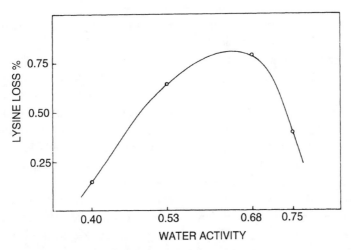

Figure 1–30 Loss of Free Lysine in Milk Powder Kept at 40°C for 10 Days as a Function of Water Activity. *Source*: From M. Loncin, J.J. Bimbenet, and J. Lenges, Influence of the Activity of Water on the Spoilage of Foodstuffs, *J. Food Technol.*, Vol. 3, pp. 131–142, 1968.

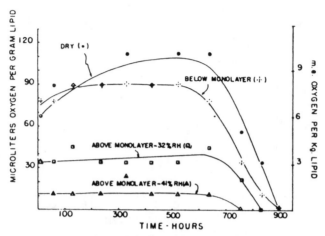

Figure 1–31 Peroxide Production in Freeze-Dried Salmon Stored at Different Relative Humidities. *Source*: From F. Martinez and T.P. Labuza, Effect of Moisture Content on Rate of Deterioration of Freeze-Dried Salmon, *J. Food Sci.*, Vol. 33, pp. 241–247, 1968.

optimal for protection against oxidation may be conducive to other spoilage reactions, such as browning.

Water activity may affect the properties of powdered dried product. Berlin et al. (1968) studied the effect of water vapor sorption on the porosity of milk powders. When the powders were equilibrated at 50 percent relative humidity (RH), the microporous structure was destroyed. The free fat content was considerably increased, which also indicates structural changes.

Other reactions that may be influenced by water activity are hydrolysis of protopectin, splitting and demethylation of pectin, autocatalytic hydrolysis of fats, and the transformation of chlorophyll into pheophytin (Loncin et al. 1968).

Rockland (1969) has introduced the concept of *local isotherm* to provide a closer relationship between sorption isotherms and stability than is possible with other methods. He suggested that the differential coefficient of moisture with respect to relative humidity ($\Delta M/\Delta RH$), calculated from sorption isotherms, is related to product stability.

The interaction between water and polymer molecules in gel formation has been reviewed by Busk (1984).

WATER ACTIVITY AND PACKAGING

Because water activity is a major factor influencing the keeping quality of a number of foods, it is obvious that packaging can do much to maintain optimal conditions for long storage life. Sorption isotherms play an important role in the selection of packaging materials. Hygroscopic products always have a steep sorption isotherm and reach the critical area of moisture content before reaching external climatic conditions. Such foods have to be packaged in glass containers with moistureproof seals or in watertight plastic (thick polyvinylchloride). For example, consider instant coffee, where the critical area is at about 50 percent RH. Under these conditions the product cakes and loses its flowability. Other products might not be hygroscopic and no unfavorable reactions occur at normal conditions of storage. Such products can be packaged in polyethylene containers.

There are some foods where the equilibrium relative humidity is above that of the external climatic conditions. The packaging material then serves the purpose of protecting the product from moisture loss. This is the case with processed cheese and baked goods.

Different problems may arise in composite foods, such as soup mixes, where several distinct ingredients are packaged together. In Figure 1–32, for example, substance B with the steep isotherm is more sensitive to moisture, and is mixed in equal quantities with substance A in an impermeable package.*

---

*The initial relative humidity of A is 65 percent and of B, 15 percent.

The initial moisture content of B is X_1, and after equilibration with A, the moisture content is X_2. The substances A and B will reach a mean relative humidity of about 40 percent, but not a mean moisture content. If this were a dry soup mix and the sensitive component was a freeze-dried vegetable with a moisture content of 2 percent and the other component, a starch or flour with a moisture content of 13 percent, the vegetable would be moistened to up to 9 percent. This would result in rapid quality deterioration due to nonenzymic browning reactions. In this case, the starch would have to be postdried.

Salwin and Slawson (1959) found that stability in dehydrated foods was impaired if several products were packaged together. A transfer of water could take place from items of higher moisture-vapor pressure to those of lower moisture-vapor pressure. These authors determined packaging compatibility by examining the respective sorption isotherms. They suggested a formula for calculation of the final equilibrium moisture content of each component from the iso-

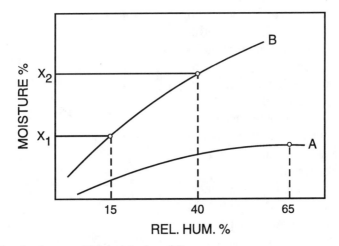

Figure 1–32 Sorption Isotherms of Materials A and B

therms of the mixed food and its equilibrium relative humidity:

$$a_w = \frac{(W_1 \cdot S_1 \cdot a_{w1'}) + (W_2 \cdot S_2 \cdot a_{w2'})}{(W_1 \cdot S_1) + (W_2 \cdot S_2)}$$

where

W_1 = gram solids of ingredient 1
S_1 = linear slope of ingredient 1
a_{w1}' = initial a_w of ingredient 1

WATER BINDING OF MEAT

According to Hamm (1962), the water-binding capacity of meat is caused by the muscle proteins. Some 34 percent of these proteins are water-soluble. The main portion of meat proteins is structural material. Only about 3 percent of the total water-binding capacity of muscle can be attributed to water-soluble (plasma) proteins. The main water-binding capacity of muscle can be attributed to actomyosin, the main component of the myofibrils. The adsorption isotherm of freeze-dried meat has the shape shown in Figure 1–33. The curve is similar to the sorption isotherms of other foods and consists of three parts. The first part corresponds to the tightly bound water, about 4 percent, which is given off at very low vapor pressures. This quantity is only about one-fifth the total quantity required to cover the whole protein with a monomolecular layer. This water is bound under simultaneous liberation of a considerable amount of energy, 3 to 6 kcal per mole of water. The binding of this water results in a volume contraction of 0.05 mL per g of protein. The binding is localized at hydrophilic groups on proteins such as polar side chains having carboxyl, amino, hydroxyl, and sulphydryl groups and also on the nondissociable carboxyl and imino groups of the peptide bonds. The binding of water is strongly influenced by the pH of meat. The effect of pH on the swelling or unswelling (that is, water-binding capacity of proteins) is schematically represented in Figure 1–34 (Honkel 1989). The second portion of the curve corresponds to multilayer adsorption, which amounts to another 4 to 6 percent of water. Hamm (1962) considered these two quantities of water to represent the real water of hydration and found them to amount to between 50 and 60 g per 100 g of protein. Muscle binds much more than this amount of water. Meat with a protein content of 20 to 22 percent contains 74 to 76 percent water, so that 100 g of protein binds about 350 to 360 g of water. This ratio is even higher in fish muscle. Most of this water is merely immobilized—retained by the net-

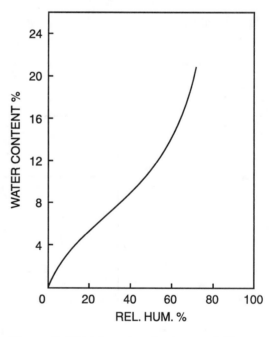

Figure 1–33 Adsorption Isotherm of Freeze-Dried Meat

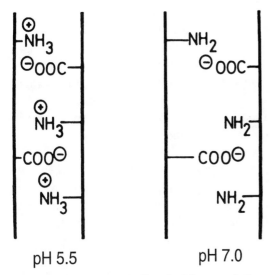

pH 5.5   pH 7.0

Figure 1–34 Water Binding in Meat as Influenced by pH

Hamm (1959a, 1959b) has proposed that during the first hour after slaughter, bivalent metal ions of muscle are incorporated into the muscle proteins at pH 6, causing a contraction of the fiber network and a dehydration of the tissue. Further changes in hydration during aging for up to seven days can be explained by an increase in the number of available carboxyl and basic groups. These result from proteolysis. Hamm and Deatherage (1960a) found that freeze-drying of beef results in a decrease in water-binding capacity in the isoelectric pH range of the muscle. The proteins form a tighter network, which is stabilized by the formation of new salt and/or hydrogen bonds. Heating beef at temperatures over 40 °C leads to strong denaturation and changes in hydration (Hamm and Deatherage 1960b). Quick freezing of beef results in a significant but small increase in the water-holding capacity, whereas slow freezing results in a significant but small decrease in water binding. These effects were thought to result from the mechanical action of ice crystals (Deatherage and Hamm 1960). The influence of heating on water binding of pork was studied by Sherman (1961b), who also investigated the effect of the addition of salts on water binding (Sherman 1961a). Water binding can be greatly affected by addition of certain salts, especially phosphates (Hellendoorn 1962). Such salt additions are used to diminish cooking losses by expulsion of water in canning hams and to obtain a better structure and consistency in manufacturing sausages. Recently, the subject of water binding has been greatly extended in scope (Katz 1997). Water binding is related to the use of water as a plasticizer and the interaction of water with the components of mixed food systems. Retaining water in mixed food systems throughout their shelf life is becoming an important

work of membranes and filaments of the structural proteins as well as by cross-linkages and electrostatic attractions between peptide chains. It is assumed that changes in water-binding capacity of meat during aging, storage, and processing relate to the free water and not the real water of hydration. The free water is held by a three-dimensional structure of the tissue, and shrinkage in this network leads to a decrease in immobilized water; this water is lost even by application of slight pressure. The reverse is also possible. Cut-up muscle can take up as much as 700 to 800 g of water per 100 g of protein at certain pH values and in the presence of certain ions. Immediately after slaughter there is a drop in hydration and an increase in rigidity of muscle with time. The decrease in hydration was attributed at about two-thirds to decomposition of ATP and at about one-third to lowering of the pH.

requirement in foods of low fat content. Such foods often have fat replacer ingredients based on proteins or carbohydrates, and their interaction with water is of great importance.

WATER ACTIVITY AND FOOD PROCESSING

Water activity is one of the criteria for establishing good manufacturing practice (GMP) regulations governing processing requirements and classification of foods (Johnston and Lin 1987). As indicated in Figure 1–35, the process requirements for

foods are governed by a_w and pH; a_w controlled foods are those with pH greater than 4.6 and a_w less than 0.85. At pH less than 4.6 and a_w greater than 0.85, foods fall into the category of low-acid foods; when packaged in hermetically sealed containers, these foods must be processed to achieve commercially sterile conditions.

Intermediate moisture foods are in the a_w range of 0.90 to 0.60. They can achieve stability by a combination of a_w with other factors, such as pH, heat, preservatives, and E_h (equilibrium relative humidity).

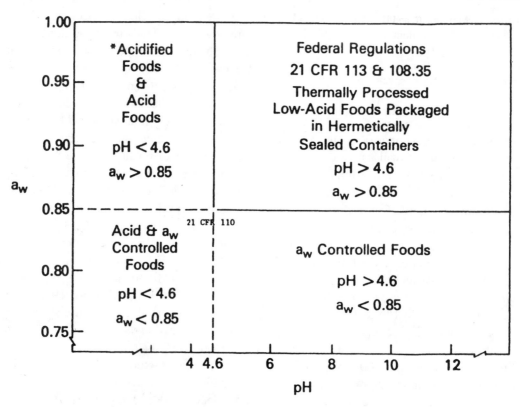

*Acidified Foods — 21 CFR 114 & 108.25

Figure 1–35 The Importance of pH and a_w on Processing Requirements for Foods. *Source*: Reprinted with permission from M.R. Johnston and R.C. Lin, FDA Views on the Importance of a_w in Good Manufacturing Practice, *Water Activity: Theory and Application to Food*, L.B. Rockland and L.R. Beuchat, eds., p. 288, 1987, by courtesy of Marcel Dekker, Inc.

REFERENCES

Acker, L. 1969. Water activity and enzyme activity. *Food Technol.* 23: 1257–1270.

Aguilera, J.M., and D.W. Stanley. 1990. *Microstructural principles of food processing and engineering.* London: Elsevier Applied Science.

Berlin, E., B.A. Anderson, and M.J. Pallansch. 1968. Effect of water vapor sorption on porosity of dehydrated dairy products. *J. Dairy Sci.* 51: 668–672.

Bone, D.P. 1987. Practical applications of water activity and moisture relations in foods. In *Water activity: Theory and application to food*, ed. L.B. Rockland and L.R. Beuchat. New York: Marcel Dekker, Inc.

Bourne, M.C. 1986. Effect of water activity on texture profile parameters of apple flesh. *J. Texture Studies* 17: 331–340.

Brunauer, S., P.J. Emmett, and E. Teller. 1938. Absorption of gasses in multimolecular layers. *J. Am. Chem. Soc.* 60: 309–319.

Bushuk, W., and C.A. Winkler. 1957. Sorption of water vapor on wheat flour, starch and gluten. *Cereal Chem.* 34: 73–86.

Busk Jr., G.C. 1984. Polymer-water interactions in gelation. *Food Technol.* 38: 59–64.

Chirife, J., and M.P. Buera. 1996. A critical review of the effect of some non-equilibrium situations and glass transitions on water activity values of food in the microbiological growth range. *J. Food Eng.* 25: 531–552.

Deatherage, F.E., and R. Hamm. 1960. Influence of freezing and thawing on hydration and charges of the muscle proteins. *Food Res.* 25: 623–629.

Hamm, R. 1959a. The biochemistry of meat aging. I. Hydration and rigidity of beef muscle (In German). *Z. Lebensm. Unters. Forsch.* 109: 113–121.

Hamm, R. 1959b. The biochemistry of meat aging. II. Protein charge and muscle hydration (In German). *Z. Lebensm. Unters. Forsch.* 109: 227–234.

Hamm, R. 1962. The water binding capacity of mammalian muscle. VII. The theory of water binding (In German). *Z. Lebensm. Unters. Forsch.* 116: 120–126.

Hamm, R., and F.E. Deatherage. 1960a. Changes in hydration and charges of muscle proteins during heating of meat. *Food Res.* 25: 573–586.

Hamm, R., and F.E. Deatherage. 1960b. Changes in hydration, solubility and charges of muscle proteins during heating of meat. *Food Res.* 25: 587–610.

Hellendoorn, E.W. 1962. Water binding capacity of meat as affected by phosphates. *Food Technol.* 16: 119–124.

Honkel, K.G. 1989. The meat aspects of water and food quality. In *Water and food quality*, ed. T.M. Hardman. New York: Elsevier Applied Science.

Johnston, M.R., and R.C. Lin. 1987. FDA views on the importance of a_w in good manufacturing practice. In *Water activity: Theory and application to food*, ed. L.B. Rockland and L.R. Beuchat. New York: Marcel Dekker, Inc.

Jouppila, K., and Y.H. Roos. 1994. The physical state of amorphous corn starch and its impact on crystallization. *Carbohydrate Polymers.* 32: 95–104.

Kapsalis, J.G. 1987. Influences of hysteresis and temperature on moisture sorption isotherms. In *Water activity: Theory and application to food*, ed. L.B. Rockland and L.R. Beuchat. New York: Marcel Dekker, Inc.

Katz, F. 1997. The changing role of water binding. *Food Technol.* 51, no. 10: 64.

Klotz, I.M. 1965. Role of water structure in macromolecules. *Federation Proc.* 24: S24–S33.

Labuza, T.P. 1968. Sorption phenomena in foods. *Food Technol.* 22: 263–272.

Labuza, T.P. 1980. The effect of water activity on reaction kinetics of food deterioration. *Food Technol.* 34, no. 4: 36–41, 59.

Labuza, T.P., S.R. Tannenbaum, and M. Karel. 1970. Water content and stability of low-moisture and intermediate-moisture foods. *Food Technol.* 24: 543–550.

Landolt-Boernstein. 1923. In *Physical-chemical tables* (In German), ed. W.A. Roth and K. Sheel. Berlin: Springer Verlag.

Leung, H.K. 1987. Influence of water activity on chemical reactivity. In *Water activity: Theory and application to food*, ed. L.B. Rockland and L.R. Beuchat. New York: Marcel Dekker, Inc.

Levine, H., and L. Slade. 1992. Glass transitions in foods. In *Physical chemistry of foods*. New York: Marcel Dekker, Inc.

Loncin, M., J.J. Bimbenet, and J. Lenges. 1968. Influence of the activity of water on the spoilage of foodstuffs. *J. Food Technol.* 3: 131–142.

Lusena, C.V., and W.H. Cook. 1953. Ice propagation in systems of biological interest. I. Effect of membranes and solutes in a model cell system. *Arch. Biochem. Biophys.* 46: 232–240.

Lusena, C.V., and W.H. Cook. 1954. Ice propagation in systems of biological interest. II. Effect of solutes at rapid cooling rates. *Arch. Biochem. Biophys.* 50: 243–251.

Lusena, C.V., and W.H. Cook. 1955. Ice propagation in systems of biological interest. III. Effect of solutes on nucleation and growth of ice crystals. *Arch. Biochem. Biophys.* 57: 277–284.

Martinez, F., and T.P. Labuza. 1968. Effect of moisture content on rate of deterioration of freeze-dried salmon. *J. Food Sci.* 33: 241–247.

Meryman, H.T. 1966. *Cryobiology.* New York: Academic Press.

Pauling, L. 1960. *The nature of the chemical bond.* Ithaca, NY: Cornell University Press.

Perry, J.H. 1963. *Chemical engineers' handbook.* New York: McGraw Hill.

Riedel, L. 1959. Calorimetric studies of the freezing of white bread and other flour products. *Kältetechn.* 11: 41–46.

Rockland, L.B. 1969. Water activity and storage stability. *Food Technol.* 23: 1241–1251.

Rockland, L.B., and S.K. Nishi. 1980. Influence of water activity on food product quality and stability. *Food Technol.* 34, no. 4: 42–51, 59.

Roos, Y.H. 1993. Water activity and physical state effects on amorphous food stability. *J. Food Process Preserv.* 16: 433–447

Roos, Y.H. 1995. Glass transition-related physicochemical changes in foods. *Food Technol.* 49, no. 10: 97–102.

Roos, Y.H., and M.J. Himberg. 1994. Nonenzymatic browning behavior, as related to glass transition of a food model at chilling temperatures. *J. Agr. Food Chem.* 42: 893–898.

Roos, Y.H, K. Jouppila, and B. Zielasko. 1996. Nonenzymatic browning-induced water plasticization. *J. Thermal. Anal.* 47: 1437–1450.

Roos, Y.H., and M. Karel. 1991a. Amorphous state and delayed ice formation in sucrose solutions. *Int. J. Food Sci. Technol.* 26: 553–566.

Roos, Y.H., and M. Karel. 1991b. Non equilibrium ice formation in carbohydrate solutions. *Cryo-Letters.* 12: 367–376.

Roos, Y.H., and M. Karel. 1991c. Phase transition of amorphous sucrose and frozen sucrose solutions. *J. Food Sci.* 56: 266–267.

Roos, Y., and M. Karel. 1991d. Plasticizing effect of water on thermal behaviour and crystallization of amorphous food models. *J. Food Sci.* 56: 38–43.

Roos, Y., and M. Karel. 1991e. Water and molecular weight effects on glass transitions in amorphous carbohydrates and carbohydrate solutions. *J. Food Sci.* 56: 1676–1681.

Salwin, H., and V. Slawson. 1959. Moisture transfer in combinations of dehydrated foods. *Food Technol.* 13: 715–718.

Saravacos, G.D. 1967. Effect of the drying method on the water sorption of dehydrated apple and potato. *J. Food Sci.* 32: 81–84.

Sherman, P. 1961a. The water binding capacity of fresh pork. I. The influence of sodium chloride, pyrophosphate and polyphosphate on water absorption. *Food Technol.* 15: 79–87.

Sherman, P. 1961b. The water binding capacity of fresh pork. III. The influence of cooking temperature on the water binding capacity of lean pork. *Food Technol.* 15: 90–94.

Speedy, R.J. 1984. Self-replicating structures in water. *J. Phys. Chem.* 88: 3364–3373.

van den Berg, C., and S. Bruin. 1981. Water activity and its estimation in food systems: Theoretical aspects. In *Water activity—Influences on food quality*, ed. L.B. Rockland and G.F. Steward. New York: Academic Press.

VandenTempel, M. 1958. Rheology of plastic fats. *Rheol. Acta* 1: 115–118.

Wierbicki, E., and F.E. Deatherage. 1958. Determination of water-holding capacity of fresh meats. *J. Agr. Food Chem.* 6: 387–392.

Lipids

INTRODUCTION

It has been difficult to provide a definition for the class of substances called lipids. Early definitions were mainly based on whether the substance is soluble in organic solvents like ether, benzene, or chloroform and is not soluble in water. In addition, definitions usually emphasize the central character of the fatty acids—that is, whether lipids are actual or potential derivatives of fatty acids. Every definition proposed so far has some limitations. For example, monoglycerides of the short-chain fatty acids are undoubtedly lipids, but they would not fit the definition on the basis of solubility because they are more soluble in water than in organic solvents. Instead of trying to find a definition that would include all lipids, it is better to provide a scheme describing the lipids and their components, as Figure 2–1 shows. The basic components of lipids (also called *derived lipids*) are listed in the central column with the fatty acids occupying the prominent position. The left column lists the lipids known as phospholipids. The right column of the diagram includes the compounds most important from a quantitative standpoint in foods. These are mostly esters of fatty acids and glycerol. Up to 99 percent of the lipids in plant and animal material consist of such esters, known as fats and oils. Fats are solid at room temperature, and oils are liquid.

The fat content of foods can range from very low to very high in both vegetable and animal products, as indicated in Table 2–1. In nonmodified foods, such as meat, milk, cereals, and fish, the lipids are mixtures of many of the compounds listed in Figure 2–1, with triglycerides making up the major portion. The fats and oils used for making fabricated foods, such as margarine and shortening, are almost pure triglyceride mixtures. Fats are sometimes divided into visible and invisible fats. In the United States, about 60 percent of total fat and oil consumed consists of invisible fats—that is, those contained in dairy products (excluding butter), eggs, meat, poultry, fish, fruits, vegetables, and grain products. The visible fats, including lard, butter, margarine, shortening, and cooking oils, account for 40 percent of total fat intake. The interrelationship of most of the lipids is represented in Figure 2–1. A number of minor components, such as hydrocarbons, fat-soluble vitamins, and pigments are not included in this scheme.

Fats and oils may differ considerably in composition, depending on their origin. Both fatty acid and glyceride composition may

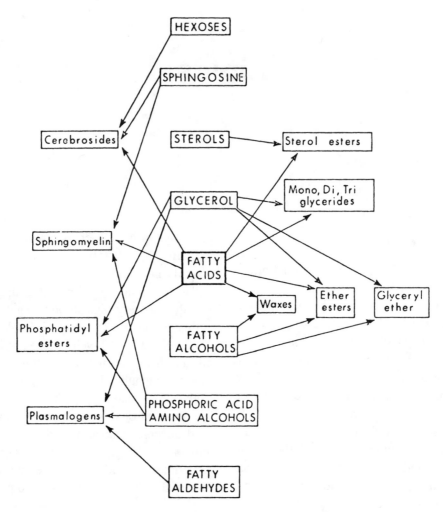

Figure 2–1 Interrelationship of the Lipids

result in different properties. Fats and oils can be classified broadly as of animal or vegetable origin. Animal fats can be further subdivided into mammal depot fat (lard and tallow) and milk fat (mostly ruminant) and marine oils (fish and whale oil). Vegetable oils and fats can be divided into seed oils (such as soybean, canola), fruit coat fats (palm and olive oils), and kernel oils (coconut and palm kernel).

The scientific name for esters of glycerol and fatty acids is acylglycerols. Triacylglycerols, diacylglycerols, and monoacylglycerols have three, two, or one fatty acid ester linkages. The common names for these compounds are glycerides, triglycerides, diglycerides, and monoglycerides. The scientific and common names are used interchangeably in the literature, and this practice is followed in this book.

Table 2–1 Fat Contents of Some Foods

Product	Fat (%)
Asparagus	0.25
Oats	4.4
Barley	1.9
Rice	1.4
Walnut	58
Coconut	34
Peanut	49
Soybean	17
Sunflower	28
Milk	3.5
Butter	80
Cheese	34
Hamburger	30
Beef cuts	10–30
Chicken	7
Ham	31
Cod	0.4
Haddock	0.1
Herring	12.5

SHORTHAND DESCRIPTION OF FATTY ACIDS AND GLYCERIDES

To describe the composition of fatty acids it is sometimes useful to use a shorthand designation. In this convention the composition of a fatty acid can be described by two numbers separated by a colon. The first number indicates the number of carbon atoms in the fatty acid chain, the second number indicates the number of double bonds. Thus, 4:0 is short for butyric acid, 16:0 for palmitic acid, 18:1 for oleic acid, etc. The two numbers provide a complete description of a saturated fatty acid. For unsaturated fatty acids, information about the location of double bonds and their stereo isomers can be given as follows: oleic acid (the *cis* isomer) is 18:1c9; elaidic acid (the *trans* isomer) is 18:1t9. The numbering of carbon atoms in fatty acids starts normally with the carboxyl carbon as number one. In some cases polyunsaturated fatty acids are numbered starting at the methyl end; for instance, linoleic acid is represented as 18:2n-6 and linolenic acid 18:3n-3. These symbols indicate straight-chain, 18-carbon fatty acids with two and three methylene interrupted *cis* double bonds that start at the sixth and third carbon from the methyl end, respectively. These have also been described as ω6 and ω3. The reason for this type of description is that the members of each group n-6 or n-3 are related biosynthetically through processes involving desaturation, chain elongation, and chain shortening (Gunstone 1986) (Figure 2–2).

Triglycerides can be abbreviated by using the first letters of the common names of the component fatty acids. SSS indicates tristearin, PPP tripalmitin, and SOS a triglyceride with two palmitic acid residues in the 1 and 3 positions and oleic acid in the 2 position. In some cases, glyceride compositions are discussed in terms of saturated and unsaturated component fatty acids. In this case, S and U are used and glycerides would be indicated as SSS for trisaturated glyceride and SUS for a glyceride with an unsaturated fatty acid in the 2 position. In other cases, the total number of carbon atoms in a glyceride is important, and this can be shortened to glycerides with carbon numbers 54, 52, and so on. A glyceride with carbon number 54 could be made up of three fatty acids with 18 carbons, most likely to happen if the glyceride originated from one of the seed oils. A glyceride with carbon number 52 could have two component fatty acids with 18 carbons and one with 16 carbons. The carbon number does not give any information about saturation and unsaturation.

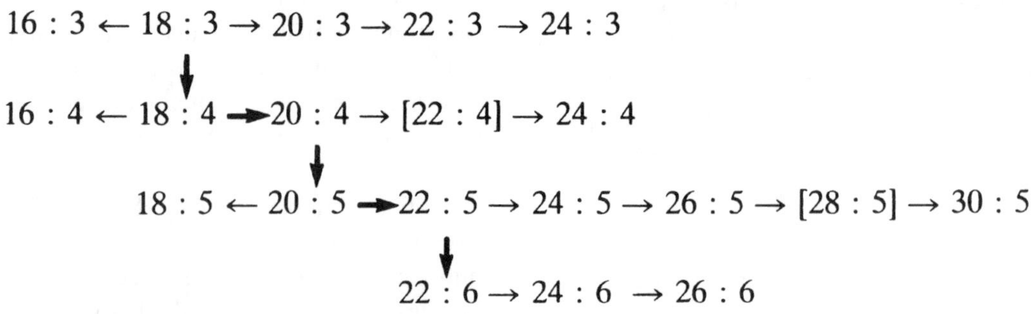

$$16 : 3 \leftarrow 18 : 3 \rightarrow 20 : 3 \rightarrow 22 : 3 \rightarrow 24 : 3$$

$$16 : 4 \leftarrow 18 : 4 \longrightarrow 20 : 4 \rightarrow [22 : 4] \rightarrow 24 : 4$$

$$18 : 5 \leftarrow 20 : 5 \longrightarrow 22 : 5 \rightarrow 24 : 5 \rightarrow 26 : 5 \rightarrow [28 : 5] \rightarrow 30 : 5$$

$$22 : 6 \rightarrow 24 : 6 \rightarrow 26 : 6$$

Figure 2–2 The *n*-3 Family Polyunsaturated Fatty Acids Based on Linolenic Acid. The heavy arrows show the relationship between the most important *n*-3 acids through desaturation (vertical arrows) and chain elongation (horizontal arrows)

COMPONENT FATTY ACIDS

Even-numbered, straight-chain saturated and unsaturated fatty acids make up the greatest proportion of the fatty acids of natural fats. However, it is now known that many other fatty acids may be present in small amounts. Some of these include odd carbon number acids, branched-chain acids, and hydroxy acids. These may occur in natural fats (products that occur in nature), as well as in processed fats. The latter category may, in addition, contain a variety of isomeric fatty acids not normally found in natural fats. It is customary to divide the fatty acids into different groups, for example, into saturated and unsaturated ones. This particular division is useful in food technology because saturated fatty acids have a much higher melting point than unsaturated ones, so the ratio of saturated fatty acids to unsaturated ones significantly affects the physical properties of a fat or oil. Another common division is into short-chain, medium-chain, and long-chain fatty acids. Unfortunately, there is no generally accepted division of these groups. Gen-erally, short-chain fatty acids have from 4 to 10 carbon atoms; medium-chain fatty acids, 12 or 14 carbon atoms; and long-chain fatty acids, 16 or more carbon atoms. However, some authors use the terms long- and short-chain fatty acid in a strictly relative sense. In a fat containing fatty acids with 16 and 18 carbon atoms, the 16 carbon acid could be called the *short-chain* fatty acid. Yet another division differentiates between essential and nonessential fatty acids.

Some of the more important saturated fatty acids are listed with their systematic and common names in Table 2–2, and some of the unsaturated fatty acids are listed in Table 2–3. The naturally occurring unsaturated fatty acids in fats are almost exclusively in the *cis*-form (Figure 2–3), although *trans*-acids are present in ruminant milk fats and in catalytically hydrogenated fats. In general, the following outline of fatty acid composition can be given:

• *Depot fats* of higher land animals consist mainly of palmitic, oleic, and stearic acid and are high in saturated fatty acids.

Table 2–2 Saturated Even- and Odd-Carbon Numbered Fatty Acids

Systematic Name	Common Name	Formula	Shorthand Description
n-Butanoic	Butyric	$CH_3 \cdot (CH_2)_2 \cdot COOH$	4:0
n-Hexanoic	Caproic	$CH_3 \cdot (CH_2)_4 \cdot COOH$	6:0
n-Octanoic	Caprylic	$CH_3 \cdot (CH_2)_6 \cdot COOH$	8:0
n-Decanoic	Capric	$CH_3 \cdot (CH_2)_8 \cdot COOH$	10:0
n-Dodecanoic	Lauric	$CH_3 \cdot (CH_2)_{10} \cdot COOH$	12:0
n-Tetradecanoic	Myristic	$CH_3 \cdot (CH_2)_{12} \cdot COOH$	14:0
n-Hexadecanoic	Palmitic	$CH_3 \cdot (CH_2)_{14} \cdot COOH$	16:0
n-Octadecanoic	Stearic	$CH_3 \cdot (CH_2)_{16} \cdot COOH$	18:0
n-Eicosanoic	Arachidic	$CH_3 \cdot (CH_2)_{18} \cdot COOH$	20:0
n-Docosanoic	Behenic	$CH_3 \cdot (CH_2)_{20} \cdot COOH$	22:0
n-Pentanoic	Valeric	$CH_3 \cdot (CH_2)_3 \cdot COOH$	5:0
n-Heptanoic	Enanthic	$CH_3 \cdot (CH_2)_5 \cdot COOH$	7:0
n-Nonanoic	Pelargonic	$CH_3 \cdot (CH_2)_7 \cdot COOH$	9:0
n-Undecanoic	—	$CH_3 \cdot (CH_2)_9 \cdot COOH$	11:0
n-Tridecanoic	—	$CH_3 \cdot (CH_2)_{11} \cdot COOH$	13:0
n-Pentadecanoic	—	$CH_3 \cdot (CH_2)_{13} \cdot COOH$	15:0
n-Heptadecanoic	Margaric	$CH_3 \cdot (CH_2)_{15} \cdot COOH$	17:0

The total content of acids with 18 carbon atoms is about 70 percent.

- *Ruminant milk fats* are characterized by a much greater variety of component fatty acids. Lower saturated acids with 4 to 10 carbon atoms are present in relatively large amounts. The major fatty acids are palmitic, oleic, and stearic.
- *Marine oils* also contain a wide variety of fatty acids. They are high in unsaturated fatty acids, especially those unsaturated acids with long chains containing 20 or 22 carbons or more. Several of these fatty acids, including eicosapentaenoic acid (EPA) and docosahexaenoic acid (DHA), have recently received a good deal of attention because of biomedical interest (Ackman 1988b).
- *Fruit coat fats* contain mainly palmitic, oleic, and sometimes linoleic acids.
- *Seedfats* are characterized by low contents of saturated fatty acids. They contain palmitic, oleic, linoleic, and linolenic acids. Sometimes unusual fatty acids may be present, such as erucic acid in rapeseed oil. Recent developments in plant breeding have made it possible to change the fatty acid composition of seed oils dramatically. Rapeseed oil in which the erucic acid has been replaced by oleic acid is known as canola oil. Low linolenic acid soybean oil can be obtained, as

Table 2–3 Unsaturated Fatty Acids

Systematic Name	Common Name	Formula	Shorthand Description
Dec-9-enoic		$CH_2=CH \cdot (CH_2)_7 \cdot COOH$	10:1
Dodec-9-enoic		$CH_3 \cdot CH_2 \cdot CH=CH \cdot (CH_2)_7 \cdot COOH$	12:1
Tetradec-9-enoic	Myristoleic	$CH_3 \cdot (CH_2)_3 \cdot CH=CH \cdot (CH_2)_7 \cdot COOH$	14:1
Hexadec-9-enoic	Palmitoleic	$CH_3 \cdot (CH_2)_5 \cdot CH=CH \cdot (CH_2)_7 \cdot COOH$	16:1
Octadec-6-enoic	Petroselinic	$CH_3 \cdot (CH_2)_{10} \cdot CH=CH \cdot (CH_2)_4 \cdot COOH$	18:1
Octadec-9-enoic	Oleic	$CH_3 \cdot (CH_2)_7 \cdot CH=CH \cdot (CH_2)_7 \cdot COOH$	18:1
Octadec-11-enoic	Vaccenic	$CH_3 \cdot (CH_2)_5 \cdot CH=CH \cdot (CH_2)_9 \cdot COOH$	18:1
Octadeca-9:12-dienoic	Linoleic	$CH_3 \cdot (CH_2)_4 \cdot (CH=CH \cdot CH_2)_2 \cdot (CH_2)_6 \cdot COOH$	18:2ω6
Octadeca-9:12:15-trienoic	Linolenic	$CH_3 \cdot CH_2 \cdot (CH=CH \cdot CH_2)_3 \cdot (CH_2)_6 \cdot COOH$	18:3ω3
Octadeca-6:9:12-trienoic	γ-Linolenic	$CH_3 \cdot (CH_2)_4 \cdot (CH=CH \cdot CH_2)_3 \cdot (CH_2)_3 \cdot COOH$	18:3ω6
Octadeca-9:11:13-trienoic	Elaeostearic	$CH_3 \cdot (CH_2)_3 \cdot (CH=CH)_3 \cdot (CH_2)_7 \cdot COOH$	20:3
Eicos-9-enoic	Gadoleic	$CH_3 \cdot (CH_2)_9 \cdot CH=CH \cdot (CH_2)_7 \cdot COOH$	20:1
Eicosa-5:8:11:14-tetraenoic	Arachidonic	$CH_3 \cdot (CH_2)_4 \cdot (CH=CH \cdot CH_2)_4 \cdot (CH_2)_2 \cdot COOH$	20:4ω6
Eicosa-5:8:11:14:17-pentaenoic acid	EPA	$CH_3 \cdot CH_2 \cdot (CH=CH \cdot CH_2)_5 \cdot (CH_2)_2 \cdot COOH$	20:5ω3
Docos-13-enoic	Erucic	$CH_3 \cdot (CH_2)_7 \cdot CH=CH \cdot (CH_2)_{11} \cdot COOH$	22:1
Docosa-4:7:10:13:16:19-hexaenoic acid	DHA	$CH_3 \cdot CH_2 (CH=CH \cdot CH_2)_6 \cdot (CH_2) \cdot COOH$	22:6ω3

can sunflower and linseed oils with more desirable fatty acid composition.

The depot fats of higher land animals, especially mammals, have relatively simple fatty acid composition. The fats of birds are somewhat more complex. The fatty acid compositions of the major food fats of this group are listed in Table 2–4. The kind of feed consumed by the animals may greatly influence the composition of the depot fats. Animal depot fats are characterized by the presence of 20 to 30 percent palmitic acid, a property shared by human depot fat. Many of the seed oils, in contrast, are very low in palmitic acid. The influence of food con-

Figure 2–3 Structures of Octadec-*cis*-9-Enoic Acid (Oleic Acid) and Octadec-*trans*-9-Enoic Acid (Elaidic Acid)

Table 2–4 Component Fatty Acids of Animal Depot Fats

Animal	Fatty Acids Wt %						
	14:0	*16:0*	*16:1*	*18:0*	*18:1*	*18:2*	*18:3*
Pig	1	24	3	13	41	10	1
Beef	4	25	5	19	36	4	Trace
Sheep	3	21	2	25	34	5	3
Chicken	1	24	6	6	40	17	1
Turkey	1	20	6	6	38	24	2

sumption applies equally for the depot fat of chicken and turkey (Marion et al. 1970; Jen et al. 1971). The animal depot fats are generally low in polyunsaturated fatty acids. The iodine value of beef fat is about 50 and of lard about 60. Iodine value is generally used in the food industry as a measure of total unsaturation in a fat.

Ruminant milk fat is extremely complex in fatty acid composition. By using gas chromatography in combination with fractional distillation of the methyl esters and adsorption chromatography, Magidman et al. (1962) and Herb et al. (1962) identified at least 60 fatty acids in cow's milk fat. Several additional minor fatty acid components have been found in other recent studies. About 12 fatty acids occur in amounts greater than 1 percent (Jensen and Newburg 1995). Among these, the short-chain fatty acids from butyric to capric are characteristic of ruminant milk fat. Data provided by Hilditch and Williams (1964) on the component fatty acids of some milk fats are listed in Table 2–5. Fatty acid compositions are usually reported in percentage by weight, but in the case of fats containing short-chain fatty acids (or very long-chain fatty acids) this method may not give a good impression of the molecular proportions of fatty acids present. Therefore, in many instances, the fatty acid composi-

tion is reported in mole percent, as is the case with the data in Table 2–5. According to Jensen (1973) the following fatty acids are present in cow's milk fat: even and odd saturated acids from 2:0 to 28:0; even and odd monoenoic acids from 10:1 to 26:1, with the exception of 11:1, and including positional and geometric isomers; even unsaturated fatty acids from 14:2 to 26:2 with some conjugated geometric isomers; polyenoic even acids from 18:3 to 22:6 including some conjugated *trans* isomers; monobranched fatty acids 9:0 and 11:0 to 25:0—some *iso* and some *ante-iso* (*iso* acids have a methyl branch on the penultimate carbon, *ante-iso* on the next to penultimate carbon [Figure 2–4]); multibranched acids from 16:0 to 28:0, both odd and even with three to five methyl branches; and a number of keto, hydroxy, and cyclic acids.

It is impossible to determine all of the constituents of milk fatty acids by a normal chromatographic technique, because many of the minor component fatty acids are either not resolved or are covered by peaks of other major fatty acids. A milk fat chromatogram of fatty acid composition is shown in Figure 2–5. Such fatty acid compositions as reported are therefore only to be considered as approximations of the major component fatty acids; these are listed in Table 2–6. This

Table 2–5 The Component Fatty Acids of Some Milk Fats in Mole %

Fatty Acid	Cow	Goat	Sheep
4:0	9.5	7.5	7.5
6:0	4.1	4.7	5.3
8:0	0.8	4.3	3.5
10:0	3.2	12.8	6.4
Total short chain	17.6	29.3	22.7
12:0	2.9	6.6	4.5
14:0	11.5	11.8	9.9
16:0	26.7	24.1	21.6
18:0	7.6	4.7	10.3
20:0	1.8	0.4	0.8
10–12 unsaturated	1.1	1.4	1.0
16:1	4.3	2.2	2.0
18:1	22.4	16.5	21.6
18:2	3.1	2.8	4.3
20–22 unsaturated	1.0	0.2	1.3

Source: From T.P. Hilditch and P.N. Williams, *The Chemical Constitution of Natural Fats*, 4th ed., 1964, John Wiley & Sons.

table reports the most recent results of the major component fatty acids in bovine milk fat as well as their distribution among the sn-1, sn-2, and sn-3 positions in the triacylglycerols (Jensen and Newburg 1995).

In most natural fats the double bonds of unsaturated fatty acids occur in the *cis* configuration. In milk fat a considerable proportion is in the *trans* configuration. These *trans* bonds result from microbial action in the rumen where polyunsaturated fatty acids of the feed are partially hydrogenated. Catalytic hydrogenation of oils in the fat industry

also results in *trans* isomer formation. The level of *trans* isomers in milk fat has been reported as 2 to 4 percent (deMan and deMan 1983). Since the total content of unsaturated fatty acids in milk fat is about 34 percent, *trans* isomers may constitute about 10 percent of total unsaturation. The complexity of the mixture of different isomers is demonstrated by the distribution of positional and geometric isomers in the monoenoic fatty acids of milk fat (Table 2–7) and in the unconjugated 18:2 fatty acids (Table 2–8). The iodine value of milk fat is

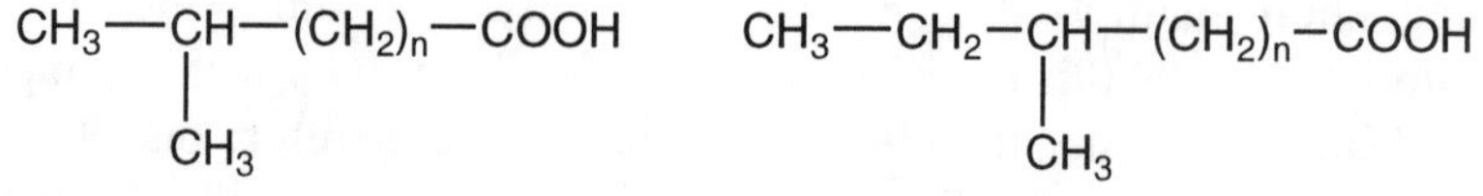

$$CH_3-CH-(CH_2)_n-COOH \qquad CH_3-CH_2-CH-(CH_2)_n-COOH$$
$$\quad\;\; | \qquad\qquad\qquad\qquad\qquad\qquad\qquad | $$
$$\quad\;\; CH_3 \qquad\qquad\qquad\qquad\qquad\qquad\quad CH_3$$

Figure 2–4 Examples of Iso- and Ante-Iso-Branched-Chain Fatty Acids

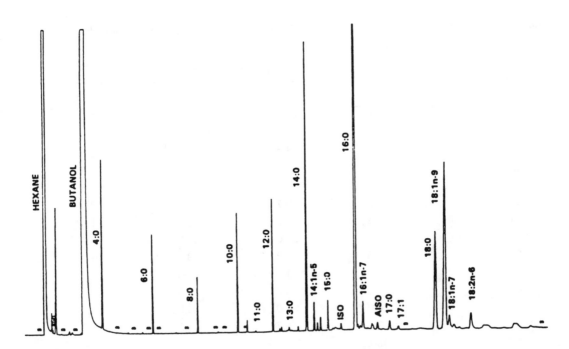

Figure 2–5 Chromatogram of Milk Fat Fatty Acid Composition Analyzed as Butyl Esters on a 30-m Capillary Column. *Source:* Reprinted from R.G. Ackman, Animal and Marine Lipids, in *Improved and Technological Advances in Alternative Sources of Lipids*, B. Kamel and Y. Kakuda, eds., p. 298, © 1994, Aspen Publishers, Inc.

in the range of 30 to 35, much lower than that of lard, shortening, or margarine, which have similar consistencies.

Marine oils have also been found to contain a large number of component fatty acids. Ackman (1972) has reported as many as 50 or 60 components. Only about 14 of these are of importance in terms of weight percent of the total. These consist of relatively few saturated fatty acids (14:0, 16:0, and 18:0) and a larger number of unsaturated fatty acids with 16 to 22 carbon atoms and up to 6 double bonds. This provides the possibility for many positional isomers.

The complexity of the fatty acid composition of marine oils is evident from the chromatogram shown in Figure 2–6 (Ackman 1994). The end structure of the polyunsatu-

rated fatty acids is of nutritional importance, especially eicosapentaenoic acid (EPA), 20:5ω3 or 20:5 n-3, and docosahexaenoic acid (DHA), 22:6ω3 or 22:6 n-3. The double bonds in marine oils occur exclusively in the *cis* configuration. EPA and DHA can be produced slowly from linolenic acid by herbivore animals, but not by humans. EPA and DHA occur in major amounts in fish from cold, deep waters, such as cod, mackerel, tuna, swordfish, sardines, and herring (Ackman 1988a; Simopoulos 1988). Arachidonic acid is the precursor in the human system of prostanoids and leukotrienes.

Ackman (1988b) has drawn attention to the view that the fatty acid compositions of marine oils are all much the same and vary

Table 2–6 Major Fatty Acids of Bovine Milk Fat and Their Distribution in the Triacylglycerols

Fatty Acids (mol%)	Bovine Milk Fat			
	TG	sn-1	sn-2	sn-3
4:0	11.8	—	—	35.4
6:0	4.6	—	0.9	12.9
8:0	1.9	1.4	0.7	3.6
10:0	37	1.9	3.0	6.2
12:0	3.9	4.9	6.2	0.6
14:0	11.2	9.7	17.5	6.4
15:0	2.1	2.0	2.9	1.4
16:0	23.9	34.0	32.3	5.4
16:1	2.6	2.8	3.6	1.4
17:0	0.8	1.3	1.0	0.1
18:0	7.0	10.3	9.5	1.2
18:1	24.0	30.0	18.9	23.1
18:2	2.5	1.7	3.5	2.3
18:3	Trace	—	—	—

Source: Reprinted with permission from R.G. Jensen and D.S. Newburg, Milk Lipids, in *Handbook of Milk Composition*, R.G. Jensen, ed., p. 546, © 1995, Academic Press.

only in the proportions of fatty acids. The previously held view was that marine oils were species-specific. The major fatty acids of marine oils from high-, medium-, and low-fat fish are listed in Table 2–9 (Ackman 1994).

The fatty acid composition of egg yolk is given in Table 2–10. The main fatty acids are palmitic, oleic, and linoleic. The yolk constitutes about one-third of the weight of the edible egg portion. The relative amounts of egg yolk and white vary with the size of the egg. Small eggs have relatively higher amounts of yolk. The egg white is virtually devoid of fat.

The vegetable oils and fats can be divided into three groups on the basis of fatty acid composition. The first group comprises oils containing mainly fatty acids with 16 or 18 carbon atoms and includes most of the seed oils; in this group are cottonseed oil, peanut oil, sunflower oil, corn oil, sesame oil, olive oil, palm oil, soybean oil, and safflower oil. The second group comprises seed oils containing erucic (docos-13-enoic) acid. These include rapeseed and mustard seed oil. The third group is the vegetable fats, comprising coconut oil and palm kernel oil, which are highly saturated (iodine value about 15), and cocoa butter, the fat obtained from cocoa beans, which is hard and brittle at room temperature (iodine value 38). The component fatty acids of some of the most common vegetable oils are listed in Table 2–11. Palmitic is the most common saturated fatty acid in vegetable oils, and only very small amounts of stearic acid are present. Oils containing linolenic acid, such

Table 2–7 Positional and Geometric Isomers of Bovine Milk Lipid Monoenoic Fatty Acids (Wt%)

Position of Double Bond	cis Isomers				trans Isomers	
	14:1	*16:1*	*17:1*	*18:1*	*16:1*	*18:1*
5	1.0	Tr			2.2	
6	0.8	1.3	3.4		7.8	1.0
7	0.9	5.6	2.1		6.7	0.8
8	0.6	Tr	20.1	1.7	5.0	3.2
9	96.6	88.7	71.3	95.8	32.8	10.2
10	—	Tr	Tr	Tr	1.7	10.5
11	—	2.6	2.9	2.5	10.6	35.7
12	—	Tr	Tr	—	12.9	4.1
13	—	—	—	—	10.6	10.5
14	—	—	—	—	—	9.0
15	—	—	—	—	—	6.8
16	—	—	—	—	—	7.5

Source: From R.G. Jensen, Composition of Bovine Milk Lipids, *J. Am. Oil Chem. Soc.*, Vol. 50, pp. 186–192, 1973.

as soybean oil, are unstable. Such oils can be slightly hydrogenated to reduce the linolenic acid content before use in foods. Another fatty acid that has received attention for its possible beneficial effect on health is the n-6 essential fatty acid, gamma-linolenic acid (18:3 n-6), which occurs at a level of 8 to 10 percent in evening primrose oil (Carter 1988).

The *Crucifera* seed oils, including rapeseed and mustard oil, are characterized by the presence of large amounts of erucic acid

Table 2–8 Location of Double Bonds in Unconjugated 18:2 Isomers of Milk Lipids

cis, cis	cis, trans or trans, cis	trans, trans
11, 15	11, 16 and/or 11, 15	12, 16
10,15	10, 16 and/or 10, 15	11, 16 and/or 11, 15
9, 15	9, 15 and/or 9, 16	10, 16 and/or 10, 15
8, 15 and/or 8, 12	8, 16 and/or 8, 15	9, 16 and/or 9, 15
7, 15 and/or 7, 12	and/or 8, 12	and/or 9, 13
6, 15 and/or 6, 12		

Source: From R.G. Jensen, Composition of Bovine Milk Lipids, *J. Am. Oil Chem. Soc.*, Vol. 50, pp. 186–192, 1973.

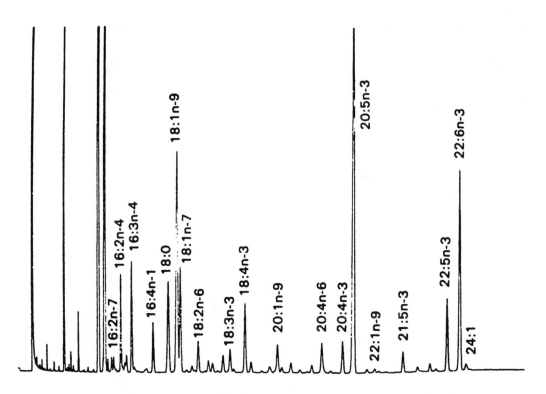

Figure 2–6 Chromatogram of the Fatty Acid Composition of Fish Oil (Menhaden). Analysis of methyl esters on a 30-m capillary column. *Source:* Reprinted from R.G. Ackman, Animal and Marine Lipids, in *Improved and Technological Advances in Alternative Sources of Lipids*, B. Kamel and Y. Kakuda, eds., p. 308, 1994, Aspen Publishers, Inc.

(docos-13-enoic) and smaller amounts of eicos-11-enoic acid. Rapeseed oil of the variety *Brassica napus* may have over 40 percent of erucic acid (Table 2–12), whereas *Brassica campestris* oil usually has a much lower erucic acid content, about 22 percent. Because of possible health problems resulting from ingestion of erucic acid, new varieties of rapeseed have been introduced in recent years; these are the so-called low-erucic acid rapeseed (LEAR) varieties, which produce LEAR oil. When the seed is also low in glucosinolates, the oil is known as canola oil. Plant breeders have succeeded in reducing the erucic acid level to less than 1 percent and as a result canola oil has a very

high level of oleic acid (Table 2–12). The breeding of these varieties has in effect resulted in the creation of a completely new oil. Removal of the erucic and eicosenoic acids results in a proportional increase in the oleic acid content. The low erucic acid oil is a linolenic acid–containing oil and is therefore similar in this respect to soybean oil. The fatty acid composition of mustard oil is given in Table 2–12. It is similar to that of *B. campestris* oil.

Vegetable fats, in contrast to the oils, are highly saturated, have low iodine values, and have high melting points. Coconut oil and palm kernel oil belong to the lauric acid fats. They contain large amounts of medium- and

Table 2–9 Total Fat Content and Major Fatty Acids in High-, Medium-, and Low-Fat Fish

	High Fat		Medium Fat		Low Fat	
	Capelin	Sprat	Blue Whiting	Capelin	Dogfish	Saith, Gutted
Total fat	14.1	12.9	7.4	4.0	1.7	0.4
Fatty acid						
14:0	7.1	5.5	3.9	7.3	1.6	1.7
16:0	9.9	17.5	11.5	9.7	15.3	12.4
16:1	11.0	5.8	6.1	8.3	4.9	2.7
18:1	13.4	18.0	14.8	14.5	20.8	13.1
20:1	16.3	7.4	10.7	13.6	11.2	5.9
22:1	12.6	12.8	12.4	10.4	7.9	3.5
20:5n-3	8.6	7.4	10.4	9.2	6.0	12.7
22:6n-3	6.7	11.7	12.6	11.0	15.5	30.6
Total	85.6	86.1	82.4	84.0	84.8	82.6

Source: Reprinted from R.G. Ackman, Animal and Marine Lipids, in *Improved and Technological Advances in Alternative Sources of Lipids*, B. Kamel and Y. Kakuda, eds., p. 302, 1994, Aspen Publishers, Inc.

short-chain fatty acids, especially lauric acid (Table 2–13). Cocoa butter is unusual in that it contains only three major fatty acids—palmitic, stearic, and oleic—in approximately equal proportions.

Table 2–10 Fatty Acid Composition of Egg Yolk

Fatty Acid	%
Total saturated	36.2
14:0	0.3
16:0	26.6
18:0	9.3
Total monounsaturated	48.2
16:1	4.0
18:1	44.1
Total polyunsaturated	14.7
18:2	13.4
18:3	0.3
20:4	1.0

COMPONENT GLYCERIDES

Natural fats can be defined as mixtures of mixed triglycerides. Simple triglycerides are virtually absent in natural fats, and the distribution of fatty acids both between and within glycerides is selective rather than random. When asymmetric substitution in a glycerol molecule occurs, enantiomorphic forms are produced (Kuksis 1972; Villeneuve and Foglia 1997). This is illustrated in Figure 2–7. Glycerol has a plane of symmetry or mirror plane, because two of the four substituents on the central carbon atom are identical. When one of the carbon atoms is esterified with a fatty acid, a monoglyceride results and two nonsuperimposable structures exist. These are called enantiomers and are also referred to as chiral. A racemic mixture is a mixture of equal amounts of enantiomers. Asymmetric or chiral compounds are formed in 1-monoglycerides; all 1, 2-diglycerides; 1,

Table 2–11 Component Fatty Acids of Some Vegetable Oils

	Fatty Acid Wt%					
Oil	16:0	18:0	18:1	18:2	18:3	Total C18
Canola	4	2	56	26	10	96
Cottonseed	27	2	18	51	Trace	73
Peanut*	13	3	38	41	Trace	83
Olive	10	2	78	7	—	90
Rice bran	16	2	42	37	1	84
Soybean	11	4	22	53	8	89
Sunflower	5	5	20	69	—	95
Sunflower high oleic	4	5	81	8	—	96
Palm	44	4	39	11	—	54
Cocoa butter	26	34	35	3	—	74

*Peanut oil also contains about 3% of 22:0 and 1% of 22:1.

3-diglycerides containing unlike substituents; and all triglycerides in which the 1- and 3- positions carry different acyl groups.

The glyceride molecule can be represented in the wedge and slash form (Figure 2–8). In this spatial representation, the wedge indicates a substituent coming out of the plane toward the observer, and the slash indicates a substituent going away from the observer. The three carbon atoms of the glycerol are then described by the stereospecific numbering (*sn*) with the three carbon atoms designated *sn*-1 from the top to *sn*-3 at the bottom.

When a fat or oil is characterized by determination of its component fatty acids, there still remains the question as to how these acids are distributed among and within the glycerides. Originally theories of glyceride distribution were attempts by means of mathematical schemes to explain the occurrence

Table 2–12 Component Fatty Acids of Some Crucifera Seed Oils (Wt%)

	Fatty Acid							
Seed Oil	16:0	18:0	18:1	18:2	18:3	20:1	22:1	Total C18
Rapeseed (*B. campestris*)	4	2	33	18	9	12	22	62
Rapeseed (*B. napus*)	3	1	17	14	9	11	45	41
Canola (LEAR)	4	2	55	26	10	2	<1	96
Mustard (*B. juncea*)	4	—	22	24	14	12	20	60

Source: Data from B.M. Craig et al., Influence of Genetics, Environment, and Admixtures on Low Erucic Acid Rapeseed in Canada, *J. Am. Oil Chem. Soc.*, Vol. 50, pp. 395–399, 1973; and M. Vaisey-Genser and N.A.M. Eskin, *Canola Oil: Properties and Performance*, 1987, Canola Council.

Table 2–13 Component Fatty Acids of Some Vegetable Fats (Wt %)

Vegetable Fat	Fatty Acid								
	6:0	*8:0*	*10:0*	*12:0*	*14:0*	*16:0*	*18:0*	*18:1*	*18:2*
Coconut	0.5	9.0	6.8	46.4	18.0	9.0	1.0	7.6	1.6
Palm kernel	—	2.7	7.0	46.9	14.1	8.8	1.3	18.5	0.7
Cocoa butter	—	—	—	—	—	26.2	34.4	37.3	2.1

of particular kinds and amounts of glycerides in natural fats. Subsequent theories have been refinements attempting to relate to the biochemical mechanisms of glyceride synthesis. Hilditch proposed the concept of even distribution (Gunstone 1967). In the rule of even (or widest) distribution, each fatty acid in a fat is distributed as widely as possible among glyceride molecules. This means that when a given fatty acid A constitutes about 35 mole percent or more of the total fatty acids (A + X), it will occur at least once in all triglyceride molecules, as represented by GAX_2. If A occurs at levels of 35 to 70 mole percent, it will occur twice in an increasing number of triglycerides GA_2X. At levels over 70 percent, simple triglycerides GA_3 are formed. In strictly random distribution the amount of GA_3 in a fat would be proportional to the cube of the percentage of A present. For example, at 30 percent A there would be 2.7 percent of GA_3, which under rules of even distribution would occur only at levels of A over 70 percent (Figure 2–9).

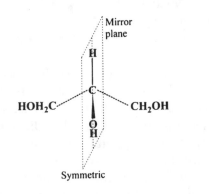

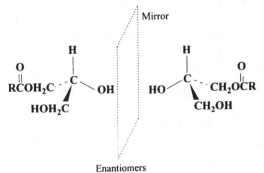

Figure 2–7 Plane of Symmetry of a Glycerol Molecule (Top) and Mirror Image of Two Enantiomers of a Mono-Acylglycerol (bottom). *Source:* Reprinted with permission from P. Villeneuve and T.A. Foglia, Lipase Specificities: Potential Application in Bioconversions, *Inform*, 8, pp. 640–650, © 1997, AOCS Press.

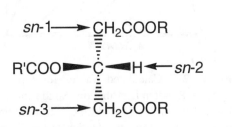

Figure 2–8 Stereospecific Numbering of the Carbons in a Triacylglycerol

The theory of restricted random distribution was proposed by Kartha (1953). In this theory the fatty acids are distributed at random, but the content of fully saturated glycerides is limited to the amount that can remain fluid *in vivo*. This theory is followed by the 1,3 random, 2 random distribution hypothesis of Vander Wal (1964). According to this theory, all acyl groups at the 2-positions of the glycerol moieties of a fat are distributed therein at random. Equally, all acyl groups at the 1- and 3-positions are distributed at random and these positions are identical. Application of this theory to the results obtained with a number of fats gave good agreement (Vander Wal 1964), as Table 2–14 shows.

In vegetable fats and oils, the saturated fatty acyl groups have a tendency to occupy the 1- and 3- positions in the glycerides and the unsaturated acyl groups occupy the 2-position (Figure 2–10). Since these fats contain a limited number of fatty acids, it is customary to show the glyceride composition in terms of saturated (S) and unsaturated (U) acids. The predominant glyceride types in these fats and oils are S-U-S and S-U-U. Lard is an exception—saturated acyl groups predominate in the 2-position. The glyceride distribution of cocoa butter results in a fat with a sharp melting point of about 30 to 34°C. It is hard and brittle below the melting point, which makes the fat useful for chocolate and confectionery manufacture. Other fats with similar fatty acid composition, such as sheep depot fat (see Table 2–4), have a greater variety of glycerides, giving the fat a higher melting point (about 45°C) and a wider melting range, and a greasy and soft appearance.

Brockerhoff et al. (1966) studied the fatty acid distribution in the 1-, 2-, and 3-positions of the triglycerides of animal depot fats by stereospecific analysis. The distribution among the three positions was nonrandom. The distribution of fatty acids seems to be governed by chain length and unsaturation. In most fats a short chain and unsaturation direct a fatty acid toward position 2. The depot fat of pigs is an exception, palmitic acid being predominant in position 2. In the fats of marine animals, chain length is the directing factor, with polyunsaturated and short-chain fatty acids accumulated in the 2-position and long chains in the 1- and 3-positions. In the fats of birds, unsaturation seems to be the only directing factor and these acids accumulate in the 2-position.

The positional distribution of fatty acids in pig fat (lard) and cocoa butter is shown in Table 2–15. Most of the unsaturation in lard is located in the 1- and 3-positions, whereas in cocoa butter the major portion of the unsaturation is located in the 2-position. This

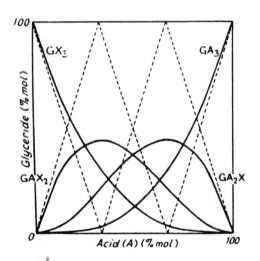

Figure 2–9 Calculated Values for Glyceride Types in Random Distribution (Solid Lines) and Even Distribution (Dotted Lines). *Source:* From F.D. Gunstone, *An Introduction to the Chemistry of Fats and Fatty Acids,* 1967, Chapman and Hall.

Table 2–14 Comparison of the Glyceride Composition of Some Natural Fats as Determined Experimentally and as Calculated by 1,3 Random, 2 Random Hypothesis

		Molecular Species					
Fat	Method	SSS (Mole %)	SUS (Mole %)	SSU (Mole %)	USU (Mole %)	UUS (Mole %)	UUU (Mole %)
Lard	Experiment	8	0	29	36	15	12
Lard	Calculated	6	2	29	36	12	15
Chicken fat	Experiment	3	10	9	12	38	28
Chicken fat	Calculated	3	10	10	9	36	32
Cocoa butter	Experiment	5	66	7	3	20	1
Cocoa butter	Calculated	5	69	2	0	22	2

Source: From R.J. Vander Wal, Triglyceride Structure, *Adv. Lipid Res.*, Vol. 2, pp. 1–16, 1964.

difference accounts for the difference in physical properties of the two fats (deMan et al. 1987).

Milk fat, with its great variety of fatty acids, also has a very large number of glycerides. It is possible, by, for example, fractional crystallization from solvents, to separate milk fat in a number of fractions with different melting points (Chen and deMan 1966). Milk fat is peculiar in some respects. Its short-chain fatty acids are classified chemically as saturated compounds but behave physically like unsaturated fatty acids. One of the unsaturated fatty acids, the so-called oleic acid, is partly *trans* and has a much higher melting point than the *cis* isomers. In the highest melting fraction from milk fat, there is very little short-chain fatty acid and little unsaturation, mostly in the *trans* configuration (Woodrow and deMan 1968). The low melting fractions are high in short-chain fatty acids and unsaturation (*cis*). The general distribution of major fatty acids in whole milk fat is as follows (Morrison

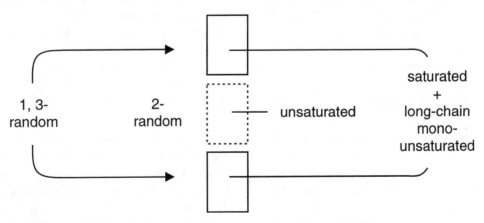

Figure 2–10 Fatty Acid Distribution in the Triacylglycerols of Vegetable Oils

Table 2–15 Positional Distribution Fatty Acids in Pig Fat and Cocoa Butter

Fat	Position	Fatty Acid (Mole %)					
		14:0	16:0	16:1	18:0	18:1	18:2
Pig fat	1	0.9	9.5	2.4	29.5	51.3	6.4
	2	4.1	72.3	4.8	2.1	13.4	3.3
	3	0	0.4	1.5	7.4	72.7	18.2
Cocoa butter	1	—	34.0	0.6	50.4	12.3	1.3
	2	—	1.7	0.2	2.1	87.4	8.6
	3	—	36.5	0.3	52.8	8.6	0.4

Source: From W.C. Breckenridge, Stereospecific Analysis of Triacylglycerols, in *Fatty Acids and Glycerides*, A. Kuksis, ed., 1978, Plenum Press.

1970): 4:0 and 6:0 are located largely in primary positions; 18:0 and 18:1 are preferentially in primary positions; 10:0, 12:0, and 16:0 are distributed randomly or with a slight preference for the secondary position; and 14:0 is predominantly in the secondary position. The distribution of milk fat triacylglycerols according to carbon number and unsaturation has been reported by Jensen and Newburg (1995) and is presented in Table 2–16.

PHOSPHOLIPIDS

All fats and oils and fat-containing foods contain a number of phospholipids. The lowest amounts of phospholipid are present in pure animal fats such as lard and beef tallow. In some crude vegetable oils, such as cottonseed, corn, and soybean oils, phospholipids may be present at levels of 2 to 3 percent. Fish, crustacea, and mollusks contain approximately 0.7 percent of phospholipids in the muscle tissue. Phospholipids are surface active, because they contain a lipophilic and hydrophilic portion. Since they can easily be hydrated, they can be removed from fats and oils during the refin-

ing process. In some cases they may be removed by separation of two phases; for example, if butter is melted and filtered, the pure oil thus obtained is free from phospholipids. The structure of the most important phospholipids is given in Figure 2–11. After refining of oils, neutralization, bleaching, and deodorization, the phospholipid content is reduced to virtually zero. The phospholipids removed from soybean oil are used as emulsifiers in certain foods, such as chocolate. Soybean phospholipids contain about 35 percent lecithin and 65 percent cephalin. The fatty acid composition of phospholipids is usually different from that of the oil in which they are present. The acyl groups are usually more unsaturated than those of the triglycerides. Phospholipids of many vegetable oils contain two oleic acid residues. The phospholipids of milk do not contain the short-chain fatty acids found in milk fat triglycerides, and they contain more long-chain polyunsaturated fatty acids than the triglycerides. The composition of cow's milk phospholipids has been reported by Jensen (1973), as shown in Table 2–17. The difference in composition of triglycerides and phospholipids in mackerel is demon-

Table 2–16 Distribution (wt %) of Milk Fat Triacylglycerols According to Carbon Number and Unsaturation

Carbon Number	Number of Double Bonds			
	0	1	2	3
34	4.8	1.4	—	—
36	5.0	4.9	2.6	—
38	4.6	6.9	2.9	3.1
40	2.0	4.6	3.1	1.2
42	1.5	2.4	2.1	1.2
44	1.0	2.8	2.9	1.0
46	1.3	2.1	2.2	1.0
48	1.6	2.2	2.2	1.0
50	2.6	3.4	2.7	0.8
52	2.7	5.7	1.9	0.4
54	2.2	1.4	0.3	—
Total	29.3	37.8	22.9	9.7

Source: Reprinted with permission from R.G. Jensen and D.S. Newburg, Milk Lipids, in *Handbook of Milk Composition*, R.G. Jensen, ed., p. 550, © 1995, Academic Press.

strated by the data reported by Ackman and Eaton (1971), as shown in Table 2–18. The phospholipids of flesh and liver in mackerel are considerably more unsaturated than the triglycerides.

The distribution of fatty acids in phospholipids is not random, with saturated fatty acids preferentially occupying position 1 and unsaturated fatty acids position 2.

UNSAPONIFIABLES

The unsaponifiable fraction of fats consists of sterols, terpenic alcohols, aliphatic alcohols, squalene, and hydrocarbons. The distribution of the various components of the unsaponifiable fraction in some fats and oils is given in Table 2–19. In most fats the major components of the unsaponifiable fraction are sterols. Animal fats contain cholesterol

and, in some cases, minor amounts of other sterols such as lanosterol. Plant fats and oils contain phytosterols, usually at least three, and sometimes four (Fedeli and Jacini 1971). They contain no or only trace amounts of cholesterol. The predominant phytosterol is β-sitosterol; the others are campesterol and stigmasterol. In rapeseed oil, brassicasterol takes the place of stigmasterol. Sterols are compounds containing the perhydrocyclopenteno-phenanthrene nucleus, which they have in common with many other natural compounds, including bile acids, hormones, and vitamin D. The nucleus and the description of the four rings, as well as the system of numbering of the carbon atoms, are shown in Figure 2–12A. The sterols are solids with high melting points. Stereochemically they are relatively flat molecules, usually with all *trans* linkages, as shown in

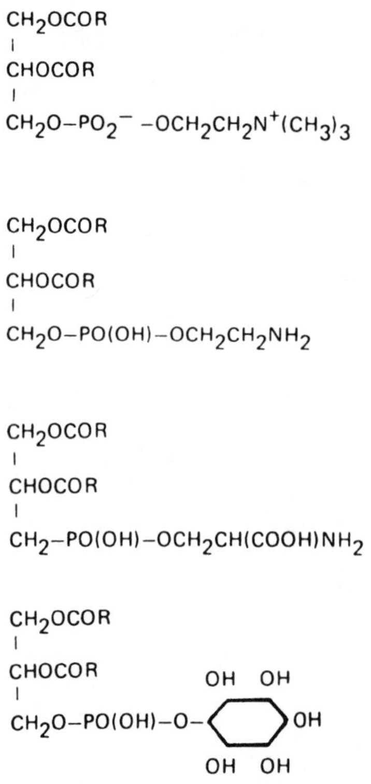

CH_2OCOR
|
$CHOCOR$
|
$CH_2O{-}PO_2^-\ {-}OCH_2CH_2N^+(CH_3)_3$

Phosphatidylcholine
(lecithin)

CH_2OCOR
|
$CHOCOR$
|
$CH_2O{-}PO(OH){-}OCH_2CH_2NH_2$

Phosphatidylethanolamine
(cephalin)

CH_2OCOR
|
$CHOCOR$
|
$CH_2{-}PO(OH){-}OCH_2CH(COOH)NH_2$

Phosphatidylserine

CH_2OCOR
|
$CHOCOR$
|
$CH_2O{-}PO(OH){-}O{-}$ OH OH

Phosphoinositides

Figure 2–11 Structure of the Major Phospholipids

Table 2–17 Composition of the Phospholipids of Cow's Milk

Phospholipid	Mole (%)
Phosphatidylcholine	34.5
Phosphatidylethanolamine	31.8
Phosphatidylserine	3.1
Phosphatidylinositol	4.7
Sphingomyelin	25.2
Lysophosphatidylcholine	Trace
Lysophosphatidylethanolamine	Trace
Total choline phospholipids	59.7
Plasmalogens	3
Diphosphatidyl glycerol	Trace
Ceramides	Trace
Cerebrosides	Trace

Source: From R.G. Jensen, Composition of Bovine Milk Lipids, *J. Am. Oil Chem. Soc.*, Vol. 50, pp. 186–192, 1973.

Table 2–18 Triglycerides and Phospholipids of Mackerel Lipids and Calculated Iodine Values for Methyl Esters of Fatty Acid from Lipids

	Triglycerides			Phospholipids		
	In Lipid (%)	In Tissue (%)	Ester Iodine Value	In Lipid (%)	In Tissue (%)	Ester Iodine Value
Light flesh	89.5	9.1	152.3	4.7	0.5	242.9
Dark flesh	74.2	10.7	144.3	11.3	1.6	208.1
Liver	79.5	14.4	130.9	9.3	1.7	242.1

Source: From R.G. Ackman and C.A. Eaton, Mackerel Lipids and Fatty Acids, *Can. Inst. Food Sci. Technol. J.*, Vol. 4, pp. 169–174, 1971.

Figure 2–12B. The ring junction between rings A and B is *trans* in some steroids, *cis* in others. The junctions between B and C and C and D are normally *trans*. Substituents that lie above the plane, as drawn in Figure 2–12C, are named β, those below the plane, α. The 3-OH group in cholesterol (Figure 2–12C) is the β-configuration, and it is this group that may form ester linkages. The composition of the plant sterols is given in Figure 2–13. Part of the sterols in natural fats are present as esters of fatty acids; for example, in milk fat, about 10 percent of the cholesterol occurs in the form of cholesterol esters.

The sterols provide a method of distinguishing between animal and vegetable fats by means of their acetates. Cholesterol acetate has a melting point of 114°C, whereas phytosterol acetates melt in the range of 126 to 137°C. This provides a way to detect adulteration of animal fats with vegetable fats.

Table 2–19 Composition of the Unsaponifiable Fraction of Some Fats and Oils

Oils	Hydrocarbons	Squalene	Aliphatic Alcohols	Terpenic Alcohols	Sterols
Olive	2.8–3.5	32–50	0.5	20–26	20–30
Linseed	3.7–14.0	1.0–3.9	2.5–5.9	29–30	34.5–52
Teaseed	3.4	2.6	—	—	22.7
Soybean	3.8	2.5	4.9	23.2	58.4
Rapeseed	8.7	4.3	7.2	9.2	63.6
Corn	1.4	2.2	5.0	6.7	81.3
Lard	23.8	4.6	2.1	7.1	47.0
Tallow	11.8	1.2	2.4	5.5	64.0

Source: From G. Jacini, E. Fedeli, and A. Lanzani, Research in the Nonglyceride Substances of Vegetable Oils, *J. Assoc. Off. Anal. Chem.*, Vol. 50, pp. 84–90, 1967.

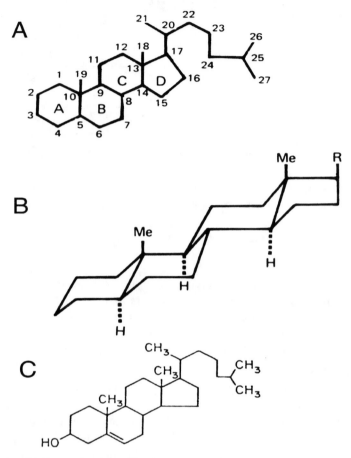

Figure 2–12 Sterols. (**A**) Structure of the Steroid Nucleus, (**B**) Stereochemical Representation, and (**C**) Cholesterol

The sterol content of some fats and oils is given in Table 2–20. Cholesterol is the main sterol of animal, fish, and marine fats and oils.

The hydrocarbons of the unsaponifiable oils are members of the *n*-paraffins as well as of the branched-chain paraffins of the *iso* and *ante-iso* configuration. The composition of hydrocarbon constituents of some vegetable oils has been reported by Jacini et al. (1967) and is listed in Table 2–21.

The structure of squalene is presented in Figure 2–14, which also gives the structure of one of the terpenic alcohols, geranyl geraniol; this alcohol has been reported to be a component of the nonglyceride fraction of vegetable oils (Fedeli et al. 1966).

AUTOXIDATION

The unsaturated bonds present in all fats and oils represent active centers that, among other things, may react with oxygen. This reaction leads to the formation of primary, secondary, and tertiary oxidation products that may make the fat or fat-containing foods unsuitable for consumption.

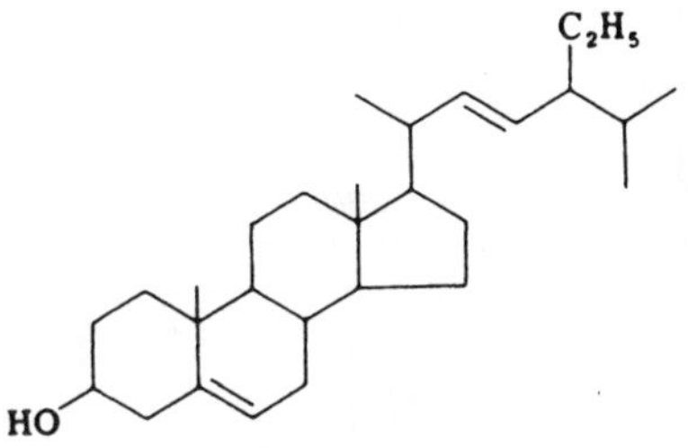

Stigmasterol

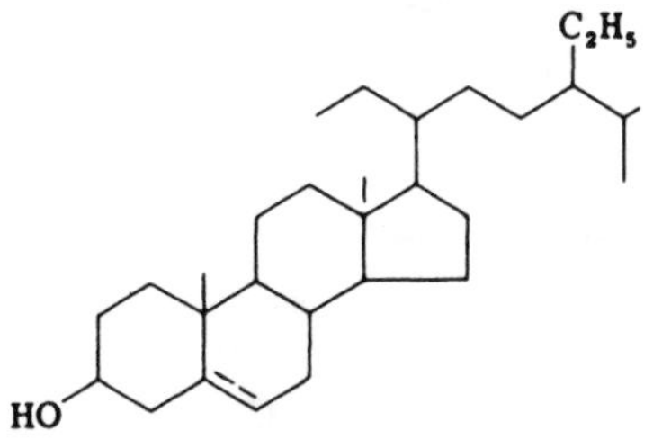

β-Sitosterol

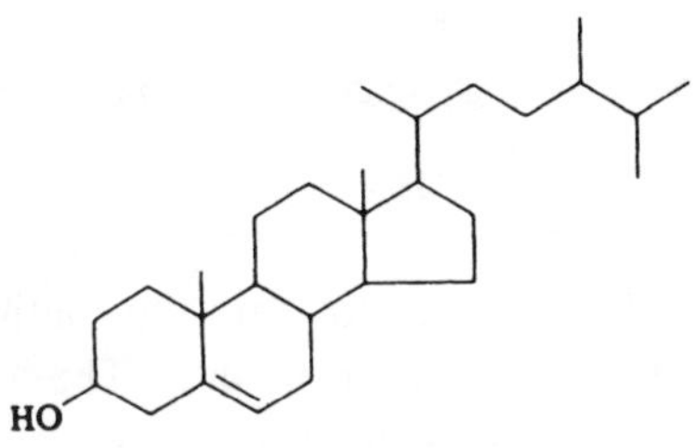

Campesterol

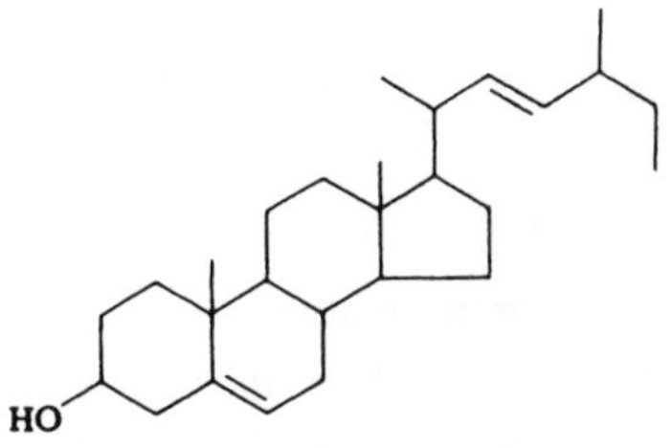

Brassicasterol

Figure 2–13 Structures of the Plant Sterols

The process of autoxidation and the resulting deterioration in flavor of fats and fatty foods are often described by the term *rancidity*. Usually rancidity refers to oxidative deterioration but, in the field of dairy science, rancidity refers usually to hydrolytic changes resulting from enzyme activity. Lundberg (1961) distinguishes several types of rancidity. In fats such as lard, common oxidative rancidity results from exposure to oxygen; this is characterized by a sweet but undesirable odor and flavor that become progressively more intense and unpleasant as oxidation progresses. Flavor reversion is the term used for the objectionable flavors that develop in oils containing linolenic acid. This type of oxidation is produced with considerably less oxygen than with common oxidation. A type of oxidation similar to reversion may take place in dairy products, where a very small amount of oxygen may

result in intense oxidation off-flavors. It is interesting to note that the linolenic acid content of milk fat is quite low.

Among the many factors that affect the rate of oxidation are the following:

Table 2–20 Sterol Content of Fats and Oils

Fat	Sterol (%)
Lard	0.12
Beef tallow	0.08
Milk fat	0.3
Herring	0.2–0.6
Cottonseed	1.4
Soybean	0.7
Corn	1.0
Rapeseed	0.4
Coconut	0.08
Cocoa butter	0.2

Table 2–21 Hydrocarbon Composition of Some Vegetable Oils

Oils	n-Paraffins	iso- and/or ante-iso Paraffins	Unidentified	Total Hydrocarbons
Corn	C_{11-31}	C_{11-21}	8	40
Peanut	C_{11-30}	C_{11-23}	7	40
Rapeseed	C_{11-31}	C_{11-17}, C_{19-21}	6	36
Linseed	C_{11-35}	C_{11-21}	7	43–45
Olive	C_{11}, C_{13-30}	—	6	29

Source: From G. Jacini, E. Fedeli, and A. Lanzani, Research in the Nonglyceride Substances of Vegetable Oils, *J. Assoc. Off. Anal. Chem.*, Vol. 50, pp. 84–90, 1967.

- amount of oxygen present
- degree of unsaturation of the lipids
- presence of antioxidants
- presence of prooxidants, especially copper, and some organic compounds such as heme-containing molecules and lipoxidase
- nature of packaging material
- light exposure
- temperature of storage

The autoxidation reaction can be divided into the following three parts: initiation, propagation, and termination. In the initiation part, hydrogen is abstracted from an olefinic compound to yield a free radical.

$$RH \rightarrow R^{\cdot} + H^{\cdot}$$

The removal of hydrogen takes place at the carbon atom next to the double bond and can be brought about by the action of, for instance, light or metals. The dissociation energy of hydrogen in various olefinic compounds has been listed by Ohloff (1973) and is shown in Table 2–22. Once a free radical has been formed, it will combine with oxygen to form a peroxy-free radical, which can in turn abstract hydrogen from another unsaturated molecule to yield a peroxide and a new free radical, thus starting the propagation reaction. This reaction may be repeated up to several thousand times and has the nature of a chain reaction.

$$R^{\cdot} + O_2 \longrightarrow RO_2^{\cdot}$$

$$RO_2^{\cdot} + RH \longrightarrow ROOH + R^{\cdot}$$

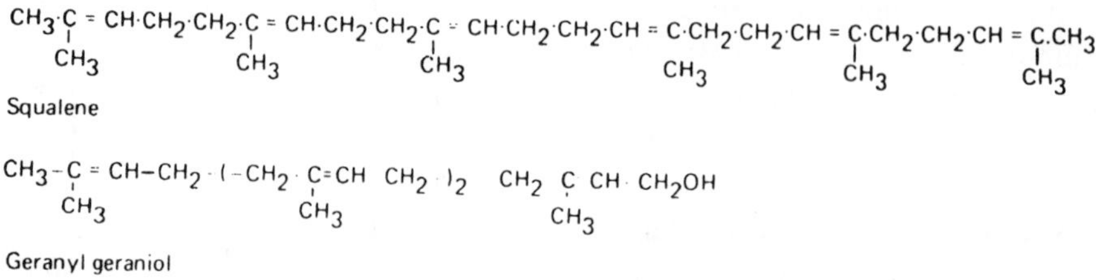

CH$_3$·C = CH·CH$_2$·CH$_2$·C = CH·CH$_2$·CH$_2$·C = CH·CH$_2$·CH$_2$·CH = C·CH$_2$·CH$_2$·CH = C·CH$_2$·CH$_2$·CH = C·CH$_3$
     CH$_3$ CH$_3$ CH$_3$ CH$_3$ CH$_3$ CH$_3$

Squalene

CH$_3$–C = CH–CH$_2$·(–CH$_2$·C=CH·CH$_2$·)$_2$·CH$_2$·C·CH·CH$_2$OH
     CH$_3$ CH$_3$ CH$_3$

Geranyl geraniol

Figure 2–14 Structure of Squalene and Geranyl Geraniol

Table 2–22 Dissociation Energy for the Abstraction of Hydrogen from Olefinic Compounds and Peroxides

Compound	ΔE (kcal/mole)
H—CH=CH$_2$	103
H—CH$_2$—CH$_2$—CH$_3$	100
H—CH$_2$—CH=CH$_2$	85

H—CH—CH=CH—CH$_2$—
    |
   CH$_3$
	77

—CH=CH—CH—CH=CH—
         |
        H
	65

H—OO—R	90

Source: From G. Ohloff, Fats as Precursors, in *Functional Properties of Fats in Foods*, J. Solms, ed., 1973, Forster Publishing.

The propagation can be followed by termination if the free radicals react with themselves to yield nonactive products, as shown here:

$$R^{\bullet} + R^{\bullet} \longrightarrow R—R$$

$$R^{\bullet} + RO_2^{\bullet} \longrightarrow RO_2R$$

$$nRO_2^{\bullet} \longrightarrow (RO_2)_n$$

The hydroperoxides formed in the propagation part of the reaction are the primary oxidation products. The hydroperoxide mechanism of autoxidation was first proposed by Farmer (1946). These oxidation products are generally unstable and decompose into the secondary oxidation products, which include a variety of compounds, including carbonyls, which are the most important. The peroxides have no importance to flavor deterioration, which is wholly caused by the secondary oxidation products. The nature of the process can be represented by the curves of Figure 2–15 (Pokorny 1971). In the initial stages of the reaction, the amount of hydroperoxides increases slowly; this stage is termed the induction period. At the end of the induction period, there is a sudden increase in peroxide content. Because peroxides are easily determined in fats, the peroxide value is frequently used to measure the progress of oxidation. Organoleptic changes are more closely related to the secondary oxidation products, which can be measured by various procedures, including the benzidine value, which is related to aldehyde decomposition products. As the aldehydes are themselves oxidized, fatty acids are formed; these free fatty acids may be considered tertiary oxidation products. The length of the induction period, therefore, depends

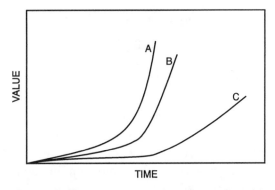

Figure 2–15 Autoxidation of Lard. (A) peroxide value, (B) benzidine value, (C) acid value. *Source:* From J. Pokorny, Stabilization of Fats by Phenolic Antioxidants, *Can. Inst. Food Sci. Technol. J.*, Vol. 4, pp. 68–74, 1971.

on the method used to determine oxidation products.

Although even saturated fatty acids may be oxidized, the rate of oxidation greatly depends on the degree of unsaturation. In the series of 18-carbon-atom fatty acids 18:0, 18:1, 18:2, 18:3, the relative rate of oxidation has been reported to be in the ratio of 1:100:1200:2500. The reaction of unsaturated compounds proceeds by the abstraction of hydrogen from the α carbon, and the resulting free radical is stabilized by resonance as follows:

$$-\overset{1}{\underset{\cdot}{C}}H-\overset{2}{C}H=\overset{3}{C}H- \;\rightleftharpoons\; -\overset{1}{C}H=\overset{2}{C}H-\overset{3}{\underset{\cdot}{C}}H-$$

If oleic acid is taken as example of a monoethenoid compound (cis-9-octadecenoic acid), the reaction will proceed by abstraction of hydrogen from carbons 8 or 11, resulting in two pairs of resonance hybrids.

$$COOH-(CH_2)_6-\overset{8}{C}H_2-\overset{9}{\underset{}{C}}H=\overset{10}{C}H-\overset{11}{C}H_2-(CH_2)_6-CH_3$$

$$-\overset{8}{\underset{\cdot}{C}}H-\overset{9}{C}H=\overset{10}{C}H-\overset{11}{C}H_2- \qquad\qquad -\overset{8}{C}H_2-\overset{9}{\underset{\cdot}{C}}H-\overset{10}{C}H=\overset{11}{C}H-$$

and   and

$$-\overset{8}{C}H=\overset{9}{C}H-\overset{10}{\underset{\cdot}{C}}H-\overset{11}{C}H_2- \qquad\qquad -\overset{8}{C}H_2-\overset{9}{C}H=\overset{10}{C}H-\overset{11}{\underset{\cdot}{C}}H-$$

This leads to the formation of the following four isomeric hydroperoxides:

$$-\overset{8}{C}H-\overset{9}{C}H=\overset{10}{C}H-\overset{11}{C}H_2$$
$$|$$
$$OOH$$

$$-\overset{8}{C}H=\overset{9}{C}H-\overset{10}{C}H-\overset{11}{C}H_2$$
$$|$$
$$OOH$$

$$-\overset{8}{C}H_2-\overset{9}{C}H-\overset{10}{C}H=\overset{11}{C}H-$$
$$|$$
$$OOH$$

$$-\overset{8}{C}H_2-\overset{9}{C}H=\overset{10}{C}H-\overset{11}{C}H-$$
$$|$$
$$OOH$$

In addition to the changes in double bond position, there is isomerization from cis to trans, and 90 percent of the peroxides formed may be in the trans configuration (Lundberg 1961).

From linoleic acid (cis-cis-9,12-octadecadienoic acid), three isomeric hydroperoxides can be formed as shown in the next formula. In this mixture of 9, 11, and 13 hydroperoxides, the conjugated ones occur in greatest

$$-\overset{9}{C}H-\overset{10}{C}H=\overset{11}{C}H-\overset{12}{C}H=\overset{13}{C}H-$$
$$|$$
$$OOH$$

$$-\overset{9}{C}H=\overset{10}{C}H-\overset{11}{C}H-\overset{12}{C}H=\overset{13}{C}H-$$
$$|$$
$$OOH$$

$$-\overset{9}{C}H=\overset{10}{C}H-\overset{11}{C}H=\overset{12}{C}H-\overset{13}{C}H-$$
$$|$$
$$OOH$$

quantity because they are the more stable forms. The hydroperoxides occur in the cis-trans and trans-trans configurations, the content of the latter being greater with higher temperature and greater extent of oxidation. From the oxidation of linolenic acid (cis, cis, cis-9,12,15-octadecatrienoic acid), six isometric hydroperoxides can be expected according to theory, as shown:

$$-\overset{9}{C}H-\overset{10}{C}H=\overset{11}{C}H-\overset{12}{C}H=\overset{14}{C}H_2-\overset{14}{C}H_2-\overset{15}{C}H=\overset{16}{C}H-$$
$$|$$
$$OOH$$

$$-CH=CH-\overset{11}{C}H-CH=CH-CH_2-CH=CH-$$
$$|$$
$$OOH$$

$$-CH=CH-CH_2-\overset{12}{C}H-CH=CH-CH=CH-$$
$$|$$
$$OOH$$

$$-CH=CH-CH=CH-\overset{13}{C}H-CH_2-CH=CH-$$
$$|$$
$$OOH$$

$$-CH=CH-CH_2-CH=CH-\overset{14}{C}H-CH=CH-$$
$$|$$
$$OOH$$

$$-CH=CH-CH_2-CH=CH-CH=CH-\overset{16}{C}H-$$
$$|$$
$$OOH$$

Hydroperoxides of linolenate decompose more readily than those of oleate and linoleate because active methylene groups are present. The active methylene groups are the ones located between a single double bond and a conjugated diene group. The hydrogen at this methylene group could readily be abstracted to form dihydroperoxides. The possibilities here for decomposition products are obviously more abundant than with oleate oxidation.

The decomposition of hydroperoxides has been outlined by Keeney (1962). The first step involves decomposition to the alkoxy and hydroxy free radicals.

$$R-CH(OOH)-R \longrightarrow R-CH-R + \cdot OH$$
$$|$$
$$O \cdot$$

The alkoxy radical can react to form aldehydes.

$$R-CH-R \longrightarrow R \cdot + RCHO$$
$$|$$
$$O \cdot$$

This reaction involves fission of the chain and can occur on either side of the free radical. The aldehyde that is formed can be a short-chain volatile compound, or it can be attached to the glyceride part of the molecule; in this case, the compound is nonvola-

tile. The volatile aldehydes are in great part responsible for the oxidized flavor of fats.

The alkoxy radical may also abstract a hydrogen atom from another molecule to yield an alcohol and a new free radical, as shown:

$$R-CH-R + R^1H \longrightarrow R-CH-R + R^{1 \cdot}$$
$$| \qquad\qquad\qquad\qquad |$$
$$O \cdot \qquad\qquad\qquad\qquad OH$$

The new free radicals formed may participate in propagation of the chain reaction. Some of the free radicals may interact with themselves to terminate the chain, and this could lead to the formation of ketones as follows:

$$R-CH-R + R^{1 \cdot} \longrightarrow R-C-R + R^1H$$
$$| \qquad\qquad\qquad\qquad\quad ||$$
$$O \cdot \qquad\qquad\qquad\qquad\quad O$$

As indicated, a variety of aldehydes have been demonstrated in oxidized fats. Alcohols have also been identified, but the presence of ketones is not as certain. Keeney (1962) has listed the aldehydes that may be formed from breakdown of hydroperoxides of oxidized oleic, linoleic, linolenic, and arachidonic acids (Table 2–23). The aldehydes are powerful flavor compounds and have very low flavor thresholds; for example, 2,4-decadienal has a flavor threshold of less than one part per billion. The presence of a double bond in an aldehyde generally lowers the flavor threshold considerably. The aldehydes can be further oxidized to carboxylic acids or other tertiary oxidation products.

When chain fission of the alkoxy radical occurs on the other side of the free radical group, the reaction will not yield volatile aldehydes but will instead form nonvolatile aldehydo-glycerides. Volatile oxidation products can be removed in the refining process

Table 2–23 Hydroperoxides and Aldehydes (with Single Oxygen Function) That May Be Formed in Autoxidation of Some Unsaturated Fatty Acids

Fatty Acid	Methylene Group Involved[*]	Isomeric Hydroperoxides Formed from the Structures Contributing to the Intermediate Free Radical Resonance Hybrid	Aldehydes Formed by Decomposition of the Hydroperoxides
Oleic	11	11-hydroperoxy-9-ene	octanal
		9-hydroperoxy-10-ene	2-decenal
	8	8-hydroperoxy-9-ene	2-undecenal
		10-hydroperoxy-8-ene	nonanal
Linoleic	11	13-hydroperoxy-9,11-diene	hexanal
		11-hydroperoxy-9,12-diene	2-octenal
		9-hydroperoxy-10,12-diene	2,4-decadienal
Linolenic	14	16-hydroperoxy-9,12,14-triene	propanal
		14-hydroperoxy-9,12,15-triene	2-pentenal
		12-hydroperoxy-9,13,15-triene	2,4-heptadienal
	11	13-hydroperoxy-9,11,15-triene	3-hexenal
		11-hydroperoxy-9,12,15-triene	2,5-octadienal
		9-hydroperoxy-10,12,15-triene	2,4,7-decatrienal
Arachidonic	13	15-hydroperoxy-5,8,11,13-tetraene	hexanal
		13-hydroperoxy-5,8,11,14-tetraene	2-octenal
		11-hydroperoxy-5,8,12,14-tetraene	2,4-decadienal
	10	12-hydroperoxy-5,8,10,14-tetraene	3-nonenal
		10-hydroperoxy-5,8,11,14-tetraene	2,5-undecadienal
		8-hydroperoxy-5,9,11,14-tetraene	2,4,7-tridecatrienal
	7	9-hydroperoxy-5,7,11,14-tetraene	3,6-dodecadienal
		7-hydroperoxy-5,8,11,14-tetraene	2,5,8-tetradecatrienal
		5-hydroperoxy-6,8,11,14-tetraene	2,4,7,10-hexadecatetraenal

[*]Only the most active methylene groups in each acid are considered.

Source: From M. Keeney, Secondary Degradation Products, in *Lipids and Their Oxidation*, H.W. Schultz et al., eds., 1962, AVI Publishing Co.

during deodorization, but the nonvolatile products remain; this can result in a lower oxidative stability of oils that have already oxidized before refining.

The rate and course of autoxidation depend primarily on the composition of the fat—its degree of unsaturation and the types of unsaturated fatty acids present. The absence, or at least a low value, of peroxides does not necessarily indicate that an oil is not oxidized. As Figure 2–16 indicates, peroxides are labile and may be transformed into secondary oxidation products. A combined index of primary and secondary oxidation products gives a better evaluation of the state of oxidation of an oil. This is expressed as Totox value: Totox value = 2 × peroxide value + anisidine value. (Anisidine value is a

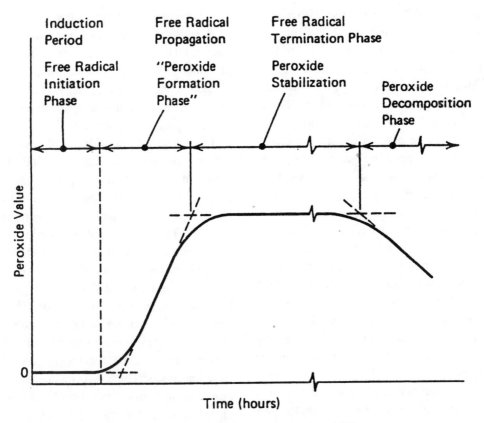

Figure 2–16 Peroxide Formation and Decomposition as a Function of Time

measure of secondary oxidation products.) Removal of oxygen from foods will prevent oxidation, but, in practice, this is not easy to accomplish in many cases. At high temperatures (100 to 140°C) such as those used in the accelerated tests for oil stability (active oxygen method), formic acid is produced, which can be used to indicate the end of the induction period. The formation of formic acid results from aldehyde decomposition. Peroxidation of aldehydes establishes a resonance equilibrium between two limiting forms.

$$R–CH_2–\overset{\cdot}{C}O \rightleftharpoons R–\overset{\cdot}{C}H–CHO$$

The second hybrid ties up oxygen at the α carbon to yield the α-hydroperoxy aldehyde as follows:

$$R–\overset{\cdot}{C}H–CHO+O_2 \rightarrow R–\overset{\overset{\displaystyle \overset{\cdot}{O}}{|}\overset{\displaystyle O}{|}}{C}H–CHO$$

Breakdown of oxygen and carbon bonds yields formic acid and a new aldehyde.

$$R–\overset{\overset{\displaystyle \overset{\cdot}{O}}{|}\overset{\displaystyle O}{|}}{C}–CH–CHO \rightarrow HCOOH+RCHO$$

deMan et al. (1987) investigated this reaction with a variety of oils and found that although formic acid was the main reaction product, other short-chain acids from acetic to caproic were also formed. Trace metals, especially copper, and to a lesser extent iron, will catalyze fat oxidation; metal deactivators such as citric acid can be used to reduce the effect. Lipoxygenase (lipoxidase) and heme compounds act as catalysts of lipid oxidation. Antioxidants can be very effective in slowing down oxidation and increasing the induction period. Many foods contain natural antioxidants; the tocopherols are the most important of these. They are present in greater amounts in vegetable oils than in animal fats, which may explain the former's greater stability.

Antioxidants such as tocopherols may be naturally present; they may be induced by processes such as smoking or roasting, or added as synthetic antioxidants. Antioxidants act by reacting with free radicals, thus terminating the chain. The antioxidant AH may react with the fatty acid free radical or with the peroxy free radical,

$$AH + R^{\cdot} \rightarrow RH + A^{\cdot}$$
$$AH + RO_2^{\cdot} \rightarrow RO_2H + A^{\cdot}$$

The antioxidant free radical deactivated by further oxidation to quinones, thus terminating the chain. Only phenolic compounds that can easily produce quinones are active as antioxidants (Pokorny 1971). At high concentrations antioxidants may have a prooxidant effect and one of the reactions may be as follows:

$$A^{\cdot} + RH \rightarrow AH + R^{\cdot}$$

Tocopherols in natural fats are usually present at optimum levels. Addition of anti-oxidant beyond optimum amounts may result in increasing the extent of prooxidant action. Lard is an example of a fat with very low natural antioxidant activity and antioxidant must be added to it, to provide protection. The effect of antioxidants can be expressed in terms of protection factor, as shown in Figure 2–17 (Pokorny 1971). The highly active antioxidants that are used in the food industry are active at about 10 to 50 parts per million (ppm). Chemical structure of the antioxidants is the most important factor affecting their activity. The number of synthetic antioxidants permitted in foods is limited, and the structure of the most widely used compounds is shown in Figure 2–18. Propyl gallate is more soluble in water than in fats. The octyl and dodecyl esters are more fat soluble. They are heat resistant and nonvolatile with steam, making them useful for frying oils and in baked products. These are considered to have *carry-through properties*. Butylated hydroxyanisole (BHA) has carry-through properties but butylated hydroxy toluene (BHT) does not, because it is volatile with steam. The compound *tert*-butyl hydroquinine (TBHQ) is used for its effectiveness in increasing oxidative stability of polyunsaturated oils and fats. It also provides carry-through protection for fried foods. Antioxidants are frequently used in combination or together with synergists. The latter are frequently metal deactivators that have the ability to chelate metal ions. An example of the combined effect of antioxidants is shown in Figure 2–19. It has been pointed out (Zambiazi and Przybylski 1998) that fatty acid composition can explain only about half of the oxidative stability of a vegetable oil. The other half can be contributed to minor components including tocopherols, metals, pigments, free fatty acids, phenols, phospholipids, and sterols.

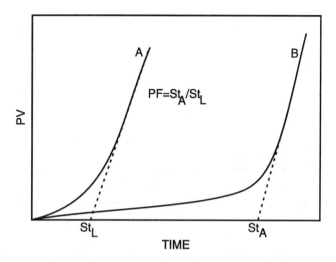

Figure 2–17 Determination of Protection Factor. (A) lard, (B) lard + antioxidant. *Source:* From J. Pokorny, Stabilization of Fats by Phenolic Antioxidants, *Can. Inst. Food Sci. Technol. J.*, Vol. 4, pp. 68–74, 1971.

PHOTOOXIDATION

Oxidation of lipids, in addition to the free radical process, can be brought about by at least two other mechanisms—photooxidation and enzymic oxidation by lipoxygenase. The latter is dealt with in Chapter 10. Light-induced oxidation or photooxidation results from the reactivity of an excited state of oxygen, known as singlet oxygen ($^1O_2$). Ground-state or normal oxygen is triplet oxygen ($^3O_2$). The activation energy for the reaction of normal oxygen with an unsaturated fatty acid is very high, of the order of 146 to 273

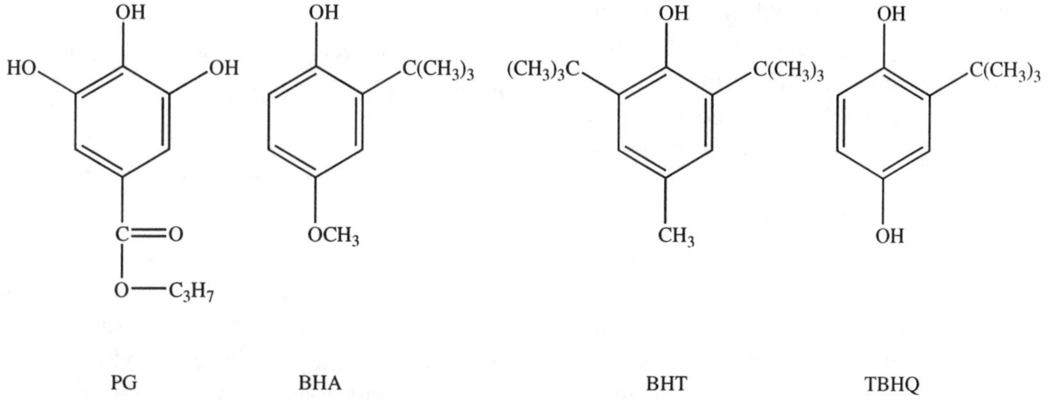

PG   BHA   BHT   TBHQ

Figure 2–18 Structure of Propyl Gallate (PG), Butylated Hydroxyanisole (BHA), Butylated Hydroxy Toluene (BHT), and Tert-Butyl Hydroquinone (TBHQ)

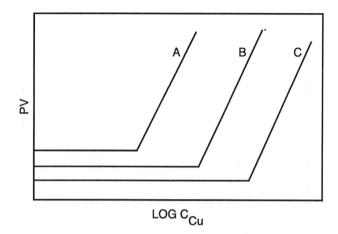

Figure 2–19 Effect of Copper Concentration on Protective Effect of Antioxidants in Lard. (A) lard + 0.01% BHT, (B) lard + 0.01% ascorbyl palmitate, (C) lard + 0.005% BHT and 0.05% ascorbyl palmitate. *Source:* From J. Pokorny, Stabilization of Fats by Phenolic Antioxidants, *Can. Inst. Food Sci. Technol. J.,* Vol. 4, pp. 68–74, 1971.

kJ/mole. When oxygen is converted from the ground state to the singlet state, energy is taken up amounting to 92 kJ/mole, and in this state the oxygen is much more reactive. Singlet-state oxygen production requires the presence of a sensitizer. The sensitizer is activated by light, and can then either react directly with the substrate (type I sensitizer) or activate oxygen to the singlet state (type II sensitizer). In both cases unsaturated fatty acid residues are converted into hydroperoxides. The light can be from the visible or ultraviolet region of the spectrum.

Singlet oxygen is short-lived and reverts back to the ground state with the emission of light. This light is fluorescent, which means that the wavelength of the emitted light is higher than that of the light that was absorbed for the excitation. The reactivity of singlet oxygen is 1,500 greater than that of ground-state oxygen. Compounds that can act as sensitizers are widely occurring food components, including chlorophyll, myoglobin, riboflavin, and heavy metals. Most of

these compounds promote type II oxidation reactions. In these reactions the sensitizer is transformed into the activated state by light. The activated sensitizer then reacts with oxygen to produce singlet oxygen.

$$sen \xrightarrow{h\nu} sen^*$$

$$sen^* + O_2 \longrightarrow sen + {}^1O_2$$

The singlet oxygen can react directly with unsaturated fatty acids.

$$^1O_2 + RH \longrightarrow ROOH$$

The singlet oxygen reacts directly with the double bond by addition, and shifts the double bond one carbon away. The singlet oxygen attack on linoleate produces four hydroperoxides as shown in Figure 2–20. Photooxidation has no induction period, but the reaction can be quenched by carotenoids

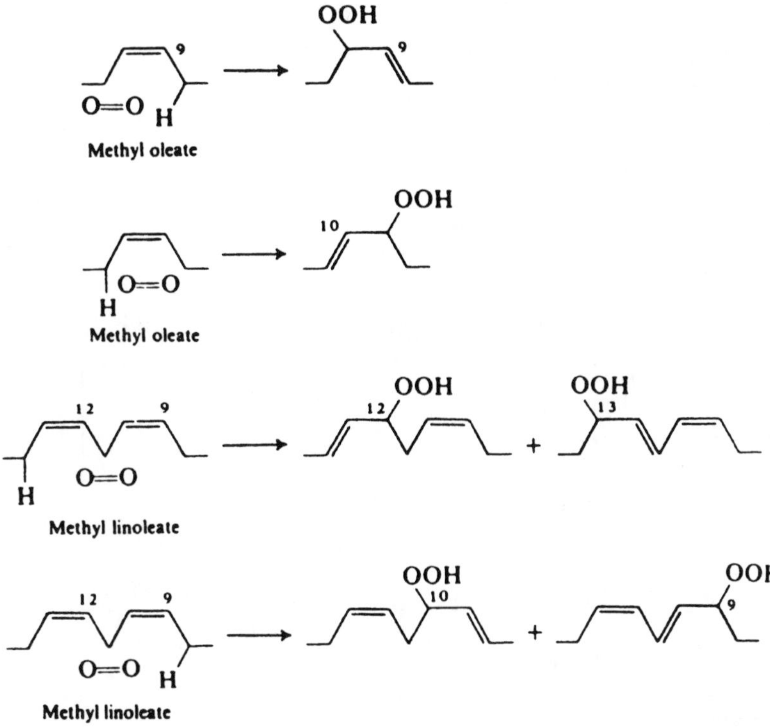

Figure 2–20 Photooxidation. Singlet-oxygen attack on oleate produces two hydroperoxides; linoleate yields four hydroperoxides

that effectively compete for the singlet oxygen and bring it back to the ground state.

Phenolic antioxidants do not protect fats from oxidation by singlet oxidation (Yasaei et al. 1996). However, the antioxidant ascorbyl palmitate is an effective singlet oxygen quencher (Lee et al. 1997). Carotenoids are widely used as quenchers. Rahmani and Csallany (1998) reported that in the photooxidation of virgin olive oil, pheophytin A functioned as sensitizer, while β-carotene acted as a quencher.

The combination of light and sensitizers is present in many foods displayed in transparent containers in brightly lit supermarkets. The light-induced deterioration of milk has been studied extensively. Sattar et al. (1976) reported on the light-induced flavor deterioration of several oils and fats. Of the five fats examined, milk fat and soybean oil were most susceptible and corn oil least susceptible to singlet oxygen attack. The effect of temperature on the rate of oxidation of illuminated corn oil was reported by Chahine and deMan (1971) (Figure 2–21). They found that temperature has an important effect on photooxidation rates, but even freezing does not completely prevent oxidation.

HEATED FATS—FRYING

Fats and oils are heated during commercial processing and during frying. Heating during

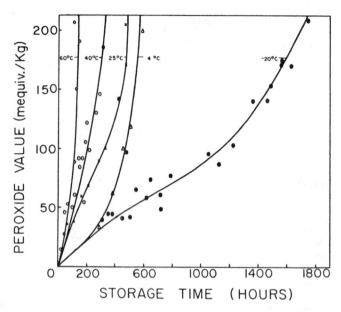

Figure 2–21 Effect of Temperature on Rate of Oxidation of Illuminated Corn Oil. *Source:* From M.H. Chahine and J.M. deMan, Autoxidation of Corn Oil under the Influence of Fluorescent Light, *Can. Inst. Food Sci. Technol. J.,* Vol. 4, pp. 24–28, 1971.

processing mainly involves hydrogenation, physical refining, and deodorization. Temperature used in these processes may range from 120°C to 270°C. The oil is not in contact with air, which eliminates the possibility of oxidation. At the high temperatures used in physical refining and deodorization, several chemical changes may take place. These include randomization of the glyceride structure, dimer formation, *cis-trans* isomerization, and formation of conjugated fatty acids (positional isomerization) of polyunsaturated fatty acids (Hoffmann 1989). The *trans* isomer formation in sunflower oil as a result of high temperature deodorization is shown in Figure 2–22 (Ackman 1994).

Conditions prevailing during frying are less favorable than those encountered in the above-mentioned processes. Deep frying, where the food is heated by immersion in hot oil, is practiced in commercial frying as well as in food service operations. The temperatures used are in the range of 160°C to 195°C. At lower temperatures frying takes longer, and at higher temperatures deterioration of the oil is the limiting factor. Deep frying is a complex process involving both the oil and the food to be fried. The reactions taking place are schematically presented in Figure 2–23. Steam is given off during the frying, which removes volatile antioxidants, free fatty acids, and other volatiles. Contact with the air leads to autoxidation and the formation of a large number of degradation products. The presence of steam results in hydrolysis, with the production of free fatty acids and partial glycerides. At lower frying temperatures the food has to be fried longer to reach the desirable color, and this results in higher oil uptake. Oil absorption by fried

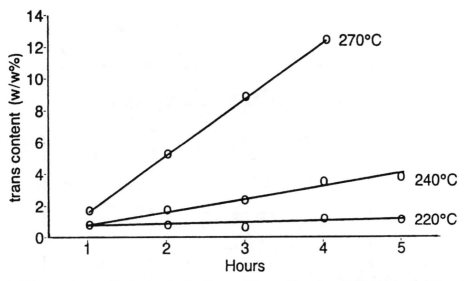

Figure 2–22 *Trans* Isomer Formation in Sunflower Oil as a Function of Deodorization Temperature. *Source:* Reprinted from R.G. Ackman, Animal and Marine Lipids, in *Improved and Technological Advances in Alternative Sources of Lipids*, B. Kamel and Y. Kakuda, eds., p. 301, 1994, Aspen Publishers, Inc.

foods may range from 10 to 40 percent, depending on conditions of frying and the nature and size of the food.

Oils used in deep frying must be of high quality because of the harsh conditions during deep frying and to provide satisfactory shelf life in fried foods. The suitability of an oil for frying is directly related to its content of unsaturated fatty acids, especially linolenic acid. This has been described by Erickson (1996) as "inherent stability" calculated from the level of each of the unsaturated fatty acids (oleic, linoleic, and linolenic) and their relative reaction rate with oxygen. The inherent stability calculated for a number of oils is given in Table 2–24. The higher the inherent stability, the less suitable the oil is for frying. The liquid seed oils, such as soybean and sunflower oil, are not suitable for deep frying and are usually partially hydrogenated for this purpose. Such hydrogenated oils can take the form of shortenings, which may be plastic solids or pourable suspensions. Through plant breeding and genetic engineering, oils with higher inherent stability can be obtained, such as high-oleic sunflower oil, low-linolenic canola oil, and low-linolenic soybean oil.

The stability of frying oils and fats is usually measured by an accelerated test known as the active oxygen method (AOM). In this test, air is bubbled through an oil sample maintained at 95°C and the peroxide value is measured at intervals. At the end point the peroxide value shows a sharp increase, and this represents the AOM value in hours. Typical AOM values for liquid seed oils range from 10 to 30 hours; heavy-duty frying shortenings range from 200 to 300 hours. AOM values of some oils and fats determined by measuring the peroxide value and using an automatic recording of volatile acids produced during the test are given in Table 2–25 (deMan et al. 1987).

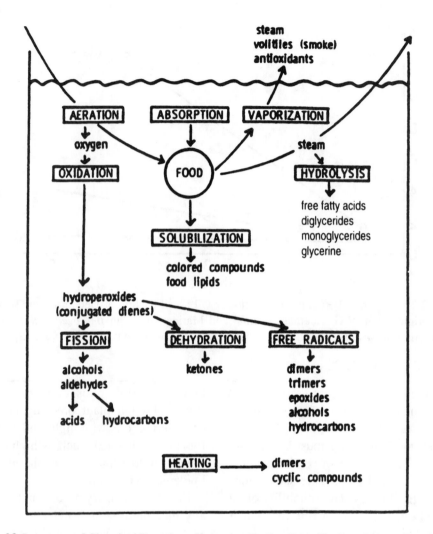

Figure 2–23 Summary of Chemical Reactions Occurring During Deep Frying. *Source:* Reprinted with permission from F.T. Orthoefer, S. Gurkin, and K. Lui, Dynamics of Frying in Deep Frying, in *Chemistry, Nutrition and Practical Applications*, E.G. Perkins and M.D. Erickson, eds., p. 224. © 1996, AOCS Press.

As shown in Figure 2–23, oil breakdown during frying can be caused by oxidation and thermal alteration. Oxidation can result in the formation of oxidized monomeric, dimeric, and oligomeric triglycerides as well as volatile compounds including aldehydes, ketones, alcohols, and hydrocarbons. In addition, oxidized sterols may be formed. Thermal degradation can result in cyclic monomeric triglycerides and nonpolar dimeric and oligomeric triglycerides. The polymerization reaction may take place by conversion of part of the *cis-cis*-1,4 diene system of linoleates to the *trans-trans* conjugated diene. The 1,4 and 1,3 dienes can combine in a Diels-Alder type addition reaction to produce a dimer as

Table 2–24 Inherent Stability of Oils for Use in Frying

Oil	Iodine Value	Inherent Stability*
Soybean	130	7.4
Sunflower	120	7.7
High-oleic sunflower	90	2.0
Corn	110	6.2
Cottonseed	98	5.2
Canola	110	5.4
Peanut	92	4.5
Lard	60	1.4
Olive	88	1.8
Palm	55	1.4
Palm olein	58	1.6
Palm stearin	35	1.0
Tallow	50	0.7
Palm kernel	17	0.5
Coconut	9	0.4

*Inherent stability calculated from decimal fraction of fatty acids multiplied by relative reaction rates with oxygen, assuming rate for oleic acid = 1, linoleic acid = 10, and linolenic acid = 25.

Table 2–25 Active Oxygen Method (AOM) Time of Several Oils and Fats as Determined by Peroxide Value and Conductivity Measurements

Oil	AOM Time (POV)[a]	AOM Time (Conductivity)[b]
Sunflower	6.2	7.1
Canola	14.0	15.8
Olive	17.8	17.8
Corn	12.4	13.8
Peanut	21.1	21.5
Soybean	11.0	10.4
Triolein	8.1	7.4
Lard	42.7	43.2
Butterfat	2.8	2.0

[a]At peroxide value 100.
[b]At intercept of conductivity curve and time axis.

Source: Reprinted with permission from J.M. deMan, et al., Formation of Short Chain Volatile Organic Acids in the Automated AOM Method, *J.A.O.C.S.*, Vol. 64, p. 996, © 1987, American Oil Chemists' Society.

shown in Figure 2–24. Other possible routes for dimer formation are through free radical reactions. As shown in Figure 2–25, this may involve combination of radicals, intermolecular addition, and intramolecular addition. From dimers, higher oligomers can be produced; the structure of these is still relatively unknown.

Another class of compounds formed during frying is cyclic monomers of fatty acids. Linoleic acid can react at either the C9 or C12 double bonds to give rings between carbons 5 and 9, 5 and 10, 8 and 12, 12 and 17, and 13 and 17. Cyclic monomers with a cyclopentenyl ring have been isolated from heated sunflower oil, and their structure is illustrated in Figure 2–26 (Le Quéré and Sébédio 1996).

Some countries such as France require that frying oils contain less than 2 percent linolenic acid. Several European countries have set maximum limits for the level of polar

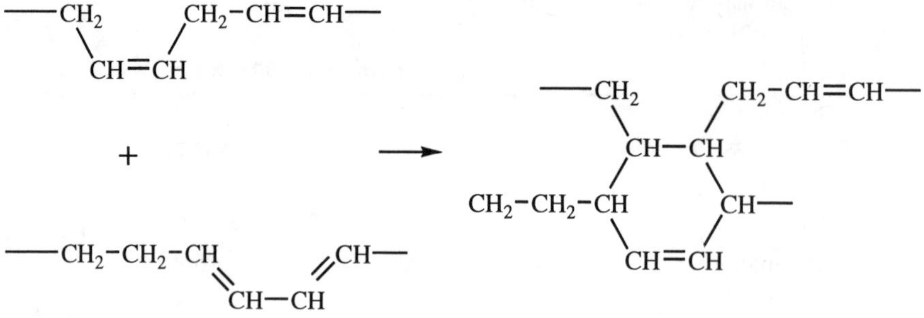

Figure 2–24 Polymerization of Diene Systems To Form Dimers

compounds or for the level of free fatty acids beyond which the fat is considered unfit for human consumption. In continuous industrial frying, oil is constantly being removed from the fryer with the fried food and replenished with fresh oil so that the quality of the oil can remain satisfactory. This is more difficult in intermittent frying operations.

FLAVOR REVERSION

Soybean oil and other fats and oils containing linolenic acid show the reversion phenomenon when exposed to air. Reversion flavor is a particular type of oxidized flavor that develops at comparatively low levels of oxidation. The off-flavors may develop in oils

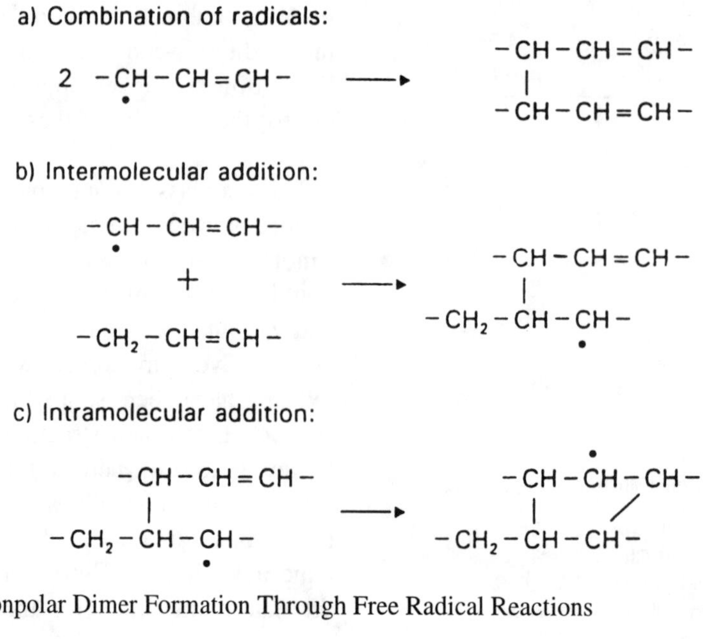

Figure 2–25 Nonpolar Dimer Formation Through Free Radical Reactions

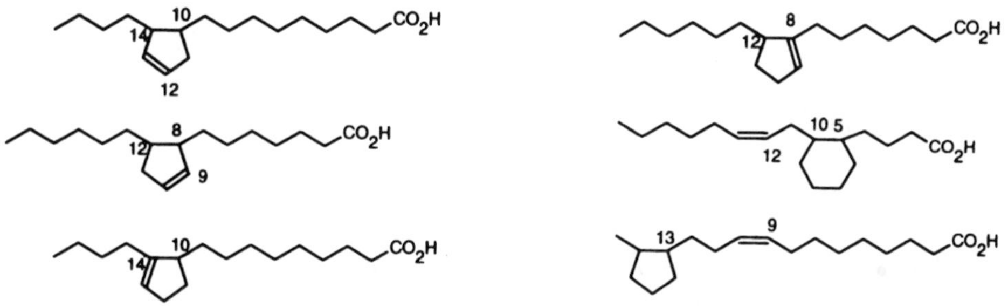

Figure 2–26 Cyclic Fatty Acid Monomers Formed from Linoleic Acid in Heated Sunflower Oil

that have a peroxide value of as little as 1 or 2. Other oils may not become rancid until the peroxide value reaches 100. Linolenic acid is generally recognized as the determining factor of inversion flavors. These off-flavors are variously described as grassy, fishy, and painty (Cowan and Evans 1961). The origin of these flavors appears to be the volatile oxidation products resulting from the terminal pentene radical of linolenic acid, $CH_3-CH_2-CH=CH-CH_2-$. Hoffmann (1962) has listed the flavor descriptions of reverted soybean oil (Table 2–26) and the volatile decomposition products isolated from reverted or oxidized soybean oil (Table 2–27).

The first perceptible reversion flavor was found to be caused by 3-*cis*-hexenal, which has a pronounced green bean odor. Other flavorful aldehydes isolated were 2-*trans*-hexenal (green, grassy), 2-*trans*-nonenal (rancid), and 2-*trans*-6-*cis*-nonadienal (cucumber flavor). These findings illustrate the complexity of the reversion flavor. Similar problems occur with other polyunsaturated oils such as fish oil and rapeseed oil.

HYDROGENATION

Hydrogenation of fats is a chemical reaction consisting of addition of hydrogen at double bonds of unsaturated acyl groups. This reaction is of great importance to industry, because it permits the conversion of liquid oils into plastic fats for the production of margarine and shortening. For some oils, the process also results in a decreased susceptibility to oxidative deterioration. In the hydrogenation reaction, gaseous hydrogen, liquid oil, and solid catalyst participate under agitation in a closed vessel. Although most industrial processes use solid nickel catalysts, interest in organometallic compounds that serve as homogeneous catalysts has increased greatly. Frankel and Dutton (1970)

Table 2–26 Flavor Descriptions Used for Crude, Processed, and Reverted Soybean Oil

State	Flavor
Crude	Grassy, beany
Freshly processed	Sweet, pleasant, nutty
Reverted	Grassy, bany, buttery, melony, tallowy, painty, fishy

Source: From G. Hoffmann, Vegetable Oils, in *Lipids and Their Oxidation*, H.W. Schultz et al., eds., 1962, AVI Publishing Co.

Table 2–27 Volatile Compounds Isolated from Reverted or Oxidized Soybean Oil

		Aldehydes		
Saturated		$\Delta 2$ Unsaturated	$\Delta 2,4$ Unsaturated	$\Delta 3$ Unsaturated
$C_1 C_2 C_3^*$ $C_4 C_5 C_6^*$		$C_4 C_5 C_6 C_7^*$ C_8	$C_6 \underbrace{C_7 C_7^*}\, C_8^* C_9$	$\underbrace{C_6 C_6}$
$C_7 C_8 C_9$		C_9^* $C_{10} C_{11}$	$\underbrace{C_{10} C_{10}^*}\, C_{12}$	

Ketones and Dicarb.	Alcohols and Esters	Acids
Methyl-pentyl ketone	1-octen-3-ol	Saturated
Di-n-propyl ketone	Ethanol	C_1–C_9 or C_{10}
Malondialdehyde	Ethyl formate	
	Ethyl acetate	

*Main products

⌣⌢ Stereo-isomers

Source: From G. Hoffmann, Vegetable Oils, in *Lipids and Their Oxidation,* H.W. Schultz et al., eds., 1962, AVI Publishing Co.

---

have represented catalytic hydrogenation by the following scheme, in which the reacting species are the olefinic substrate (S), the metal catalyst (M), and H_2:

$$S + M \rightleftharpoons [S{-}M]$$
$$(1)$$
$$[S{-}M{-}H_2] \longrightarrow SH_2 + M$$
$$(3)$$
$$M + H_2 \rightleftharpoons [M{-}H_2]$$
$$(2)$$

The intermediates 1, 2, and 3 are organo-metallic species. If the reaction involves heterogeneous catalysis, the olefins and hydrogen are bound to the metal by chemisorption. If homogeneous catalysis takes place, the intermediates are organometallic complexes. The intermediates are labile and short-lived and cannot usually be isolated. In heterogeneous catalysis, the surface of the metal performs the function of catalyst and

the preparation of the catalyst is of major importance. When hydrogen is added to double bonds in a natural fat consisting of many component glycerides and different component unsaturated fatty acids, the result depends on many factors, if the reaction is not carried to completion. Generally, hydrogenation of fats is not carried to completion and fats are hydrogenated only partially. Under these conditions, hydrogenation may be selective or nonselective. Selectivity means that hydrogen is added first to the most unsaturated fatty acids. Selectivity is increased by increasing hydrogenation temperature and decreased by increasing pressure and agitation. Table 2–28 shows the effect of selectivity on the properties of soybean oil. The selectively hydrogenated oil is more resistant to oxidation because of the preferential hydrogenation of the linolenic

Table 2–28 Differences in Selective and Nonselective Hydrogenation of Soybean Oil

Characteristic	Selective	Non-selective
Induction period AOM (hr)	240	31
Micropenetration	70 (more plastic)	30
Capillary mp (°C)	39	55
Condition:		
Temp (°C)	177	121
Pressure (psi)	5	50
Ni catalyst (%)	0.05	0.05

Source: From W.O. Lundberg, *Autoxidation and Antioxidants*, 1961, John Wiley & Sons.

acid. The influence of selectivity conditions on the fatty acids of hydrogenated cottonseed and peanut oil is demonstrated by the data presented in Table 2–29. The higher the selectivity, the lower the level of polyunsaturated fatty acids will be and the higher the level of monounsaturates.

Another important factor in hydrogenation is the formation of positional and geometrical isomers. Formation of *trans* isomers is rapid and extensive. The isomerization can be understood by the reversible character of

chemisorption. When the olefinic bond reacts, two carbon-metal bonds are formed as an intermediate stage (represented by an asterisk in Figure 2–27). The intermediate may react with an atom of adsorbed hydrogen to yield the "half-hydrogenated" compound, which remains attached by only one bond. Additional reaction with hydrogen results in formation of the saturated compound. There is also the possibility that the half-hydrogenated olefin may again attach itself to the catalyst surface at a carbon on either side of the existing bond, with simultaneous loss of hydrogen. Upon desorption of this species, a positional or geometrical isomer may result. The proportion of *trans* acids is high because this is the more stable configuration.

Double bond migration occurs in both directions and more extensively in the direction away from the ester group. This is true not only for the *trans* isomers that are formed but also for the *cis* isomers. The composition of the positional isomers in a partially hydrogenated margarine fat is shown in Figures 2–28 and 2–29 (Craig-Schmidt 1992). In a partially hydrogenated fat, the analysis of component fatty acids by gas-liquid chromatography is difficult because of the presence of many isomeric

Table 2–29 Fatty Acid Composition of Cottonseed and Peanut Oil Hydrogenated under Different Conditions of Selectivity to Iodine Value 65

Oil	Hydrogenation Conditions	Fatty Acids		
		Saturated (%)	Oleic (%)	Linoleic (%)
Cottonseed	Moderately selective	31.5	64.5	4.0
Peanut	Moderately selective	27.5	72.5	—
Cottonseed	Nonselective	36.0	56.0	8.0
Peanut	Nonselective	30.0	67.0	3.0
Cottonseed	Very nonselective	39.5	48.5	12.0
Peanut	Very nonselective	33.0	61.0	6.0

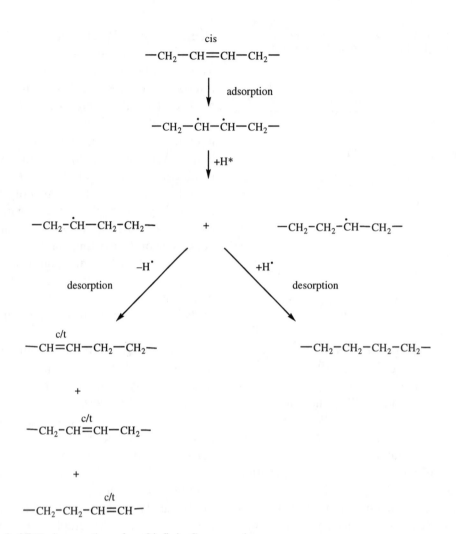

Figure 2–27 Hydrogenation of an Olefinic Compound

fatty acids. This is shown in the chromatogram in Figure 2–30 (Ratnayake 1994). The hydrogenation of oleate can be represented as follows:

The change from oleate to *iso* oleate involves no change in unsaturation but does result in a considerably higher melting point. Hydrogenation of linoleate first produces some conjugated dienes, followed by formation of positional and geometrical isomers of oleic acid and, finally, stearate.

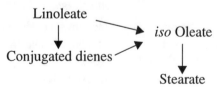

Hydrogenation of linolenate is more complex and greatly dependent on reaction con-

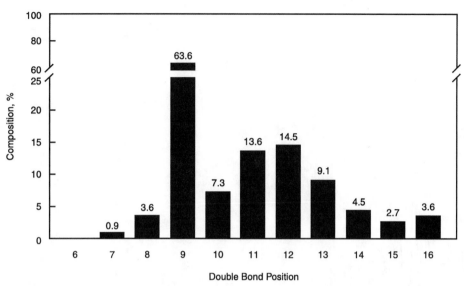

Figure 2–28 Positional Isomers of 18:1 *cis* Formed in the Partial Hydrogenation of a Margarine Fat. *Source:* Reprinted with permission from M.C. Craig-Schmidt, Fatty Acid Isomers in Foods, in *Fatty Acids in Foods and Their Health Implications*, C.K. Chow, ed., p. 369, 1992, by courtesy of Marcel Dekker, Inc.

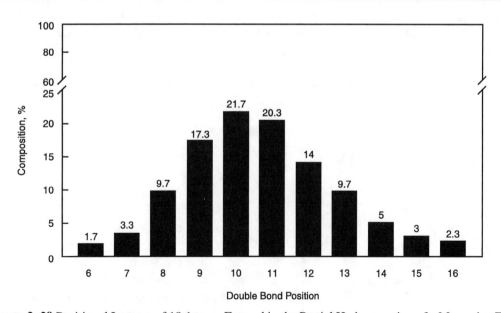

Figure 2–29 Positional Isomers of 18:1 *trans* Formed in the Partial Hydrogenation of a Margarine Fat. *Source:* Reprinted with permission from M.C. Craig-Schmidt, Fatty Acid Isomers in Foods, in *Fatty Acids in Foods and Their Health Implications*, C.K. Chow, ed., p. 369, 1992, by courtesy of Marcel Dekker, Inc.

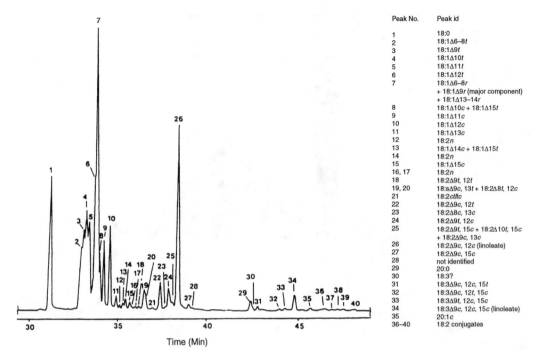

Peak No.	Peak id
1	18:0
2	18:1Δ6–8t
3	18:1Δ9t
4	18:1Δ10t
5	18:1Δ11t
6	18:1Δ12t
7	18:1Δ6–8r
	+ 18:1Δ9r (major component)
	+ 18:1Δ13–14r
8	18:1Δ10c + 18:1Δ15t
9	18:1Δ11c
10	18:1Δ12c
11	18:1Δ13c
12	18:2n
13	18:1Δ14c + 18:1Δ15t
14	18:2n
15	18:1Δ15c
16, 17	18:2n
18	18:2Δ9t, 12t
19, 20	18:sΔ9c, 13t + 18:2Δ8t, 12c
21	18:2ctltc
22	18:2Δ9c, 12t
23	18:2Δ8c, 13c
24	18:2Δ9t, 12c
25	18:2Δ9t, 15c + 18:2Δ10t, 15c
	+ 18:2Δ9c, 13c
26	18:2Δ9c, 12c (linoleate)
27	18:2Δ9c, 15c
28	not identified
29	20:0
30	18:3?
31	18:3Δ9c, 12c, 15t
32	18:3Δ9c, 12t, 15c
33	18:3Δ9t, 12c, 15c
34	18:3Δ9c, 12c, 15c (linoleate)
35	20:1c
36–40	18:2 conjugates

Figure 2–30 Gas Chromatogram of the Fatty Acid Methyl Esters from Partially Hydrogenated Soybean Oil, Using a 100-m Fused Silica Capillary Column Coated with SP2560. *Source:* Reprinted with permission from W.M.N. Ratnayake, Determination of Trans Unsaturation by Infrared Spectrophotometry and Determination of Fatty Acid Composition of Partially Hydrogenated Vegetable Oils and Animal Fats by Gas Chromatography/Infrared Spectrophotometry: Collaborative Study, *J.A.O.A.C. Intern.*, Vol. 78, pp. 783–802, © 1994.

ditions. The reactions can be summarized as follows:

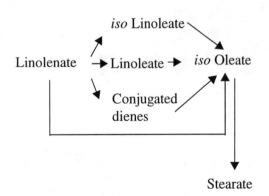

During hydrogenation, solid isomers of oleic acids are formed by partial hydrogenation of polyunsaturated acid groups or by isomerization of oleic acid. This has an important effect on the consistency of partially hydrogenated oils. For example, compare olive oil, which has an iodine value of 80 and is liquid at room temperature, with soybean oil hydrogenated to the same iodine value, which is a fat with a consistency like lard.

Nickel catalysts are poisoned by sulfur and phosphorous compounds, free fatty acids, and residual soaps. Oils are refined and sometimes bleached before hydrogenation. Sulfur compounds are not easily removed from the oil. Oils that contain sulfur compounds are rapeseed oil, canola oil, and fish oils (Wijesundera et al. 1988). High-erucic rapeseed oil is very difficult to hydrogenate unless it is deodorized (deMan et al. 1995). Canola oils of the double-zero variety that are low in erucic acid and glucosinolates still contain traces of sulfur in the form of isothiocyanates (Abraham and deMan 1985).

When catalysts are poisoned by sulfur the hydrogenation reaction is slowed down and the formation of *trans* isomers is increased. U.S. stick margarines are reported to contain 24 percent of *trans* fatty acids, and soft margarines contain 14 to 18 percent. Shortenings contain 22.5 percent, and fats in snack foods contain up to 46 percent of *trans* fatty acids (Craig-Schmidt 1992).

It is difficult to eliminate oxidation-sensitive polyunsaturated fatty acids by partial hydrogenation of fish oils. This has been demonstrated by Ackman (1973) in the progressive hydrogenation of anchovetta oil. The original eicosapentaenoic acid (20:5 ω 3) is not completely removed until an iodine value of 107.5 is reached. Even at this point there are other polyunsaturated fatty acids present that may be susceptible to flavor reversion.

In the nonselective hydrogenation of typical seed oils, polyunsaturated fatty acids are rapidly reduced and *trans*-isomer levels increase to high values. Figure 2–31 shows the hydrogenation of canola oil (deMan et al. 1982).

INTERESTERIFICATION

It is possible to change the position of fatty acid radicals on the glycerides in a fat by the process known as interesterification, randomization, or ester interchange. This is because, in the presence of certain catalysts, the fatty acid radicals can be made to move between hydroxyl positions so that an essentially random fatty acid distribution results, according to the following reaction pattern (Formo 1954):

$$RCOOR^2 + R^1 COOR^3 \rightleftharpoons RCOOR^3 + R^1COOR^2$$

This process is used in industry to modify the crystallization behavior and the physical properties of fats. It can also be used to produce solid fats for margarine and shortening that are low in *trans* fatty acids. An additional advantage is that polyunsaturated fatty acids, which are destroyed during hydrogenation, are not affected. Several types of interesterification are possible. A fat can be randomized by carrying out the reaction at temperatures above its melting point, several raw materials may be interesterified together so that a new product results, or a fat can be interesterified at a temperature below its melting point so that only the liquid fraction reacts (this is known as directed interesterification). The effect of randomization can be demonstrated with the case of a mixture of equal amounts of two simple glycerides, such as triolein and tristearin (Figure 2–32). After randomization, six possible triglycerides are found in quantities that can be calculated. When the blend of the two glycerides is other than in equal quantities the results can be derived from a graph such as the one in Figure 2–33. The graph indicates that the maximum levels of the intermediate glycerides A_2B and AB_2 are formed at molar fractions of one-third A or one-third B.

The theoretical number of glycerides formed by interesterification of mixtures

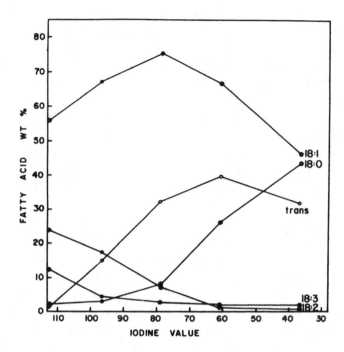

Figure 2–31 Change in Fatty Acid Composition During Hydrogenation of Canola Oil

containing different fatty acids has been described by Rozenaal (1992) and is shown in Table 2–30. The table also gives the formula for calculating the total number of glycerides formed. For example, for $n = 4$ the number of glycerides formed is 20 and for $n = 6$ the number is 56. Thus, interesterification results in increased complexity of the oil.

In directed interesterification, one of the reaction components is removed from the reaction mixture. This can be achieved by selecting a reaction temperature at which the trisaturated glycerides become insoluble and precipitate. The equilibrium is then disturbed and more trisaturates are formed, which can then be precipitated. Because of the low temperature employed, the reaction is up to 10 to 20 times slower than the random process. Another procedure of directed interesterification involves the continuous distilling of low molecular weight fatty acids, such as

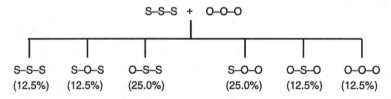

Figure 2–32 Interesterification of a Mixture of Equal Quantities of Triolein and Tristearin

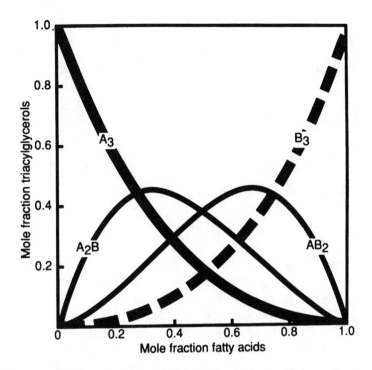

Figure 2–33 Triacylglycerol Distribution in a Randomized Binary Mixture. *Source:* Reprinted with permission from A. Rozenaal, Interesterification of Oils and Fats, *Inform*, 3, pp. 1233–1235, © 1992, AOCS Press.

those present in coconut oil with high free fatty acid content (Hustedt 1976).

The reaction mechanism of interesterification using sodium methoxide as a catalyst is a two-step process (Sreenivasan 1978; Rozenaal 1992). First, the catalyst combines with the glyceride at one of the carbonyl locations (Figure 2–34). Then the anion of the catalyst and the alkoxy group of the ester are exchanged. The catalyst has changed but remains active. At the end of the reaction there remains an amount of fatty acid methyl ester equivalent to the amount of sodium methoxide catalyst used. The randomization reaction continues until equilibrium has been reached. The reaction

Table 2–30 Theoretical Triacylglycerol Composition after Interesterification for n Fatty Acids (A, B, C, D) with Molar Fractions a, b, c, d

Type	Number	Amount
Mono acid (AAA, BBB)	n	a^3, b^3, c^3
Diacid (AAB, AAC)	$n(n-1)$	$3a^2b$, $3ab^2$, $3a^2c$
Triacid (ABC, BCD)	$1/6\, n(n-1)(n-2)$	$6abc$, $6acd$
Total	$n^3/6 + n^2/2 + n/3$	

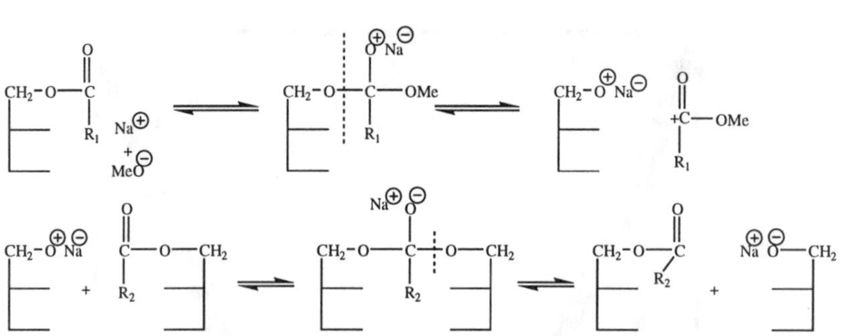

Figure 2–34 Reaction Mechanism of the Interesterification Process

is terminated by destroying the catalyst through addition of water or organic acid, which converts the fatty acid methyl ester into free fatty acid. The reaction is intramolecular as well as intermolecular. Freeman (1968) has reported that the intramolecular rearrangement occurs at a faster rate than the general randomization.

Interesterification can occur without the use of a catalyst at high temperature (300°C or higher). However, this process is slow and a number of other reactions occur such as polymerization and isomerization. The process is usually carried out at temperatures of about 100°C or lower. A number of catalysts can be used, including alkali metals and alkali metal alkylates. These catalysts are extremely susceptible to destruction by water and free fatty acids and are also affected by peroxides, carbon dioxide, and oxygen. A useful and convenient catalyst is the combination of sodium or potassium hydroxide and glycerol (Figure 2–35). The first step is the formation of sodium glycerate, which then reacts with a triacylglycerol to form the active catalyst and a monoacylglycerol as byproduct. The first step takes place at 60°C under vacuum to neutralize free fatty acids and remove water. The second step and the interesterification occurs at 130°C. The presence of glycerol results in the formation of some partial glycerides.

Rozenaal determined the reaction rate for the randomization of palm oil (Figure 2–36). The reaction rate, which was measured by determination of the solid fat content, increased with temperature. There is evidence of an induction period at lower temperatures.

Random interesterification can result in either an increase or a decrease in melting point and solid fat content, depending on the composition of the original fat or fat blend. When cocoa butter is interesterified, the unique melting properties are completely changed (Figure 2–37). Interesterification of lard has been used extensively. Lard produces coarse crystals because it tends to crystallize in the β form. Palmitic acid is mostly located in the sn-2 position of the disaturated glycerides (S_2U). When lard is randomized, the level of palmitic acid in the sn-2 position drops from 64 to 24 percent. The result is a smooth-textured fat that crystallizes in the β' form. Randomized lard has an improved plastic range and makes a better shortening. Palm oil shows the phenomenon of post-hardening or post-crystallization. This is a disadvantage in a number of applica-

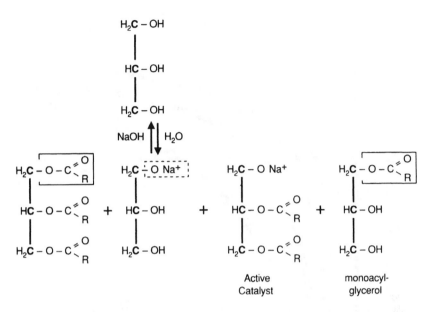

Figure 2–35 Formation of Interesterification Catalyst from Sodium Hydroxide and Glycerol in a Two-Step Process. *Source:* Reprinted with permission from A. Rozenaal, Interesterification of Oils and Fats, *Inform*, 3, pp. 1233–1235, © 1992, AOCS Press.

tions. Interesterification eliminates this problem.

In the formulation of margarines and shortenings, a hardstock is often combined with unmodified liquid oil. A useful hardstock for the formulation of soft margarines is an interesterified blend of palm stearin and palm kernel oil or fully hydrogenated palm kernel oil. Interesterification is used to produce *trans* free fats for making margarines and shortenings. The traditional method for transforming oils into fats involves hydrogenation and this results in high *trans* levels. The physical and chemical properties of *trans* free fats made by interesterfication have been described by Petrauskaite et al. (1998).

Interesterification can also be carried out by using lipase enzymes as a catalyst. This type of application is described in Chapter 10.

Ester interchange of fats with a large excess of glycerol, at high temperature, under vacuum, and in the presence of a catalyst, results in an equilibrium mixture of mono-, di-, and triglycerides. After removal of excess glycerol, the mixture is called technical monoglyceride and contains about 40 percent of 1-monoglyceride. Technical monoglycerides are used as emulsifying agents in foods. Molecular distillation yields products with well over 90 percent 1-monoglycerides; the distilled monoglycerides are also widely used in foods.

PHYSICAL PROPERTIES

Fats and oils are mixtures of mixed triglycerides. Fats are semisolid at room temperature; they are known as plastic fats. The solid character of fats is the result of the presence

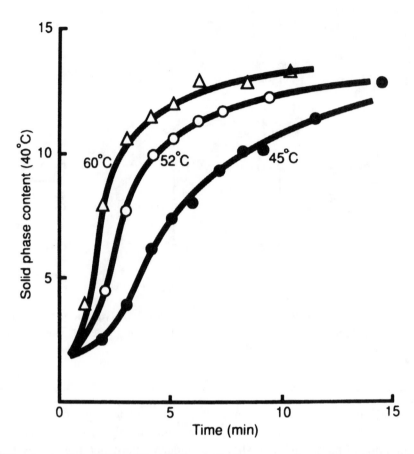

Figure 2–36 Randomization of Palm Oil at Different Temperatures. *Source:* Reprinted with permission from A. Rozenaal, Interesterification of Oils and Fats, *Inform*, 3, pp. 1233–1235, © 1992, AOCS Press.

of a certain proportion of crystallized triglycerides. Most fats contain a range of triglycerides of different melting points from very high to very low. When a fat is liquefied by heating, all glycerides are in the liquid state; upon cooling, some of the higher melting fractions become insoluble and crystallize. As the temperature is lowered, more glycerides become insoluble and crystallize, thereby increasing the solid fat content.

Crystallization of a fat is a slow process, whereas melting is instantaneous. When a fat crystallizes, the latent heat of crystallization is liberated and a volume contraction takes place. When a fat melts, the heat effect is negative and volume expands (Figure 2–38). These changes have been used for the measurement of some fat properties such as melting point, solidification temperature, and solid fat content. Because fats contain a range of glycerides of different melting points, there is no distinct melting point but rather a melting range. Nevertheless, melting points of fats are often determined. These are not real melting points, but some arbitrary temperature at which virtually all of the fat

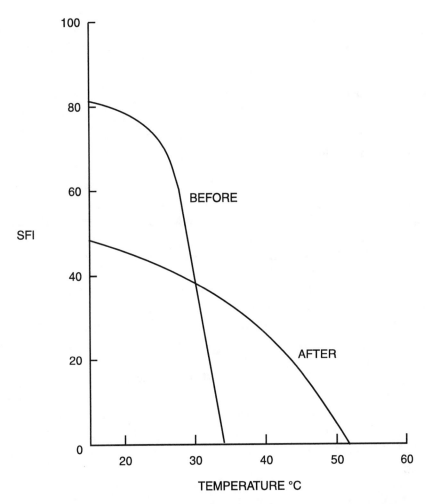

Figure 2–37 Solid Fat Index (SFI) of Cocoa Butter Before and After Interesterification

has become liquid. The value of these melting points depends on the measurement technique employed (Mertens and deMan 1972). The melting point of a fat is basically determined by the melting points of its constituent fatty acids. As shown in Table 2–31, chain length and unsaturation affect melting point. In addition, the configuration around the double bond is important. The arrangement of the fatty acids in different glyceride types also affects the melting point, as shown in Table 2–32. One of the most important properties of fats is the solid fat content. Because the solid fat content is dependent on temperature, it is common to determine the solid fat profile, which is a graph representing the relationship between solid fat content and temperature. Earlier methods for measuring solid fat were based on dilatometry, a technique using the melting expansion of fats on heating. Modern methods employ pulsed nuclear magnetic resonance. The dilatometer

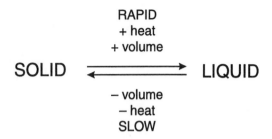

Figure 2–38 Melting and Solidification of Fats

technique gives an approximation of the true solid fat content and is reported as solid fat index (SFI). The nuclear magnetic resonance results are true solid fat measurements and reported as solid fat content (SFC). The relationship between solid fat content and physical properties is demonstrated in Figure 2–39 (Bracco 1994), where the difference in properties of cocoa butter and tallow is demonstrated. The steep solid fat curve of cocoa butter provides coolness and flavor release, and its high solids content at room temperature ensures hardness and heat resistance. Tallow has a less steep solids curve with solids beyond 37°C, which gives rise to a waxy mouthfeel.

The rate at which a liquid fat is cooled is important in establishing solid fat content and crystal size. In contrast to the glassy state forming systems described in Chapter 1, fats do not form a glassy state. The rate

of supercooling determines whether nucleation or crystal growth will predominate. At low supercooling, nucleation is at a minimum and large crystals are formed. At high supercooling, nucleation is high and small crystals result (Figure 2–40). Another result of high supercooling is the formation of solid solutions or mixed crystals. When fats are cooled quickly to well below their melting point (e.g., 0°C), the high melting glycerides crystallize together with some of the lower melting ones into mixed crystals. When the fat is subsequently tempered at a temperature close to the melting point, the mixed crystals recrystallize into crystals of more uniform composition. This means that a rapidly cooled fat contains more solid fat than the same fat that has been tempered after the same cooling treatment (Table 2–33).

Size and amount of fat crystals may vary from one product to another and even between products of the same composition but of dissimilar temperature history (that is,

Table 2–31 Melting Points of Some Fatty Acids

Fatty Acid	MP (°C)
Oleic (*cis*)	13
Elaidic (*trans*)	44
Stearic	70
Linoleic (*cis-cis*)	–5
Linelaidic (*trans-trans*)	28
Butyric	–8

Table 2–32 Melting Points of Some Triglycerides

Triglycerides		MP (°C)
Trisaturated	S-S-S	72
	P-P-P	65
	S-P-P	62
Disaturated	P-0-P	37
	O-P-P	34
	S-0-P	35
	S-P-O	39
	P-S-O	36
Diunsaturated	O-O-P	19
	O-O-S	23
Triunsaturated	O-O-O	5
	E-E-E	42
	L-L-L	–10

Note: S = stearic, P = palmitic, O = oleic, E = elaidic, L = linoleic.

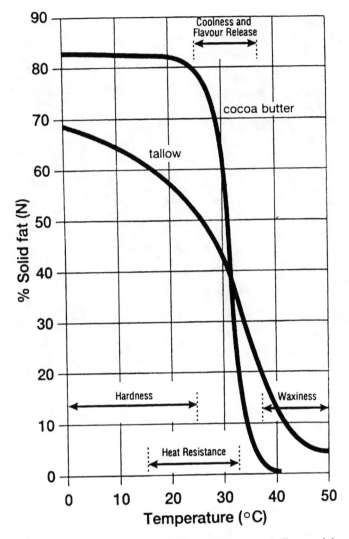

Figure 2–39 Physical Properties of Tallow and Cocoa Butter as Influenced by Solid Fat Profile. *Source:* Reprinted with permission from U. Bracco, Effect of Triglyceride Structure on Fat Absorption, *American Journal of Clinical Nutrition,* Vol. 60, (Suppl.) p. 1008S, © 1994, American Society for Clinical Nutrition.

a slightly different manufacturing method). Crystal size in fats is usually from 0.1 to 5 μm and occasionally up to 50 μm. Fats with such large crystals become grainy to the taste and even to the eye. Fat crystals form a three-dimensional network that lends rigidity to the product and holds the liquid portion of the fat. The proportion of solids in a fat is very important in determining the physical properties of a product. Fats may retain their solid character with solid fat contents as low as 12 to 15 percent. Below this level, fats become pourable and lose their plastic character. The relationship between hardness and

A   B

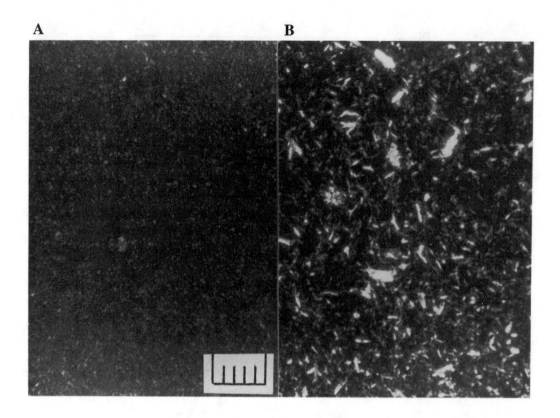

Figure 2–40 Effect of Supercooling on Fat Crystal Size. Crystals are seen in the polarizing microscope. (**A**) high supercooling, (**B**) low supercooling. One scale unit equals 10 micrometers.

solid fat content, as shown in Figure 2–41, indicates that there is only a relatively narrow range of solids that results in a product being neither too hard nor too soft. This is called the plastic range of fats. Shortening is an example of a product that requires an extended plastic range.

In addition to crystal content and crystal size and shape, there is another phenomenon in the crystallization of fats know as polymorphism. Long-chain organic compounds including glycerides exhibit polymorphism, the existence of more than one crystal form. Polymorphism results from different patterns of

molecular packing in fat crystals. It is generally assumed that, in a liquid, molecules move around at random. There is evidence (Hernqvist 1984) that in melted fat a limited arrangement of ordered lamellar units exists (short-range order). The size of these lamellar units decreases with increasing temperature. It has been suggested that these lamellar units are the beginning of a crystal nucleus (Figure 2–42), which will develop into a crystal (long-range order). Fats may occur in three basic polymorphs, designated α (alpha), β' (beta prime), and β (beta). The α form is the least stable and has the lowest melting point; the β

Table 2–33 Solid Fat Content of a Soft Margarine Fat with and Without Tempering at °C

Measurement Temperature (°C)	Solid Fat Content	
	Tempered	Not Tempered
10	22.3	30.7
20	11.7	16.0
30	4.1	6.3
35	2.8	1.9

form is the most stable and has the highest melting point; the β' form is intermediate in stability and melting point. Polymorphic transformations occur from α to β' to β and are irreversible. When a fat is cooled rapidly the α polymorph is produced, which is usually quickly converted to the β' form. In commer- cially produced fats such as shortenings and margarines, β' is the desirable form because β' crystals are small and result in a smooth tex- ture and good functionality. Depending on fatty acid and glyceride composition, fats may remain stable in the β' form or may convert to the β form. The latter will result in a large

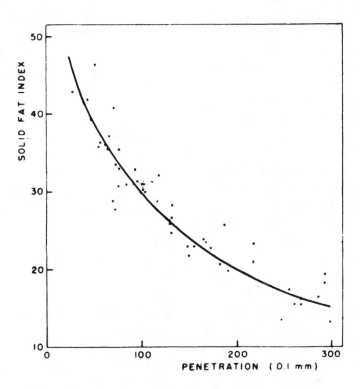

Figure 2–41 Relationship Between Hardness and Solid Fat Content

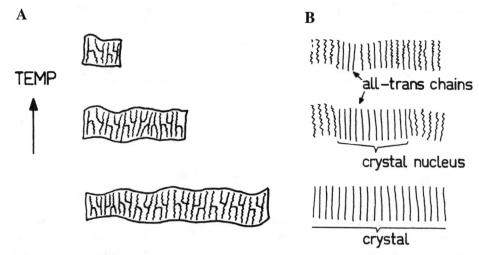

Figure 2–42 Lamellar Units and Crystal Formation. (**A**) Existence of lamellar units in a liquid oil just above its melting point. (**B**) Proposed mechanism for crystal formation in a triglyceride melt. *Source:* Reprinted with permission from L. Hernqvist, On the Structure of Triglycerides in the Liquid State and Fat Crystallization, *Fette Seifen Anstrichm*, Vol. 86, p. 299, Copyright © 1984, Lars Hernqvist.

increase in crystal size and grainy texture. The conversion of the β' to the β form is not an all-or-nothing phenomenon. Different ratios of β' and β crystals can exist in fats. The difference in melting points of the β' and β forms of some simple triglycerides are given in Table 2–34. The triglyceride PSP is unusual because it does not crystallize in the β form but remains stable in the β' form.

Assignment of the polymorphic form of a fat can be done unequivocally only by X-ray diffraction, sometimes supported by differential scanning calorimetry. The characteristics of the three polymorphs are as follows:

- α *form*—one strong, short spacing in the X-ray diffraction pattern at 0.42 nm. The chains are arranged in a hexagonal crystal structure (H) with no order of the zig-zag chain planes. The chains have no angle of tilt (Lutton 1972) (Figure 2–43).
- β' *form*—two strong, short spacings in the X-ray diffraction pattern at 0.38 and

0.42 nm. The chains are arranged in an orthorhombic crystal structure. The zig-zag planes are arranged in a perpendicular crystal structure. The chains have an angle of tilt between 50 and 70 degrees.
- β *form*—one strong, short spacing at 0.46 nm and two spacings at 0.37 and 0.39 nm. This form is the most densely packed and is in a triclinic structure. The zigzag planes are in a parallel arrangement, and the chains have an angle of tilt of 50 to 70 degrees.

The smallest repeating unit in a crystal is known as a subcell. Figure 2–44 summarizes the crystal arrangements of the three polymorphic forms. The chain length of the fatty acids in the triglycerides influences the packing of the successive layers in the different polymorphs. Examples of glyceride structures of 8-8-8 and 6-8-6 are given in Figure 2–45. Fats that consist of triglycerides with fatty acids of uniform chain length are more

Table 2–34 Melting Points (°C) of the β′ and β Forms of Some Simple Triglycerides

Triglyceride	β′	β
PPP	56.7	66.2
SSS	64.2	73.5
PSP	68.8	—
POP	30.5	35.3
POS	33.2	38.2
SOS	36.7	41.2

easily transformed from the β′ to the β form. Riiner (1971) has suggested a mechanism for the β′ to β transition (Figure 2–46). This mechanism explains only the change in crystal packing. The large change in crystal size is an additional effect of this polymorphic transformation.

In the lengthwise direction of molecules in the crystal, two possible arrangements have been described (Figure 2–47). Recognized by the long spacings in the X-ray diffraction pattern, these arrangements are known as double-chain length and triple-chain length forms. Both β′ and β polymorphs may exist in double- or triple-chain length forms and are distinguished by the suffix β-2 or β-3. The double-chain length forms are the most stable.

Hoerr (1960) has described the appearance of the different crystal forms when observed by polarized light microscopy. Alpha crystals are fragile platelets of about 5 μm in size. Beta prime crystals are tiny needles of about 1 to 2 μm. Beta crystals are large and irregularly shaped, ranging from 20 to 50 μm in size.

Sherbon (1974) has proposed a comprehensive scheme for the possible transitions between the different crystal forms of triglycerides. The scheme (Figure 2–48) indicates whether the transition is endothermic or exothermic. The different crystal forms are given the subscript L, referring to the nomenclature proposed by Lutton (1972).

The tendency of fats to remain stable in the β′ form for a long time or to quickly convert to the β form is more difficult to predict than that of pure triglycerides. In margarines and shortenings it is desirable that the fat crystals are in the β′ form. In margarines this results in a smooth texture, shiny surface, and good meltdown properties. The β′ crystals in shortenings give good aeration in cakes and creams. The solids or fat crystals in margarines and shortenings consist of triglycerides with melting points considerably higher than those of the whole fat. It is thought that the triglycerides of the solids with the highest melting points set the trend of the polymorphic behavior of margarines and shortenings. These triglycerides are referred to as high melting glycerides (HMGs) and can be obtained by fractionation from acetone (D'Souza et al. 1991). When the HMGs con-

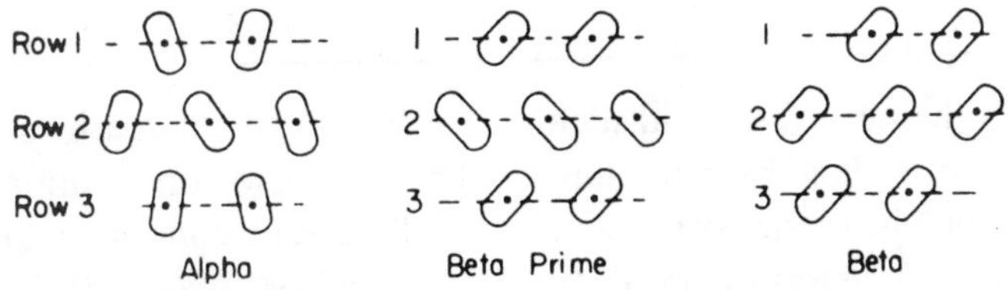

Figure 2–43 Cross-Sectional Structures of Long-Chain Compounds. *Source:* From E.S. Lutton, Technical Lipid Structures, *J. Am. Oil Chem. Soc.*, Vol. 49, pp. 1–9, 1972.

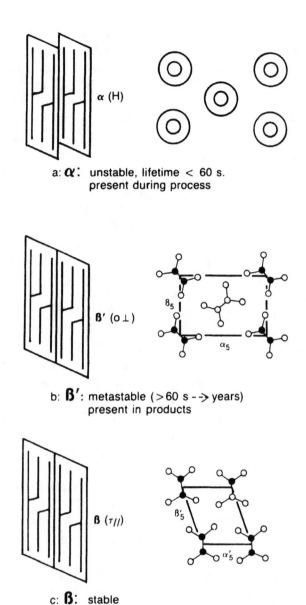

a: α: unstable, lifetime < 60 s.
present during process

b: β': metastable (>60 s --> years)
present in products

c: β: stable

Figure 2–44 Representation of the Packing of Triacylglycerols in the Three Main Polymorphic Forms

sist of a high level of the same fatty acid such as palmitic acid or stearic acid or a mixture of stearic and any *trans* isomeric form 18:1, the fat crystals quickly transform from β' to β form. This happens when canola oil or sunflower oil is hydrogenated and used as a sole

hardstock in margarines and shortenings. The fatty acid composition of canola oil (see Table 2–11) consists mainly of 18-carbon fatty acids such as 18:0, 18:1, 18:2, and 18:3; it contains only 4 percent palmitic acid. When canola oil is hydrogenated, the solids

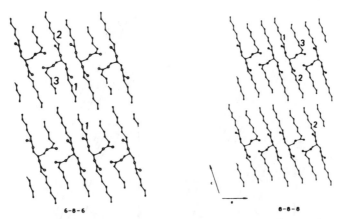

Figure 2–45 Tilted Arrangement of Glycerides in the Crystal Structure. Examples of postulated 8-8-8 and 6-8-6 triglyceride structures. *Source:* From E.S. Lutton, Technical Lipid Structures, *J. Am. Oil Chem. Soc.*, Vol. 49, pp. 1–9, 1972.

that are produced consist of triglycerides of 18:0 or 18:1 *trans* fatty acid, which results in a high level of 54 (3 × 18) carbon triglycerides. These triglycerides are liable to undergo polymorphic transition to the β form. Triglycerides consisting of only palmitic acid or tripalmitin (48-carbon triglycerides) are equally likely to convert to the β form. This can happen when palm stearin, which is obtained from fractionated palm oil, is used as the sole hardstock for a margarine or shortening (deMan and deMan 1995). Soybean oil contains 11 percent palmitic acid (see Table 2–11). When soybean oil is hydrogenated, the solids are more diverse in fatty acid chain length than those of hydrogenated canola or sunflower oil of similar consistency. The triglyceride composition (car-

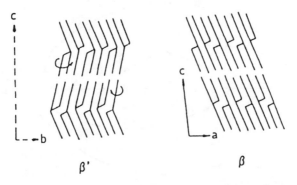

Figure 2–46 Suggested Mechanism for the β′ to β Polymorphic Transition. *Source:* Reprinted with permission from U. Riiner, Phase Behaviour of Hydrogenated Fats II, *Lebensm. Wiss. Tehnol.*, Vol. 4, p. 117, © 1971, Academic Press Ltd.

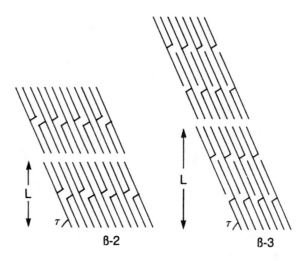

Figure 2–47 Double-Chain Length and Triple-Chain Length Arrangements in Triacylglycerol Crystals

bon number) of common vegetable oils is displayed in Table 2–35. Canola and sunflower oil contain high levels of 54-carbon triacylglycerol (TAG), while palm oil has the lowest level. When these oils are hydrogenated, the solids in the early stages of hydrogenation contain less 54-carbon and more 50-carbon triglycerides than the original oils, because the 50-carbon triglycerides already contain two saturated fatty acids and are likely to be included in the solids at the early stage. The 50-carbon triglycerides of the solids consist mainly of PSP or PEP (E stands for elaidic acid or any form of isomeric 18:1 *trans*). These triglycerides are very stable in the β' form, as mentioned earlier, and prefer-

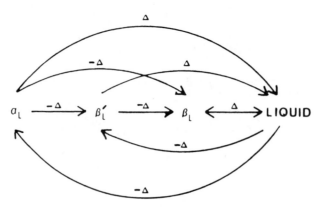

Figure 2–48 Polymorphic Transitions Between Different Crystal Forms of Triglycerides. *Source:* From J.W. Sherbon, Crystallization and Fractionation of Milk Fat, *J. Am. Oil Chem. Soc.*, Vol. 51, pp. 22–25, 1974.

Table 2–35 Triacylglycerol (TAG) Composition (%) by Carbon Number of Common Vegetable Oils

	TAG (Carbon Number)				
Oil	48	50	52	54	56
Canola	—	1.1	13.0	76.8	5.6
Soybean	—	3.3	27.6	66.7	1.7
Sunflower	—	2.8	20.2	75.1	0.7
Sunflower high olein	—	2.0	15.0	80.6	1.0
Corn	—	4.6	30.4	64.2	0.8
Olive	—	4.7	27.7	66.7	0.9
Peanut	—	5.5	30.9	54.2	5.3
Cottonseed	0.9	13.6	43.5	40.5	1.3
Palm	8.0	42.5	40.5	9.0	—

ably should be part of the solids. The solid 52-carbon triglycerides are mainly made up of PSS or PEE. Palmitic acid in vegetable oils is mainly located in the 1 or the 3 position. Palm oil has a high level of 50- and 52-carbon triglycerides and a low level of 54-carbon TAG (Table 2–36). The 50-carbon TAG consist mainly of POP (26 percent) and PLP (8 percent). When palm oil is hydrogenated, POP and PLP change to PSP or PEP, which provides a high level of desirable β′ triglycerides. In addition, the solid 52-carbon triglycerides that are produced are also β′ tending. The β′ stability of hydrogenated canola oil can be greatly enhanced by incorporation of hydrogenated palm oil at a level

Table 2–36 Major Triacylglycerols (TAGs) of Palm Oil and Their Melting Points

TAG	Carbon Number	%	Melting Point (°C)	
			β′	β
*PPP	48	6	56.7	66.2
*PPS	50	1	59.9	62.9
*PSP	50	.5	68.8	—
POP	50	26	30.5	35.3
*PPO	50	6	35.4	40.4
PLP	50	7	18.6	NA
POO	52	19	14.2	19.2
POS	52	3	33.2	38.2
PLO	52	4	NA	NA
OOO	54	3	−11.8	5.1

*Likely to be in the palm stearin fraction.
NA = Not available.

of 10 percent. For margarines where it is desirable to have low levels of *trans* fatty acids, palm midfraction that contains very high levels of POP and PLP (up to 60 percent) can be hydrogenated and incorporated at lower levels to secure a desirable level of PSP and PEP.

Diversification of the fatty acid composition of the solids can also be achieved by including lauric oils such as palm kernel or coconut oil, preferably in the fully hydrogenated form (Nor Aini et al. 1996; 1997). A β′ stable hardstock can, upon dilution with liquid oil, crystallize in the β form, as is done in soft margarines and to a certain extent in stick margarines. Elevated storage temperatures can also induce a polymorphic transition. β crystals are initially small but upon storage can grow into large, needle-like agglomerates that make the product grainy, make the texture brittle, and give the surface a dull appearance. Shortenings always contain about 10 to 12 percent hydrogenated palm oil—usually in the fully hydrogenated form, which extends the plastic range so that the product is still plastic at higher temperatures. The melting point of shortenings is in the low range of 40 to 45°C.

FRACTIONATION

Fats can be separated into fractions with different physical characteristics by fractional crystallization from a solvent or by fractionation from the melt. The former process gives sharply defined fractions but is only used for production of high-value fats; the latter process is much simpler and more cost-effective. Fractionation from the melt or dry fractionation is carried out on a very large scale with palm oil and other fats including beef tallow, lard, and milk fat.

There are several reasons for employing fractional crystallization (Hamm 1995):

- To remove small quantities of high melting components that might result in cloudiness of an oil. This can be either a triacylglycerol fraction or non-triacylglycerol compound. The former happens when soybean oil is lightly hydrogenated to convert it to a more stable oil. The resulting solid triglycerides have to be removed to yield a clear oil. The latter occurs when waxes crystallize from oils such as sunflower oil. This type of fractionation is known as winterization.
- To change a fat or oil into two or more fractions with different melting characteristics. In simple dry fractionation, a hard fraction (stearin) and a liquid fraction (olein) are obtained. This is by far the most common application of fractionation.
- To produce well-defined fractions with special physical properties that can be used as specialty fats or confectionery fats. This is often done by solvent fractionation.

The process of fractionation involves the controlled and limited crystallization of a melted fat. By careful management of the rate of cooling and the intensity of agitation, it is possible to produce a slurry of relatively large crystals that can be separated from the remaining liquid oil by filtration. The major application of fractionation is with palm oil; many millions of tons of palm oil are fractionated into palm stearin and palm olein every year. Palm oil is unusual among vegetable oils. It has a high level of palmitic acid; contains a substantial amount of a trisaturated simple glyceride, tripalmitin; and has a high level of SUS glycerides. The major glycerides present in palm oil are listed in Table 2–36, together with their melting points. When a liquid fat is cooled, a crystalline solid is formed; its composition and yield are determined by the final temperature of the oil. From Table 2–36 it is obvious that the glycerides most likely to crystallize are PPP, PPS, and PSP. It is theoretically impossible to separate these glycerides sharply from the rest of the glycerides. There are two reasons: (1) the formation of solid solutions and (2) the problem of entrainment.

The formation of solid solutions can be explained by the phase diagram of Figure 2–49 (Timms 1984; 1995). This relates to two triglycerides A and B, which form a solid solution. In other words, they crystallize together into mixed crystals. At temperature T_1 the solid phase has the composition indicated by c and the liquid phase composition is a. The fraction of the solid phase is equal to ab/ac. As the crystallization temperature differs from T_1, the composition of the solid and liquid phases will also differ. Solid solutions are formed when two or more solutes have melting points that are not very different. If the solute and solvent molecules have widely differing melting points, no solid solutions will be formed and the solubility is dependent only on the solute. This is known as ideal solubility. A plot of the solubility of tripalmitin in 2-oleo-dipalmitin is presented in Figure 2–50. At all temperatures the actual solubility is higher than the ideal solubility, and the resulting solid phase that crystallizes is composed not only of tripalmitin but a solid solution of tripalmitin and 2-oleo-dipalmitin.

Entrainment results from mechanical entrapment of liquid phase in the crystal cake obtained by filtration. The composition of the filter cake is highly dependent on the degree of pressure applied in the filtration process. The composition of palm oil and its main fractions are given in Table 2–37 (Siew et al. 1992; 1993). The triglycerides in palm oil and palm olein are listed in Table 2–38. Depending on the process used, the stearin can be obtained in yields of 20 to 40 percent, with iodine values ranging from 29 to 47. This is important because the olein is considered the more valuable commodity. Palm olein is widely used as a liquid oil in tropical climates. However, in moderate climates, it crystallizes at lower temperatures, just as olive oil and peanut oil do. Palm stearin is finding increasing application as a nonhydrogenated hard fat and as a component in interesterification for the production of no-*trans* margarines.

The fractionation of palm oil can be carried out in a number of ways to yield a variety of products as shown in Figure 2–51. In this multistage process, a palm midfraction is obtained that can be further fractionated to yield a cocoa butter equivalent (Kellens 1994). Palm oil can be double fractionated to yield a so-called super olein with iodine value of 65.

Milk fat fractionation has been described by Deffense (1993). By combining multistep fractionation and blending, it is possible to produce modified milk fats with improved functional properties.

COCOA BUTTER AND CONFECTIONERY FATS

Cocoa butter occupies a special place among natural fats because of its unusual and highly valued physical properties. Products containing cocoa butter, such as chocolate, are solid at room temperature; have a desirable "snap"; and melt smoothly and rapidly in the mouth, giving a cooling effect with no

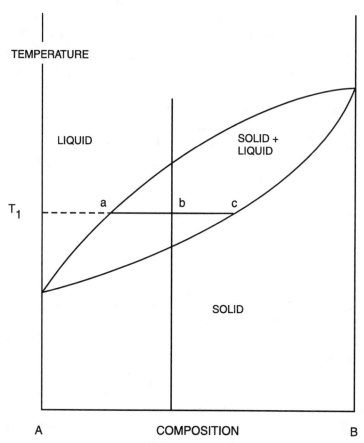

Figure 2–49 Phase Diagram of Mixture of Triglycerides A and B, Showing a Continuous Solid Solution. *Source:* Reprinted from R.E. Timms, Crystallization of Fats, in *Developments in Oils and Fats*, R.J. Hamilton, ed., p. 206, © 1995, Aspen Publishers, Inc.

greasy impression on the palate. Confectionery fats or specialty fats are often designed to have many of the positive traits of cocoa butter—properties that make them more suitable for special applications. The physical properties of cocoa butter are demonstrated by its solid fat profile (Figure 2–37). The solid fat curve indicates a high solid fat content at room temperature, which steeply declines as the temperature is raised to just below human body temperature; at this temperature no solid fat remains, so there is no waxy mouthfeel. These desirable properties are com-

pletely lost when cocoa butter is subjected to randomization of its triglyceride structure. The main characteristic of cocoa butter is the presence of a high content of symmetrical monounsaturated triglycerides. Cocoa butter contains only three major fatty acids—palmitic, stearic, and oleic. The three main glyceride types are of the SUS type, with the unsaturated fatty acid in the 2-position. Lard, which has a similar fatty acid composition, has mostly saturated fatty acids in the 2-position and very different physical properties. Table 2–39 presents the triglyceride compo-

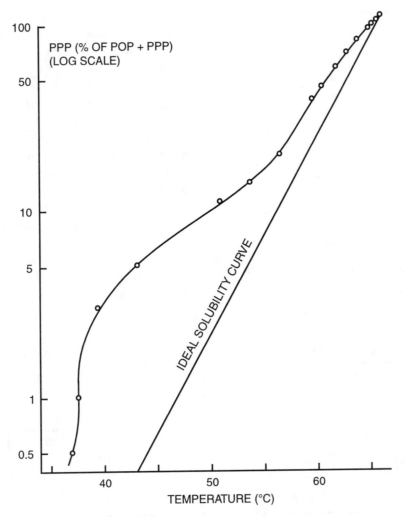

Figure 2–50 Solubility Diagram of Tripalmitin in 2-oleo-dipalmitin. *Source:* Reprinted from R.E. Timms, Crystallization of Fats, in *Developments in Oils and Fats*, R.J. Hamilton, ed., p. 209, © 1995, Aspen Publishers, Inc.

Table 2–37 Composition and Slip Melting Point (SMP) of Palm Oil and Its Fractions

	Fatty Acid (Wt %)				
Product	16:0	18:0	18:1	18:2	SMP (°C)
Palm oil	44.1	4.4	39.0	10.6	36.7
Palm olein	40.9	4.2	41.5	11.6	21.5
Palm stearin	56.8	4.9	29.0	7.2	51.4

Source: Reprinted with permission from W.L. Siew, et al., Identity Characteristics of Malaysian Palm Oil Products: Fatty Acid and Triglyceride Composition and Solid Fat Content, *E.L.A.E.I.S.*, Vol. 5, No. 1, p. 40, © 1993, Palm Oil Research Institute of Malaysia.

Table 2–38 Composition of Palm Oil and Palm Olein

Triglyceride	Palm Oil	Palm Olein
SSS (mainly PPP)	8	0.5
SUS (mainly POP)	50	48
SUU (mainly POO)	37	44
UUU	5	7
Iodine value	51–53	57–58

Source: Reprinted with permission from W.L. Siew, et al., Identity Characteristics of Malaysian Palm Oil Products: Fatty Acid and Triglyceride Composition and Solid Fat Content, *E.L.A.E.I.S.*, Vol. 4, No. 1, p. 40, © 1993, Palm Oil Research Institute of Malaysia.

sition of cocoa butter from the three main producing countries and demonstrates the natural variability of this fat. The melting points of the main triglycerides POP, POS, and SOS are given in Table 2–34. The polymorphic behavior of cocoa butter is more complex than that of its component glycerides, and a specific system for cocoa butter is often used. This was introduced by Wille and Lutton (1966) and recognizes six different polymorphs—I, for the lowest melting form, through VI, for the highest melting form (Table 2–40). Another system in use recognizes only five polymorphs, designated γ, α, β′, β2, and β1, in order of increasing stability and melting point (Loisel et al. 1998; van Malssen et al. 1996a, b, c).

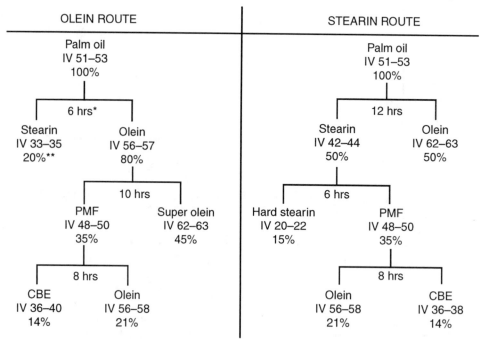

* Cycle times in a De Smet fractionation
** Results from 6 bar membrane press filtration

Figure 2–51 Products Obtained by Multistage Fractionation of Palm Oil. *Source:* Reprinted with permission from Developments in Fat Fractionation Technology, Paper No. 0042, 1994, p. 29, © Mark Kellens, PhD.

Table 2–39 Triglyceride Composition of Cocoa Butter from the Main Producing Countries (Mole %)

	Ghana	Malaysia	Brazil
Symmetrical monounsaturated	80	87	72
Symmetrical diunsaturated	9	7	8
Unsymmetrical diunsaturated	8	5	17
Polyunsaturated	3	1	3

The desirable physical properties of cocoa butter and chocolate—snap, gloss, melting in the mouth, and flavor release—depend on the formation of polymorph V or β_2. Cocoa butter is a tempering fat, which means that a special tempering procedure has to be followed to produce the desired polymorphic form. This involves keeping the molten chocolate mass at 50 to 60°C for one hour, then cooling to 25 to 27°C to initiate crystallization, and then heating to 29 to 31°C before molding and final cooling to 5 to 10°C.

After long storage or unfavorable storage conditions such as extreme temperatures, chocolate may show "bloom." This is a grayish covering of the surface caused by crystals of the most stale β phase (phase VI). Eventually the change progresses to the interior of the chocolate. The resulting change in crystal structure and melting point makes the product unsuitable for consumption. The tendency of chocolate, especially dark chocolate, to bloom has been attributed to the high level of POS glycerides. Milk fat is miscible with cocoa butter without greatly affecting its melting properties, but this is not the case for most other fats. Several kinds of fats that are intended to replace cocoa butter are known as specialty fats or confectionery fats. These are mainly cocoa butter substitutes (CBSs) and cocoa butter improvers and equivalents (CBIs and CBEs).

Miscibility is an important characteristic of specialty fats. When fats of different composition are mixed, they may show eutectic effects. This means that the melting point or solid fat content of the blend is lower than that of the individual components. This happens when cocoa butter is mixed with a CBS. The resulting isosolids diagram of Figure 2–52 shows the strong eutectic effect that can lead to unacceptable softening. Mixing of cocoa butter and CBE gives no eutectic effect, and this type of fat can be used in any proportion

Table 2–40 Polymorphic Forms of Cocoa Butter and Their Melting Points (°C)

Phase	Melting Point	Phase	Melting Point
I	17.3	γ	16–18
II	23.3	α	21–24
III	25.5	β'	27–29
IV	27.5		
V	33.8	β_2	34–35
VI	36.3	β_1	36–37

with cocoa butter (Figure 2–53). CBEs are generally based on three raw materials—shea oil, illipe butter, and palm oil–and processed by fractionation. CBEs also require the same tempering procedures as cocoa butter. It is possible to tailor make CBEs to higher solid fat content and melting point than some of the softer types of cocoa butter; these fats are described as CBIs (Figure 2–54).

Cocoa butter substitutes are available in two types, lauric and nonlauric. Lauric CBSs are not compatible with cocoa butter and must only be used with low-fat cocoa powder. Lauric CBSs are based on palm kernel oil or coconut oil. Fractionation of palm kernel oil yields a stearin containing mainly saturated triglycerides composed mainly of lauric and myristic acids. The stearin itself after hydrogenation is an excellent CBS (Figure 2–55). It has a very high solid fat

content at 20°C, and the molded products are resistant to fat bloom. They do not need tempering, and the crystals formed are stable.

Nonlauric CBSs are produced by hydrogenation of liquid oils, frequently followed by fractionation and/or blending. The objective of hydrogenation is to achieve a product with a steep melting curve. This can be accomplished by using a sulfur-modified nickel catalyst, which produces a very high level of *trans* isomers. These products, especially those made from palm olein, are very stable in the β′ form. The nonlauric CBS can be significantly improved by fractionation, which results in a narrower melting range. Nonfractionated CBSs are used in compound coating fats for cookies. The fractionated, hydrogenated CBSs have better eating quality and can tolerate up to 25 percent cocoa butter when used in coatings. Lauric CBS

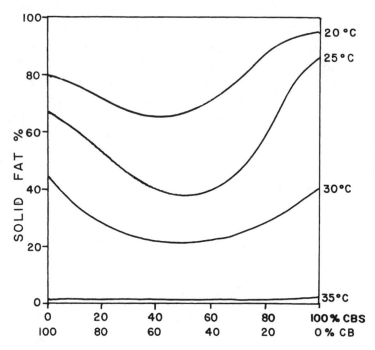

Figure 2–52 Isosolids Diagram of Blends of Cocoa Butter (CB) and Cocoa Butter Substitute (CBS), Showing Strong Eutectic Effect

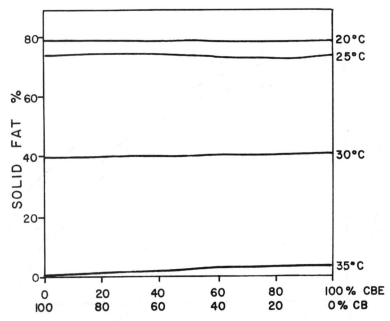

Figure 2–53 Isosolids Diagram of Blends of Cocoa Butter (CB) and Cocoa Butter Equivalent (CBE). There is no eutectic.

can tolerate no more than 6 percent admixture of cocoa butter.

EMULSIONS AND EMULSIFIERS

A satisfactory definition of emulsions has been provided by Becher (1965), who states that an emulsion is a heterogeneous system, consisting of one immiscible liquid intimately dispersed in another one, in the form of droplets with a diameter generally over 0.1 μm. These systems have a minimal stability, which can be enhanced by surface-active agents and some other substances. In foods, emulsions usually contain the two phases, oil and water. If water is the continuous phase and oil the disperse phase, the emulsion is of the oil in water (O/W) type. In the reverse case, the emulsion is of the water in oil (W/O) type. A third material or combination of several materials is required to confer stability upon the emulsion. These are surface-active agents called emulsifiers. The action of emulsifiers can be enhanced by the presence of stabilizers. Emulsifiers are surface-active compounds that have the ability to reduce the interfacial tension between air-liquid and liquid-liquid interfaces. This ability is the result of an emulsifier's molecular structure: the molecules contain two distinct sections, one having polar or hydrophilic character, the other having nonpolar or hydrophobic properties. The extent of the lowering of interfacial tension by surface-active agents is shown in Figure 2–56. Most surface-active agents reduce the surface tension from about 50 dynes/cm to less than 10 dynes/cm when used in concentrations below 0.2 percent.

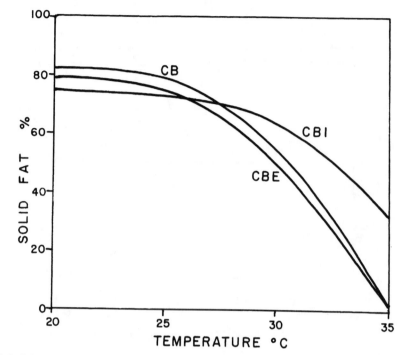

Figure 2–54 Solid Fat Profiles of Cocoa Butter (CB), Cocoa Butter Equivalent (CBE), and Cocoa Butter Improver (CBI)

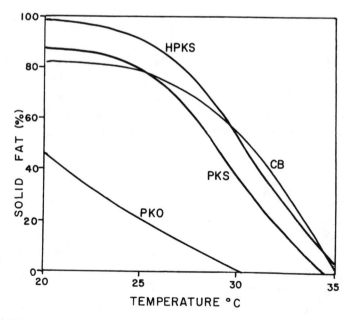

Figure 2–55 Solid Fat Profiles of Cocoa Butter (CB), Palm Kernel Oil (PKO), Palm Kernel Stearin (PKS), and Hydrogenated Palm Kernel Stearin (HPKS)

The relative size of the hydrophilic and hydrophobic sections of an emulsifier molecule mostly determines its behavior in emulsification. To make the selection of the proper emulsifier for a given application, the so-called HLB (hydrophile-lipophile balance) system was developed. It is a numerical expression for the relative simultaneous attraction of an emulsifier for water and for oil. The HLB of an emulsifier is an indication of how it will behave but not how efficient it is. Emulsifiers with low HLB tend to form W/O emulsions, those with intermediate HLB form O/W emulsions, and those with high HLB are solubilizing agents. HLB values can be calculated from the saponification number of the ester (S) and the acid value of the fatty acid radical (A_v) as follows:

$$HLB = 20\left(1 - \frac{S}{A_v}\right)$$

In many cases the HLB value is determined experimentally and various methods have been used (Friberg 1976). The HLB scale goes from 0 to 20, in theory at least, since at each end of the scale the compounds would have little emulsifying activity. The HLB values of some commercial nonionic emulsifiers are given in Table 2–41 (Griffin 1965).

Foods contain many natural emulsifiers, of which phospholipids are the most common. Crude phospholipid mixtures obtained by degumming of soya oil are utilized extensively as food emulsifiers and are known as soya-lecithin. This product contains a variety of phospholipids, not just lecithin.

Emulsifiers can be tailor-made to serve in many food emulsion systems. Probably the most widely used is the group of monoglycerides obtained by glycerolysis of fats. Reaction of an excess of glycerol with a fat under vacuum at high temperature and in the pres-

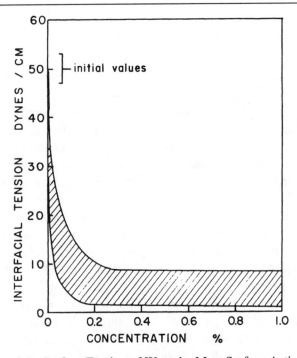

Figure 2–56 Lowering of the Surface Tension of Water by Most Surface-Active Compounds. *Source: From P. Becher, Emulsions—Theory and Practice, 1965, Van Nostrand Reinhold Publishing Co.*

Table 2–41 HLB Values of Some Commercial Nonionic Emulsifiers

Trade Name	Chemical Designation	HLB
Span 85	Sorbitan trioleate	1.8
Span 65	Sorbitan tristearate	2.1
Atmos 150	Mono- and diglycerides from the glycerolysis of edible fats	3.2
Atmul 500	Mono- and diglycerides from the glycerolysis of edible fats	3.5
Atmul 84	Glycerol monostearate	3.8
Span 80	Sorbitan monooleate	4.3
Span 60	Sorbitan monostearate	4.7
Span 40	Sorbitan monopalmitate	6.7
Span 20	Sorbitan monolaurate	8.6
Tween 61	Polyoxyethylene sorbitan monostearate	9.6
Tween 81	Polyoxyethylene sorbitan monooleate	10.0
Tween 85	Polyoxyethylene sorbitan trioleate	11.0
Arlacel 165	Glycerol monostearate (acid stable, self-emulsifying)	11.0
Myrj 45	Polyoxyethylene monostearate	11.1
Atlas G-2127	Polyoxyethylene monolaurate	12.8
Myrj 49	Polyoxyethylene monostearate	15.0
Myrj 51	Polyoxyethylene monostearate	16.0

Source: From W.C. Griffin, Emulsions, in *Kirk-Othmer Encyclopedia of Chemical Technology*, 2nd ed., Vol. 8, pp. 117–154, 1965, John Wiley & Sons.

ence of a catalyst results in the formation of so-called technical monoglycerides. These are mixtures of mono-, di-, and triglycerides. Only the 1-monoglycerides are active as emulsifiers. By molecular distillation under high vacuum, a product can be obtained in which the 1-monoglyceride content exceeds 90 percent.

The ability of emulsifier molecules to orient themselves at interfaces is exemplified in the phase behavior of monoglycerides. The emulsifier molecules in aqueous systems show mesomorphism (formation of liquid crystalline phases). In such systems, several mesophases may exist, as Krog and Larsson (1968) have shown. When 1-monopalmitin crystals in 20 to 30 percent of water are heated to 60°C, a mesophase, called the *neat* phase, is formed (Figure 2–57A). This structure consists of bimolecular lipid layers separated by water, with the chains in a disordered state. If this phase is cooled, a gel is formed (Figure 2–57B). The structure is lamellar with the hydrocarbon chains extended and tilted in a 54-degree angle toward the water layers. When the neat phase is heated, a stiff cubic phase is formed, called *viscous isotropic* (Figure 2–57C). It consists of small water spheres arranged in a face-centered lattice with the polar groups pointed toward the water. On further heating, an isotropic fluid is obtained (Figure 2–57D), in which polar groups and water form disk-like arrangements. With excess water, a dispersion is formed (Figure

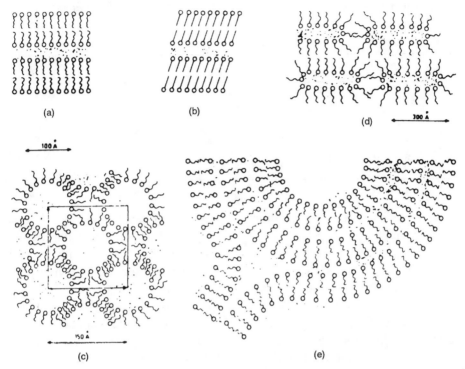

Figure 2–57 Structure of the Mesophases of Monoglyceride-Water Systems. **(A)** neat phase, **(B)** gel, **(C)** viscous isotropic, **(D)** fluid isotropic, and **(E)** dispersion. *Source:* From N. Krog and K. Larsson, Phase Behavior and Rheological Properties of Aqueous Systems of Industrial Distilled Monoglycerides, *Chem. Phys. Lipids*, Vol. 2, pp. 129–143, 1968.

2–57E), which consists of concentric bimolecular shells of monoglyceride molecules alternating with water.

Emulsions are stabilized by a variety of compounds, mostly macromolecules such as proteins and starches.

Emulsifiers have many additional functions in foods. They form complexes with food components, resulting in modified physical properties of the food system. For example, consider the amylose-complexing effect of emulsifiers (Krog 1971). This effect is useful for improving the shelf life of bread (antifirming effect) and modifying the physical characteristics of potato products, pasta, and similar foods.

The amount of emulsifier required to provide a monomolecular layer for an emulsion of any particle size can be calculated from the relationship A/V = 6/D, where A, V, and D are area, volume, and diameter, respectively, of the spherical particles. The total area for 1 mL of oil dispersed as uniform spheres thus calculated is given in Table 2–42. If sodium stearate is used as the emulsifier, the amount required can be calculated, since this molecule has a surface area of about 0.2 nm^2.

$$W = \frac{\text{Total area (nm}^2) \times 306}{0.2 \times 6 \times 10^{23}}$$

where

W =   weight of emulsifier

Table 2–42 Total Surface Area of 1 mL of Oil Dispersed as Uniform Spheres in an Emulsion

Diameter		Total Surface Area		
μm	cm	m^2	cm^2	nm^2
10	10^{-3}	0.6	6×10^3	6×10^{17}
1	10^{-4}	6	6×10^4	6×10^{18}
0.1	10^{-5}	60	6×10^5	6×10^{19}
0.01	10^{-6}	600	6×10^6	6×10^{20}

306 = molecular weight of sodium stearate

6×10^{23} = Avogadro's number

The amount of emulsifier required increases sharply with decreasing particle size of the emulsion, as shown in Table 2–43. This constitutes a practical limitation to the lower limit of particle size that can be obtained in emulsification. It also indicates that a substance that consists of molecules with a large ratio of surface area to molecular weight and that yields a strong film should be an efficient emulsifier.

NOVEL OILS AND FATS

The first oil modified by plant breeding was canola, which is now a major oil produced worldwide. Through plant breeding and increasingly through genetic engineering, new types of oil are being created. Oils with different fatty acid composition are being developed to produce oils with improved nutritional properties, improved oxidative stability, and improved physical or technological properties. To obtain canola oil with improved oxidative stability, low-linolenic canola has been developed. A low-linolenic soybean oil has also been produced, as well as high-oleic sunflower, high-oleic safflower, and low-linolenic linseed oil. These oils with improved oxidative stability are recommended for use as frying oils (Erickson and Frey 1994).

The rapeseed plant is amenable to genetic modification for the production of oils with a variety of compositions. One example is laurate canola, which produces an oil with about 40 percent lauric acid. This makes it somewhat similar to the lauric oils, coconut and palm kernel. The difference is in the triacylglycerol composition, as it contains LOL and LOO glycerides (L = lauric acid). When these are hydrogenated triacylglycerols LSL and LSS are formed, which are not available from other sources.

Genetic engineering of oilseed plants can be used to control fatty acid chain length and

Table 2–43 Amount of Sodium Stearate Emulsifier Required To Provide a Monomolecular Layer in Emulsions of Different Particle Size

Particle Diameter Micrometer	Sodium Stearate (as % of oil)
10	0.15
1	1.5
0.1	15
0.01	150

unsaturation as well as triglyceride structure (Lassner 1997). It is also used to control resistance to herbicides and to improve yield. Another example of the use of genetic engineering is the development of a mustard plant with oil composition identical to canola oil. This allows oilseed production in areas that are too dry for the cultivation of canola because the mustard plant is more drought tolerant.

FAT REPLACERS

Fat replacers are substances that are meant to replace the calories provided by fat (9 kcal/g) either partly or completely. They may consist of fat-like substances that are not or only partly absorbed by humans or of proteinaceous or carbohydrate compounds that mimic the gustatory qualities of fats. The latter will be dealt with in Chapters 3 and 4.

Olestra has been developed for use as a frying fat that has no calories. Olestra is a sucrose polyester with six to eight acyl groups, derived from soybean, corn, cottonseed, or sunflower fatty acids. It is not absorbed in the human digestive system and, therefore, yields no calories. Other fat replacers are based on the fact that the level of 9 kcal/g does not apply when short-chain

fatty acids are present. It is also known that long-chain fatty acids, especially stearic acid, are incompletely metabolized and yield calorie values less than 9 kcal/g. By combining these two types of fatty acid into glycerides, fats are obtained that have energy values of 5 kcal/g. A product of this type is Salatrim (an acronym for short and long acyl triglyceride molecules). It contains at least one stearic acid and one short-chain fatty acid in each of the component glycerides. It is produced by fully hydrogenating a liquid oil (canola or soybean) and then trans-esterifying this with short-chain acids such as acetic, propionic, or butyric acid. By varying the composition, a family of products is obtained that range from a solid to a liquid at room temperature (Kosmark 1996). A problem in the use of these low-calorie fats is that they contain only saturated fatty acids, and the claim of reduced calorie content is not permitted on nutritional labels (in contrast to nondigestible carbohydrates, see Chapter 4).

Medium-chain triacylglycerols are another type of structured fat made by trans-esterification of caprylic and capric fatty acids derived from coconut or palm kernel oil (Akoh 1997).

REFERENCES

Abraham, V., and J.M. deMan. 1985. Determination of volatile sulfur compounds in canola oil. *J. Am. Oil Chem. Soc.* 62: 1025–1028.

Ackman, R.G. 1972. The analysis of fatty acids and related materials by gas-liquid chromatography. In *Progress in the chemistry of fats and other lipids.* Vol. 12, ed. R.T. Holman. Oxford, UK: Pergamon Press.

Ackman, R.G. 1973. *Marine lipids and fatty acids in human nutrition.* Food and Agriculture Organization Technical Conference on Fishery Products, Japan.

Ackman, R.G. 1988a. Some possible effects on lipid biochemistry of differences in the distribution on glycerol of long-chain n-3 fatty acids in the fats of marine fish and marine mammals. *Atherosclerosis* 70: 171–173.

Ackman, R.G. March 1988b. The year of the fish oils. *Chem. Ind.* 139–145.

Ackman, R.G. 1994. Animal and marine lipids. In *Improved and technological advances in alternative sources of lipids,* ed. B. Kamel and Y. Kakuda. London: Blackie Academic and Professional.

Ackman, R.G., and C.A. Eaton. 1971. Mackerel lipids and fatty acids. *Can. Inst. Food Sci. Technol. J.* 4: 169–174.

Akoh, C.C. 1997. Making new structured fats by chemical reaction and enzymatic modification. *Lipid Technol.* 9: 61–66.

Becher, P. 1965. *Emulsions: Theory and practice.* New York: Van Nostrand Reinhold Publishing Co.

Bracco, U. 1994. Effect of triglycerides structure on fat absorption. *Am. J. Clin. Nutr.* 60: 1002S–1009S.

Brockerhoff, H., R.J. Hoyle, and N. Wolmark. 1966. Positional distribution of fatty acids in triglycerides of animal depot fats. *Biochim. Biophys. Acta* 116: 67–72.

Carter, J.P. 1988. Gamma-linolenic acid as a nutrient. *Food Technol.* 81: 72–82.

Chahine, M.H., and J.M. deMan. 1971. Autoxidation of corn oil under the influence of fluorescent light. *Can. Inst. Food Sci. Technol. J.* 4: 24–28.

Chen, P.C., and J.M. deMan. 1966. Composition of milk fat fractions obtained by fractional crystallization from acetone. *J. Dairy Sci.* 49: 612–616.

Cowan, J.C., and C.D. Evans. 1961. Flavor reversion. In *Autoxidation and antioxidants,* ed. W.O. Lundberg. New York: John Wiley & Sons.

Craig-Schmidt, M.C. 1992. Fatty acid isomers in foods. In *Fatty acids in foods and their health implications,* ed. C.K. Chow. New York: Marcel Dekker.

Deffense, E. 1993. Milk fat fractionation today: A review. *J. Am. Oil Chem. Soc.* 70: 1193–1201.

deMan, J.M., and L. deMan. 1995. Palm oil as a component for high quality margarine and shortening formulations. *Malaysian Oil Sci. Technol.* 4: 56–60.

deMan, J.M., Y. El-Shattory, and L. deMan. 1982. Hydrogenation of canola oil (Tower). *Chem. Mikrobiol. Technol. Lebensm.* 7: 117–124.

deMan, L., and J.M. deMan. 1983. *Trans* fatty acids in milk fat. *J. Am. Oil Chem. Soc.* 60: 1095–1098.

deMan, L., F. Tie, and J.M. deMan. 1987. Formation of short chain volatile organic acids in the automated AOM method. *J. Am. Oil Chem. Soc.* 64: 993–996.

deMan, L., et al. 1995. Physical and chemical properties of hydrogenated high erucic rapeseed oil. *Fat Sci. Technol.* 97: 485–490.

D'Souza, V., L. deMan, and J.M. deMan. 1991. Chemical and physical properties of the high melting fractions of commercial margarines. *J. Am. Oil Chem. Soc.* 68: 153–162.

Erickson, D.R. 1996. Production and composition of frying fats. In *Deep frying: Chemistry, nutrition and practical applications,* ed. E.G. Perkins and M.D. Erickson. Champaign, IL: AOCS Press.

Erickson, M.D., and N. Frey. 1994. Property enhanced oils in food applications. *Food Technol.* 11: 63–68.

Farmer, E.H. 1946. Peroxidation in relation to olefinic structure. *Trans. Faraday Soc.* 42: 228–236.

Fedeli, E., et al. 1966. Isolation of geranyl geraniol from the unsaponifiable fraction of linseed oil. *J. Lipid Res.* 7: 437–441.

Fedeli, E., and G. Jacini. 1971. Lipid composition of vegetable oils. *Adv. Lipid Res.* 9: 335–382.

Formo, M.W. 1954. Ester reactions of fatty materials. *J. Am. Oil Chem. Soc.* 31: 548–559.

Frankel, E.N., and H.J. Dutton. 1970. Hydrogenation with homogeneous and heterogeneous catalysts. In *Topics in lipid chemistry.* Vol. 1, ed. F.D. Gunstone. London: Logos Press, Ltd.

Freeman, I.P. 1968. Interesterification. I. Change of glyceride composition during the course of interesterification. *J. Am. Oil Chem. Soc.* 45: 456–460.

Friberg, S. 1976. *Food emulsions.* New York: Marcel Dekker, Inc.

Griffin, W.C. 1965. Emulsions. In *Kirk-Othmer encyclopedia of chemical technology,* 2nd ed., 8: 117–154. New York: John Wiley & Sons.

Gunstone, F.D. 1967. *An introduction to the chemistry of fats and fatty acids.* London: Chapman and Hall.

Gunstone, F.D. 1986. Fatty acid structure. In *The lipid handbook,* ed. F.D. Gunstone, J.L. Harwood, and F.B. Padley. London: Chapman and Hall.

Hamm, W. 1995. Trends in edible oil fractionation. *Trends Food Sci. Technol.* 6: 121–126.

Herb, S.F., et al. 1962. Fatty acids of cows' milk. B. Composition by gas-liquid chromatography aided by other methods of fractionation. *J. Am. Oil Chem. Soc.* 39: 142–146.

Hernqvist, L. 1984. On the structure of triglycerides in the liquid state and fat crystallization. *Fette Seifen Anstrichm.* 86: 297–300.

Hilditch, T.P., and P.N. Williams. 1964. *The chemical constitution of natural fats.* 4th ed. New York: John Wiley & Sons.

Hoerr, C.W. 1960. Morphology of fats, oils and shortenings. *J. Am. Oil Chem. Soc.* 37: 539–546.

Hoffmann, G. 1989. *The chemistry and technology of edible oils and fats and their high fat products.* London: Academic Press.

Hustedt, H.H. 1976. Interesterification of edible oils. *J. Am. Oil Chem. Soc.* 53: 390–392.

Jacini, G., E. Fedeli, and A. Lanzani. 1967. Research in the nonglyceride substances of vegetable oils. *J. Assoc. Off. Anal. Chem.* 50: 84–90.

Jen, J.J., et al. 1971. Effects of dietary fats on the fatty acid contents of chicken adipose tissue. *J. Food Sci.* 36: 925–929.

Jensen, R.G. 1973. Composition of bovine milk lipids. *J. Am. Oil Chem. Soc.* 50: 186–192.

Jensen, R.G., and D.S. Newburg. 1995. Milk lipids. In *Handbook of milk composition*, ed. R.G. Jensen. New York: Academic Press.

Kartha, A.R.S. 1953. The glyceride structure of natural fats. II. The rule of glyceride type distribution of natural fats. *J. Am. Oil Chem. Soc.* 30: 326–329.

Keeney, M. 1962. Secondary degradation products. In *Lipids and their oxidation*, ed. H.W. Schultz, E.A. Day, and R.O. Sinnhuber. Westport, CT: AVI Publishing Co.

Kellens, M. 1994. Developments in fat fractionation technology. Paper No. 0042. London: Society of Chemical Industry.

Kosmark, R. 1996. Salatrim: Properties and applications. *Food Technol.* 50, no. 4: 98–101.

Krog, N. 1971. Amylose complexing effect of food grade emulsifiers. *Stärke* 23: 206–210.

Krog, N., and K. Larsson. 1968. Phase behavior and rheological properties of aqueous systems of industrial distilled monoglycerides. *Chem. Phys. Lipids* 2: 129–143.

Kuksis, A. 1972. Newer developments in determination of structure of glycerides and phosphoglycerides. In *Progress in the chemistry of fats and other lipids*. Vol. 12, ed. R.T. Holman. Oxford, UK: Pergamon Press.

Lassner, M. 1997. Transgenic oilseed crops: A transition from basic research to product development. *Lipid Technol.* 9: 5–9.

Lee, K.H., M.Y. Yung, and S.Y. Kim. 1997. Quenching mechanism and kinetics of ascorbyl palmitate for the reduction of photosensitized oxidation of oils. *J. Am. Oil Chem. Soc.* 74: 1053–1057.

Le Quéré, J.L., and J.L. Sébédio. 1996. Cyclic monomers of fatty acids. In *Deep frying: Chemistry, nutrition and practical applications*, ed. E.G. Perkins and M.D. Erickson. Champaign, IL: AOCS Press.

Loisel, C., et al. 1998. Phase transitions and polymorphism of cocoa butter. *J. Am. Oil Chem. Soc.* 75: 425–439.

Lundberg, W.O. 1961. Autoxidation and antioxidants. New York: John Wiley & Sons.

Lutton, E.S. 1972. Technical lipid structures. *J. Am. Oil Chem. Soc.* 49: 1–9.

Magidman, P., et al. 1962. Fatty acids of cows' milk. A. Techniques employed in supplementing gas-liquid chromatography for identification of fatty acids. *J. Am. Oil Chem. Soc.* 39: 137–142.

Marion, W.W., S.T. Maxon, and R.M. Wangen. 1970. Lipid and fatty acid composition of turkey liver, skin and depot tissue. *J. Am. Oil Chem. Soc.* 47: 391–392.

Mertens, W.G., and J.M. deMan. 1972. Automatic melting point determination of fats. *J. Am. Oil Chem. Soc.* 49: 366–370.

Morrison, W.R. 1970. Milk lipids. In *Topics in lipid chemistry*. Vol. 1, ed. F.D. Gunstone. London: Logos Press.

Nor Aini, I., et al. 1996. Chemical composition and physical properties of soft (tub) margarines sold in Malaysia. *J. Am. Oil Chem. Soc.* 73: 995–1001.

Nor Aini, I., et al. 1997. Chemical and physical properties of plastic fat products sold in Malaysia. *J. Food Lipids* 4: 145–164.

Ohloff, G. 1973. Fats as precursors (In German). In *Functional properties of fats in foods*, ed. J. Solms. Zurich, Switzerland: Forster Publishing.

Petrauskaite, V., et al. 1998. Physical and chemical properties of *trans*-free fats produced by chemical interesterification of vegetable oil blends. *J. Am. Oil Chem. Soc.* 75: 489–493.

Pokorny, J. 1971. Stabilization of fats by phenolic antioxidants. *Can. Inst. Food Sci. Technol. J.* 4: 68–74.

Rahmani, M., and A.S. Csallany. 1998. Role of minor constituents in the photooxidation of virgin olive oil. *J. Am. Oil Chem Soc.* 75: 837–843.

Ratnayake, W.M.N. 1994. Determination of *trans* unsaturation by infrared spectrophotometry and determination of fatty acid composition of partially hydrogenated vegetable oils and animal fats by gas chromatography/infrared spectrophotometry: Collaborative study. *JAOAC Intern.* 78: 783–802.

Riiner, U. 1971. Phase behavior of hydrogenated fats. II. Polymorphic transitions of hydrogenated sunflowerseed oil and solutions. *Lebensm. Wiss. Technol.* 4: 113–117.

Rozenaal, A. 1992. Interesterification of oils and fats. *Inform* 3: 1232–1237.

Sattar, A., J.M. deMan, and J.C. Alexander. 1976. Light induced oxidation of edible oils and fats. *Lebensm. Wiss. Technol.* 9: 149–152.

Sherbon, J.W. 1974. Crystallization and fractionation of milk fat. *J. Am. Oil Chem. Soc.* 51: 22–25.

Siew, W.L., et al. 1992. Identity characteristics of Malaysian palm oil products: Fatty acid and triglyceride composition and solid fat content. *Elaeis* 4: 79–85.

Siew, W.L., et al. 1993. Identity characteristics of Malaysian palm oil products: Fatty acid and triglyceride composition and solid fat content. *Elaeis* 5: 38–46.

Simopoulos, A.P. March/April 1988. ω-3 fatty acids in growth and development and in health and disease. *Nutr. Today* 10–19.

Sreenivasan, B. 1978. Interesterification of fats. *J. Am. Oil Chem. Soc.* 45: 456–460.

Timms, R.E. 1984. Phase behaviour of fats and their mixtures. *Progr. Lipid Res.* 23: 1–38.

Timms, R.E. 1995. Crystallization of fats. In *Developments in oils and fats*, ed. R.J. Hamilton. New York: Chapman and Hall.

Vander Wal, R.J. 1964. Triglyceride structure. *Adv. Lipid Res.* 2: 1–16.

van Malssen, K., R. Peschar, and H. Schenk. 1996a. Real-time X-ray powder diffraction investigations on cocoa butter. I. Temperature dependent crystallization behaviour. *J. Am. Oil Chem. Soc.* 73: 1209–1215.

van Malssen, K., R. Peschar, and H. Schenk. 1996b. Real-time X-ray powder diffraction investigations on cocoa butter. II. The relationship between melting behaviour and composition of β-cocoa butter. *J. Am. Oil Chem. Soc.* 73: 1217–1223.

van Malssen, K., R. Peschar, and H. Schenk. 1996c. Real-time X-ray powder diffraction investigations on cocoa butter. III. Direct β-crystallization of cocoa butter: Occurrence of a memory effect. *J. Am. Oil Chem. Soc.* 73: 1223–1230.

Villeneuve, P., and T.A. Foglia. 1997. Lipase specificities: Potential application in lipid bioconversions. *Inform* 8: 640–650.

Wijesundera, R.C., et al. 1988. Determination of sulfur contents of vegetable and marine oils by ion chromatography and indirect ultraviolet photometry of their combustion products. *J. Am. Oil Chem. Soc.* 65: 1526–1530.

Wille, R.L., and E.S. Lutton. 1966. Polymorphism of cocoa butter. *J. Am. Oil Chem. Soc.* 43: 491–496.

Woodrow, I.L., and J.M. deMan. 1968. Distribution of *trans* unsaturated fatty acids in milk fat. *Biochim. Biophys. Acta* 152: 472–478.

Yasaei, P.M., et al. 1996. Singlet oxygen oxidation of lipids resulting from photochemical sensitizers in the presence of antioxidants. *J. Am. Oil Chem. Soc.* 73: 1177–1181.

Zambiazi, R.C., and R. Przybylski. 1998. Effect of endogenous minor components on the oxidative stability of vegetable oils. *Lipid Technol.* 10: 58–62.

CHAPTER 3

Proteins

INTRODUCTION

Proteins are polymers of some 21 different amino acids joined together by peptide bonds. Because of the variety of side chains that occur when these amino acids are linked together, the different proteins may have different chemical properties and widely different secondary and tertiary structures. The various amino acids joined in a peptide chain are shown in Figure 3–1. The amino acids are grouped on the basis of the chemical nature of the side chains (Krull and Wall 1969). The side chains may be polar or nonpolar. High levels of polar amino acid residues in a protein increase water solubility. The most polar side chains are those of the basic and acidic amino acids. These amino acids are present at high levels in the soluble albumins and globulins. In contrast, the wheat proteins, gliadin and glutenin, have low levels of polar side chains and are quite insoluble in water. The acidic amino acids may also be present in proteins in the form of their amides, glutamine and asparagine. This increases the nitrogen content of the protein. Hydroxyl groups in the side chains may become involved in ester linkages with phosphoric acid and phosphates. Sulfur amino acids may form disulfide cross-links between neighboring peptide chains or between different parts of the same chain. Proline and hydroxyproline impose significant structural limitations on the geometry of the peptide chain.

Proteins occur in animal as well as vegetable products in important quantities. In the developed countries, people obtain much of their protein from animal products. In other parts of the world, the major portion of dietary protein is derived from plant products. Many plant proteins are deficient in one or more of the essential amino acids. The protein content of some selected foods is listed in Table 3–1.

AMINO ACID COMPOSITION

Amino acids joined together by peptide bonds form the primary structure of proteins. The amino acid composition establishes the nature of secondary and tertiary structures. These, in turn, significantly influence the functional properties of food proteins and their behavior during processing. Of the 20 amino acids, only about half are essential for human nutrition. The amounts of these essential amino acids present in a protein and their availability determine the nutritional quality of the protein. In general, animal proteins are of higher quality than plant proteins. Plant

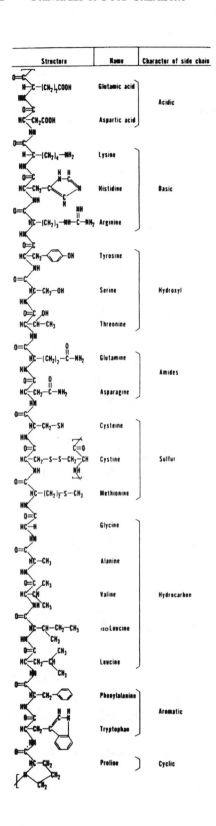

Structure	Name	Character of side chain
	Glutamic acid	Acidic
	Aspartic acid	
	Lysine	
	Histidine	Basic
	Arginine	
	Tyrosine	
	Serine	Hydroxyl
	Threonine	
	Glutamine	Amides
	Asparagine	
	Cysteine	
	Cystine	Sulfur
	Methionine	
	Glycine	
	Alanine	
	Valine	Hydrocarbon
	isoLeucine	
	Leucine	
	Phenylalanine	Aromatic
	Tryptophan	
	Proline	Cyclic

Figure 3–1 Component Amino Acids of Proteins Joined by Peptide Bonds and Character of Side Chains. *Source:* From Northern Regional Research Laboratory, U.S. Department of Agriculture.

proteins can be upgraded nutritionally by judicious blending or by genetic modification through plant breeding. The amino acid composition of some selected animal and vegetable proteins is given in Table 3–2.

Egg protein is one of the best quality proteins and is considered to have a biological value of 100. It is widely used as a standard, and protein efficiency ratio (PER) values sometimes use egg white as a standard. Cereal proteins are generally deficient in lysine and threonine, as indicated in Table

Table 3–1 Protein Content of Some Selected Foods

Product	Protein (g/100 g)
Meat: beef	16.5
pork	10.2
Chicken (light meat)	23.4
Fish: haddock	18.3
cod	17.6
Milk	3.6
Egg	12.9
Wheat	13.3
Bread	8.7
Soybeans: dry, raw	34.1
cooked	11.0
Peas	6.3
Beans: dry, raw	22.3
cooked	7.8
Rice: white, raw	6.7
cooked	2.0
Cassava	1.6
Potato	2.0
Corn	10.0

Table 3–2 Amino Acid Content of Some Selected Foods (mg/g Total Nitrogen)

Amino Acid	Meat (Beef)	Milk	Egg	Wheat	Peas	Corn
Isoleucine	301	399	393	204	267	230
Leucine	507	782	551	417	425	783
Lysine	556	450	436	179	470	167
Methionine	169	156	210	94	57	120
Cystine	80	—	152	159	70	97
Phenylalanine	275	434	358	282	287	305
Tyrosine	225	396	260	187	171	239
Threonine	287	278	320	183	254	225
Valine	313	463	428	276	294	303
Arginine	395	160	381	288	595	262
Histidine	213	214	152	143	143	170
Alanine	365	255	370	226	255	471
Aspartic acid	562	424	601	308	685	392
Glutamic acid	955	1151	796	1866	1009	1184
Glycine	304	144	207	245	253	231
Proline	236	514	260	621	244	559
Serine	252	342	478	281	271	311

3–3. Soybean is a good source of lysine but is deficient in methionine. Cottonseed protein is deficient in lysine and peanut protein in methionine and lysine. The protein of potato although present in small quantity (Table 3–1) is of excellent quality and is equivalent to that of whole egg.

Table 3–3 Limiting Essential Amino Acids of Some Grain Proteins

Grain	First Limiting Amino Acid	Second Limiting Amino Acid
Wheat	Lysine	Threonine
Corn	Lysine	Tryptophan
Rice	Lysine	Threonine
Sorghum	Lysine	Threonine
Millet	Lysine	Threonine

PROTEIN CLASSIFICATION

Proteins are complex molecules, and classification has been based mostly on solubility in different solvents. Increasingly, however, as more knowledge about molecular composition and structure is obtained, other criteria are being used for classification. These include behavior in the ultracentrifuge and electrophoretic properties. Proteins are divided into the following main groups: simple, conjugated, and derived proteins.

Simple Proteins

Simple proteins yield only amino acids on hydrolysis and include the following classes:

- *Albumins.* Soluble in neutral, salt-free water. Usually these are proteins of relatively low molecular weight. Examples are egg albumin, lactalbumin, and serum albumin in the whey proteins of milk, leucosin of cereals, and legumelin in legume seeds.

- *Globulins*. Soluble in neutral salt solutions and almost insoluble in water. Examples are serum globulins and β-lactoglobulin in milk, myosin and actin in meat, and glycinin in soybeans.
- *Glutelins*. Soluble in very dilute acid or base and insoluble in neutral solvents. These proteins occur in cereals, such as glutenin in wheat and oryzenin in rice.
- *Prolamins*. Soluble in 50 to 90 percent ethanol and insoluble in water. These proteins have large amounts of proline and glutamic acid and occur in cereals. Examples are zein in corn, gliadin in wheat, and hordein in barley.
- *Scleroproteins*. Insoluble in water and neutral solvents and resistant to enzymic hydrolysis. These are fibrous proteins serving structural and binding purposes. Collagen of muscle tissue is included in this group, as is gelatin, which is derived from it. Other examples include elastin, a component of tendons, and keratin, a component of hair and hoofs.
- *Histones*. Basic proteins, as defined by their high content of lysine and arginine. Soluble in water and precipitated by ammonia.
- *Protamines*. Strongly basic proteins of low molecular weight (4,000 to 8,000). They are rich in arginine. Examples are clupein from herring and scombrin from mackerel.

Conjugated Proteins

Conjugated proteins contain an amino acid part combined with a nonprotein material such as a lipid, nucleic acid, or carbohydrate. Some of the major conjugated proteins are as follows:

- *Phosphoproteins*. An important group that includes many major food proteins. Phosphate groups are linked to the hydroxyl groups of serine and threonine. This group includes casein of milk and the phosphoproteins of egg yolk.
- *Lipoproteins*. These are combinations of lipids with protein and have excellent emulsifying capacity. Lipoproteins occur in milk and egg yolk.
- *Nucleoproteins*. These are combinations of nucleic acids with protein. These compounds are found in cell nuclei.
- *Glycoproteins*. These are combinations of carbohydrates with protein. Usually the amount of carbohydrate is small, but some glycoproteins have carbohydrate contents of 8 to 20 percent. An example of such a mucoprotein is ovomucin of egg white.
- *Chromoproteins*. These are proteins with a colored prosthetic group. There are many compounds of this type, including hemoglobin and myoglobin, chlorophyll, and flavoproteins.

Derived Proteins

These are compounds obtained by chemical or enzymatic methods and are divided into primary and secondary derivatives, depending on the extent of change that has taken place. Primary derivatives are slightly modified and are insoluble in water; rennet-coagulated casein is an example of a primary derivative. Secondary derivatives are more extensively changed and include proteoses, peptones, and peptides. The difference between these breakdown products is in size and solubility. All are soluble in water and not coagulated by heat, but proteoses can be precipitated with saturated ammonium sulfate solution. Peptides contain two or more amino acid residues. These breakdown products are formed during the processing of many foods, for example, during ripening of cheese.

PROTEIN STRUCTURE

Proteins are macromolecules with different levels of structural organization. The primary structure of proteins relates to the peptide bonds between component amino acids and also to the amino acid sequence in the molecule. Researchers have elucidated the amino acid sequence in many proteins. For example, the amino acid composition and sequence for several milk proteins is now well established (Swaisgood 1982).

Some proteolytic enzymes have quite specific actions; they attack only a limited number of bonds, involving only particular amino acid residues in a particular sequence. This may lead to the accumulation of well-defined peptides during some enzymic proteolytic reactions in foods.

The secondary structure of proteins involves folding the primary structure. Hydrogen bonds between amide nitrogen and carbonyl oxygen are the major stabilizing force. These bonds may be formed between different areas of the same polypeptide chain or between adjacent chains. In aqueous media, the hydrogen bonds may be less significant, and van der Waals forces and hydrophobic interaction between apolar side chains may contribute to the stability of the secondary structure. The secondary structure may be either the α-helix or the sheet structure, as shown in Figure 3–2. The helical structures are stabilized by intramolecular hydrogen bonds, the sheet structures by intermolecular hydrogen bonds. The requirements for maximum stability of the helix structure were established by Pauling et al. (1951). The helix model involves a translation of 0.54 nm per turn along the central axis. A complete turn is made for every 3.6 amino acid residues. Proteins do not necessarily have to occur in a complete α-helix configuration; rather, only parts of the peptide chains may be helical, with other areas of the chain in a more or less unordered configuration. Proteins with α-helix structure may be either globular or fibrous. In the parallel sheet structure, the polypeptide chains are almost fully extended and can form hydrogen bonds between adjacent chains. Such structures are generally insoluble in aqueous solvents and are fibrous in nature.

The tertiary structure of proteins involves a pattern of folding of the chains into a compact unit that is stabilized by hydrogen bonds, van der Waals forces, disulfide bridges, and hydrophobic interactions. The tertiary structure results in the formation of a tightly packed unit with most of the polar amino acid residues located on the outside and hydrated. This leaves the internal part with most of the apolar side chains and virtually no hydration. Certain amino acids, such as proline, disrupt the α-helix, and this causes fold regions with random structure (Kinsella 1982). The nature of the tertiary structure varies among proteins as does the ratio of α-helix and random coil. Insulin is loosely folded, and its tertiary structure is stabilized by disulfide bridges. Lysozyme and glycinin have disulfide bridges but are compactly folded.

Large molecules of molecular weights above about 50,000 may form quaternary structures by association of subunits. These structures may be stabilized by hydrogen bonds, disulfide bridges, and hydrophobic interactions. The bond energies involved in

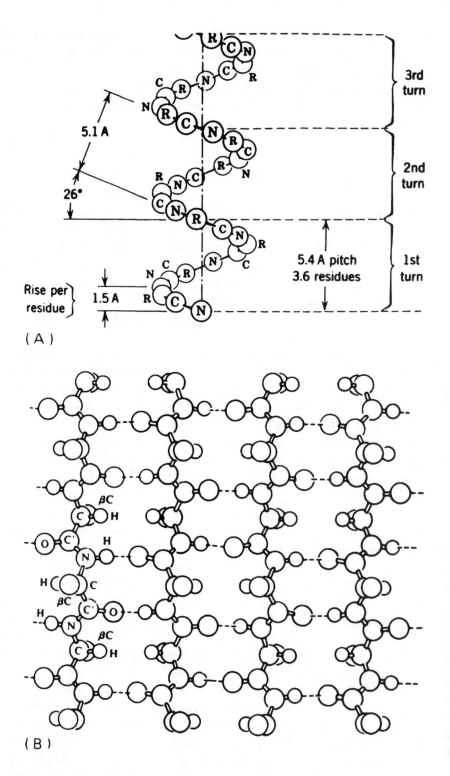

Figure 3–2 Secondary Structures of Proteins, (**A**) Alpha Helix, (**B**) Antiparallel Sheet

forming these structures are listed in Table 3–4.

The term *subunit* denotes a protein chain possessing an internal covalent and noncovalent structure that is capable of joining with other similar subunits through noncovalent forces or disulfide bonds to form an oligomeric macromolecule (Stanley and Yada 1992). Many food proteins are oligomeric and consist of a number of subunits, usually 2 or 4, but occasionally as many as 24. A listing of some oligomeric food proteins is given in Table 3–5. The subunits of proteins are held together by various types of bonds: electrostatic bonds involving carboxyl, amino, imidazole, and guanido groups; hydrogen bonds involving hydroxyl, amide, and phenol groups; hydrophobic bonds involving long-chain aliphatic residues or aromatic groups; and covalent disulfide bonds involving cystine residues. Hydrophobic bonds are not true bonds but have been described as interactions of nonpolar groups. These nonpolar groups or areas have a tendency to orient themselves to the interior of the protein molecule. This tendency depends on the relative number of nonpolar amino acid residues and their location in the peptide chain. Many food proteins, especially plant storage proteins, are highly hydrophobic—so much so that not all of the hydrophobic areas can be oriented toward the inside and have to be located on the surface. This is a possible factor in subunits association and in some cases may result in aggregation. The hydrophobicity values of some food proteins as reported by Stanley and Yada (1992) are listed in Table 3–6.

The well-defined secondary, tertiary, and quaternary structures are thought to arise directly from the primary structure. This means that a given combination of amino acids will automatically assume the type of structure that is most stable and possible given the considerations described by Pauling et al. (1951).

Table 3–4 Bond Energies of the Bonds Involved in Protein Structure

Bond	Bond Energy* (kcal/mole)
Covalent C-C	83
Covalent S-S	50
Hydrogen bond	3–7
Ionic electrostatic bond	3–7
Hydrophobic bond	3–5
Van der Waals bond	1–2

*These refer to free energy required to break the bonds: in the case of a hydrophobic bond, the free energy required to unfold a nonpolar side chain from the interior of the molecule into the aqueous medium.

Table 3–5 Oligomeric Food Proteins

Protein	Molecular Weight (d)	Subunits
Lactoglobulin	35,000	2
Hemoglobin	64,500	4
Avidin	68,300	4
Lipoxygenase	108,000	2
Tyrosinase	128,000	4
Lactate dehydrogenase	140,000	4
7S soy protein	200,000	9
Invertase	210,000	4
Catalase	232,000	4
Collagen	300,000	3
11S soy protein	350,000	12
Legumin	360,000	6
Myosin	475,000	6

Source: Reprinted with permission from D.W. Stanley and R.Y. Yada, Thermal Reactions in Food Protein Systems, *Physical Chemistry of Foods*, H.G. Schwartzberg and R.H. Hartel, eds., p. 676, 1992, by courtesy of Marcel Dekker, Inc.

DENATURATION

Denaturation is a process that changes the molecular structure without breaking any of the peptide bonds of a protein. The process is peculiar to proteins and affects different proteins to different degrees, depending on the structure of a protein. Denaturation can be brought about by a variety of agents, of which the most important are heat, pH, salts, and surface effects. Considering the complexity of many food systems, it is not surprising that denaturation is a complex process that cannot easily be described in simple terms. Denaturation usually involves loss of biological activity and significant changes in some physical or functional properties such as solubility. The destruction of enzyme activity by heat is an important operation in food processing. In most cases, denaturation is nonreversible; however, there are some exceptions, such as the recovery of some types of enzyme activity after heating. Heat denaturation is sometimes desirable—for example, the denaturation of whey proteins for the production of milk powder used in baking. The relationship among temperature, heating time, and the extent of whey protein denaturation in skim milk is demonstrated in Figure 3–3 (Harland et al. 1952).

The proteins of egg white are readily denatured by heat and by surface forces when egg white is whipped to a foam. Meat proteins are denatured in the temperature range 57 to 75°C, which has a profound effect on texture, water holding capacity, and shrinkage.

Denaturation may sometimes result in the flocculation of globular proteins but may also lead to the formation of gels. Foods may be denatured, and their proteins destabilized, during freezing and frozen storage. Fish proteins are particularly susceptible to destabilization. After freezing, fish may become tough and rubbery and lose moisture. The caseinate micelles of milk, which are quite stable to heat, may be destabilized by freezing. On frozen storage of milk, the stability of the caseinate progressively decreases, and this may lead to complete coagulation.

Protein denaturation and coagulation are aspects of heat stability that can be related to the amino acid composition and sequence of the protein. Denaturation can be defined as *a major change in the native structure that does not involve alteration of the amino acid sequence*. The effect of heat usually involves a change in the tertiary structure, leading to a less ordered arrangement of the polypeptide chains. The temperature range in which denaturation and coagulation of most proteins take place is about 55 to 75°C, as indicated in Table 3–7. There are some notable exceptions to this general pattern. Casein and gelatin are examples of proteins that can be

Table 3–6 Hydrophobicity Values of Some Food Proteins

Protein	Hydrophobicity cal/residue
Gliadin	1300
Bovine serum albumin	1120–1000
α-Lactalbumin	1050
β-Lactoglobulin	1050
Actin	1000
Ovalbumin	980
Collagen	880
Myosin	880
Casein	725
Whey protein	387
Gluten	349

Source: Reprinted with permission from D.W. Stanley and R.Y. Yada, Thermal Reactions in Food Protein Systems, *Physical Chemistry of Foods*, H.G. Schwartzberg and R.H. Hartel, eds., p. 677, 1992, by courtesy of Marcel Dekker, Inc.

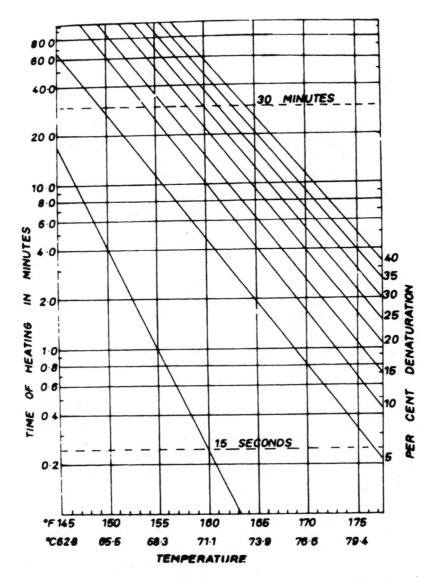

Figure 3–3 Time-Temperature Relationships for the Heat Denaturation of Whey Proteins in Skim Milk. *Source*: From H.A. Harland, S.T. Coulter, and R. Jenness, The Effects of Various Steps in the Manufacture on the Extent of Serum Protein Denaturation in Nonfat Dry Milk Solids. *J. Dairy Sci.* 35: 363–368, 1952.

boiled without apparent change in stability. The exceptional stability of casein makes it possible to boil, sterilize, and concentrate milk, without coagulation. The reasons for this exceptional stability have been discussed by Kirchmeier (1962). In the first place, restricted formation of disulfide bonds due to low content of cystine and cysteine results in increased stability. The relationship between coagulation temperature as a measure of stability and sulfur amino acid content is shown in Tables 3–7 and 3–8. Peptides, which are low in these particular amino acids, are less likely to become involved in the type of sulfhydryl agglomeration shown in Figure 3–4. Casein, with its extremely low content of sulfur amino acids, exemplifies this behavior. The heat stability of casein is also explained by the restraints against forming a folded tertiary structure. These restraints are due to the relatively high content of proline and hydroxyproline in the heat stable proteins (Table 3–9). In a peptide chain free of proline, the possibility of forming inter- and intramolecular hydrogen bonds is better than in a chain containing many proline residues (Figure 3–5). These considerations show how amino acid composition directly relates to secondary and tertiary structure of proteins; these structures are, in turn, responsible for some of the physical properties of the protein and the food of which it is a part.

NONENZYMIC BROWNING

The nonenzymic browning or *Maillard reaction* is of great importance in food manufacturing and its results can be either desir-

able or undesirable. For example, the brown crust formation on bread is desirable; the brown discoloration of evaporated and sterilized milk is undesirable. For products in which the browning reaction is favorable, the resulting color and flavor characteristics are generally experienced as pleasant. In other products, color and flavor may become quite unpleasant.

The browning reaction can be defined as the sequence of events that begins with the reaction of the amino group of amino acids, peptides, or proteins with a glycosidic hydroxyl group of sugars; the sequence terminates with the formation of brown nitrogenous polymers or melanoidins (Ellis 1959).

The reaction velocity and pattern are influenced by the nature of the reacting amino acid or protein and the carbohydrate. This means that each kind of food may show a different browning pattern. Generally, lysine is the most reactive amino acid because of the free ε-amino group. Since lysine is the limiting essential amino acid in many food proteins, its destruction can substantially reduce the nutritional value of the protein. Foods that are rich in reducing sugars are very reactive, and this explains why lysine in milk is destroyed more easily than in other

Table 3–7 Heat Coagulation Temperatures of Some Albumins and Globulins and Casein

Protein	Coagulation Temp. (°C)
Egg albumin	56
Serum albumin (bovine)	67
Milk albumin (bovine)	72
Legumelin (pea)	60
Serum globulin (human)	75
β-Lactoglobulin (bovine)	70–75
Fibrinogen (human)	56–64
Myosin (rabbit)	47–56
Casein (bovine)	160–200

Table 3–8 Cysteine and Cystine Content of Some Proteins (g Amino Acid/100 g Protein)

Protein	Cysteine (%)	Cystine (%)
Egg albumin	1.4	0.5
Serum albumin (bovine)	0.3	5.7
Milk albumin	6.4	—
β-Lactoglobulin	1.1	2.3
Fibrinogen	0.4	2.3
Casein	—	0.3

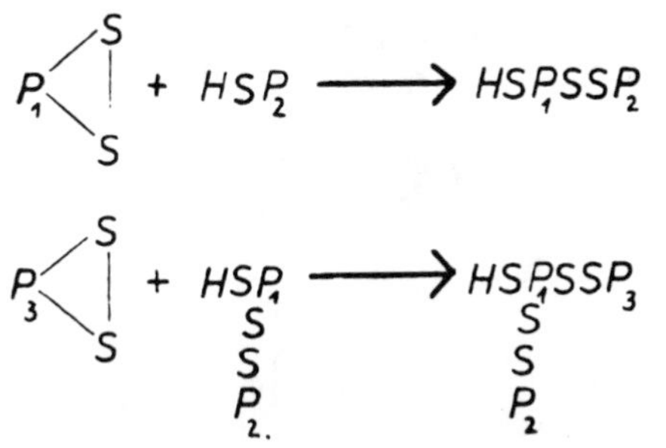

$$P_1\begin{matrix}S\\|\\S\end{matrix} + HSP_2 \longrightarrow HSP_1SSP_2$$

$$P_3\begin{matrix}S\\|\\S\end{matrix} + HSP_1 \longrightarrow HSP_3SSP_1$$
$$\begin{matrix}S\\S\\P_2.\end{matrix} \qquad \begin{matrix}S\\S\\P_2\end{matrix}$$

Figure 3–4 Reactions Involved in Sulfhydryl Polymerization of Proteins. *Source*: From O. Kirchmeier, The Physical-Chemical Causes of the Heat Stability of Milk Proteins, *Milchwissenschaft* (German), Vol. 17, pp. 408–412, 1962.

foods (Figure 3–6). Other factors that influence the browning reaction are temperature, pH, moisture level, oxygen, metals, phosphates, sulfur dioxide, and other inhibitors.

The browning reaction involves a number of steps. An outline of the total pathway of melanoidin formation has been given by Hodge (1953) and is shown in Figure 3–7. According to Hurst (1972), the following five steps are involved in the process:

1. The production of an *N*-substituted glycosylamine from an aldose or ketose reacting with a primary amino group of an amino acid, peptide, or protein.
2. Rearrangement of the glycosylamine by an Amadori rearrangement type of reaction to yield an aldoseamine or ketoseamine.
3. A second rearrangement of the ketoseamine with a second mole of aldose to form a diketoseamine, or the reaction

Table 3–9 Amino Acid Composition of Serum Albumin, Casein, and Gelatin (g Amino Acid/100 g Protein)

Amino Acid	Serum Albumin	Casein	Gelatin
Glycine	1.8	1.9	27.5
Alanine	6.3	3.1	11.0
Valine	5.9	6.8	2.6
Leucine	12.3	9.2	3.3
Isoleucine	2.6	5.6	1.7
Serine	4.2	5.3	4.2
Threonine	5.8	4.4	2.2
Cystine 1/2	6.0	0.3	0.0
Methionine	0.8	1.8	0.9
Phenylalanine	6.6	5.3	2.2
Tyrosine	5.1	5.7	0.3
Proline	4.8	13.5	16.4
Hydroxyproline	—	—	14.1
Aspartic acid	10.9	7.6	6.7
Glutamic acid	16.5	24.5	11.4
Lysine	12.8	8.9	4.5
Arginine	5.9	3.3	8.8
Histidine	4.0	3.8	0.8

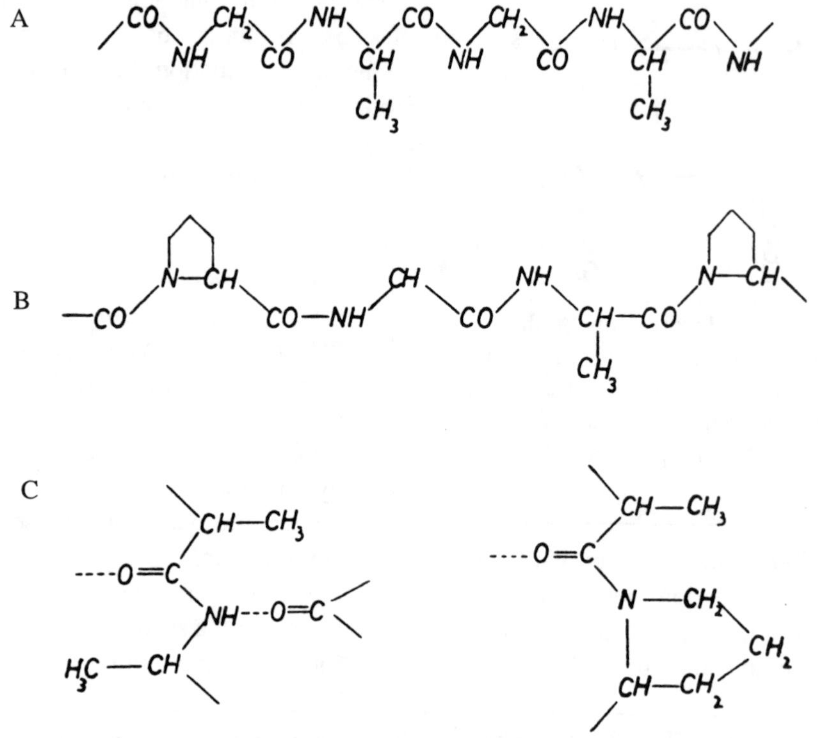

Figure 3–5 Effect of Proline Residues on Possible Hydrogen Bond Formation in Peptide Chains. (A) Proline-free chain; (B) proline-containing chain; (C) hydrogen bond formation in proline-free and proline-containing chains. *Source*: From O. Kirchmeier, The Physical-Chemical Causes of the Heat Stability of Milk Proteins, *Milchwissenschaft* (German), Vol. 17, pp. 408–412, 1962.

of an aldoseamine with a second mole of amino acid to yield a diamino sugar.

4. Degradation of the amino sugars with loss of one or more molecules of water to give amino or nonamino compounds.

5. Condensation of the compounds formed in Step 4 with each other or with amino compounds to form brown pigments and polymers.

The formation of glycosylamines from the reaction of amino groups and sugars is reversible (Figure 3–8) and the equilibrium is highly dependent on the moisture level. The mechanism as shown is thought to involve addition of the amine to the carbonyl group of the open-chain form of the sugar, elimination of a molecule of water, and closure of the ring. The rate is high at low water content; this explains the ease of browning in dried and concentrated foods.

The Amadori rearrangement of the glycosylamines involves the presence of an acid catalyst and leads to the formation of ketoseamine or 1-amino-1-deoxyketose according

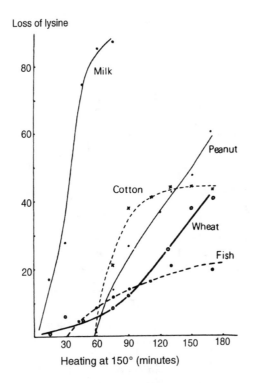

Figure 3–6 Loss of Lysine Occurring as a Result of Heating of Several Foods. *Source*: From J. Adrian, The Maillard Reaction. IV. Study on the Behavior of Some Amino Acids During Roasting of Proteinaceous Foods, *Ann. Nutr. Aliment.* (French), Vol. 21, pp. 129–147, 1967.

to the scheme shown in Figure 3–9. In the reaction of D-glucose with glycine, the amino acid reacts as the catalyst and the compound produced is 1-deoxy-1-glycino-β-D-fructose (Figure 3–10). The ketoseamines are relatively stable compounds, which are formed in maximum yield in systems with 18 percent water content. A second type of rearrangement reaction is the Heyns rearrangement, which is an alternative to the Amadori rearrangement and leads to the same type of transformation. The mechanism of the Amadori rearrangement (Figure 3–9) involves protonation of the nitrogen atom at carbon 1. The Heyns rearrangement (Figure 3–11) involves protonation of the oxygen at carbon 6.

Secondary reactions lead to the formation of diketoseamines and diamino sugars. The formation of these compounds involves complex reactions and, in contrast to the formation of the primary products, does not occur on a mole-for-mole basis.

In Step 4, the ketoseamines are decomposed by 1,2-enolization or 2,3-enolization. The former pathway appears to be the more important one for the formation of brown color, whereas the latter results in the formation of flavor products. According to Hurst (1972), the 1,2-enolization pathway appears mainly to lead to browning but also contributes to formation of off-flavors through hydroxymethylfurfural, which may be a fac-

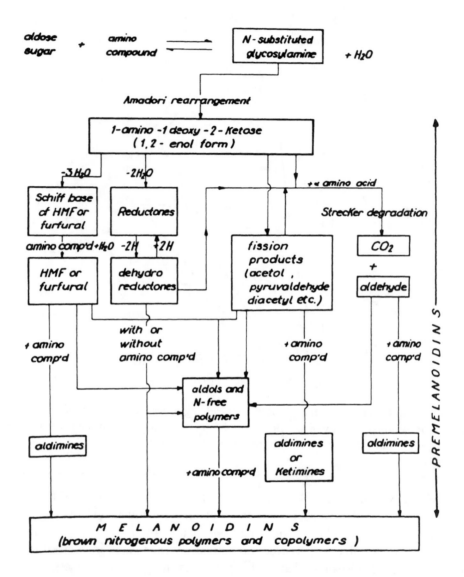

Figure 3–7 Reaction Pattern of the Formation of Melanoidins from Aldose Sugars and Amino Compounds. *Source*: From J.E. Hodge, Chemistry of Browning Reactions in Model Systems, *Agr. Food Chem.*, Vol. 1, pp. 928–943, 1953.

tor in causing off-flavors in stored, overheated, or dehydrated food products. The mechanism of this reaction is shown in Figure 3–12 (Hurst 1972). The ketoseamine (1) is protonated in acid medium to yield (2). This is changed in a reversible reaction into the 1,2-enolamine (3) and this is assisted by the N substituent on carbon 1. The following steps involve the β-elimination of the hydroxyl group on carbon 3. In (4) the enolamine is in the free base form and converts to the Schiff base (5). The Schiff base may

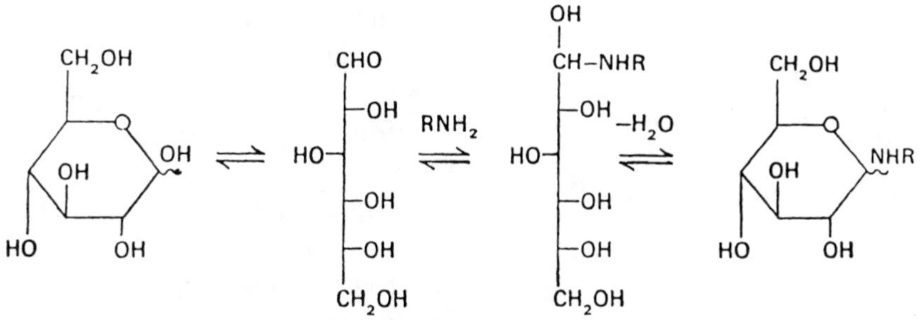

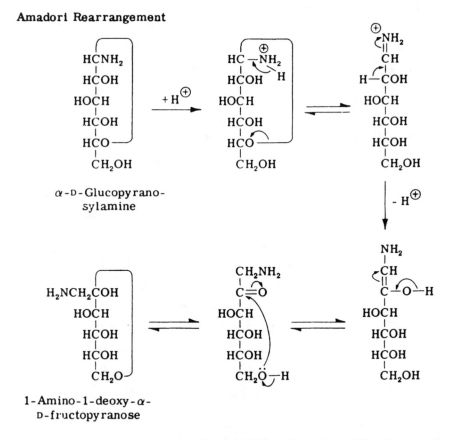

Figure 3–8 Reversible Formation of Glycosylamines in the Browning Reaction. *Source*: From D.T. Hurst, *Recent Development in the Study of Nonenzymic Browning and Its Inhibition by Sulpher Dioxide*, BFMIRA Scientific and Technical Surveys, No. 75, Leatherhead, England, 1972.

Amadori Rearrangement

α-D-Glucopyrano-sylamine

1-Amino-1-deoxy-α-D-fructopyranose

Figure 3–9 Amadori Rearrangement. *Source*: From M.J. Kort, Reactions of Free Sugars with Aqueous Ammonia, *Adv. Carbohydrate Chem. Biochem.*, Vol. 25, pp 311–349, 1970.

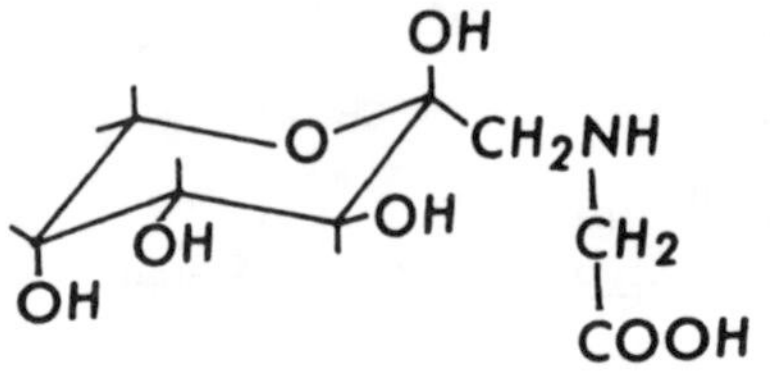

Figure 3–10 Structure of 1-Deoxy-1-Glycino-β-D-Fructose

undergo hydrolysis and form the enolform (7) of 3-deoxyosulose (8). In another step the Schiff base (5) may lose a proton and the hydroxyl from carbon 4 to yield a new Schiff base (6). Both this compound and the 3-deox-

yosulose may be transformed into an unsaturated osulose (9), and by elimination of a proton and a hydroxyl group, hydroxymethylfurfural (10) is formed.

Following the production of 1,2-enol forms of aldose and ketose amines, a series of degradations and condensations results in the formation of melanoidins. The α-β-dicarbonyl compounds enter into aldol type condensations, which lead to the formation of polymers, initially of small size, highly hydrated, and in colloidal form. These initial products of condensation are fluorescent, and continuation of the reaction results in the formation of the brown melanoidins. These polymers are of nondistinct composition and contain

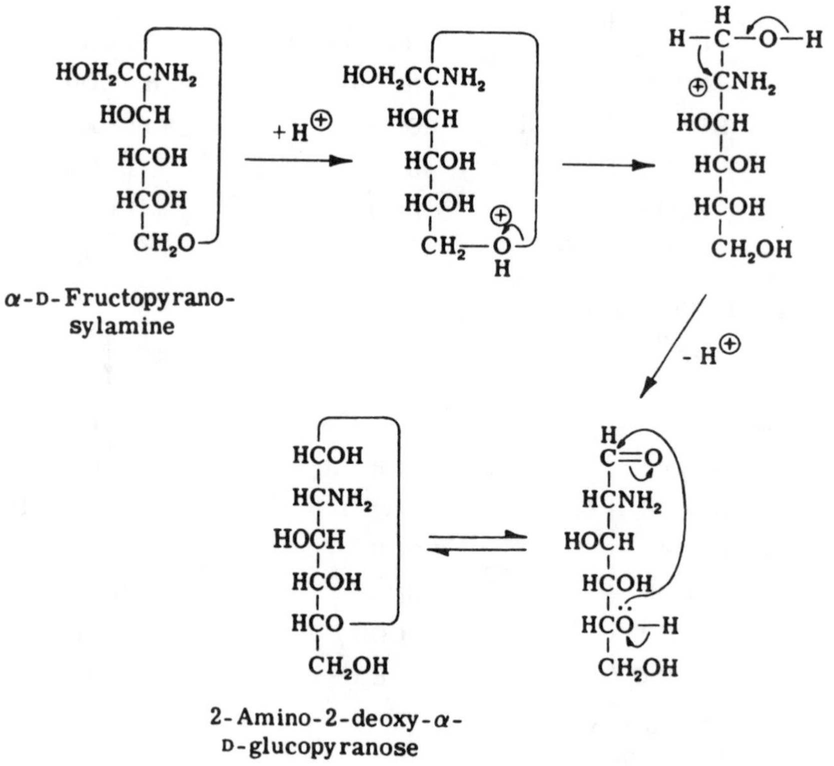

Figure 3–11 Heyns Rearrangement. *Source*: From M.J. Kort, Reactions of Free Sugars with Aqueous Ammonia, *Adv. Carbohydrate Chem. Biochem.*, Vol. 25, pp. 311–349, 1970.

varying levels of nitrogen. The composition varies with the nature of the reaction partners, pH, temperature, and other conditions.

The flavors produced by the Maillard reaction also vary widely. In some cases, the flavor is reminiscent of caramelization. The Strecker degradation of α-amino acids is a reaction that also significantly contributes to the formation of flavor compounds. The dicarbonyl compounds formed in the previously described schemes react in the following manner with α-amino acids:

$$R-\overset{O}{\overset{\|}{C}}-\overset{O}{\overset{\|}{C}}-R + R^1 CHNH_2 \longrightarrow$$

$$R^1 CHO + CO_2 + R-CHNH_2-\overset{O}{\overset{\|}{C}}-R$$

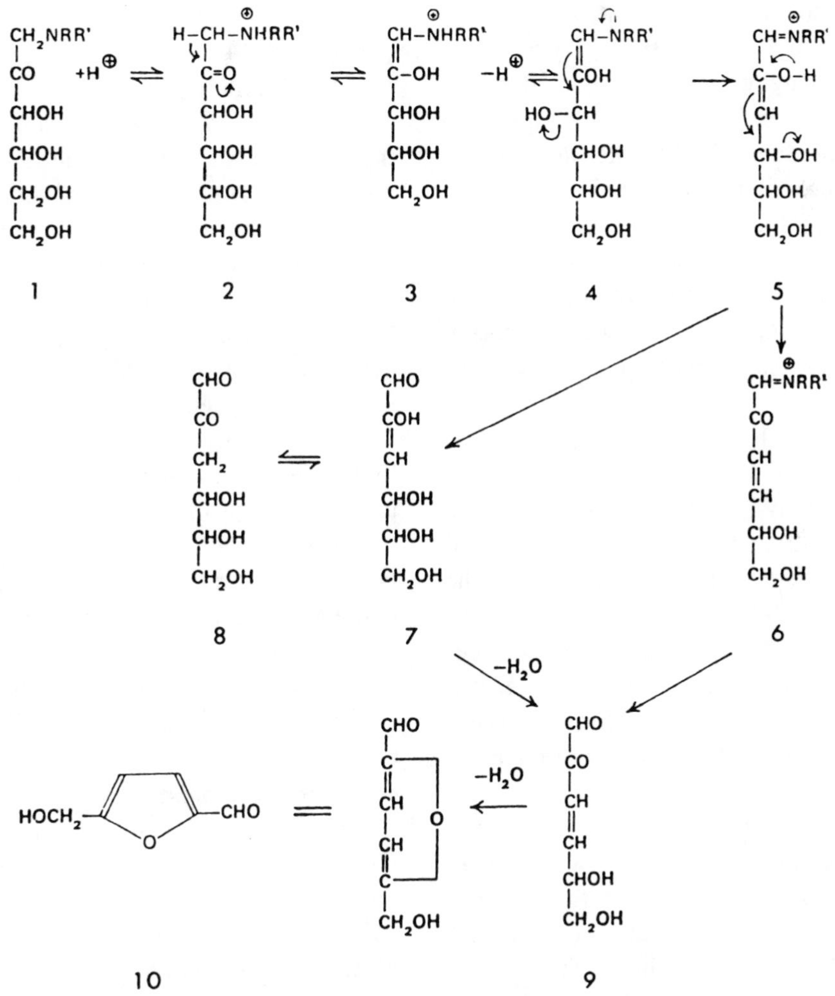

Figure 3–12 1,2-Enolization Mechanism of the Browning Reaction. *Source*: From D.T. Hurst, Recent Developments in the Study of Nonenzymic Browning and Its Inhibition by Sulphur Dioxide, BFMIRA Scientific and Technical Surveys, No. 75, Leatherhead, England, 1972.

The amino acid is converted into an aldehyde with one less carbon atom (Schönberg and Moubacher 1952). Some of the compounds of browning flavor have been described by Hodge et al. (1972). Corny, nutty, bready, and crackery aroma compounds consist of planar unsaturated heterocyclic compounds with one or two nitrogen atoms in the ring. Other important members of this group are partially saturated N-heterocyclics with alkyl or acetyl group substituents. Compounds that contribute to pungent, burnt aromas are listed in Table 3–10. These are mostly vicinal polycarbonyl compounds and α,β-unsaturated aldehydes. They condense rapidly to form melanoidins. The Strecker degradation aldehydes contribute to the aroma of bread, peanuts, cocoa, and other roasted foods. Although acetic, phenylacetic, isobutyric, and isovaleric aldehydes are prominent in the aromas of bread, malt, peanuts, and cocoa, they are not really characteristic of these foods (Hodge et al. 1972).

A somewhat different mechanism for the browning reaction has been proposed by Burton and McWeeney (1964) and is shown in Figure 3–13. After formation of the aldosylamine, dehydration reactions result in the production of 4- to 6-membered ring compounds. When the reaction proceeds under conditions of moderate heating, fluorescent nitrogenous compounds are formed. These react rapidly with glycine to yield melanoidins.

The influence of reaction components and reaction conditions results in a wide variety of reaction patterns. Many of these conditions are interdependent. Increasing temperature results in a rapidly increasing rate of browning; not only reaction rate, but also the pattern of the reaction may change with temperature. In model systems, the rate of browning increases two to three times for each 10° rise in temperature. In foods containing fructose, the increase may be 5 to 10 times for each 10° rise. At high sugar contents, the rate may be even more rapid. Temperature also affects the composition of the pigment formed. At higher temperatures, the carbon content of the pigment increases and more pigment is formed per mole of carbon dioxide released. Color intensity of the pigment increases with increasing temperature. The effect of temperature on the reaction rate of D-glucose with DL-leucine is illustrated in Figure 3–14.

In the Maillard reaction, the basic amino group disappears; therefore, the initial pH or the presence of a buffer has an important effect on the reaction. The browning reaction is slowed down by decreasing pH, and the browning reaction can be said to be self-inhibitory since the pH decreases with the loss of the basic amino group. The effect of pH on the reaction rate of D-glucose with DL-leucine is demonstrated in Figure 3–15. The effect of pH on the browning reaction is highly dependent on moisture content. When a large amount of water is present, most of the browning is caused by caramelization, but at low water levels and at pH greater than 6, the Maillard reaction is predominant.

The nature of the sugars in a nonenzymic browning reaction determines their reactivity. Reactivity is related to their conformational stability or to the amount of open-chain structure present in solution. Pentoses are more reactive than hexoses, and hexoses more than reducing disaccharides. Nonreducing disaccharides only react after hydrolysis has taken place. The order of reactivity of some of the aldohexoses is: mannose is more reactive than galactose, which is more reactive than glucose.

The effect of the type of amino acid can be summarized as follows. In the α-amino acid

Table 3–10 Aroma and Structure Classification of Browned Flavor Compounds

Aromas:	Burnt (pungent, empyreumatic)	Variable (aldehydic, ketonic)
Structure:	Polycarbonyls(α,β-Unsat'd aldehydes–C:O–C:O–,=C–CHO)	Monocarbonyls (R–CHO, R–C:O–CH$_3$)
Examples of compounds:	Glyoxal Pyruvaldehyde Diacetyl Mesoxalic dialdehyde	Strecker aldehydes Isobutyric Isovaleric Methional
	Acrolein Crotonaldehyde	2-Furaldehydes 2-Pyrrole aldehydes C$_3$–C$_6$ Methyl ketones

Source: From J.E. Hodge, F.D. Mills, and B.E. Fisher, Compounds of Browned Flavor from Sugar-Amine Reactions, *Cereal Sci. Today*, Vol. 17, pp. 34–40, 1972.

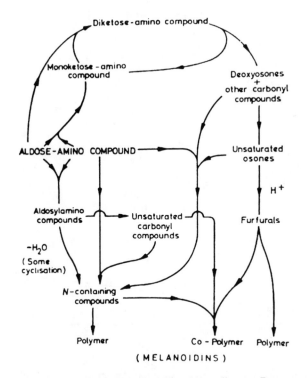

Figure 3–13 Proposed Browning Reaction Mechanism According to Burton and McWeeney. *Source*: From H.S. Burton and D.J. McWeeney, Non-Enzymatic Browning: Routes to the Production of Melanoidins from Aldoses and Amino Compounds, *Chem. Ind.*, Vol. 11, pp. 462–463, 1964.

series, glycine is the most reactive. Longer and more complex substituent groups reduce the rate of browning. In the ω-amino acid series, browning rate increases with increasing chain length. Ornithine browns more rapidly than lysine. When the reactant is a protein, particular sites in the molecule may react faster than others. In proteins, the ε-amino group of lysine is particularly vulnerable to attack by aldoses and ketoses.

Moisture content is an important factor in influencing the rate of the browning reaction. Browning occurs at low temperatures and intermediate moisture content; the rate increases with increasing water content. The rate is extremely low below the glass transition temperature, probably because of limited diffusion (Roos and Himberg 1994; Roos et al. 1996a, b).

Methods of preventing browning could consist of measures intended to slow reaction rates, such as control of moisture, temperature, or pH, or removal of an active intermediate. Generally, it is easier to use an inhibitor. One of the most effective inhibitors of browning is sulfur dioxide. The action of sulfur dioxide is unique and no other suitable inhibitor has been found. It is known that sulfite can combine with the carbonyl group of an aldose to give an addition compound:

$$NaHSO_3 + RCHO \rightarrow RCHOHSO_3Na$$

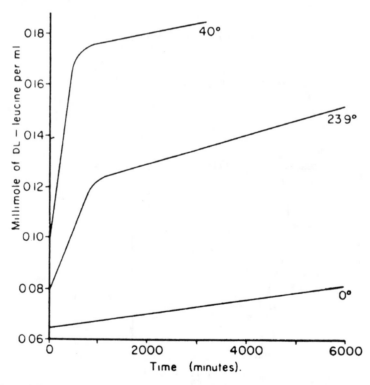

Figure 3–14 Effect of Temperature on the Reaction Rate of D-Glucose with DL-Leucine. *Source*: From G. Haugard, L. Tumerman, and A. Sylvestri, A Study on the Reaction of Aldoses and Amino Acids, *J. Am. Chem. Soc.*, Vol. 73, pp. 4594–4600, 1951.

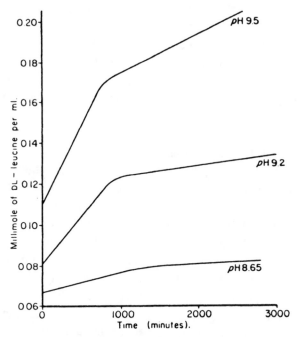

Figure 3–15 Effect of pH on the Reaction Rate of D-Glucose with DL-Leucine. *Source*: From G. Haugard, L. Tumerman, and A. Sylvestri, A Study on the Reaction of Aldoses and Amino Acids, *J. Am. Chem. Soc.*, Vol. 73, pp. 4594–4600, 1951.

However, this reaction cannot possibly account for the inhibitory effect of sulfite. It is thought that sulfur dioxide reacts with the degradation products of the amino sugars, thus preventing these compounds from condensing into melanoidins. A serious drawback of the use of sulfur dioxide is that it reacts with thiamine and proteins, thereby reducing the nutritional value of foods. Sulfur dioxide destroys thiamine and is therefore not permitted for use in foods containing this vitamin.

CHEMICAL CHANGES

During processing and storage of foods, a number of chemical changes involving proteins may occur (Hurrell 1984). Some of these may be desirable, others undesirable.

Such chemical changes may lead to compounds that are not hydrolyzable by intestinal enzymes or to modifications of the peptide side chains that render certain amino acids unavailable. Mild heat treatments in the presence of water can significantly improve the protein's nutritional value in some cases. Sulfur-containing amino acids may become more available and certain antinutritional factors such as the trypsin inhibitors of soybeans may be deactivated. Excessive heat in the absence of water can be detrimental to protein quality; for example, in fish proteins, tryptophan, arginine, methionine, and lysine may be damaged. A number of chemical reactions may take place during heat treatment including decomposition, dehydration of serine and threonine, loss of sulfur from cysteine, oxidation of cysteine and methionine, cyclization of glutamic and aspartic acids and threonine (Mauron 1970; 1983).

The nonenzymic browning, or Maillard, reaction causes the decomposition of certain amino acids. For this reaction, the presence of a reducing sugar is required. Heat damage may also occur in the absence of sugars. Bjarnason and Carpenter (1970) demonstrated that the heating of bovine plasma albumin for 27 hours at 115°C resulted in a 50 percent loss of cystine and a 4 percent loss of lysine. These authors suggest that amide-type bonds are formed by reaction between the ε-amino group of lysine and the amide groups of asparagine or glutamine, with the reacting units present either in the same peptide chain or in neighboring ones (Figure 3–16). The Maillard reaction leads to the formation of brown pigments, or melanoidins, which are not well defined and may result in numerous flavor and odor compounds. The browning reaction may also result in the blocking of lysine. Lysine becomes unavailable when it is involved in the Amadori reaction, the first stage of browning. Blockage of lysine is nonexistent in pasteurization of milk products, and is at 0 to 2 percent in UHT sterilization, 10 to 15 percent in conventional sterilization, and 20 to 50 percent in roller drying (Hurrell 1984).

Some amino acids may be oxidized by reacting with free radicals formed by lipid oxidation. Methionine can react with a lipid peroxide to yield methionine sulfoxide. Cysteine can be decomposed by a lipid free radical according to the following scheme:

$$L^{\cdot} + Cys{-}SH \longrightarrow Cys{-}S^{\cdot} \longrightarrow Cys{-}S{-}S{-}Cys$$
$$\longrightarrow Cys^{\cdot} + H_2S$$
$$\overset{H^{\cdot}}{\longrightarrow} Alanine$$

The decomposition of unsaturated fatty acids produces reactive carbonyl compounds that may lead to reactions similar to those involved in nonenzymic browning. Methionine can be oxidized under aerobic conditions in the presence of SO_2 as follows:

$$R{-}S{-}CH_3 + 2SO_3^= \rightarrow R{-}SO{-}CH_3 + 2SO_4^=$$

This reaction is catalyzed by manganese ions at pH values from 6 to 7.5. SO_2 can also react with cystine to yield a series of oxidation products. Some of the possible reaction products resulting from the oxidation of sulfur amino acids are listed in Table 3–11. Nielsen et al. (1985) studied the reactions between protein-bound amino acids and oxidizing lipids. Significant losses occurred of the amino acids lysine, tryptophan, and histidine. Methionine was extensively oxidized to its sulfoxide. Increasing water activity increased losses of lysine and tryptophan but had no effect on methionine oxidation.

Alkali treatment of proteins is becoming more common in the food industry and may result in several undesirable reactions. When cystine is treated with calcium hydroxide, it is transformed into amino-acrylic acid, hydrogen sulfide, free sulfur, and 2-methyl thiazolidine-2, 4-dicarboxylic acid as follows:

This can also occur under alkaline conditions, when cystine is changed into amino-

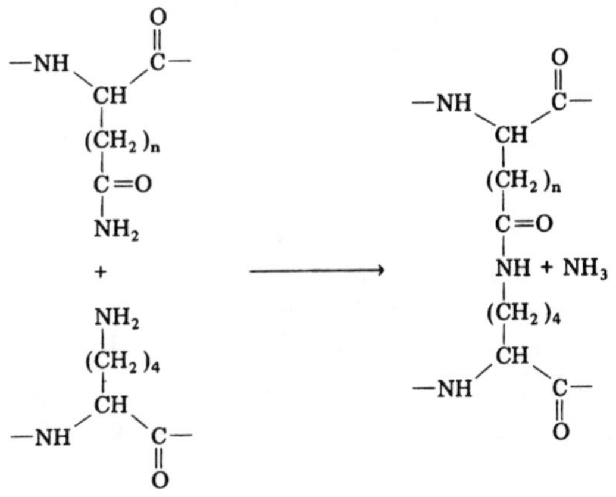

$$-NH-\underset{\underset{(CH_2)_n}{\overset{\overset{O}{\parallel}}{\underset{|}{CH-C-}}}}{} \quad \begin{matrix} C=O \\ | \\ NH_2 \end{matrix} \\ + \\ \begin{matrix} NH_2 \\ | \\ (CH_2)_4 \\ | \\ -NH-\underset{\overset{||}{O}}{CH-C-} \end{matrix} \longrightarrow \quad -NH-\underset{\underset{(CH_2)_n}{\underset{|}{CH}}}{\overset{\overset{O}{\parallel}}{C-}} \\ \begin{matrix} C=O \\ | \\ NH + NH_3 \\ | \\ (CH_2)_4 \\ | \\ -NH-\underset{\overset{||}{O}}{CH-C-} \end{matrix}$$

Figure 3–16 Formation of Amide-Type Bonds from the Reaction Between ε-amine Groups of Lysine and Amide Groups of Asparagine ($n = 1$) Glutamine ($n = 2$). *Source*: From J. Bjarnason and K. J. Carpenter, Mechanisms of Heat Damage in Proteins. 2 Chemical Changes in Pure Proteins, *Brit. J. Nutr.*, Vol. 24, pp. 313–329, 1970.

---

acrylic acid and thiocysteine by a β-elimination mechanism, as follows:

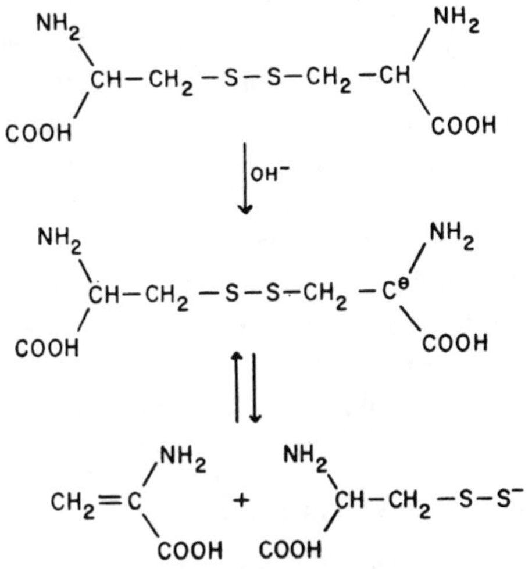

$$\underset{COOH}{\overset{NH_2}{\diagdown}}CH-CH_2-S-S-CH_2-CH\underset{\diagdown COOH}{\overset{\diagup NH_2}{}}$$

$$\Big\downarrow OH^-$$

$$\underset{COOH}{\overset{NH_2}{\diagdown}}CH-CH_2-S-S-CH_2-C^{\ominus}\underset{\diagdown COOH}{\overset{\diagup NH_2}{}}$$

$$\Updownarrow$$

$$CH_2=C\underset{\diagdown COOH}{\overset{\diagup NH_2}{}} \quad + \quad \underset{COOH}{\overset{NH_2}{\diagdown}}CH-CH_2-S-S^-$$

Amino-acrylic acid (dehydroalanine) is very reactive and can combine with the ε-amino group of lysine to yield lysinoalanine (Ziegler 1964) as shown:

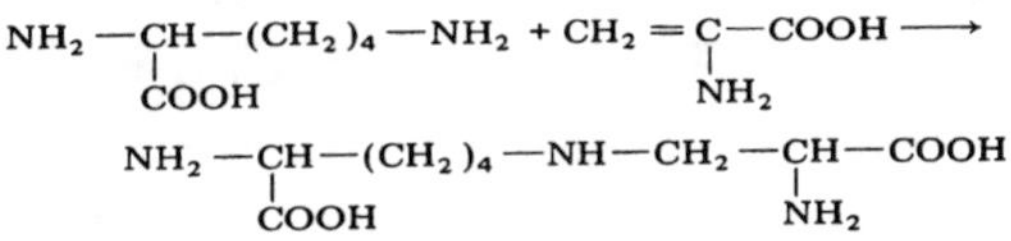

$$NH_2-\underset{\underset{COOH}{|}}{CH}-(CH_2)_4-NH_2 + CH_2=\underset{\underset{NH_2}{|}}{C}-COOH \longrightarrow$$

$$NH_2-\underset{\underset{COOH}{|}}{CH}-(CH_2)_4-NH-CH_2-\underset{\underset{NH_2}{|}}{CH}-COOH$$

Lysinoalanine formation is not restricted to alkaline conditions—it can also be formed by prolonged heat treatment. Any factor favoring lower pH and less drastic heat treatment will reduce the formation of lysinoalanine. Hurrell (1984) found that dried whole milk and UHT milk contained no lysinoalanine and that evaporated and sterilized milk contained 1,000 ppm. More severe treatment with alkali can decompose arginine into ornithine and urea. Ornithine can combine with dehydroalanine in a reaction similar to the one giving lysinoalanine and, in this case, ornithinoalanine is formed.

Treatment of proteins with ammonia can result in addition of ammonia to dehydroalanine to yield β-amino-alanine as follows:

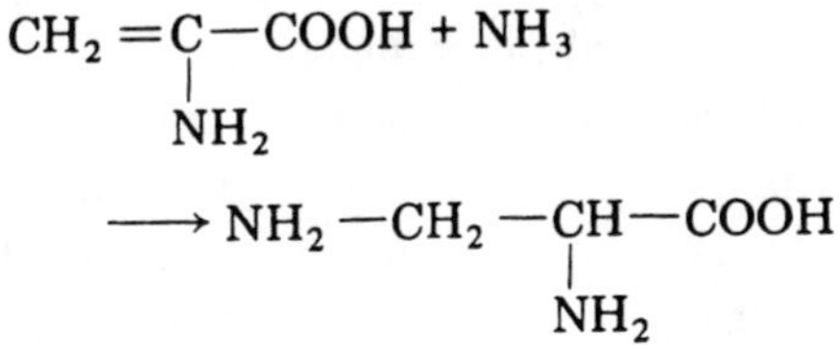

$$CH_2=\underset{\underset{NH_2}{|}}{C}-COOH + NH_3$$

$$\longrightarrow NH_2-CH_2-\underset{\underset{NH_2}{|}}{CH}-COOH$$

Light-induced oxidation of proteins has been shown to lead to off-flavors and destruction of essential amino acids in milk. Patton (1954) demonstrated that sunlight attacks methionine and converts it into methional (β-methylmercaptopropionaldehyde), which can cause a typical sunlight off-flavor at a level of 0.1 ppm. It was later demonstrated by Finley and Shipe (1971) that the source of the light-induced off-flavor in milk resides in a low-density lipoprotein fraction.

Table 3–11 Oxidation Products of the Sulfur-Containing Amino Acids

Name		Formula
Methionine		$R\text{-}S\text{-}CH_3$
	Sulfoxide	$R\text{-}SO\text{-}CH_3$
	Sulfone	$R\text{-}SO_2\text{-}CH_3$
Cystine		$R\text{-}S\text{-}S\text{-}R$
	Disulfoxide	$R\text{-}SO\text{-}SO\text{-}R$
	Disulfone	$R\text{-}SO_2\text{-}SO_2\text{-}R$
Cysteine		$R\text{-}SH$
	Sulfenic	$R\text{-}SOH$
	Sulfinic	$R\text{-}SO_2H$
	Sulfonic (or cysteic acid)	$R\text{-}SO_3H$

Proteins react with polyphenols such as phenolic acids, flavonoids, and tannins, which occur widely in plant products. These reactions may result in the lowering of available lysine, protein digestibility, and biological value (Hurrell 1984).

Racemization is the result of heat and alkaline treatment of food proteins. The amino acids present in proteins are of the L-series. The racemization reaction starts with the abstraction of an α-proton from an amino acid residue to give a negatively charged planar carbanion. When a proton is added back to this optically inactive intermediate, either a D- or L-enantiomer may be formed (Masters and Friedman 1980). Racemization leads to reduced digestibility and protein quality.

FUNCTIONAL PROPERTIES

Increasing emphasis is being placed on isolating proteins from various sources and using them as food ingredients. In many applications functional properties are of great importance. Functional properties have been defined as those physical and chemical properties that affect the behavior of proteins in food systems during processing, storage, preparation, and consumption (Kinsella 1982). A summary of these properties is given in Table 3–12.

Even when protein ingredients are added to food in relatively small amounts, they may significantly influence some of the physical properties of the food. Hermansson (1973) found that addition of 4 percent of a soybean protein isolate to processed meat significantly affected firmness, as measured by extrusion force, compression work, and sensory evaluation.

The emulsifying and foaming properties of proteins relate to their adsorption at interfaces and to the structure of the protein film formed there (Mitchell 1986). The emulsifying and emulsion stabilizing capacity of protein meat additives is important to the production of sausages. The emulsifying properties of proteins are also involved in the production of whipped toppings and coffee whiteners. The whipping properties of proteins are essential in the production of whipped toppings. Paulsen and Horan (1965) determined the functional characteristics of edible soya flours, especially in relation to their use in bakery products. They evaluated the measurable parameters of functional properties such as water dispersibility, wettability, solubility, and foaming characteristics as those properties affected the quality of baked products containing added soya flour.

Some typical functional properties of food proteins are listed in Table 3–13.

Surface Activity of Proteins

Proteins can act as surfactants in stabilizing emulsions and foams. To perform this function proteins must be amphiphilic just

Table 3–12 Functional Properties of Food Proteins

General Property	Functional Criteria
Organoleptic	Color, flavor, odor
Kinesthetic	Texture, mouth feel, smoothness, grittiness, turbidity
Hydration	Solubility, wettability, water absorption, swelling, thickening, gelling, syneresis, viscosity
Surface	Emulsification, foaming (aeration, whipping), film formation
Binding	Lipid-binding, flavor-binding
Structural	Elasticity, cohesiveness, chewiness, adhesion, network cross-binding, aggregation, dough formation, texturizability, fiber formation, extrudability
Rheological	Viscosity, gelation
Enzymatic	Coagulation (rennet), tenderization (papain), mellowing ("proteinases")
"Blendability"	Complementarity (wheat-soy, gluten-casein)
Antioxidant	Off-flavor prevention (fluid emulsions)

Source: From J.E. Kinsella, Structure and Functional Properties of Food Proteins, in *Food Proteins*, P.F. Fox and J.J. Condon, eds., 1982, Applied Science Publishers.

Table 3–13 Functional Properties of Proteins in Food Systems

Functional Property	Mode of Action	Food System
Solubility	Protein solvation	Beverages
Water absorption and binding	Hydrogen bonding of water; entrapment of water (no drip)	Meat, sausages, bread, cakes
Viscosity	Thickening; water binding	Soups, gravies
Gelation	Protein acts as adhesive material	Meat, sausages, baked goods, pasta products
Elasticity	Hydrophobic bonding in gluten; disulfide links in gels	Meats, bakery products
Emulsification	Formation and stabilization of fat emulsion	Sausages, bologna, soup, cakes
Fat absorption	Binding of free fat	Meats, sausages, doughnuts
Flavor-binding	Adsorption, entrapment, release	Simulated meats, bakery products, etc.
Foaming	Forms stable films to entrap gas	Whipped toppings, chiffon, desserts, angel cakes

Source: From J.E. Kinsella, Structure and Functional Properties of Food Proteins, in *Food Proteins*, P.F. Fox and J.J. Condon, eds., 1982, Applied Science Publishers.

like the emulsifiers discussed in Chapter 2. This is achieved when part of the protein structure contains predominantly amino acids with hydrophobic side chains and another part contains mostly hydrophilic side chains. The molecule is then able to orient itself in the oil-water interface. Thus, the ability of proteins to serve as emulsifiers varies greatly among proteins. The emulsifying capacity of a protein depends not only on its overall hydrophobicity but, more importantly, on the distribution of the hydrophobic or charged groups along the polypeptide chain and the manner in which the chain is folded (Dalgleish 1989). Hydrophobic side

chains are likely to be folded into the inside of the molecule leaving the outside more hydrophilic. To be effective surfactants, proteins need to have flexible polypeptide chains, so that they are able to orient at the interface. Only proteins that have little secondary structure and are able to unfold at the interface are effective emulsifiers. Nakai and Powrie (1981) have shown the relationship between solubility, charge frequency, and hydrophobicity in graphical manner (Figure 3–17) and in the form of a table relating these parameters to the functional properties of proteins (Table 3–14).

There are two important considerations in

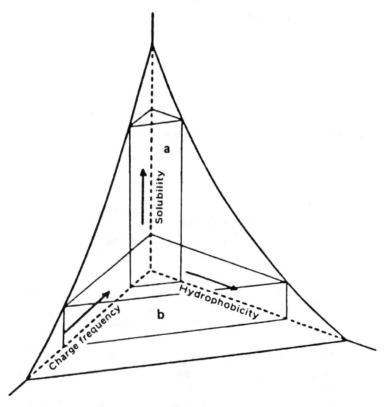

Figure 3–17 Relationship Between Solubility, Charge Frequency, and Hydrophobicity of Proteins. *Source*: Reprinted with permission from S. Nakai and W.D. Powrie, Modification of Proteins for Functional and Nutritional Improvements, in *Cereals—A Renewable Resource, Theory and Practice*, Y. Pomeranz and L. Munck, eds., p. 225, © 1981, American Association of Cereal Chemists.

Table 3–14 Contribution of Hydrophobicity, Charge Frequency, and Structural Parameters to Functionality of Proteins

	Hydrophobicity	Charge frequency	Structure
Solubility	−	+	−
Emulsification	+sur	(−)	+
Foaming	+tot	−	+
Fat binding	+sur	(−)	−
Water holding	−	+	?
Heat coagulation	+tot	−	+
Dough making	(+)	−	+

+: positive contribution; −: negative contribution; sur: surface hydrophobicity; tot: total hydrophobicity; ( ): contributes to a lesser extent.

Source: Reprinted with permission from S. Nakai and W.D. Powrie, Modification of Proteins for Functional and Nutritional Improvements, *Cereals—Renewable Resource, Theory and Practice*, Y. Pomeranz and L. Munck, eds., p. 235, © 1981, American Association of Cereal Chemists.

emulsion formation: the binding of the protein to the oil-water interface and the stability of the emulsion. For an emulsion to possess stability, the proteins have to form a cohesive film. The cohesiveness of such films is stabilized by intramolecular disulfide bonds. In addition, emulsion particles can be stabilized by steric factors. This happens when disordered protein molecules at the interface prevent emulsion droplets from approaching one another closely enough to permit coagulation as a result of attraction by van der Waals forces (Dalgleish 1989).

The amount of protein required to form a stable emulsion depends on the size of emulsion droplets produced and the nature of the protein. The size of the droplets, assuming the presence of sufficient protein to cover the interfacial area, is determined by the input of mechanical energy by a device such as a homogenizer or colloid mill. According to Dalgleish (1989), the protein load for monolayer coverage of the oil-water interface is in the order of a few mg/m². β-casein is an effective emulsifier protein and has been reported to give an interfacial loading of 2 to 3 mg/m².

Gel Formation

Proteins can form gels by acid coagulation, action of enzymes, heat, and storage. A gel is a protein network that immobilizes a large amount of water. The network is formed by protein-protein interactions. Gels are characterized by having relatively high non-Newtonian viscosity, elasticity, and plasticity. Examples of protein gels are a variety of dairy products including yogurt, soybean curd (tofu), egg protein gels including mayonnaise, and meat and fish protein gels. Some types of gel formation are reversible, especially those produced by heat. Gelatin gels are produced when a heated solution of gelatin is cooled. This sol-gel transformation is reversible. Most other types of gel formation are not reversible. Gelation has been described as a two-stage process (Pomeranz 1991). The first stage is a denaturation of the native protein into unfolded polypeptides, and the second stage is a gradual association to form the gel matrix. The type of association and, therefore, the nature of the gel depends on a variety of covalent and noncovalent interactions involving disulfide bonds, hydrogen bonds, ionic and hydrophobic interactions, or combinations of these.

Protein gels can be divided into two types: aggregated gels and clear gels (Barbut 1994). Aggregated gels are formed from casein and from egg white proteins and are opaque because of the relatively large size of the protein aggregates. Clear gels are formed from smaller particles, such as those formed from whey protein isolate, and have high water-holding capacity. The formation of such gels from ovalbumin is illustrated in Figure 3–18 (Hatta and Koseki 1988), showing the influence of protein concentration, pH, and ionic strength. Many dairy products involve gel formation through the action of acid or by combined activity of acid and enzymes. Mayonnaise is an oil-in-water emulsion, in which egg yolk protein acts as the emulsifier. The presence of acetic acid in the form of vinegar or citric acid from lemon juice leads to interaction of the proteins covering the emulsion droplets, resulting in a gel-type emulsion.

ANIMAL PROTEINS

Milk Proteins

The proteins of cow's milk can be divided into two groups: the caseins, which are phosphoproteins and comprise 78 percent of the total weight, and the milk serum proteins, which make up 17 percent of the total weight. The latter group includes β-lactoglobulin (8.5 percent), a lactalbumin (5.1 percent), immune globulins (1.7 percent), and serum albumins. In addition, about 5 percent of milk's total weight is nonprotein nitrogen (NPN)-containing substances, and these include peptides

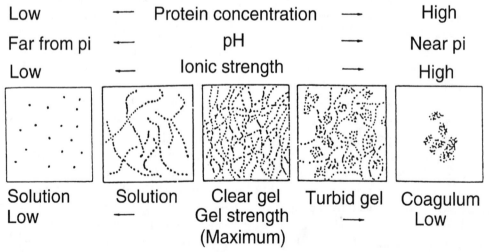

Figure 3–18 Model for the Formation of a Gel Network by Heated Ovalbumin, pi = isoelectric point. *Source:* Reprinted with permission from H. Hatta and T. Koseki, Relationship of SH Groups to Functionality of Ovalbumin, in *Food Proteins*, J.E. Kinsella and W.G. Soucie, eds., p. 265, © 1988, American Oil Chemists' Society.

and amino acids. Milk also contains very small amounts of enzymes, including peroxidase, acid phosphatase, alkaline phosphatase, xanthine oxidase, and amylase. The protein composition of bovine herd milk is listed in Table 3–15 (Swaisgood 1995), and the amino acid composition of the milk proteins is shown in Table 3–16.

Casein is defined as the heterogeneous group of phosphoproteins precipitated from skim milk at pH 4.6 and 20°C. The proteins remaining in solution, the serum or whey proteins, can be separated into the classic lactoglobulin and lactalbumin fractions by half saturation with ammonium sulfate or by full saturation with magnesium sulfate, as is shown in Figure 3–19. However, this separation is possible only with unheated milk. After heating by, for example, boiling, about 80 percent of the whey proteins will precipitate with the casein at pH 4.6; this property has been used to develop a method for measuring the degree of heat exposure of milk and milk products.

Casein exists in milk as relatively large, nearly spherical particles of 30 to 300 nm in diameter (Figure 3–20). In addition to acid precipitation, casein can be separated from milk by rennet action or by saturation with sodium chloride. The composition of the casein depends on the method of isolation. In the native state, the caseinate particles contain relatively large amounts of calcium and phosphorus and smaller quantities of magnesium and citrate and are usually referred to as calcium caseinatephosphate or calcium phosphocaseinate particles. When adding acid to

Table 3–15 Protein Composition of Mature Bovine Herd Milk

Protein	g/liter
Total protein	36
Total casein	29.5
Whey protein	6.3
α_{s1}-casein	11.9
α_{s2}-casein	3.1
β-casein	9.8
κ-casein	3.5
γ-casein	1.2
α-lactalbumin	1.2
β-lactoglobulin	3.2
Serum albumin	0.4
Immunoglobulin	0.8
Proteose-peptones	1.0

Source: Reprinted with permission from H.E. Swaisgood, Protein and Amino Acid Composition of Bovine Milk, in *Handbook of Milk Composition*, R.G. Jensen, ed., p. 465, © 1982, Academic Press.

Table 3–16 Amino Acid Composition of Milk Proteins

Amino Acid	g/kg protein
Essential amino acids	
Threonine	46
Valine	66
Methionine	26
Cystine	8
Isoleucine	59
Leucine	97
Phenylalanine	49
Lysine	81
Histidine	27
Arginine	35
Tryptophan	17
Nonessential amino acids	
Aspartic acid, asparagine	79
Serine	56
Glutamic acid, glutamine	219
Proline	99
Glycine	20
Alanine	34
Tyrosine	51

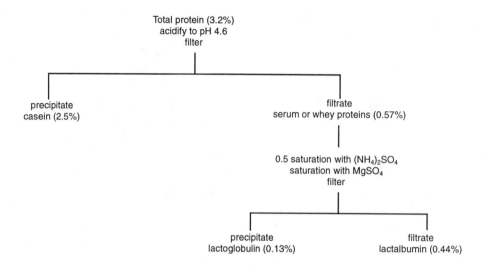

Figure 3–19 Separation of Milk Proteins by Precipitation

milk, the calcium and phosphorus are progressively removed until, at the isoelectric point of pH 4.6, the casein is completely free of salts. Other methods of preparing casein yield other products; for example, salt precipitation does not remove the calcium and

Figure 3–20 Electron Photomicrograph of Casein Micelles in Milk. *Source*: From I. Vujicic, J.M. deMan, and I.L. Woodrow, Interaction of Polyphosphates and Citrate with Skim Milk Proteins, *Can. Inst. Food Technol. J.*, Vol. 1, pp. 17–21, 1968.

phosphorus, and rennet action involves limited proteolysis. The rennet casein is named paracasein.

Casein is a nonhomogeneous protein that consists of four components, identified as α_{s1}-, α_{s2}-, β-, and κ-casein. The γ-casein mentioned in the literature (see Table 3–15) has been identified as proteolytic fragments of β-casein (Wong et al. 1996). The four casein components occur as genetic variants. Such genetically determined variants or polymorphs differ from one another by one or more amino acid substitutions and/or deletions. The complete amino acid sequence has been established, and the exact nature of the genetic variants has been determined. The genetic variants are described by the suffix CN (for casein) and a capital A, B, C, etc., as well as by the number of phosphorylations. Genetic variants of the four caseins are listed in Table 3–17 (Wong et al. 1996). The caseins have sites of phosphorylation that have a unique amino acid sequence. The three–amino acid sequence is Ser-X-A, with X

Table 3–17 Genetic Variants of Caseins

Variant	Total Amino Acid Residues	Molecular Weight
α_{s1}-CN B-8P	199	23,614
α_{s2}-CN A-11P	207	24,350
β-CN A1-5P	209	23,982
κ-CN B-1P	169	19,023

being any amino acid and A either glutamic acid or serine-phosphate. One part of the α-casein is precipitable by calcium ions and has been designated calcium-sensitive casein or α_s. The non–calcium-sensitive fraction, κ-casein, is the protein assumed to confer stability on the casein micelle; this has been found to be removed by the action of rennin, thereby leaving the remaining casein precipitable by calcium ions. κ-casein is the fraction with the lowest phosphate content. The two α_s-caseins show strong association. The association of β-casein is temperature dependent. At 4°C only monomers exist; at temperatures greater than 8°C association will occur. α_{s1}-casein has more acidic than basic amino acids and has a net negative charge of 22 at pH 6.5. The polypeptide chain contains 8.5 percent proline that is distributed uniformly, resulting in no apparent secondary structure. α_{s2}-casein has the highest number of phosphorylations and a low proline content. β-casein is a single polypeptide chain with a total of 209 amino acids, and has seven genetic variants. The distribution of amino acids in the polypeptide chain is quite specific. The N-terminal segment has a high negative charge, giving it hydrophilic properties, the C-terminal portion is highly hydrophobic. This arrangement lends surfactant properties to this protein.

Casein contains 0.86 percent phosphorus, and it is assumed that this is present exclusively in the form of monophosphate esters with the hydroxyl groups of serine and threonine. Limited and specific hydrolysis of casein with proteolytic enzymes has produced a number of large polypeptides that resist further hydrolysis. An electrophoretically homogeneous phosphopeptone has been isolated from a trypsin digest of β-casein by Peterson et al. (1958). This phosphopeptone has a molecular weight of about 3,000 and consists of 24 amino acid residues of 10 different amino acids. The peptone contains 5 phosphate residues linked to 4 serine and 1 threonine groups. This constitutes essentially all of the phosphorus of β-casein and it appears that the phosphate residues are localized in a relatively small region of the casein molecule.

In addition to ester phosphate, casein contains calcium phosphate in the colloidal form. It appears that the presence of this colloidal calcium phosphate helps maintain the structural integrity of the casein micelle. Although the composition and structure of most of the casein fractions are now well established, the exact nature of the arrangement of the caseins and calcium phosphate into a micelle is not well known. Many models of micelle structure have been proposed (Farrell 1973; Farrell and Thompson 1974). These can be divided into three groups: coat-core models, internal structure models, and subunit models. In the most popular, the coat-core model, it is assumed that the core contains the calcium-sensitive α_s-caseins and that this core is covered by a layer of κ-casein. The function of the κ-casein coat is to protect the micelle from insolubilization by calcium ions. The κ-casein is readily attacked by the enzyme rennin, thus removing the coat and resulting in coagulation of the micelles. This model most readily accounts for the action of rennin but does not explain the position of the colloidal calcium phosphate.

It appears that micelles are formed by cross-linking of some of the ester phosphate groups by calcium. Chelation of calcium results in dissociation and solubilization of the micelles, and the rate at which this happens corresponds to the ester phosphate content of the monomers (Aoki et al. 1987).

The whey proteins of milk were originally thought to be composed of two main components, lactalbumin and lactoglobulin, as indicated in Figure 3–19. Then it was found that the lactalbumin contains a protein with the characteristics of a globulin. This protein, known as β-lactoglobulin, is the most abundant of the whey proteins. It has a molecular weight of 36,000. In addition to β-lactoglobulin, the classic lactalbumin fraction contains α-lactalbumin, serum albumin, and at least two minor components.

β-lactoglobulin is rich in lysine, leucine, glutamic acid, and aspartic acid. It is a globular protein with five known genetic variants. Variants A and B have 162 amino acids and molecular weights of 18,362 and 18,276, respectively. β-lactoglobulin has a tightly packed structure and consists of eight strands of antiparallel β sheets. The interior of the molecule is hydrophobic. The molecular structure also contains a certain amount of α helix, which plays a role in the formation of the usually occurring dimer. The association is pH dependent. β-lactoglobulin A will form octamers at low temperature and high concentration and at pH values between 3.5 and 5.2. Below pH 3.5 the protein dissociates into monomers. This protein is the only milk protein containing cysteine and, therefore, contains free sulfhydryl groups, which play a role in the development of cooked flavor in heated milk. The cysteine group is also involved in thermal denaturation. At pH 6.7 and above 67°C, β-lactoglobulin denatures, followed by aggregation. The first step in the denaturation is a series of reversible conformational changes that result in exposure of cysteine. The next step involves association through sulfhydryl-disulfide exchange.

The differences between genetic variants, although minor, may result in marked changes in some properties (Aschaffenburg 1965). The two chains of β-lactoglobulin C differ from the chains of the B variant in that a histidine residue has taken the place of a glutamic acid or glutamine residue. The A chains differ in two places from the B chains: aspartic acid replaces glycine and valine replaces alanine. Because of these minute differences, A is less soluble and more stable when heated than B. Variant A has a tendency to form tetramers at pH 4.5, whereas this tendency is absent in B. These differences are thought to be the result of differences in the three-dimensional folding or tertiary structure of the amino acid chains.

α-lactalbumin occurs as genetic variants A and B, each with 123 amino acid residues. The molecular weight of A is 14,147 and B is 14,175. The amino acid sequence of α-lactalbumin is very similar to that of hen egg-white lysozyme. α-lactalbumin has a high binding capacity for calcium and some other metals. It is insoluble at the isoelectric range from pH 4 to 5. The calcium in α-lactalbumin is bound very strongly and protects the stability of the molecule against thermal denaturation.

The immune globulins were previously divided into euglobulin and pseudoglobulin. The level of these proteins in colostrum is very high and they have been shown to be transferred to the blood of the young calf, indicating that they are absorbed unchanged. The three classes of immunoglobulins in milk are designated IgM(γM), IgA(γA), and IgG(γG) (Gordon and Kalan 1974). IgG is subdivided into IgG1 and IgG2. The serum albumin has been shown to be identical to the blood serum albumin.

Nonfat milk (skim milk) is the raw material from which a number of milk protein products are manufactured. A schematic diagram of the various products obtained by processing of skim milk is given in Figure 3–21 (Wong et al. 1996). These products are used as raw materials in many manufactured foods and include caseins, caseinates, and coprecipitates (Morr 1984). Acid casein results from isoelectric precipitation of casein at pH 4.6 to 4.7. The curd is recovered by centrifugation, then washed and

dried. Alkali neutralization of the wet curd yields caseinate, which is spray dried. Rennet casein is made by rennet coagulation followed by washing and drying of the curd. Coprecipitates include both casein and whey proteins and are made from heated skim milk. The heating denatures the whey proteins, which can then be precipitated with acid together with the casein.

Whey protein concentrate is made from whey, the by-product of cheese making. Removal of lactose and minerals requires reverse osmosis end ultrafiltration processing (Figure 3–22).

An up-to-date coverage of our present knowledge of milk proteins is given by Wong et al. (1996).

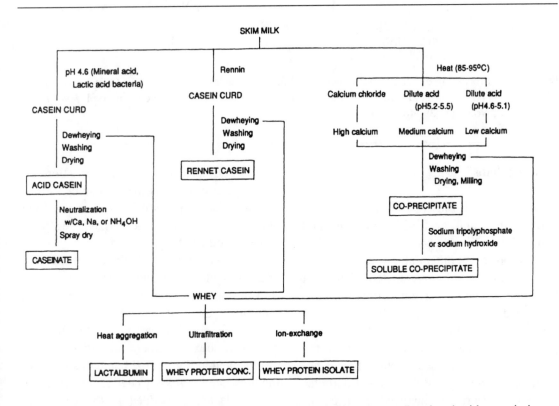

Figure 3–21 Production of Protein Products from Skim Milk. *Source*: Reprinted with permission from D.W.S. Wong, et al., Structures and Functionalities of Milk Proteins, *Crit. Rev. Food Sci. Nutr.*, Vol. 36, p. 834, © 1996, CRC Press, Inc.

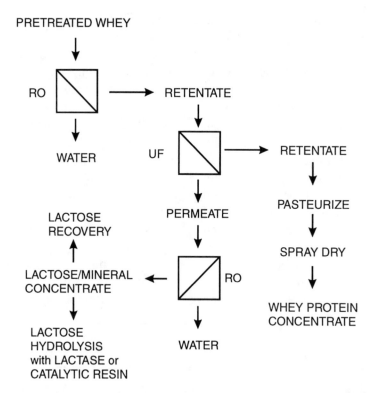

Figure 3–22 Processing of Whey To Produce Whey Protein Concentrate by Reverse Osmosis (RO) and Ultrafiltration (UF). *Source*: From C. V. Morr, Production and Use of Milk Proteins in Food, *Food Technol.*, Vol. 38, No. 7, pp. 39–48, 1984.

Meat Proteins

The proteins of muscle consist of about 70 percent structural or fibrillar proteins and about 30 percent water-soluble proteins. The fibrillar proteins contain about 32 to 38 percent myosin, 13 to 17 percent actin, 7 percent tropomyosin, and 6 percent stroma proteins. Meat and fish proteins contribute to highly organized structures that lend particular properties to these products. Some of the other proteins discussed in this chapter are more or less globular and consist of particles that are not normally involved in an extensive structural array. Examples are milk proteins and the protein bodies in cereals and

oilseeds. Extensive structure formation involving these proteins may occur in various technological processes such as making cheese from milk or texturized vegetable protein products from oilseeds.

Muscle is made up of fibers that are several centimeters long and measure 0.01 to 0.1 mm in diameter. The fibers are enclosed in membranes called sarcolemma and are arranged in bundles that enclose fat and connective tissue. The fibers are cross-striated, as indicated in Figure 3–23, and this is due to the presence of cross-striated myofibrils. The myofibrils are embedded in the cell cytoplasm called sarcoplasm. The fibers contain peripherally distributed nuclei; a diagram of

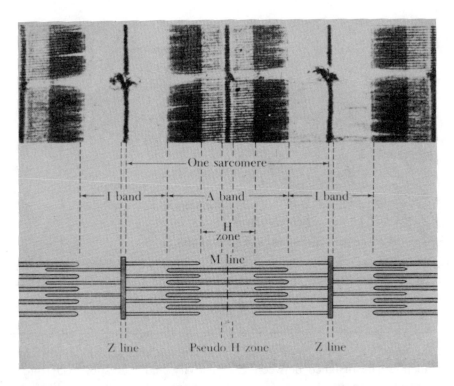

Figure 3–23 Microscopic Structure of Striated Muscle in Longitudinal Section. The top shows an electron micrograph, and a schematic representation of the composing units is given in the lower part. Magnification is about 23,000 times. Band I is light because it consists of thin filaments. The dark part of the A band is dense because it consists of overlapping thick and thin filaments. The thick filaments have a bulge in the center. This gives rise to the M line. Next to this is a bare region called the pseudo H zone. *Source*: From H.E. Huxley, The Mechanism of Muscular Contraction, *Sci. Am.*, Vol. 213, No. 6, pp. 18–27, 1965.

the arrangement of the various constituents of a muscle fiber is given in Figure 3–24 (Cassens 1971). In addition to the constituents mentioned, the muscle fibers contain other components including mitochondria, ribosomes, lysosomes, and glycogen granules. The fibers make up the largest part of the muscle volume, but there is from 12 to 18 percent of extracellular space.

The fibrils are optically nonuniform, which accounts for the striated appearance. Compounds with different refractive indexes are arranged along the fiber. The A bands are anisotropic and show up in polarized light, in contrast to the isotropic I bands, which look dark. The arrangement is schematically represented in Figure 3–23. The A bands contain thick filaments of 10 to 11 nm diameter. The I bands are composed of thinner filaments (5 nm). The A and I bands are the contractile elements and consist of myosin and actin respectively. The I band is dissected by a dark line called the Z line. The A band contains a light area named the H zone. It repre-

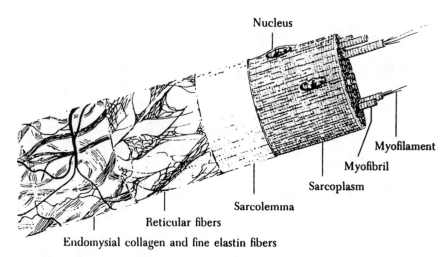

Figure 3–24 Diagrammatic Representation of a Muscle Fiber. *Source*: From R.G. Cassens, Microscopic Structure of Animal Tissues, in *The Science of Meat and Meat Products*, J.F. Price and B.S. Schweigert, eds., 1971, W.H. Freeman and Co.

sents the area where no actin is opposite the myosin filaments. The distance between Z lines is the sarcomere length.

Meat contains three general types of proteins: the soluble proteins, which can easily be removed by extraction with weak salt solutions (ionic strength ≤ 0.1); the contractile proteins; and the stroma proteins of the connective tissue. The soluble proteins are classed as myogens and myoalbumins. The myogens are a heterogeneous group of metabolic enzymes. After extraction of the soluble proteins, the fibril and stroma proteins remain. They can be extracted with buffered 0.6M potassium chloride to yield a viscous gel of actomyosin.

Myosin is the most abundant of the muscle proteins and makes up about 38 percent of the total. Myosin is a highly asymmetric molecule with a molecular weight of about 500,000 that contains about 60 to 70 percent α-helix structure. The molecule has a relatively high charge and contains large amounts of glutamic and aspartic acids and

dibasic amino acids. Myosin has enzyme activity and can split ATP into ADP and monophosphate, thereby liberating energy that is used in muscle contraction. The myosin molecule is not a single entity. It can be separated into two subunits by means of enzymes or action of the ultracentrifuge. The subunits with the higher molecular weight, about 220,000, are called heavy meromyosin. Those with low molecular weight, about 20,000, are called light meromyosin. Only the heavy meromyosin has ATP-ase activity.

Actin makes up about 13 percent of the muscle protein, so the actin-myosin ratio is about 1:3. Actin occurs in two forms: G-actin and F-actin (G and F denote globular and fibrous). G-actin is a monomer that has a molecular weight of about 47,000 and is a molecule of almost spherical shape. Because of its relatively high proline content, it has only about 30 percent of α-helix configuration. F-actin is a large polymer and is formed when ATP is split from G-actin. The units of actin combine to form a double helix of indefinite length, and molecular weights of actin have been reported to be in the order of several millions. Bodwell and McClain (1971) indicate that actin polymers may have an apparent molecular weight of over 14,000,000, with a length of 1,160 nm and a diameter of 6 nm. The transformation of G-actin into F-actin is schematically represented in Figure 3–25.

Actomyosin is a complex of F-actin and myosin and is responsible for muscle contraction and relaxation. Contraction occurs when myosin ATP-ase activity splits ATP to form phosphorylated actin and ADP. For this reaction, the presence of K^+ and Mg^{++} is required. Relaxation of muscle depends on regeneration of ATP from ADP by phosphorylation from creatine phosphate. The precise mechanism of contraction is still unknown, although a working hypothesis is available (Bailey 1982).

The composition of various cuts of meats and their structural implications have been described by Ranken (1984).

Collagen

The contractile meat proteins are separated and surrounded by layers of connective tissues. The amount and nature of this connective tissue is an important factor in the tenderness or toughness and the resulting eating quality of meat. Collagens form the most widely occurring group of proteins in the animal body. They are part of the connective tissues in muscle and organs, skin, bone, teeth, and tendons. The collagens are a distinct class of proteins as can be demonstrated by X-ray diffraction analysis. This technique shows that collagen fibrils have regular periodicity of 64 nm, which can be increased under tension to 400 nm. Collagen exists as a triple helix; the formation of collagen into the triple helix is shown in Figure 3–26 (Gross 1961). The triple helix is the tropocollagen molecule; these are lined up in a staggered array, overlapping by one-quarter of their length to form a fibril. The fibrils are stacked in layers to form connective tissue. Important in the formation of these structures is the high content of hydroxyproline and hydroxylysine. The content of dibasic and diacidic amino acids is also high, but tryptophan and cystine are absent. As a result of this particular amino acid composition, there are few interchain cross-bonds, and collagen swells readily in acid or alkali.

Heating of collagen fibers in water to 60 to 70°C shortens them by one-third or one-fourth of the original length. This temperature is characteristic of the type of collagen and is called shrink temperature (T_s). The T_s of fish skin collagen is very low, 35°C. When the temperature is increased to about 80°C, mammalian collagen changes into gelatin. Certain amino acid sequences are common in collagen, such as Gly-Pro-Hypro-Gly. In a triple helix, only certain sequences are permissible. The structural unit of the collagen fibrils is tropocollagen with a length of 280 nm, a diameter of 1.5 nm, and a molecular weight of 360,000. Gelatin is a soluble protein derived from insoluble collagen. Although it can be made from different animal by-products, hide is the common source of gelatin production. The process of trans-

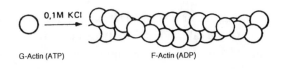

Figure 3–25 Transformation of G-Actin into F-Actin

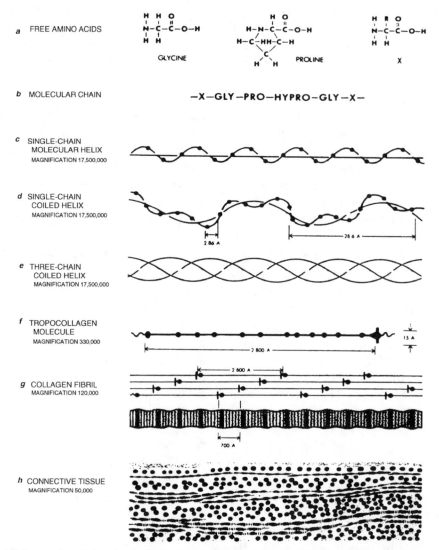

Figure 3–26 Successive Stages in the Formation of Collagen. Free amino acids are joined in a peptide chain, which twists itself into a left-handed helix. Three of these intertwine to form a right-handed superhelix. This is the tropocollagen molecule. Molecules line up in a staggered fashion to overlap by one-quarter length to form a fibril. *Source*: From J. Gross, Collagen, *Sci. Am.,* Vol. 204, No. 5, pp. 120–130, 1961.

forming collagen into gelatin involves the following three changes:

1. rupture of a limited number of peptide bonds to reduce the length of the chains

2. rupture or disorganization of a number of the lateral bonds between chains

3. change in chain configuration

The last of these is the only change essential for the conversion of collagen to gelatin. The conditions employed during the production of gelatin determine its characteristics. If there is extensive breakdown of peptide bonds, many lateral bonds may remain intact and soluble fragments are produced. If many lateral bonds are destroyed, the gelatin molecules may have relatively long chain lengths. Thus, there is a great variety of gelatins. In normal production, the hides or bones are extracted first, under relatively mild conditions, followed by successive extractions under more severe conditions. The first extraction yields the best-quality gelatin. The term *gelatin* is used for products derived from mammalian collagen that can be dispersed in water and show a reversible sol-gel change with temperature. The gels formed by gelatin can be considered as a partial return of the molecules to an ordered state. However, the return to the highly ordered state of collagen is not possible. High-quality gelatin has an average chain length of 60,000 to 80,000, whereas the value for native collagen is infinite.

The process of gel formation is probably closely associated with the presence of guanidino groups of arginine. Hypobromite has the ability to destroy guanidino groups and, when added to gelatin, it inhibits gelation.

There are three types of gelatin: alpha, with a molecular weight of 80,000 to 125,000; beta, with molecular weight of 160,000 to 250,000; and gamma, with

molecular weight of 240,000 to 375,000 (Poppe 1992). The typical amino acid sequence in gelatin is Gly-X-Y, where X is mostly proline and Y is mostly hydroxyproline (Figure 3–27).

When gelatin is placed in cold water it will absorb 5 to 10 times its own weight in water and swell. When this material is heated to above the melting point, between 27 and 34°C, the swollen gelatin dissolves. This sol-gel transformation is reversible. Poppe (1992) has described the mechanism of gel formation. It involves the reversion from a random coil to a helix structure. Upon cooling, the imino acid–rich regions of different chains form a helical structure, which is stabilized by hydrogen bonding. This then forms the three-dimensional gel matrix.

Commercial gelatin is available in two types, A and B. Type A gelatin has undergone an acid pretreatment and type B a lime pretreatment. They differ in their viscosity and their ability to combine with negatively charged hydrocolloids such as carrageenan.

Fish Proteins

The proteins of fish flesh can be divided into three groups on the basis of solubility, as indicated in Table 3–18. The skeletal muscle of fish consists of short fibers arranged between sheets of connective tissue, although the amount of connective tissue in fish mus-

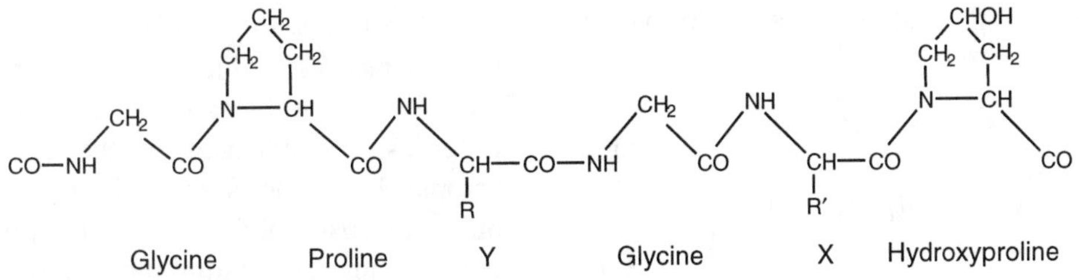

Figure 3–27 Amino Acid Sequence in Gelatin

Table 3–18 Division of the Proteins of Fish Flesh According to Solubility

Ionic Strength at Which Soluble	Name of Group	Location
Equal to or greater than 0	"Myogen" easily soluble	Mainly sarcoplasm, muscle cell juice
Greater than about 0.3	"Structural" less soluble	Mainly myofibrils, contractile elements
Insoluble	"Stroma"	Mainly connective tissues, cell walls, etc.

cle is less than that in mammalian tissue and the fibers are shorter. The myofibrils of fish muscle have a striated appearance similar to that of mammalian muscle and contain the same major proteins, myosin, actin, actomyosin, and tropomyosin. The soluble proteins include most of the muscle enzymes and account for about 22 percent of the total protein. The connective tissue of fish muscle is present in lower quantity than in mammalian muscle; the tissue has different physical properties, which result in a more tender texture of fish compared with meat. The structural proteins consist mainly of actin and myosin, and actomyosin represents about three-quarters of the total muscle protein. Fish actomyosin has been found to be quite labile and easily changed during processing and storage. During frozen storage, the actomyosin becomes progressively less soluble, and the flesh becomes increasingly tough. Connell (1962) has described the changes that may occur in cod myosin. When stored in dilute neutral solution, myosin rapidly denatures and forms aggregates in a step-wise manner as follows:

$$M \longrightarrow M_D \tag{1}$$

$$\left. \begin{array}{l} M_D + M_D \rightarrow 2M_D \\[1em] 2M_D + M_D \rightarrow 3M_D \end{array} \right\} \tag{2}$$

Equation 1 represents the change from native to denatured protein and follows first-order reaction kinetics. In successive steps represented by equations 2, dimers, trimers, and higher polymers are formed in a concentration-dependent reaction pattern. The aggregation is assumed to be mostly in a lateral fashion with only little end-to-end aggregation. The instability of fish myosin appears to be one of the major factors causing the lability of fish muscle.

The interest in using fish for the production of fish protein concentrate has waned because the product lacks satisfactory functional properties. More promising ways of using fish resources involve fish protein gels (surimi), which can be made into attractive consumer products (Mackie 1983).

Egg Proteins

The proteins of eggs are characterized by their high biological value and can be divided into the egg white and egg yolk proteins. The egg white contains at least eight different proteins, which are listed in Table 3–19. Some of these proteins have unusual properties, as indicated in the table; for example, lysozyme is an antibiotic, ovomucoid is a trypsin inhibitor, ovomucin inhibits hemagglutination, avidin binds biotin, and conalbumin binds iron. The antimicrobial

Table 3–19 Protein Composition of Egg White

Constituent	Approximate Amount (%)	Approximate Isoelectric Point (pH)	Unique Properties
Ovalbumin	54	4.6	Denatures easily, has sulfhydryls
Conalbumin	13	6.0	Complexes iron, anti-microbial
Ovomucoid	11	4.3	Inhibits enzyme trypsin
Lysozyme	3.5	10.7	Enzyme for polysaccha-rides antimicrobial
Ovomucin	1.5	?	Viscous, high sialic acid, reacts with viruses
Flavoprotein-apoprotein	0.8	4.1	Binds riboflavin
"Proteinase inhibitor"	0.1	5.2	Inhibits enzyme (bacterial proteinase)
Avidin	0.05	9.5	Binds biotin, antimicrobial
Unidentified proteins	8	5.5, 7.5 8.0, 9.0	Mainly globulins
Nonprotein	8		Primarily half glucose and salts (poorly characterized)

Source: From R.R. Feeney and R.M. Hill, Protein Chemistry and Food Research, in *Advances in Food Research*, Vol. 10, C.O. Chichester, E.M. Mrak, and G.F. Stewart, eds., 1960, Academic Press.

properties help to protect the egg from bacterial invasion.

Liquid egg white contains 10 to 11 percent of protein, and the dried form contains about 83 percent. The most abundant protein is ovalbumin, a phosphoprotein with a molecular weight of 45,000 that contains a small proportion of carbohydrate. The carbohydrate is present as a polysaccharide composed of two glucosamine and four mannose groups. Ovalbumin can be separated by electrophoresis into two components, one component with two phosphate groups and another component with one phosphate group. Some of the diphosphate changes to monophosphate on storage. Ovalbumin is readily denatured by heat.

Conalbumin has a molecular weight of 70,000 and has iron-binding and antimicrobial properties. It can render iron unavailable to microorganisms; this property is lost after heat denaturation. Iron is bound in the ferric form by coordination. The groups involved in the binding of iron are amino, carboxyl, guanidine, and amides. When these groups are blocked, the iron-binding property is lost.

Ovomucoid is a trypsin inhibitor and a glycoprotein with a molecular weight of 27,000 to 29,000, containing mannose and glucosamine. This protein is highly resistant to denaturation.

Lysozyme is classed as a globulin and has the ability to cause lysis of bacterial cells. There are three fractions, designated G_1, G_2, and G$_3$. The activity resides in the G$_1$ fraction. The protein has a molecular weight of 14,000 to 17,000 and is a basic protein with unusually high content of histidine, arginine, and lysine. It is stable to many agents that denature other proteins, such as heat, cold, and denaturing reagents. It is also quite resistant to proteolysis by enzymes such as papain and trypsin.

Ovomucin is an insoluble protein, which precipitates from egg white on dilution with water. It is not well known, has a high molecular weight (7,600,000), and is a mucoprotein. The ability of certain viruses to agglutinate red blood cells, called *hemagglutination*, is inhibited by ovomucin.

Avidin is a protein characterized by its ability to bind biotin and render it unavailable. Heat denaturation destroys this property.

Egg yolk proteins precipitate when the yolk is diluted with water. The protein components of egg yolk are listed in Table 3–20. The yolk contains a considerable amount of lipid, part of which occurs in bound form as lipoproteins. Lipoproteins are excellent emulsifiers, and egg yolk is widely used in foods for that reason. The two lipoproteins are lipovitellin, which has 17 to 18 percent lipid, and lipovitellenin, which has 36 to 41 percent lipid. The protein portions of these compounds after removal of the lipid are named vitellin and vitellenin. The former contains 1 percent phosphorus, the latter 0.29 percent.

The membranes of eggs consist of keratins and mucins.

When fluid egg yolk is frozen, changes take place, causing the thawed yolk to form a gel (Powrie 1984). Gelation increases as the freezing temperature is lowered from –6°C to –14°C. Gradual aggregation of lipoprotein is postulated as the cause of gelation.

PLANT PROTEINS

As with animal proteins, plant proteins occur in wide variety. Many plant proteins have until recently received much less study than the animal proteins. This is gradually changing, and more information is now becoming available on many nontraditional food proteins. Proteins can be obtained from leaves, cereals, oilseeds, and nuts. Leaf proteins have been extracted from macerated leaves and are very labile. They are readily denatured at about 50°C and undergo surface denaturation in the pH range 4.5 to 6.0. Cereal seed proteins are generally low in lysine. Peanut protein is poor in lysine, tryptophan, methionine, and threonine. Legume seeds are low in cyst(e)ine and methionine. Great improvement in nutritional value can

Table 3–20 Protein Components of Egg Yolk

Constituent	Approximate Amount (%)	Particular Properties
Livetin	5	Contains enzymes—poorly characterized
Phosvitin	7	Contains 10% phosphorus
Lipoproteins	21	Emulsifiers
(Total protein)	(33)	

Source: From R.R. Feeney and R.M. Hill, Protein Chemistry and Food Research, in *Advances in Food Research*, Vol. 10, C.O. Chichester, E.M. Mrak, and G.F. Stewart, eds., 1960, Academic Press.

sometimes be obtained by judicious mixing of different products.

The proteins of cereal grains are very important to their physical properties, even though the protein content of grains is not very high. Protein levels vary within wide limits, depending on species, soil, fertilizer, and climate. The protein is nonuniformly distributed throughout the kernel, with the center having the lowest protein content. Wheat has a protein content of 8 to 14 percent, rye about 12 percent, barley 10 percent, and rice 9 percent.

Wheat Proteins

Wheat proteins are unique among plant proteins and are responsible for bread-making properties of wheat. The classic method of fractionation based on solubility characteristics indicates the presence of four main fractions (Figure 3–28): albumin, which is water-soluble and coagulated by heat; globulin, soluble in neutral salt solution; gliadin, a prolamine soluble in 70 percent ethanol; and glutenin, a glutelin insoluble in alcohol but soluble in dilute acid or alkali.

The methods of gel electrophoresis and iso-electric focusing now provide highly efficient tools for separation of these proteins. By using these techniques, both gliadin and glutenin have been shown to be complex mixtures. Gliadin and glutenin are the storage, or gluten-forming, proteins of wheat. The formation of gluten takes place when flour is mixed with water. The gluten is a coherent elastic mass, which holds together other bread components such as starch and gas bubbles,

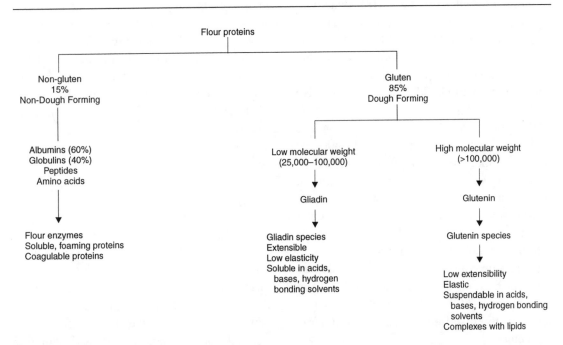

Figure 3–28 Schematic Representation of the Main Protein Fractions of Wheat Flour. *Source*: From J. Holme, A Review of Wheat Flour Proteins and Their Functional Properties, *Bakers' Dig.*, Vol. 40, No. 6, pp. 38–42, 78, 1966.

thus providing the basis for the crumb structure of bread. The hydration of the gluten proteins results in the formation of fibrils
(Simmonds and Orth 1973), with gliadins
forming films and glutenins forming strands.

Gluten proteins have a high content of
glutamine but are low in the essential amino
acids lysine, methionine, and tryptophan.
The insolubility of gluten proteins can be
directly related to their amino acid composition. High levels of nonpolar side chains
result from the presence of glutamic and
aspartic acids as the amides. Because these
are not ionized, there is a high level of apolar
(hydrogen) bonding. This contributes to
aggregation of the molecules and results in
low solubility.

Heat damage to gluten can result from
excessive air temperatures used in the drying
of wet grain. The gluten becomes tough and
is more difficult to extract. Heat-denatured
wheat gives bread poor texture and loaf volume (Schofield and Booth 1983).

Gliadin and glutenin are composed of many
different molecular species. Gliadin proteins
consist mostly of single-chain units and have
molecular weights near 36,500 (Bietz and
Wall 1972). Whole gliadin also contains
polypeptides of molecular weight 11,400 that
may be albumins, a major polypeptide of
molecular weight 44,200, and ω-gliadins of
molecular weights 69,300 and 78,100. The
polypeptides of 44,200 and 36,500 molecular
weight are joined through disulfide bonds
into higher molecular weight proteins. Glutenin consists of a series of at least 15 polypeptides with molecular weights ranging from
11,600 to 133,000. These units are bound
together by disulfide bonds to form large
molecules.

The nongluten albumin and globulin proteins represent from 13 to 35 percent of the
total protein of cereal flours. This protein

fraction contains glycoproteins, nucleoproteins, lipoproteins, and a variety of enzymes.

Chemical modification of gluten proteins
plays an important role in the industrial use
of cereals. Especially reactions that lead to
splitting or formation of an SS bond can
greatly influence solubility and rheological
properties such as extensibility and elasticity.

An example of the reduction of an SS bond
by means of an SH containing reagent is as
follows:

The disulfide bonds in wheat gluten play
an important role in cross-linking polypeptide chains. Some of the bonds present in
cereal proteins are shown in Figure 3–29
(Wall 1971). Reduction of the disulfide
bonds in gliadin and glutenin results in the
unfolding of the peptide chains (Krull and
Wall 1969), as indicated in Figure 3–30. This
type of change has a profound effect on the
rheological properties of dough (Pomeranz
1968).

Maize Proteins

Maize (corn), wheat, and rice are the three
main cereal crops of the world. The protein
content of maize varies widely depending on
variety, climate, and other factors; it is generally in the 9 to 10 percent range. The main
proteins of maize are the storage proteins of
the endosperm, namely zein and glutelin.
There are also minor amounts of albumin
and globulin. Maize proteins are low in levels of the essential amino acids lysine and

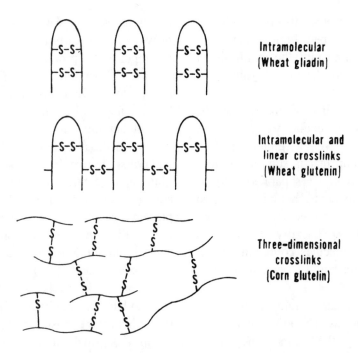

Figure 3–29 Types of Disulfide Bonds in Cereal Proteins. *Source*: From J.S. Wall, Disulfide Bonds: Determination, Location, and Influence on Molecular Properties of Proteins, *Agr. Food Chem.*, Vol. 19, pp. 619–625, 1971.

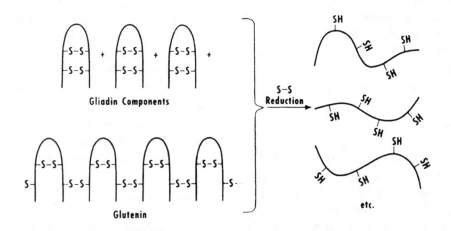

Figure 3–30 Reduction of Disulfide Bonds in Gliadin and Glutenin. *Source*: From L.H. Krull and J.S. Wall, Relationship of Amino Acid Composition and Wheat Protein Properties, *Bakers' Dig.*, Vol. 43, No. 4, pp. 30–39, 1969.

tryptophan. To overcome this problem Mertz et al. (1964) developed high-lysine corn, Opaque-2. In this mutant the synthesis of zein, the protein with the lowest lysine content, is suppressed.

The storage proteins of maize can be divided into low molecular weight (zeins) and high molecular weight (glutelins). Zein is the protein group that is soluble in alcohols. The zeins can be separated into as many as 30 components, belonging to two major groups with molecular weights of about 22 and 24kDa (Lasztity 1996). Zeins are asymmetric molecules containing about 45 percent α-helix and 15 percent β-sheet; the rest is aperiodic.

The high molecular weight storage proteins are the glutelins. The glutelins are less well-defined than the zeins. They have higher lysine, arginine, histidine, and tryptophan and lower glutamic acid content than the zeins. Glutelin consists of several subunits joined together by disulfide bonds.

The remaining proteins, present in amounts of about 8 percent each, are albumins and globulins. These proteins are soluble in water and/or salt solutions, and are characterized by higher levels of essential amino acids and lower levels of glutamic acid. Albumins have higher contents of aspartic acid, proline, glycine, and alanine, and lower levels of glutamic acid and arginine than the globulins.

Rice Proteins

Rice and wheat are the two staple cereals for much of the world population. The wheat kernel is almost never eaten as such, but only after extensive processing such as milling and baking. In contrast, rice is eaten mostly as the intact kernel after the bran has been removed. In some parts of the world, however, rice is also consumed in the form of rice noodles. The protein content of polished or white rice is lower than that of wheat. Values reported for protein content of rice range from 6 to 9 percent. The nutritional value of rice protein is high because of its relatively high content of lysine, the first limiting essential amino acid. Up to 18 percent of the protein in the rice kernel is lost in the bran and polish. In rice the main storage protein is glutelin, in contrast to wheat, whose main storage protein is gliadin. The approximate protein distribution in rice proteins is as follows: albumin 5 percent, globulin 10 percent, prolamin 5 percent, and glutenin 80 percent (Lookhart 1991).

The protein in rice is present in the form of encapsulated protein bodies, which are found throughout the endosperm. The protein bodies may be small or large with the former containing primarily glutelin and the latter containing prolamin and glutelin (Hamaker 1994). The protein bodies are insoluble and remain intact during cooking. Rice has little or no intergranular matrix protein. This characteristic is different than in most other cereals, and it may have an effect in the process of noodle making.

Glutelin is the principal protein in the whole grain as well as in milled rice and rice polish. The major proteins in rice bran are albumin and globulin. This indicates that glutelin is concentrated in the milled rice and albumin and globulin are enriched in the bran and rice polish. These byproducts are mainly used as animal feed and would constitute a valuable food source if properly processed.

Soybean Proteins

The proteins in soybeans are contained in protein bodies, or aleurone grains, which measure from 2 to 20 μm in diameter. The protein bodies can be visualized by electron microscopy (Figure 3–31). Soy protein is a good source of all the essential amino acids except methionine and tryptophan. The high lysine content makes it a good complement to cereal proteins, which are low in lysine. Soybean proteins have neither gliadin nor glutenin, the unique proteins of wheat gluten. As a result, soy flour cannot be incorporated into bread without the use of special additives that improve loaf volume. The soy proteins have a relatively high solubility in water or dilute salt solutions at pH values below or above the isoelectric point. This

means they are classified as globulins. There is as yet no generally accepted nomenclature for the soy proteins, and only some of the common terminology is used here. The complex character of the mixture of proteins in soybeans is indicated by the fact that starch gel electrophoresis of acid-precipitated globulins in 5M urea with alkaline buffer reveals 14 protein bands, and in acid buffer 15 bands appear (Puski and Melnychyn 1968).

Generally, soybean proteins are differentiated on the basis of their behavior in the ultracentrifuge. Water-extractable proteins

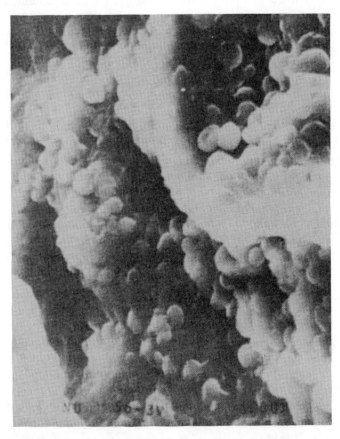

Figure 3–31 Electron Photomicrograph of the Protein Bodies in Soybeans. *Source:* From W.J. Wolf, Scanning Electron Microscopy of Soybean Protein Bodies, *J. Am. Oil Chem. Soc.*, Vol. 47, pp. 107–108, 1970.

are separated into four fractions with approximate sedimentation rates of 2, 7, 11, and 15S. The percentage of these fractions is indicated in Table 3–21, along with their components and their molecular weights. Several of the ultracentrifuge fractions can be further separated into a number of components. The 2S fraction contains trypsin inhibitors, cytochrome c, allantoinase, and two globulins. The 7S fraction contains β-amylase, hemagglutinin, lipoxygenase, and 7S globulin. The 11S fraction consists mainly of 11S globulin. This compound has been separated by electrophoresis into 18 bands in alkaline gels and 10 bands in acid gels. The 11S protein is usually named glycinin, and there are various proposals for naming other protein fractions conglycinin (Wolf 1969). The detailed subunit structures of 7S and 11S globulins, the most important of the soy proteins, have been described by Wolf (1972b). The 11S globulin has a quaternary structure consisting of 12 subunits. According to Catsimpoolas et al. (1967), these have the following amino-terminal residues: 8 glycine, 2 phenylalanine, and either 2 leucine or 2 isoleucine. It appears that the 11S protein is a dimer of two identical monomers, each consisting of six subunits, three of which are acidic and three basic. Interactions among these subunits may be a factor in stabilizing the molecule. The 7S globulin consists of 9 subunits of single polypeptide chains. The protein is a glycoprotein and the polysaccharide is attached as a single unit to one of the polypeptide chains. The carbohydrate consists of 38 mannose and 12 glucosamine residues.

Current information on soy protein fractionation and nomenclature has been given by Brooks and Moor (1985). The 7S globulins are classified into three major types. Type I is β-conglycinin (B_1–B_6), type II is β-conglycinin (B_0), and type III is γ-conglycinin.

Application of heat to soybeans or defatted soy meal makes the protein progressively more insoluble. Hydrogen bonds and hydrophobic bonds appear to be responsible for the

Table 3–21 Ultracentrifuge Fractions of Soybean Proteins

Protein Fraction	Percentage of Total	Components	Molecular Weight
2S	22	Trypsin inhibitors	8,000
			21,500
		Cytochrome c	12,000
		2.3S Globulin	18,200
		2.8S Globulin	32,000
		Allantoinase	50,000
7S	37	Beta-amylase	61,700
		Hemagglutinins	110,000
		Lipoxygenases	108,000
		7S Globulin	186,000–210,000
11S	31	11S Globulin	350,000
15S	11	–	600,000

Source: From W.J. Wolf, What Is Soy Protein, Food Technol., Vol. 26, No. 5, pp. 44–54, 1972.

decrease in solubility of the proteins during heating.

Both 7S and 11S proteins show a complicated pattern of association and dissociation reactions. The 7S globulin at 0.5 ionic strength and pH 7.6 is present as a monomer with molecular weight of 180,000 to 210,000. At 0.1 ionic strength, the molecule forms the 9S dimer. This reaction is reversible. At pH 2 and low ionic strength, the 7S globulin forms 2S and 5S compounds; these are the result of dissociation into subunits. This reaction is reversed at higher ionic strengths.

Changes in the quaternary structure of the 11S globulin have been summarized by Wolf (1972a). Secondary and tertiary structures of this protein involve no alpha helix structure but instead consist of antiparallel beta-structure and disordered regions. The structure appears to be compact and is stabilized by hydrophobic bonds. The changes in quaternary structure that occur as a function of experimental conditions are represented in Figure 3–32. At ionic strength of 0.1, the protein partially associates into agglomerates with a higher sedimentation velocity. Increasing the ionic strength reverses this reaction. Various conditions promote dissociation of the 11S protein into half-molecules with a sedimentation velocity of 7S. Further breakdown of the half-molecules may occur and will result in the formation of

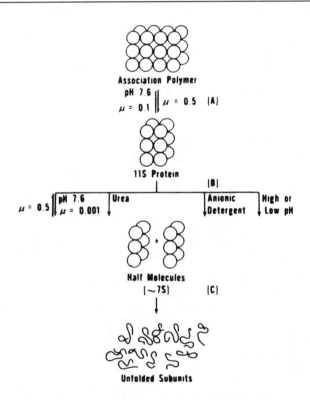

Figure 3–32 Association and Dissociation of Soybean 11S Protein as Influenced by Experimental Conditions. *Source:* From W.J. Wolf, Purification and Properties of the Proteins, in *Soybeans: Chemistry and Technology,* A.K. Smith and S.J. Circle, eds., 1972, AVI Publishing Co.

unfolded subunits. It has been suggested (Catsimpoolas 1969) that the 11S molecule is made up of two annular-hexagonal structures, each containing six alternating acidic and basic subunits. This feature would result in a stabilization of the structure by ionic bonds.

Soybean whey proteins are obtained in the solution left after acid precipitation of protein from an aqueous extract. The solution contains an unknown number of albumins and globulins, in addition to water-soluble carbohydrates, nonprotein nitrogen, salts, vitamins, and phytates. In the production of soy protein isolates, the whey proteins may create a disposal problem.

For direct human consumption, soybeans are mainly used as soymilk and tofu or bean curd. In the preparation of soymilk the extractability of proteins is related to the age and storage conditions of soybeans (Thomas et al. 1989). Adverse storage conditions result in low protein extraction and poor quality of tofu. Tofu is manufactured from soymilk with 10 percent of total solids by coagulation of the proteins with the following coagulants: $CaSO_4$, $MgSO_4$, $CaCl_2$, $MgCl_2$, or GDL (glucono delta lactone) (deMan et al. 1986). When using the salts, the Ca and Mg ions are the prime reactants. $CaSO_4$ is only slightly soluble in water; the reaction is slow, but the resulting curd is cohesive and, when pressed, has a soft texture. With soluble salts such as $CaCl_2$ the reaction is fast and the resulting curd is fragmented. When pressed, the tofu has a higher protein content and the texture is firm. Coagulation with GDL is a different process. Freshly prepared, GDL slowly decreases the pH; coagulation occurs at the isoelectric point of the proteins. The texture of GDL tofu is very homogeneous.

REFERENCES

Aoki, T., et al. 1987. Caseins are cross-linked through their ester phosphate groups by colloidal calcium phosphate. *Biochim. Biophys. Acta* 911: 238–243.

Aschaffenburg, R. 1965. Variants of milk proteins and their pattern of inheritance. *J. Dairy Sci.* 48: 128–132.

Bailey, A.J. 1982. Muscle proteins and muscle structure. In *Food proteins*, ed. P.F. Fox and J.J. Condon. New York: Applied Science Publishers.

Barbut, S. 1994. Protein gel ultrastructure and functionality. In *Protein functionality in food systems*, ed. N.S. Hettiarachachy and G.R. Ziegler. New York: Marcel Dekker Inc.

Bietz, J.A., and J.S. Wall. 1972. Wheat gluten subunits: Molecular weights determined by sodium dodecyl sulfate-polyacrylamide gel electrophoresis. *Cereal Chem.* 49: 416–430.

Bjarnason, J., and K.J. Carpenter. 1970. Mechanisms of heat damage in proteins. 2. Chemical changes in pure proteins. *Brit. J. Nutr.* 24: 313–329.

Bodwell, C.E., and P.E. McClain. 1971. Proteins. In *The science of meat and meat products*, ed. J.F. Price and B.S. Schweigert. San Francisco: W.H. Freeman and Co.

Brooks, J.R., and C.V. Moor. 1985. Current aspects of soy protein fractionation and nomenclature. *J. Am. Oil Chem. Soc.* 62: 1347–1354.

Burton, H.S., and D.J. McWeeney. 1964. Non-enzymatic browning: Routes to the production of melanoidins from aldoses and amino compounds. *Chem. Ind.* 11: 462–463.

Cassens, R.G. 1971. Microscopic structure of animal tissues. In *The science of meat and meat products*, ed. J.F. Price and B.S. Schweigert. San Francisco: W.H. Freeman and Co.

Catsimpoolas, N. 1969. Isolation of glycinin subunits by isoelectric focussing in ureamercaptoethanol. *FEBS Letters* 4: 259–261.

Catsimpoolas, N., et al. 1967. Purification and structural studies of the 11S component of soybean proteins. *Cereal Chem.* 44: 631–637.

Connell, J.J. 1962. Fish muscle proteins. In *Recent advances in food science*. Vol. 1, ed. J. Hawthorn and J.M. Leitch. London: Butterworth.

Dalgleish, D.G. 1989. Protein-stabilized emulsions and their properties. In *Water and food quality*, ed. T.M. Hardman. London: Elsevier Applied Science Publishers.

deMan, J.M., L. deMan, and S. Gupta. 1986. Texture and microstructure of soybean curd (tofu) as affected by different coagulants. *Food Microstructure* 5: 83–89.

Ellis, G.P. 1959. The Maillard reaction. In *Advances in carbohydrate chemistry*. Vol. 14, ed. M.L. Wolfrom and R.S. Tipson. New York: Academic Press.

Farrell, H.M. 1973. Models for casein micelle formation. *J. Dairy Sci.* 56: 1195–1206.

Farrell, H.M., and M.P. Thompson. 1974. Physical equilibria: Proteins. In *Fundaments of dairy chemistry*, ed. B.H. Webb et al. Westport, CT: AVI Publishing Co.

Finley, J.W., and W.F. Shipe. 1971. Isolation of a flavor producing fraction from light exposed milk. *J. Dairy Sci.* 54: 15–20.

Gordon, W.G., and E.B. Kalan. 1974. Proteins of milk. In *Fundamentals of dairy chemistry*, ed. B.H. Webb et al. Westport, CT: AVI Publishing Co.

Gross, J. 1961. *Collagen Sci. Am.* 204, no. 5: 120–130.

Hamaker, B.R. 1994. The influence of rice protein on rice quality. In *Rice science and technology*, ed. W.E. Marshall and J.I. Wadsworth. New York: Marcel Dekker.

Harland, H.A., S.T. Coulter, and R. Jenness. 1952. The effect of various steps in the manufacture on the extent of serum protein denaturation in nonfat dry milk solids. *J. Dairy Sci.* 35: 363–368.

Hatta, H., and T. Koseki. 1988. Relationship of SH groups to functionality of ovalbumin. In *Food proteins*, ed. J.E. Kinsella and W.G. Soucie. Champaign, IL: American Oil Chemistry Society.

Hermansson, A.M. 1973. Determination of functional properties of protein foods. In *Proteins in human nutrition*, ed. J.W.G. Porter and B.A. Rolls. London: Academic Press.

Hodge, J.E. 1953. Chemistry of browning reactions in model systems. *Agr. Food Chem.* 1: 928–943.

Hodge, J.E., F.D. Mills, and B.E. Fisher. 1972. Compounds of browned flavor from sugar-amine reactions. *Cereal Sci. Today* 17: 34–40.

Hurrell, R.F. 1984. Reactions of food proteins during processing and storage and their nutritional consequences. In *Developments in food proteins*, ed. B.J.F. Hudson. New York: Elsevier Applied Science Publishers.

Hurst, D.T. 1972. Recent developments in the study of nonenzymic browning and its inhibition by sulphur dioxide. BFMIRA Scientific and Technical Surveys No. 75. Leatherhead, England.

Kinsella, J.E. 1982. Structure and functional properties of food proteins. In *Food proteins*, ed. P.F. Fox and J.J. Condon. New York: Applied Science Publishers.

Kirchmeier, O. 1962. The physical-chemical causes of the heat stability of milk proteins (in German). *Milchweissenschaft* 17: 408–412.

Krull, L.H., and J.S. Wall. 1969. Relationship of amino acid composition and wheat protein properties. *Bakers' Dig.* 43, no. 4: 30–39.

Lasztity, R. 1996. *The chemistry of cereal protein*. 2nd ed. Boca Raton, FL: CRC Press.

Lookhart, G.L. 1991. Cereal proteins: Composition of their major fractions and methods of identification. In *Handbook of cereal science and technology*, ed. K.J. Lorenz and K. Kulp. New York: Marcel Dekker.

Mackie, I.M. 1983. New approaches in the use of fish proteins. In *Developments in food proteins*, ed. B.J.F. Hudson. New York: Applied Science Publishers.

Masters, P.M., and M. Friedman. 1980. Amino acid racemization in alkali-treated food proteins: Chemistry, toxicology and nutritional consequences. In *Chemical deterioration of proteins*, ed J.R. Whitaker and M. Fujimaki. Americal Chemical Society Symposium Series 123. Washington, DC: American Chemical Society.

Mauron, J. 1970. The chemical behavior of proteins during food preparation and its biological effect (in French). *J. Vitamin Res.* 40: 209–227.

Mauron, J. 1983. Interaction between food constituents during processing. In *Proceedings of the Sixth International Conference on Food Science Technology*, 301–321. Dublin, Ireland.

Mertz, E., O. Nelson, and L.S. Bates. 1964. Mutant gene that changes composition and increases lysine content of maize endosperm. *Science* 154: 279–280.

Mitchell, J.R. 1986. Foaming and emulsifying properties of proteins. In *Developments in food proteins*, ed. B.J.F. Hudson. New York: Elsevier Applied Science Publishers.

Morr, C.V. 1984. Production and use of milk proteins in food. *Food Technol.* 38, no. 7: 39–48.

Nakai, S., and W.D. Powrie. 1981. Modification of proteins for functional and nutritional improvements. In *Cereals: A renewable resource, theory and practice*, ed. Y. Pomeranz and L. Munck. St. Paul, MN: American Association of Cereal Chemistry.

Nielsen, H.K., J. Loliger, and R.F. Hurrell. 1985. Reactions of proteins with oxidizing lipids. *Brit. J. Nutr.* 53: 61–73.

Patton, S. 1954. The mechanism of sunlight flavor formation in milk with special reference to methionine and riboflavin. *J. Dairy Sci.* 37: 446–452.

Pauling, L., R.B. Corey, and H.R. Branson. 1951. The structure of proteins: Two hydrogen bonded helical configurations of the polypeptide chain. *Proc. Natl. Acad. Sci.* (US) 37: 205–211.

Paulsen, T.M., and F.E. Horan. 1965. Functional characteristics of edible soya flours. *Cereal Sci. Today* 10, no. 1: 14–17.

Peterson, R.F., L.W. Nauman, and T.L. McMeekin. 1958. The separation and amino acid composition of a pure phosphopeptone prepared from β-casein by the action of trypsin. *J. Am. Chem. Soc.* 80: 95–99.

Pomeranz, Y. 1968. Relationship between chemical composition and bread making potentialities of wheat flour. *Adv. Food Research* 16: 335–455.

Pomeranz, Y. 1991. *Functional properties of food components.* San Diego, CA: Academic Press.

Poppe, J. 1992. Gelatin. In *Thickening and gelling agents for food*, ed. A. Imeson. London: Blackie Academic and Professional.

Powrie, W.D. 1984. Chemical effects during storage of frozen foods. *J. Chem. Educ.* 61: 340–347.

Puski, G., and P. Melnychyn. 1968. Starch gel electrophoresis of soybean globulins. *Cereal Chem.* 45: 192–197.

Ranken, M.D. 1984. Composition of meat: Some structural and analytical implications. In *Developments in food proteins*, ed. B.J.F. Hudson. New York: Elsevier Applied Science Publishers.

Roos, Y.H., and M. Himberg. 1994. Non-enzymatic browning behavior, as related to glass transition, of a model at chilling temperatures. *J. Agr. Food Chem.* 42: 893–898.

Roos, Y.H., K. Jouppila, and B. Zielasko. 1996a. Non-enzymatic browning-induced water plasticization. *J. Thermal Anal.* 47: 1437–1450.

Roos, Y.H., M. Karel, and J.L. Kokini. 1996b. Glass transitions in low moisture and frozen foods: Effect on shelf life and quality. *Food Technol.* 50 (10): 95–108.

Schofield, J.D., and M.R. Booth. 1983. Wheat proteins and their technological significance. In *Develop-* *ments in food proteins*, ed. B.J.F. Hudson. New York: Elsevier Applied Science Publishers.

Schönberg, A., and R. Moubacher. 1952. The Strecker degradation of α-amino acids. *Chem. Rev.* 50: 261–277.

Simmonds, D.H., and R.A. Orth. 1973. Structure and composition of cereal proteins as related to their potential industrial utilization. In *Industrial uses of cereals*, ed. Y. Pomeranz. St. Paul, MN: American Association of Cereal Chemists.

Stanley, D.W., and R.Y. Yada. 1992. Thermal reactions in food protein systems. In *Physical chemistry of foods*, ed. H.G. Schwartzberg and R.W. Hartel. New York: Marcel Dekker, Inc.

Swaisgood, H.E. 1982. Chemistry of milk protein. In *Developments in dairy chemistry*, ed. P. F. Fox. New York: Elsevier Applied Science Publishers.

Swaisgood, H.E. 1995. Protein and amino acid composition of bovine milk. In *Handbook of milk composition*, ed. R.G. Jensen. New York: Academic Press, Inc.

Thomas, R., J.M. deMan, and L. deMan. 1989. Soymilk and tofu properties as influenced by soybean storage conditions. *J. Am. Oil Chem. Soc.* 66: 777–782.

Wall, J.S. 1971. Disulfide bonds: Determination, location and influence on molecular properties of proteins. *Agr. Food Chem.* 19: 619–625.

Wolf, W.J. 1969. Soybean protein nomenclature: A progress report. *Cereal Sci. Today* 14, no. 3: 75 –78, 129.

Wolf, W.J. 1970. Scanning electron microscopy of soybean protein bodies. *J. Am. Oil Chem. Soc.* 47: 107–108.

Wolf, W.J. 1972a. Purification and properties of the proteins. In *Soybeans: Chemistry and technology*, ed. A.K. Smith and S.J. Circle. Westport, CT: AVI Publishing Co.

Wolf, W.J. 1972b. What is soy protein. *Food Technol.* 26, no. 5: 44–54.

Wong, D.W.S., W.M. Camirand, and A.E. Pavlath. 1996. Structures and functionalities of milk proteins. *Crit. Rev. Food Sci. Nutr.* 36: 807–844.

Ziegler, K. 1964. New cross links in alkali treated wool. *J. Biol. Chem.* 239: 2713–2714.

CHAPTER 4

Carbohydrates

INTRODUCTION

Carbohydrates occur in plant and animal tissues as well as in microorganisms in many different forms and levels. In animal organisms, the main sugar is glucose and the storage carbohydrate is glycogen; in milk, the main sugar is almost exclusively the disaccharide lactose. In plant organisms, a wide variety of monosaccharides and oligosaccharides occur, and the storage carbohydrate is starch. The structural polysaccharide of plants is cellulose. The gums are a varied group of polysaccharides obtained from plants, seaweeds, and microorganisms. Because of their useful physical properties, the gums have found widespread application in food processing. The carbohydrates that occur in a number of food products are listed in Table 4–1.

MONOSACCHARIDES

D-glucose is the most important monosaccharide and is derived from the simplest sugar, D-glyceraldehyde, which is classed as an aldotriose. The designation of aldose and ketose sugars indicates the chemical character of the reducing form of a sugar and can be indicated by the simple or open-chain formula of Fischer, as shown in Figure 4–1. This type of formula shows the free aldehyde group and four optically active secondary hydroxyls. Since the chemical reactions of the sugars do not correspond to this structure, a ring configuration involving a hemiacetal between carbons 1 and 5 more accurately represents the structure of the monosaccharides. The five-membered ring structure is called furanose; the six-membered ring, pyranose. Such rings are heterocyclic because one member is an oxygen atom. When the reducing group becomes involved in a hemiacetal ring structure, carbon 1 becomes asymmetric and two isomers are possible; these are called anomers.

Most natural sugars are members of the D series. The designation D or L refers to two series of sugars. In the D series, the highest numbered asymmetric carbon has the OH group directed to the right, in the Fischer projection formula. In the L series, this hydroxyl points to the left. This originates from the simplest sugars, D- and L-glyceraldehyde (Figure 4–2).

After the introduction of the Fischer formulas came the use of the Haworth representation, which was an attempt to give a more accurate spatial view of the molecule. Because the Haworth formula does not account for the actual bond angles, the modern con-

163

Table 4–1 Carbohydrates in Some Foods and Food Products

Product	Total Sugar (%)	Mono- and Disaccharides (%)	Polysaccharides (%)
Fruits			
Apple	14.5	glucose 1.17; fructose 6.04; sucrose 3.78; mannose trace	starch 1.5; cellulose 1.0
Grape	17.3	glucose 5.35; fructose 5.33; sucrose 1.32; mannose 2.19	cellulose 0.6
Strawberry	8.4	glucose 2.09; fructose 2.40; sucrose 1.03; mannose 0.07	cellulose 1.3
Vegetables			
Carrot	9.7	glucose 0.85; fructose 0.85; sucrose 4.25	starch 7.8; cellulose 1.0
Onion	8.7	glucose 2.07; fructose 1.09; sucrose 0.89	cellulose 0.71
Peanuts	18.6	sucrose 4–12	cellulose 2.4
Potato	17.1		starch 14; cellulose 0.5
Sweet corn	22.1	sucrose 12–17	cellulose 0.7; cellulose 60
Sweet potato	26.3	glucose 0.87; sucrose 2–3	starch 14.65; cellulose 0.7
Turnip	6.6	glucose 1.5; fructose 1.18; sucrose 0.42	cellulose 0.9
Others			
Honey	82.3	glucose 28–35; fructose 34–41; sucrose 1–5	
Maple syrup	65.5	sucrose 58.2–65.5; hexoses 0.0–7.9	
Meat		glucose 0.01	glycogen 0.10
Milk	4.9	lactose 4.9	
Sugarbeet	18–20	sucrose 18–20	
Sugar cane juice	14–28	glucose + fructose 4–8; sucrose 10–20	

formational formulas (Figure 4–1) more accurately represent the sugar molecule. A number of chair conformations of pyranose sugars are possible (Shallenberger and Birch 1975) and the two most important ones for glucose are shown in Figure 4–1. These are named the *CI* D and the *IC* D forms (also described as *O-outside* and *O-inside*, respectively). In the *CI* D form of β-D-gluco-pyranose, all hydroxyls are in the equatorial

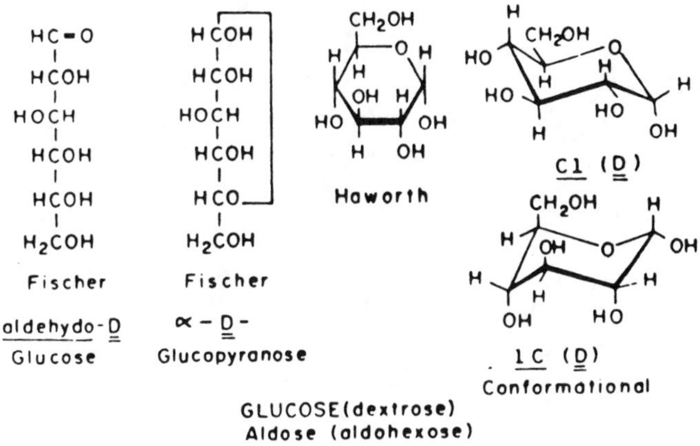

Figure 4–1 Methods of Representation of D-Glucose. *Source:* From M.L. Wolfrom, Physical and Chemical Structures of Carbohydrates, in *Symposium on Foods: Carbohydrates and Their Roles,* H.W. Schultz, R.F. Cain, and R.W. Wrolstad, eds., 1969, AVI Publishing Co.

position, which represents the highest thermodynamic stability.

The two possible anomeric forms of monosaccharides are designated by Greek letter prefix α or β. In the α-anomer the hydroxyl group points to the right, according to the Fischer projection formula; the hydroxyl group points to the left in the β-anomer. In Figure 4–1 the structure marked *C*1 D represents the α-anomer, and 1*C* D represents the β-anomer. The anomeric forms of the sugars are in tautomeric equilibrium in solution; and this causes the change in optical rotation when a sugar is placed in

Figure 4–2 Structure of D- and L-Glyceraldehyde. *Source:* From R.S. Shallenberger and G.G. Birch, *Sugar Chemistry,* 1975, AVI Publishing Co.

solution. Under normal conditions, it may take several hours or longer before the equilibrium is established and the optical rotation reaches its equilibrium value. At room temperature an aqueous solution of glucose can exist in four tautomeric forms (Angyal 1984): β-furanoside—0.14 percent, acyclic aldehyde—0.0026 percent, β-pyranoside—62 percent, and α-pyranoside—38 percent (Figure 4–3). Fructose under the same conditions also exists in four tautomeric forms as follows: α-pyranoside—trace, β-pyranoside—75 percent, α-furanoside—4 percent, and β-furanoside—21 percent (Figure 4–4) (Angyal 1976).

When the monosaccharides become involved in condensation into di-, oligo-, and polysaccharides, the conformation of the bond on the number 1 carbon becomes fixed and the different compounds have either an all-α or all-β structure at this position.

Naturally occurring sugars are mostly hexoses, but sugars with different numbers of carbons are also present in many products. There are also sugars with different func-

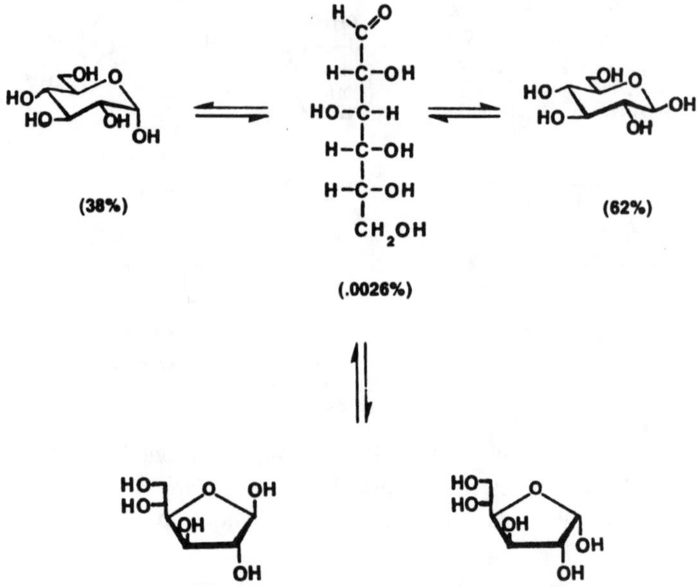

Figure 4–3 Tautomeric Forms of Glucose in Aqueous Solution at Room Temperature

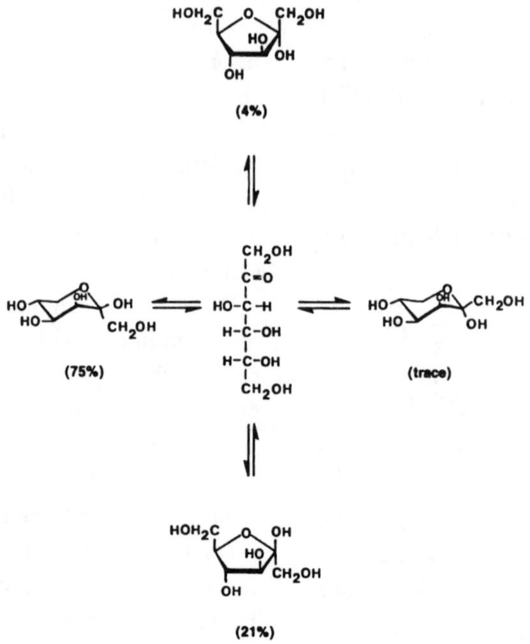

Figure 4–4 Tautomeric Forms of Fructose in Aqueous Solution at Room Temperature

tional groups or substituents; these lead to such diverse compounds as aldoses, ketoses, amino sugars, deoxy sugars, sugar acids, sugar alcohols, acetylated or methylated sugars, anhydro sugars, oligo- and polysaccharides, and glycosides. Fructose is the most widely occurring ketose and is shown in its various representations in Figure 4–5. It is the sweetest known sugar and occurs bound to glucose in sucrose or common sugar. Of all the other possible hexoses only two occur widely—D-mannose and D-galactose. Their formulas and relationship to D-glucose are given in Figure 4–6.

RELATED COMPOUNDS

Amino sugars usually contain D-glucosamine (2-deoxy-2-amino glucose). They occur as components of high molecular weight compounds such as the chitin of crustaceans and mollusks, as well as in certain mushrooms and in combination with the ovomucin of egg white.

Glycosides are sugars in which the hydrogen of an anomeric hydroxy group has been replaced by an alkyl or aryl group to form a mixed acetal. Glycosides are hydrolyzed by acid or enzymes but are stable to alkali. Formation of the full acetal means that glycosides have no reducing power. Hydrolysis of glycosides yields sugar and the aglycone. Amygdalin is an example of one of the cyanogetic glycosides and is a component of bitter almonds. The glycone moiety of this compound is gentiobiose, and complete hydrolysis yields benzaldehyde, hydrocyanic acid, and glucose (Figure 4–7). Other important glycosides are the flavonone glycosides of citrus rind, which include hesperidin and naringin, and the mustard oil glycosides, such as sinigrin, which is a component of mustard and horseradish. Deoxy sugars occur as components of nucleotides; for example, 2-deoxyribose constitutes part of deoxyribonucleic acid.

Sugar alcohols occur in some fruits and are produced industrially as food ingredients.

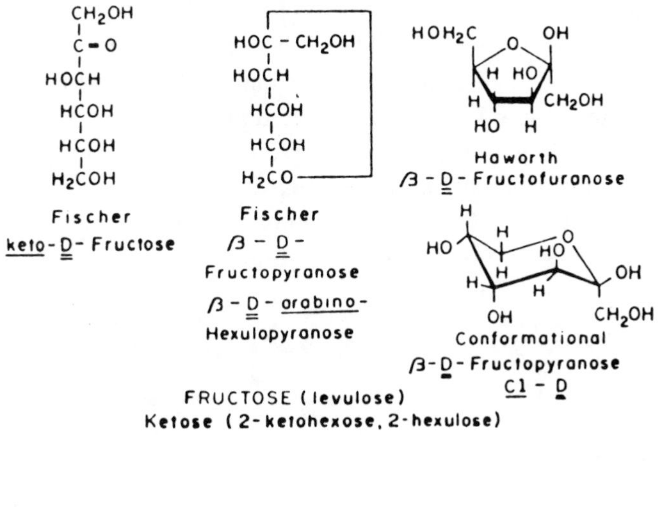

Figure 4–5 Methods of Representation of D-Fructose. *Source:* From M.L. Wolfrom, Physical and Chemical Structures of Carbohydrates, in *Symposium on Foods: Carbohydrates and Their Roles,* H.W. Schultz, R.F. Cain, and R.W. Wrolstad, eds., 1969, AVI Publishing Co.

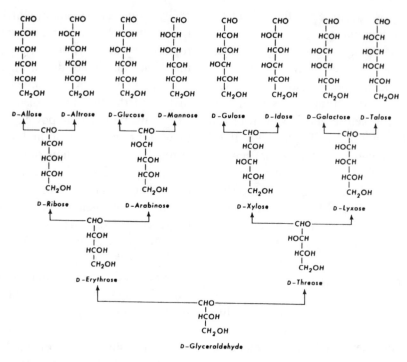

Figure 4–6 Relationship of D-Aldehyde Sugars. *Source:* From M.L. Wolfrom, Physical and Chemical Structures of Carbohydrates, in *Symposium on Foods: Carbohydrates and Their Roles,* H.W. Schultz, R.F. Cain, and R.W. Wrolstad, eds., 1969, AVI Publishing Co.

They can be made by reduction of free sugars with sodium amalgam and lithium aluminum hydride or by catalytic hydrogenation. The resulting compounds are sweet as sugars, but are only slowly absorbed and can, therefore, be used as sweeteners in diabetic foods. Reduction of glucose yields glucitol (Figure 4–8), which has the trivial name sorbitol. Another commercially produced sugar alcohol is xylitol, a five-carbon compound, which is also used for diabetic foods (Figure 4–8). Pentitols and hexitols are widely distributed in many foods, especially fruits and vegetables (Washüttl et al. 1973), as is indicated in Table 4–2.

Anhydro sugars occur as components of seaweed polysaccharides such as alginate and agar. Sugar acids occur in the pectic sub-

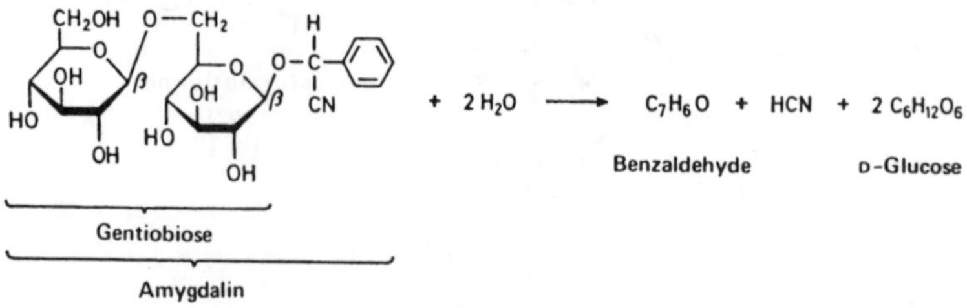

$$+ \ 2\,H_2O \longrightarrow C_7H_6O \ + \ HCN \ + \ 2\,C_6H_{12}O_6$$

Benzaldehyde   D-Glucose

Gentiobiose

Amygdalin

Figure 4–7 Hydrolysis of the Glycoside Amygdalin

Table 4–2 Occurrence of Sugar-Alcohols in Some Foods (Expressed as mg/100g of Dry Food)

Product	Arabitol	Xylitol	Mannitol	Sorbitol	Galactitol
Bananas	—	21	—	—	—
Pears	—	—	—	4600	—
Raspberries	—	268	—	—	—
Strawberries	—	362	—	—	—
Peaches	—	—	—	960	—
Celery	—	—	4050	—	—
Cauliflower	—	300	—	—	—
White mushrooms	340	128	476	—	48

Source: From J. Washüttl, P. Reiderer, and E. Bancher, A Qualitative and Quantitative Study of Sugar-Alcohols in Several Foods: A Research Role, *J. Food Sci.*, Vol. 38, pp. 1262–1263, 1973.

stances. When some of the carboxyl groups are esterified with methanol, the compounds are known as pectins. By far the largest group of saccharides occurs as oligo- and polysaccharides.

OLIGOSACCHARIDES

Polymers of monosaccharides may be either of the homo- or hetero-type. When the number of units in a glycosidic chain is in the range of 2 to 10, the resulting compound is an oligosaccharide. More than 10 units are usually considered to constitute a polysac-

charide. The number of possible oligosaccharides is very large, but only a few are found in large quantities in foods; these are listed in Table 4–3. They are composed of the monosaccharides D-glucose, D-galactose, and D-fructose, and they are closely related to one another, as shown in Figure 4–9.

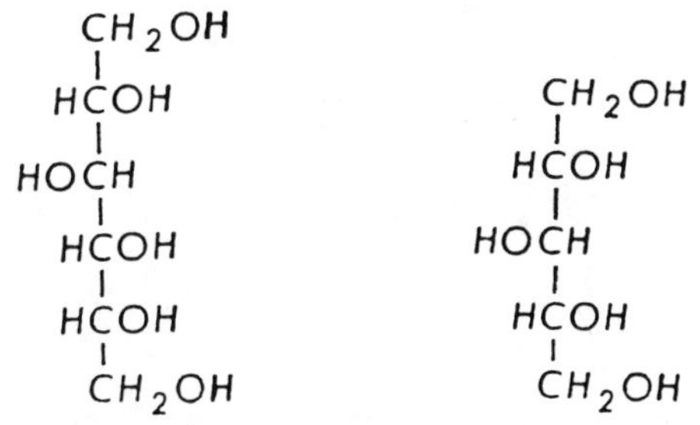

Sucrose or ordinary sugar occurs in abundant quantities in many plants and is commercially obtained from sugar cane or sugar beets. Since the reducing groups of the monosaccharides are linked in the glycosidic bond, this constitutes one of the few nonreducing disaccharides. Sucrose, therefore, does not reduce Fehling solution or form osazones and it does not undergo mutarotation in solution. Because of the unique carbonyl-to-carbonyl linkage, sucrose is highly labile in acid medium, and acid hydrolysis is more rapid than with other oligosaccharides. The structure of sucrose is shown in Figure 4–10. When sucrose is heated to 210°C, partial decomposition takes place and caramel is formed. An important reaction of sucrose,

$$
\begin{array}{cc}
CH_2OH & \\
| & \\
HCOH & CH_2OH \\
| & | \\
HOCH & HCOH \\
| & | \\
HCOH & HOCH \\
| & | \\
HCOH & HCOH \\
| & | \\
CH_2OH & CH_2OH
\end{array}
$$

Figure 4–8 Structure of Sorbitol and Xylitol

Table 4–3 Common Oligosaccharides Occurring in Foods

Sucrose	(α-D-glucopyranosyl β-D-fructofuranoside)
Lactose	(4-O-β-D-galactopyranosyl-D-glucopyranose)
Maltose	(4-O-α-D-glucopyranosyl-D-glucopyranose)
α,α-Trehalose	(α--D-glucopyranosyl-α-D-glycopyranoside)
Raffinose	[O-α-D-galactopyranosyl-(1→6)-O-α-D-glucopyranosyl-(1→2)-β-D-fructofuranoside]
Stachyose	[O-α-D-galactopyranosyl-(1→6)-O-α-D-galactopyranosyl-(1→6)-O-α-D-glucopyranosyl-(1→2)-β-D-fructofuranoside]
Verbascose	[O-α-D-galactopyranosyl-(1→6)-O-α-D-galactopyranosyl-(1→6)-O-α-D-galactopyranosyl-(1→6)-O-α-D-glucopyranosyl-(1→2)-β-D-fructofuranoside]

Source: From R.S. Shallenberger and G.G. Birch, *Sugar Chemistry,* 1975, AVI Publishing Co.

which it has in common with other sugars, is the formation of insoluble compounds with calcium hydroxide. This reaction results in the formation of tricalcium compounds $C_{12}H_{22}O_{11} \cdot 3 \ Ca(OH)_2$ and is useful for recovering sucrose from molasses. When the calcium saccharate is treated with CO_2, the sugar is liberated.

Hydrolysis of sucrose results in the formation of equal quantities of D-glucose and D-

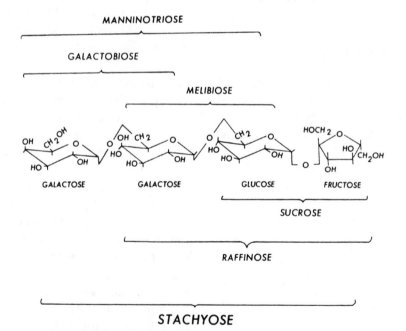

Figure 4–9 Composition of Some Major Oligosaccharides Occurring in Foods. *Source:* From R.S. Shallenberger and G.G. Birch, *Sugar Chemistry,* 1975, AVI Publishing Co.

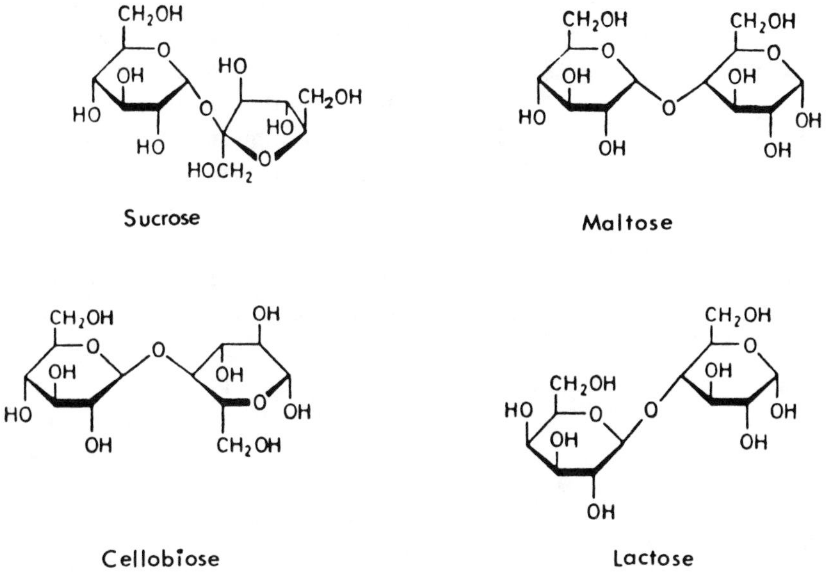

Figure 4–10 Structure of Some Important Disaccharides

fructose. Since the specific rotation of sucrose is +66.5°, of D-glucose +52.2°, and of D-fructose –93°, the resulting invert sugar has a specific rotation of –20.4°. The name invert sugar refers to the inversion of the direction of rotation.

Sucrose is highly soluble over a wide temperature range, as is indicated in Figure 4–11. This property makes sucrose an excellent ingredient for syrups and other sugar-containing foods.

The characteristic carbohydrate of milk is lactose or milk sugar. With a few minor exceptions, lactose is the only sugar in the milk of all species and does not occur elsewhere. Lactose is the major constituent of the dry matter of cow's milk, as it represents close to 50 percent of the total solids. The lactose content of cow's milk ranges from 4.4 to 5.2 percent, with an average of 4.8 percent expressed as anhydrous lactose. The lactose content of human milk is higher, about 7.0 percent.

Lactose is a disaccharide of D-glucose and D-galactose and is designated as 4-*O*-β-D-galactopyranosyl-D-glucopyranose (Figure 4–10). It is hydrolyzed by the enzyme β-D-galactosidase (lactase) and by dilute solutions of strong acids. Organic acids such as citric acid, which easily hydrolyze sucrose, are unable to hydrolyze lactose. This difference is the basis of the determination of the two sugars in mixtures.

Maltose is 4-α-D-glucopyranosyl-β-D-glucopyranose. It is the major end product of the enzymic degradation of starch and glycogen by β-amylase and has a characteristic flavor of malt. Maltose is a reducing disaccharide, shows mutarotation, is fermentable, and is easily soluble in water.

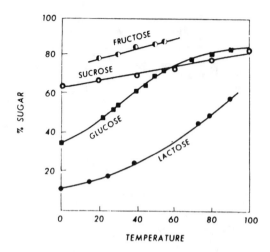

Figure 4–11 Approximate Solubility of Some Sugars at Different Temperatures. *Source:* From R.S. Shallenberger and G.G. Birch, *Sugar Chemistry*, 1975, AVI Publishing Co.

Cellobiose is 4-β-D-glucopyranosyl-β-D-glucopyranose, a reducing disaccharide resulting from partial hydrolysis of cellulose.

Legumes contain several oligosaccharides, including raffinose and stachyose. These sugars are poorly absorbed when ingested, which results in their fermentation in the large intestine. This leads to gas production and flatulence, which present a barrier to wider food use of such legumes. deMan et al. (1975 and 1987) analyzed a large number of soybean varieties and found an average content of 1.21 percent stachyose, 0.38 percent raffinose, 3.47 percent sucrose, and very small amounts of melibose. In soy milk, total reducing sugars after inversion amounted to 11.1 percent calculated on dry basis.

Cow's milk contains traces of oligosaccharides other than lactose. They are made up of two, three, or four units of lactose, glucose, galactose, neuraminic acid, mannose, and acetyl glucosamine. Human milk contains about 1 g/L of these oligosaccharides, which are referred to as the bifidus factor. The oligosaccharides have a beneficial effect on the intestinal flora of infants.

Fructooligosaccharides (FOSs) are oligomers of sucrose where an additional one, two, or three fructose units have been added by a β-(2-1)-glucosidic linkage to the fructose unit of sucrose. The resulting FOSs, therefore, contain two, three, or four fructose units. The FOSs occur naturally as components of edible plants including banana, tomato, and onion (Spiegel et al. 1994). FOSs are also manufactured commercially by the action of a fungal enzyme from *Aspergillus niger*, β-fructofuranosidase, on sucrose. The three possible FOSs are 1^F-(1-β-fructofuranosyl)$_{n-1}$ sucrose oligomers with abbreviated and common names as follows: GF_2 (1-kestose), GF_3 (nystose), and GF_4 (1^F-β-fructofuranosyl-nystose). The commercially manufactured product is a mixture of all three FOSs with sucrose, glucose, and fructose. FOSs are nondigestible by humans and are suggested to have some dietary fiber-like function.

Chemical Reactions

Mutarotation

When a crystalline reducing sugar is placed in water, an equilibrium is established between isomers, as is evidenced by a relatively slow change in specific rotation that eventually reaches the final equilibrium value. The working hypothesis for the occurrence of mutarotation has been described by Shallenberger and Birch (1975). It is assumed that five structural isomers are possible for any given reducing sugar (Figure 4–12), with pyranose and furanose ring structures being generated from a central straight-chain inter-

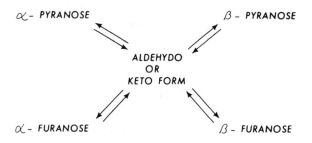

Figure 4–12 Equilibria Involved in Mutarotation. *Source:* From R.S. Shallenberger and G.G. Birch, *Sugar Chemistry,* 1975, AVI Publishing Co.

mediate. When all of these forms are present, the mutarotation is complex. When only the pyranose forms are present, the mutarotation is simple. Aldoses that have the *gluco, manno, gulo,* and *allo* configurations (Figure 4–6) exhibit simple mutarotation. D-glucose, for example, shows simple mutarotation and in aqueous solution only two forms are present, 36 percent α-D-glucopyranose and 64 percent β-D-glucopyranose. The amount of aldehyde form of glucose in solution has been estimated at 0.003 percent. The distribution of isomers in some mutarotated monosaccharides at 20°C is shown in Table 4–4. The distribution of α- and β-anomers in solutions of lactose and maltose is nearly the same as in glucose, about 32 percent α- and 64 percent β-anomer. Simple mutarotation is a first-order reaction characterized by uniform values of the reaction constants k_1 and k_2 in the equation

$$\alpha\text{-D-glucopyranose} \underset{k_2}{\overset{k_1}{\rightleftharpoons}} \beta\text{-D-glucopyranose}$$

The velocity of the reaction is greatly accelerated by acid or base. The rate is at a minimum for pyranose-pyranose interconversions in the pH range 2.5 to 6.5. Both acids and bases accelerate mutarotation rate, with bases being more effective. This was expressed by Hudson (1907) in the following equation:

$$K_{25}° = 0.0096 + 0.258 \, [H^+] + 9.750 \, [OH^-]$$

Table 4–4 Percentage Distribution of Isomers of Mutarotated Sugars at 20°C

Sugar	α-Pyranose	β-Pyranose	α-Furanose	β-Furanose
D-Glucose	31.1–37.4	64.0–67.9	—	—
D-Galactose	29.6–35.0	63.9–70.4	1.0	3.1
D-Mannose	64.0–68.9	31.1–36.0	—	—
D-Fructose	4.0?	68.4–76.0	—	28.0–31.6

Source: From R.S. Shallenberger and G.G. Birch, *Sugar Chemistry,* AVI Publishing Co.

This indicates that the effect of the hydroxyl ion is about 40,000 times greater than that of the hydrogen ion. The rate of mutarotation is also temperature dependent; increases from 1.5 to 3 times occur for every 10°C rise in temperature.

Other Reactions

Sugars in solution are unstable and undergo a number of reactions. In addition to mutarotation, which is the first reaction to occur when a sugar is dissolved, enolization and isomerization, dehydration and fragmentation, anhydride formation and polymerization may all take place. These reactions are outlined in Figure 4–13, using glucose as an example. Compounds (1) and (2) are the α and β forms in equilibrium during mutarotation with the aldehydo form (5). Heating results in dehydration of the *IC* conformation of β-D glucopyranose (3) and formation of levoglucosan (4), followed by the

sequence of reactions described under caramelization. Enolization is the formation of an enediol (6). These enediols are unstable and can rearrange in several ways. Since the reactions are reversible, the starting material can be regenerated. Other possibilities include formation of keto-D-fructose (10) and β-D-fructopyranose (11), and aldehydo-D-mannose (8) and α-D-mannopyranose (9). Another possibility is for the double bond to move down the carbon chain to form another enediol (7). This compound can give rise to saccharinic acids (containing one carboxyl group) and to 5-(hydroxy)-methylfurfural (13). All these reactions are greatly influenced by pH. Mutarotation, enolization, and formation of succharic acid (containing two carboxyl groups) are favored by alkaline pH, formation of anhydrides, and furaldehydes by acid pH.

It appears from the aforementioned reactions that on holding a glucose solution at alkaline pH, a mixture of glucose, mannose,

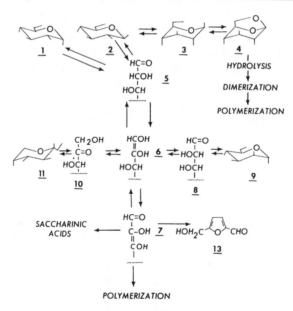

Figure 4–13 Reactions of Reducing Sugars in Solution. *Source:* From R.S. Shallenberger and G.G. Birch, *Sugar Chemistry,* 1975, AVI Publishing Co.

and fructose will be formed and, in general, any one sugar will yield a mixture of sugars. When an acid solution of sugar of high concentration is left at ambient temperature, reversion takes place. This is the formation of disaccharides. The predominant linkages in the newly formed disaccharides are α-D-1→6, and β-D-1→6. A list of reversion disaccharides observed by Thompson et al. (1954) in a 0.082N hydrochloric acid solution or in D-glucose is shown in Table 4–5.

Caramelization

The formation of the caramel pigment can be considered a nonenzymatic browning reaction in the absence of nitrogenous compounds. When sugars are subjected to heat in the absence of water or are heated in concentrated solution, a series of reactions occurs that finally leads to caramel formation. The initial stage is the formation of anhydro sugars (Shallenberger and Birch 1975). Glucose yields glucosan (1,2-anhydro-α-D-glucose) and levoglucosan (1,6-anhydro-β-D-glucose); these have widely differing specific rotation, +69° and −67°, respectively. These compounds may dimerize to form a number of reversion disaccharides, including gentio-

biose and sophorose, which are also formed when glucose is melted.

Caramelization of sucrose requires a temperature of about 200°C. At 160°C, sucrose melts and forms glucose and fructose anhydride (levulosan). At 200°C, the reaction sequence consists of three distinct stages well separated in time. The first step requires 35 minutes of heating and involves a weight loss of 4.5 percent, corresponding to a loss of one molecule of water per molecule of sucrose. This could involve formation of compounds such as isosacchrosan. Pictet and Stricker (1924) showed that the composition of this compound is 1,3′; 2,2′-dianhydro-α-D-glucopyranosyl-β-D-glucopyranosyl-β-D-fructofuranose (Figure 4–14). After an additional 55 minutes of heating, the weight loss amounts to 9 percent and the pigment formed is named caramelan. This corresponds approximately to the following equation:

$$2C_{12}H_{22}O_{11} - 4H_2O \rightarrow C_{24}H_{36}O_{18}$$

The pigment caramelan is soluble in water and ethanol and has a bitter taste. Its melting point is 138°C. A further 55 minutes of heating leads to the formation of caramelen. This compound corresponds to a weight loss of

Table 4–5 Reversion Disaccharides of Glucose in 0.082N HCl

β, β-trehalose (β-D-glucopyranosyl β-D-glucopyranoside)	0.1%
β-sophorose (2-O-β-D-glucopyranosyl-β-D-glucopyranose)	0.2%
β-maltose (4-O-α-D-glycopyranosyl-β-D-glycopyranose)	0.4%
α-cellobiose (4-O-β-D-glucopyranosyl-α-D-glucopyranose)	0.1%
β-cellobiose (4-O-β-D-glucopyranosyl-β-D-glucopyranose)	0.3%
β-isomaltose (6-O-α-D-glucopyranosyl-β-D-glucopyranose)	4.2%
α-gentiobiose (6-O-β-D-glucopyranosyl-α-D-glucopyranose)	0.1%
β-gentiobiose (6-O-β-D-glucopyranosyl-β-D-glucopyranose)	3.4%

Source: From A. Thompson et al., Acid Reversion Products from D-Glucose, *J. Am. Chem. Soc.*, Vol. 76, pp. 1309–1311, 1954.

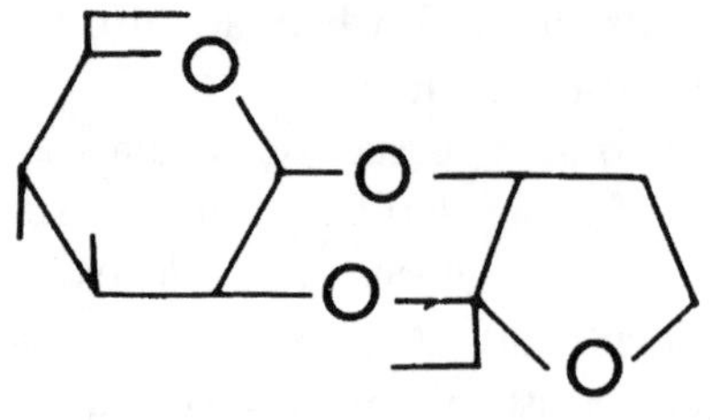

Figure 4–14 Structure of Isosacchrosan. *Source:* From R.S. Shallenberger and G.G. Birch, *Sugar Chemistry,* 1975, AVI Publishing Co.

about 14 percent, which is about eight molecules of water from three molecules of sucrose, as follows:

$$3C_{12}H_{22}O_{11} - 8H_2O \rightarrow C_{36}H_{50}O_{25}$$

Caramelen is soluble in water only and melts at 154°C. Additional heating results in the formation of a very dark, nearly insoluble pigment of average molecular composition $C_{125}H_{188}O_{80}$. This material is called humin or caramelin.

The typical caramel flavor is the result of a number of sugar fragmentation and dehydration products, including diacetyl, acetic acid, formic acid, and two degradation products reported to have typical caramel flavor by Jurch and Tatum (1970), namely, acetylformoin (4-hydroxy-2,3,5-hexane-trione) and 4-hydroxy-2,5-dimethyl-3(2H)-furanone.

Crystallization

An important characteristic of sugars is their ability to form crystals. In the commercial production of sugars, crystallization is an important step in the purification of sugar. The purer a solution of a sugar, the easier it will crystallize. Nonreducing oligosaccharides crystallize relatively easily. The fact that certain reducing sugars crystallize with more difficulty has been ascribed to the presence of anomers and ring isomers in solution, which makes these sugars intrinsically "impure" (Shallenberger and Birch 1975). Mixtures of sugars crystallize less easily than single sugars. In certain foods, crystallization is undesirable, such as the crystallization of lactose in sweetened condensed milk or ice cream.

Factors that influence growth of sucrose crystals have been listed by Smythe (1971). They include supersaturation of the solution, temperature, relative velocity of crystal and solution, nature and concentration of impurities, and nature of the crystal surface. Crystal growth of sucrose consists of two steps: (1) the mass transfer of sucrose molecules to the surface of the crystal, which is a first-order process; and (2) the incorporation of the molecules in the crystal surface, a second-order process. Under usual conditions, overall growth rate is a function of the rate of both processes, with neither being rate-controlling. The effect of impurities can be of two kinds. Viscosity can increase, thus reducing the rate of mass transfer, or impurities can involve adsorption on specific surfaces of the crystal, thereby reducing the rate of surface incorporation.

The crystal structure of sucrose has been established by X-ray diffraction and neutron diffraction studies. The packing of sucrose molecules in the crystal lattice is determined mainly by hydrogen bond formation between hydroxyl groups of the fructose moiety. As an example of the type of packing of molecules in a sucrose crystal, a projection of the crystal structure along the *a* axis is shown in Figure 4–15. The dotted square represents one unit cell. The crystal faces indicated in this figure follow planes between adjacent sucrose molecules in such a way that the

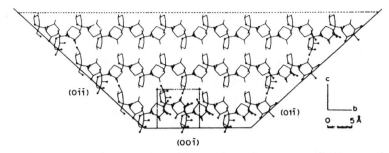

Figure 4–15 Projection of a Sucrose Crystal Along the *a* Axis. *Source:* From B.M. Smythe, Sucrose Crystal Growth, *Sugar Technol. Rev.,* Vol. 1, pp. 191–231, 1971.

furanose and pyranose rings are not intersected.

Lactose can occur in two crystalline forms, the α-hydrate and the β-anhydrous forms and can occur in an amorphous or glassy state. The most common form is the α-hydrate $(C_{12}H_{22}O_{11}\cdot H_2O)$, which can be obtained by crystallization from a supersaturated solution below 93.5°C. When crystallization is carried out above 93.5°C, the crystals formed are of β-anhydrous type. Some properties of these forms have been listed by Jenness and Patton (1959) (Table 4–6). Under normal conditions the α-

hydrate form is the stable one, and other solid forms spontaneously change to that form provided sufficient water is present. At equilibrium and at room temperature, the β-form is much more soluble and the amount of α-form is small. However, because of its lower solubility, the α-hydrate crystallizes out and the equilibrium shifts to convert β- into α-hydrate. The solubility of the two forms and the equilibrium mixture is represented in Figure 4–16.

The solubility of lactose is less than that of most other sugars, which may present problems in a number of foods containing lac-

Table 4–6 Some Physical Properties of the Two Common Forms of Lactose

Property	α-Hydrate	β-Anhydride
Melting point[1]	202°C (dec.)	252°C (dec.)
Specific rotation[2] $[\alpha]_D^{20}$	+89.4°	+35°
Solubility (g/100 mL) Water at 20°C	8	55
Water at 100°C	70	95
Specific gravity (20°C)	1.54	1.59
Specific heat	0.299	0.285
Heat of combustion (cal/g^{-1})	3761.6	3932.7

[1] Values vary with rate of heating, α-hydrate losses H_2O (120°C).

[2] Values on anhydrous basis, both forms mutarotate to +55.4°.

Source: From R. Jenness and S. Patton, *Principles of Dairy Chemistry,* 1959, John Wiley and Sons.

tose. When milk is concentrated 3:1, the concentration of lactose approaches its final solubility. When this product is cooled or when sucrose is added, crystals of α-hydrate may develop. Such lactose crystals are very hard and sharp; when left undisturbed they may develop to a large size, causing a sensation of grittiness or sandiness in the mouth. This same phenomenon limits the amount of milk solids that can be incorporated into ice cream.

The crystals of α-hydrate lactose usually occur in a prism or tomahawk shape. The latter is the basic shape and all other shapes are derived from it by different relative growth rates of the various faces. The shape of an α-hydrate lactose crystal is shown in Figure 4–17. The crystal has been character-

ized by X-ray diffraction, and the following constants for the dimensions of the unit cell and one of the axial angles have been established: $a = 0.798$ nm, $b = 2.168$ nm, $c = 0.4836$ nm, and $\beta = 109°\ 47'$. The crystallographic description of the crystal faces is indicated in Figure 4–17. These faces grow at different rates; the more a face is oriented toward the β direction, the slower it grows and the $(0\bar{1}0)$ face does not grow at all.

Amorphous or glassy lactose is formed when lactose-containing solutions are dried quickly. The dry lactose is noncrystalline and contains the same ratio of alpha/beta as the

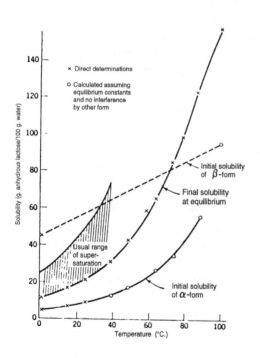

Figure 4–16 Solubility of Lactose in Water. *Source:* From E.O. Whittier, Lactose and Its Utilization: A Review, *J. Dairy Sci.,* Vol. 27, p. 505, 1944.

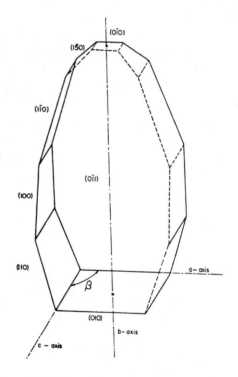

Figure 4–17 Crystallographic Representation of a Tomahawk Crystal of α-Lactose Monohydrate. *Source:* From A. Van Kreveld and A.S. Michaels, Measurement of Crystal Growth of α=Lactose, *J. Dairy Sci.,* Vol. 48, pp. 259–265, 1965.

original product. This holds true for spray or roller drying of milk products and also during drying for moisture determination. The glassy lactose is extremely hygroscopic and takes up moisture from the atmosphere. When the moisture content reaches about 8 percent, the lactose molecules recrystallize and form α-hydrate crystals. As these crystals grow, powdered products may cake and become lumpy.

Both lactose and sucrose have been shown to crystallize in an amorphous form at moisture contents close to the glass transition temperature (Roos and Karel 1991a,b; Roos and Karel 1992). When amorphous lactose is held at constant water content, crystallization releases water to the remaining amorphous material, which depresses the glass transition temperature and accelerates crystallization. These authors have done extensive studies on the glass transition of amorphous carbohydrate solutions (Roos 1993; Roos and Karel 1991d).

Seeding is a commonly used procedure to prevent the slow crystallization of lactose and the resulting sandiness in some dairy products. Finely ground lactose crystals are introduced into the concentrated product, and these provide numerous crystal nuclei. Many small crystals are formed rapidly; therefore, there is no opportunity for crystals to slowly grow in the supersaturated solution until they would become noticeable in the mouth.

Starch Hydrolyzates—Corn Sweeteners

Starch can be hydrolyzed by acid or enzymes or by a combination of acid and enzyme treatments. A large variety of products can be obtained from starch hydrolysis

using various starches such as corn, wheat, potato, and cassava (tapioca) starch. Glucose syrups, known in the United States as corn syrup, are hydrolysis products of starch with varying amounts of glucose monomer, dimer, oligosaccharides, and polysaccharides. Depending on the method of hydrolysis used, different compositions with a broad range of functional properties can be obtained. The degree of hydrolysis is expressed as dextrose equivalent (DE), defined as the amount of reducing sugars present as dextrose and calculated as a percentage of the total dry matter. Glucose syrups have a DE greater than 20 and less than 80. Below DE 20 the products are referred to as maltodextrins and above DE 80 as hydrolyzates. The properties of maltodextrins are influenced by the nature of the starch used; those of hydrolyzates are not affected by the type of starch.

The initial step in starch hydrolysis involves the use of a heat-stable endo-α-amylase. This enzyme randomly attacks α-1, 4 glycosidic bonds resulting in rapid decrease in viscosity. These enzymes can be used at temperatures as high as 105°C. This reaction produces maltodextrins (Figure 4–18), which can be used as important functional food ingredients—fillers, stabilizers, and thickeners. The next step is saccharification by using a series of enzymes that hydrolyze either the α-1,4 linkages of amylose or the α-1,6 linkages of the branched amylopectin. The action of the various starch-degrading enzymes is shown in Figure 4–19 (Olsen 1995). In addition to products containing high levels of glucose (95 to 97 percent), sweeteners with DE of 40 to 45 (maltose), 50 to 55 (high maltose), and 55 to 70 (high conversion syrup) can be produced. High dextrose syrups can be obtained by saccharification with amyloglucosidase. At the beginning of the reaction

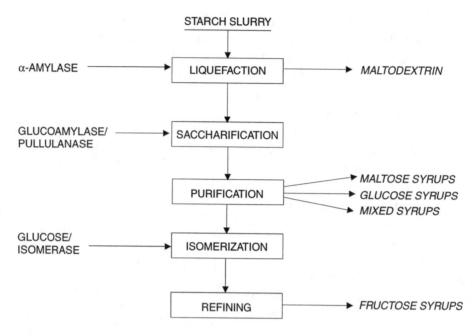

Figure 4–18 Major Steps in Enzymic Starch Conversion. *Source:* Reprinted from H.S. Olsen, Enzymic Production of Glucose Syrups, in *Handbook of Starch Hydrolysis Products and Their Derivatives*, M.W. Kearsley and S.Z. Dziedzic, eds., p. 30, © 1995, Aspen Publishers, Inc.

dextrose formation is rapid but gradually slows down. This slowdown is caused by formation of branched dextrins and because at high dextrose level the repolymerization of dextrose into isomaltose occurs.

The isomerization of glucose to fructose opened the way for starch hydrolyzates to replace cane or beet sugar (Dziezak 1987). This process is done with glucose isomerase in immobilized enzyme reactors. The conversion is reversible and the equilibrium is at 50 percent conversion. High-fructose corn syrups are produced with 42 or 55 percent fructose. These sweeteners have taken over one-third of the sugar market in the United States (Olsen 1995).

The acid conversion process has a practical limit of 55 DE; above this value, dark color and bitter taste become prominent. Depending on the process used and the reaction con-

ditions employed, a variety of products can be obtained as shown in Table 4–7 (Commerford 1974). There is a fairly constant relationship between the composition of acid-converted corn syrup and its DE. The composition of syrups made by acid-enzyme or dual-enzyme processes cannot be as easily predicted from DE.

Maltodextrins (DE below 20) have compositions that reflect the nature of the starch used. This depends on the amylose/amylopectin ratio of the starch. A maltodextrin with DE 12 shows retrogradation in solution, producing cloudiness. A maltodextrin from waxy corn at the same DE does not show retrogradation because of the higher level of α-1, 6 branches. As the DE decreases, the differences become more pronounced. A variety of maltodextrins with different functional properties, such as gel formation, can be

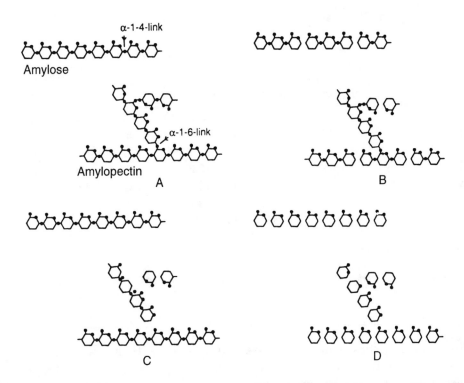

Figure 4–19 Schematic Representation of the Action of Starch-Degrading Enzymes. (A) Amylose and amylopectin, (B) action of α-amylase on amylose and amylopectin, (C) action of a debranching enzyme on amylose and amylopectin, (D) action of amyloglucosidase and debranching enzyme on amylose and amylopectin. *Source:* Reprinted from H.S. Olsen, Enzymic Production of Glucose Syrups, in *Handbook of Starch Hydrolysis Products and Their Derivatives*, M.W. Kearsley and S.Z. Dziedzic, eds., p. 36, © 1995, Aspen Publishers, Inc.

---

obtained by using different starch raw materials.

Maltodextrins of varying molecular weights are plasticized by water and decrease the glass transition temperature. Maltodextrins retard the crystallization of amorphous sucrose and at high concentrations totally inhibit sucrose crystallization (Roos and Karel 1991c).

Maltodextrins with low DE and with little or no remaining polysaccharide can be produced by using two enzymes. Alpha-amylase randomly hydrolyzes $1 \rightarrow 4$ linkages to reduce the viscosity of the suspension. Pullu-

lanase is specific for $1 \rightarrow 6$ linkages and acts as a debranching enzyme. The application of these two enzymes makes it possible to produce maltodextrins in high yield (Kennedy et al. 1985).

Polyols

Polyols or sugar alcohols occur in nature and are produced industrially from the corresponding saccharides by catalytic hydrogenation. Sorbitol, the most widely distributed natural polyol, is found in many fruits such

Table 4–7 Composition of Representative Corn Syrups

Type of Conversion	Dextrose Equivalent	Saccharides (%)							
		Mono-	Di-	Tri-	Tetra-	Penta-	Hexa-	Hepta-	Higher
Acid	30	10.4	9.3	8.6	8.2	7.2	6.0	5.2	45.1
Acid	42	18.5	13.9	11.6	9.9	8.4	6.6	5.7	25.4
Acid-enzyme	43	5.5	46.2	12.3	3.2	1.8	1.5	—	29.5[1]
Acid	54	29.7	17.8	13.2	9.6	7.3	5.3	4.3	12.8
Acid	60	36.2	19.5	13.2	8.7	6.3	4.4	3.2	8.5
Acid-enzyme	63	38.8	28.1	13.7	4.1	4.5	2.6	—	8.2[1]
Acid-enzyme	71	43.7	36.7	3.7	3.2	0.8	4.3	—	7.6[1]

[1] Includes heptasaccharides.

Source: From J.D. Commerford, Corn Sweetener Industry, in *Symposium: Sweeteners*, I.E. Inglett, ed., 1974, AVI Publishing Co.

as plums, berries, cherries, apples, and pears. It is a component of fruit juices, fruit wines, and other fruit products. It is commercially produced by catalytic hydrogenation of D-glucose. Mannitol, the reduced form of D-mannose, is found as a component of mushrooms, celery, and olives. Xylitol is obtained from saccharification of xylan-containing plant materials; it is a pentitol, being the reduced form of xylose. Sorbitol, mannitol, and xylitol are monosaccharide-derived polyols with properties that make them valuable for specific applications in foods: they are suitable for diabetics, they are noncariogenic, they possess reduced physiological caloric value, and they are useful as sweeteners that are nonfermentable by yeasts.

In recent years disaccharide alcohols have become important. These include isomalt, maltitol, lactitol, and hydrogenated starch hydrolyzates (HSH). Maltitol is hydrogenated maltose with the structure shown in Figure 4–20. It has the highest sweetness of the disaccharidepolyols compared to sugar (Table 4–8) (Heume and Rapaille 1996). It has a low negative heat of solution and, therefore, gives no cooling effect in contrast to sorbitol and xylitol. It also has a very high viscosity in solution. Sorbitol and maltitol are derived from starch by the production process illustrated in Figure 4–21. Lactitol is a disaccharide alcohol, 1,4-galactosyl-glucitol, produced by hydrogenation of lactose. It has low sweetness and a lower energy value than other polyols. It has a calorie value of 2 kcal/g and is noncariogenic (Blankers 1995). It can be used in combination with intense sweeteners like aspartame or acesulfame-K to produce sweetening

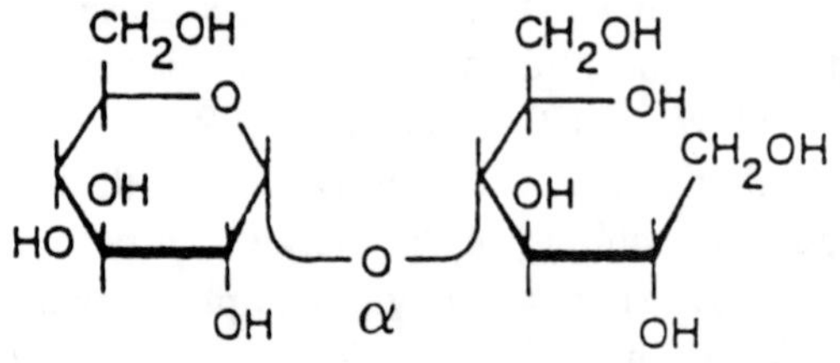

Figure 4–20 Structure of Maltitol

Table 4–8 Relative Sweetness of Polyols and Sucrose Solutions at 20°C

Compound	Relative Sweetness
Xylitol	80–100
Sorbitol	50–60
Mannitol	50–60
Maltitol	80–90
Lactitol	30–40
Isomalt	50–60
Sucrose	100

Source: Reprinted from H. Schiweck and S.C. Ziesenitz, Physiological Properties of Polyols in Comparison with Easily Metabolizable Saccharides, in *Advances in Sweeteners*, T.H. Grenby, ed., p. 87, © 1996, Aspen Publishers, Inc.

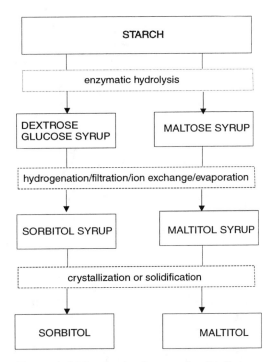

Figure 4–21 Production Process for the Conversion of Starch to Sorbitol and Maltitol. *Source:* Reprinted from H. Schiweck and S.C. Ziesenitz, Physiological Properties of Polyols in Comparison with Easily Metabolizable Saccharides, *Advances in Sweeteners*, T.H. Grenby, ed., p. 90, © 1996, Aspen Publishers, Inc.

power similar to sucrose. These combinations provide a milky, sweet taste that allows good perception of other flavors. Isomalt, also known as hydrogenated isomaltulose or hydrogenated palatinose, is manufactured in a two-step process: (1) the enzymatic transglycosylation of the nonreducing sucrose to the reducing sugar isomaltulose; and (2) hydrogenation, which produces isomalt—an equimolar mixture of D-glucopyranosyl-α-(1-1)-D-mannitol and D-glucopyranosyl-α-(1-6)-D-sorbitol. Isomalt is extremely stable and has a pure, sweet taste. Because it is only half as sweet as sucrose, it can be used as a versatile bulk sweetener (Ziesenitz 1996).

POLYSACCHARIDES

Starch

Starch is a polymer of D-glucose and is found as a storage carbohydrate in plants. It occurs as small granules with the size range and appearance characteristic to each plant species. The granules can be shown by ordi-

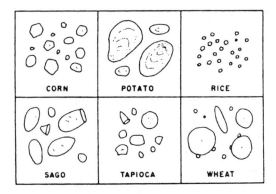

Figure 4–22 Appearance of Starch Granules as Seen in the Microscope

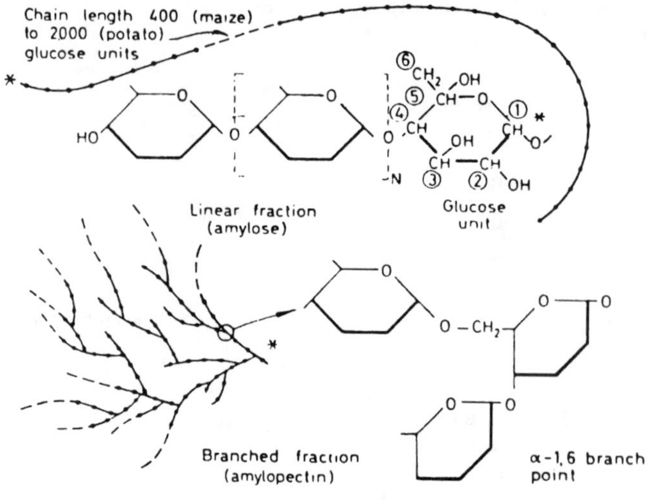

Figure 4–23 Structure of the Linear and Branched Fractions of Starch. *Source:* From J.A. Radley, Technical Properties of Starch as a Function of Its Structural Chemistry, in *Recent Advances in Food Science,* Vol. 3, J.M. Leitch and D.N. Rhodes, eds., 1963, Butterworth.

nary and polarized light microscopy and by X-ray diffraction to have a highly ordered crystalline structure (Figure 4–22).

Starch is composed of two different polymers, a linear compound, amylose, and a branched component, amylopectin (Figure 4–23). In the linear fraction the glucose units are joined exclusively by α-1→4 glucosidic bonds. The number of glucose units may range in various starches from a few hundred to several thousand units. In the most common starches, such as corn, rice, and potato, the linear fraction is the minor component and represents about 17 to 30 percent of the total. Some varieties of pea and corn starch may have as much as 75 percent amylose. The characteristic blue color of starch produced with iodine relates exclusively to the linear fraction. The polymer chain takes the form of a helix, which may form inclusion compounds with a variety of materials such as iodine. The inclusions of iodine are due to an induced dipole effect and consequent resonance along the helix. Each turn of the helix is made up of six glucose units and encloses one molecule of iodine. The length of the chain determines the color produced (Table 4–9).

Starch granules are partly crystalline; native starches contain between 15 and 45 percent crystallite material (Oates 1997). The

Table 4–9 The Color Produced by Reaction of Iodine with Amyloses of Different Chain Length

Chain Length	No. of Helix Turns	Color Produced
12	2	None
12–15	2	Brown
20–30	3–5	Red
35–40	6–7	Purple
<45	9	Blue

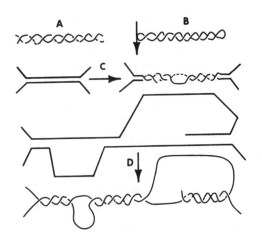

Figure 4–24 Double-Helix Formation in Starch. (A) Double helix from two molecules, (B) double helix from a single molecule, (C) alternate helix formation by central winding, (D) helix formation in large molecules. *Source:* Reprinted from L.H. Kruger and R. Murray, Starch Texture, in *Rheology and Texture in Food Quality,* J.M. deMan, P.W. Voisey, V.R. Rasper, and D.W. Stanley, eds., p. 436, © 1976, Aspen Publishers, Inc.

crystallinity can be demonstrated by X-ray diffraction techniques. Two polymorphic forms, A and B polymorphs, have been described. There is also an intermediate C form. Crystallinity results from intertwining of amylopectin chains with a linear component of over 10 glucose units to form a double helix (Figure 4–24). The double helices can associate in pairs to give either the A or B polymorphic structure. The A form is a face-centered monoclinic unit cell with 12 glucose residues in two left-handed chains containing four water molecules between the helices. The B form contains two left-handed, parallel-stranded, double helices, forming a hexagonal unit cell. The unit cell contains 12 glucose residues and 36 water molecules

(Gidley and Bociek 1985). Most cereal starches contain the A polymorph.

Amylopectin is branched because of the occurrence of α-1→6 linkages at certain points in the molecule. The branches are relatively short and contain about 20 to 30 glucose units. The outer branches can, therefore, give a red color with iodine. Certain types of cereal starch, such as waxy corn, contain only amylopectin.

The starch granule appears to be built up by deposition of layers around a central nucleus or hilum. Buttrose (1963) established that in some plants, shell formation of the starch granules is controlled by an endogenous rhythm (such as in potato starch), whereas in other plants (such as wheat starch), granule structure is controlled by environmental factors such as light and temperature. The starch granules differ in size and appearance: potato starch consists of relatively large egg-shaped granules with a diameter range of 15 to 100 μm, corn starch contains small granules of both round and angular appearance, and wheat starch also contains a diversity of sizes ranging from 2 to 35 μm. The granules show optical birefringence; that is, they appear light in the polarizing microscope between crossed filters. This property indicates some orderly orientation or crystallinity. The granules are completely insoluble in cold water and, upon heating, they suddenly start to swell at the so-called gelatinization temperature. At this point the optical birefringence disappears, indicating a loss of crystallinity.

Generally, starches with large granules swell at lower temperatures than those with small granules; potato starch swells at 59 to 67°C and corn starch at 64 to 72°C, although there are many exceptions to this rule. The swelling temperature is influenced by a variety of factors, including pH, pretreatment, heating rate, and presence of salts and sugar. Continuation of heating above the gelatinization temperature results in further swelling of the granule, and the mixture becomes viscous and translucent. In a boiled starch paste, the swollen granules still retain their identity although the birefringence is lost and the particle cannot be easily seen under the microscope. When such a paste is agitated, the granule structure breaks down and the viscosity is greatly reduced. When a cooked starch paste is cooled, it may form a gel or, under conditions of slow cooling, the linear component may form a precipitate of spherocrystals (Figure 4–25). This phenomenon, called *retrogradation*, is dependent on the size of the linear molecules. Linear molecules in potato starch have about 2,000 glucose units and have a low tendency to retrogradation. The smaller corn starch molecule, with about 400 glucose units, shows much greater tendency for association.

Hydrolysis of the chains to about 20 to 30 units completely eliminates the tendency to association and precipitation. Retrogradation of a starch paste is accelerated by freezing. After thawing a frozen starch paste, a spongy mass results, which easily loses a large part of its water under slight pressure. Swelling is inhibited by the presence of fatty acids, presumably through formation of insoluble complexes with the linear fraction. Cereal starches contain fatty acids at levels of 0.5 to 0.7 percent. All starches contain 0.06 to 0.07 percent phosphorus, in the form of glucose-6-phosphate.

The staling of bread is generally ascribed to retrogradation of starch. It is now assumed that the linear fraction is already retrograded during the baking process and that this gives the bread its elastic and tender crumb structure. Upon storage, the linear sections of the branched starch fraction slowly associate, resulting in a hardening of the crumb; this is

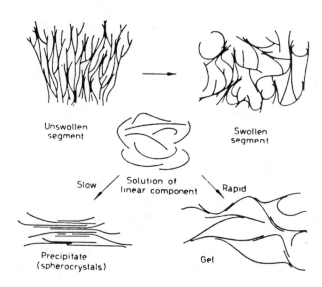

Figure 4–25 Schematic Representation of the Behavior of Starch on Swelling, Dissolving, and Retrograding. *Source:* From J.A. Radley, Technical Properties of Starch as a Function of Its Structural Chemistry, in *Recent Advances in Food Science,* Vol. 3, J.M. Leitch and D.N. Rhodes, eds., 1963, Butterworth.

known as staling. The rate of staling is temperature-dependent. Retrogradation is faster at low (although above-freezing) temperature, and bread stales more quickly when refrigerated than at room temperature. Freezing almost completely prevents staling and retrogradation.

Starches can be classified on the basis of the properties of the cooked pastes. Cereal starches (corn, wheat, rice, and sorghum) form viscous, short-bodied pastes that set to opaque gels on cooling. Root and tuber starches (potato, cassava, and tapioca) form highly viscous, long-bodied pastes. These pastes are usually clear and form only weak gels on cooling. Waxy starches (waxy corn, sorghum, and rice) form heavy-bodied, stringy pastes. These pastes are clear and have a low tendency for gel formation. High amylose starch (corn) requires high temperatures for gelatinization and gives short-bodied paste that forms a very firm, opaque gel on cooling (Luallen 1985).

Modified Starches

The properties of starches can be modified by chemical treatments that result in products suitable for specific purposes in the food industry (Whistler and Paschall 1967). Starches are used in food products to produce viscosity, promote gel formation, and provide cohesiveness in cooked starches. When a slurry of starch granules is heated, the granules swell and absorb a large amount of water; this happens at the gelatinization temperature (Figure 4–25), and the viscosity increases to a maximum. The swollen granules then start to collapse and break up, and viscosity decreases. Starch can be modified by acid treatment, enzyme treatment, cross-bonding, substitution, oxidation, and heat. Acid treatment results in thin boiling starch.

The granule structure is weakened or completely destroyed as the acid penetrates into the intermicellar areas, where a small number of bonds are hydrolyzed. When this type of starch is gelatinized, a solution or paste of low viscosity is obtained. A similar result may be obtained by enzyme treatment. The thin boiling starches yield low-viscosity pastes but retain the ability to form gels on cooling. Acid-converted waxy starches, those with low amylose levels, produce stable gels that remain clear and fluid when cooled. Acid-converted starches with higher amylose levels are more likely to form opaque gels on cooling. The acid conversion is carried out on aqueous granular starch slurries with hydrochloric or sulfuric acid at temperatures of 40 to 60°C. The action of acid is a preferential hydrolysis of linkages in the noncrystalline areas of the granules. The granules are weakened and no longer swell; they take up large amounts of water and produce pastes of low fluidity.

Cross-bonding of starch involves the formation of chemical bonds between different areas in the starch granule. This makes the granules more resistant to rupture and degradation on swelling and provides a firmer texture. The number of cross-bonds required to modify the starch granule is low; a large change in viscosity can be obtained by as few as 1 cross-bond per 100,000 glucose units. Increasing the number of cross-bonds to 1 per 10,000 units results in a product that does not swell on cooking. There are two ways to cross-link starch. The first, which gives a product known as distarch adipate, involves treating an aqueous slurry of starch with a mixture of adipic and acetic anhydrides under mildly alkaline conditions. After the reaction the starch is neutralized, washed, and dried. The second method, which produces distarch phosphate, involves treating a starch slurry with phosphorous oxychloride or sodium trimetaphosphate under alkaline conditions. Since the extent of cross-linking is low, the amount of reaction product in the modified starch is low. Free and combined adipate in cross-linked starch is below 0.09 percent. In distarch phosphate, the free and combined phosphate, expressed as phosphorus, is below 0.04 percent when made from cereal starch other than wheat, 0.11 percent if made from wheat starch, and 0.14 percent if made from potato starch (Wurzburg 1995).

Substitution of starch is achieved by reacting some of the hydroxyl groups in the starch molecules with monofunctional reagents that introduce different substituents. The action of the substituents lowers the ability of the modified starch to associate and form gels. This is achieved by preventing the linear portions of the starch molecules to form crystalline regions. The different types of substituted starch include starch acetates, starch monophosphates, starch sodium octenyl succinate, and hydroxypropyl starch ether. These substitution reactions can be performed on unmodified starch or in combination with other treatments such as acid hydrolysis or cross-linking.

Acetylation is carried out on suspensions of granular starch with acetic anhydride or vinyl acetate. Not more than 2.5 percent of acetyl groups on a dry starch basis are introduced, which equates to a degree of substitution of about 0.1 percent. Acetyl substitution reduces the ability of starch to produce gels on cooling and also increases the clarity of the cooled sol.

Starch phosphates are monophosphate esters, meaning that only one hydroxyl group is substituted in contrast to the two hydroxyl groups involved in production of cross-bonded starch. They are produced by mixing an aqueous solution of ortho-, pyro, or tripolyphosphate with granular starch; drying the mixture; and subjecting this to dry heat at 120 to 170°C. The level of phosphorus introduced into the starch does not exceed 0.4 percent. The introduction of phosphate groups as shown in Figure 4–26 gives the product an anionic charge (Wurzburg 1995). Starch monophosphates give dispersions with higher viscosity, better clarity, and better stability than the unmodified starch. They also have higher stability at low temperatures and during freezing.

Starch sodium octenyl succinate is a lightly substituted half ester produced by

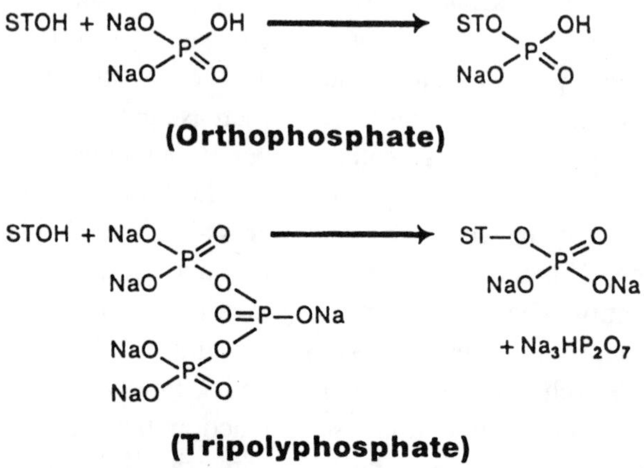

(Orthophosphate)

(Tripolyphosphate)

Figure 4–26 Phosphorylation of Starch with Sodium Ortho- or Tripolyphosphate

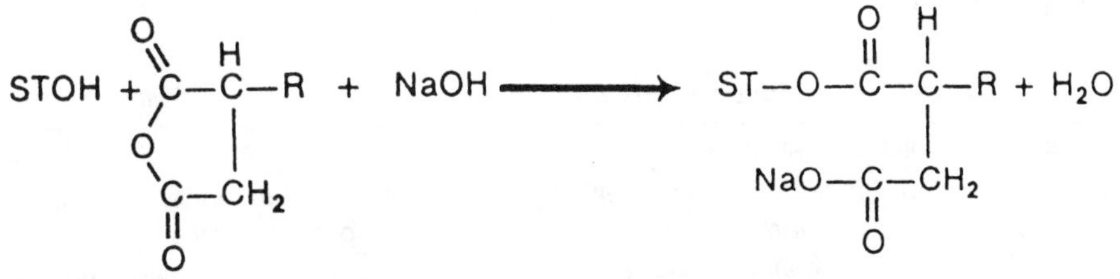

Figure 4–27 Reaction of Starch with Octenyl Succinic Anhydride

reacting an aqueous starch slurry with octenyl succinic anhydride as shown in Figure 4–27. The level of introduction of substituent groups is limited to 1 for about 50 anhydroglucose units. The treatment may be combined with other methods of conversion. The introduction of the hydrophilic carboxyl group and the lipophilic octenyl group makes this product amphiphilic and gives it the functionality of an emulsifier (Wurzburg 1995).

Hydroxypropylated starch is prepared by reacting an aqueous starch suspension with propylenol oxide under alkaline conditions at temperatures from 38 to 52°C. The reaction (Figure 4–28) is often combined with the introduction of distarch cross-links (Wurzburg 1995).

Oxidized starch is prepared by treating starch with hypochlorite. Although this starch is sometimes described as chlorinated starch, no chlorine is introduced into the molecule. The reaction is carried out by combining a starch slurry with a solution of sodium hypochlorite. Under alkaline conditions carboxyl groups are formed that modify linear portions of the molecule so that association and retrogradation are minimized. In addition to the formation of carboxyl groups, a variety of other oxidative reactions may occur including the formation of aldehydic and ketone groups. Oxidation increases the hydrophilic character of starch and lessens the tendency toward gel formation.

Dextrinization or pyroconversion is brought about by the action of heat on dry, powdered starch. Usually the heat treatment is carried out with added hydrochloric or phosphoric acid at levels of 0.15 and 0.17 percent, respectively. After addition of the acid, the starch is dried and heated in a cooker at temperatures ranging from 100 to 200°C. Two types of reaction occur, hydrolysis and transglucosidation. At low degree of conversion, hydrolysis is the main reaction and the resulting product is known as white dextrin. Transglucosidation involves initial hydrolysis of α 1-4 glucosidic bonds

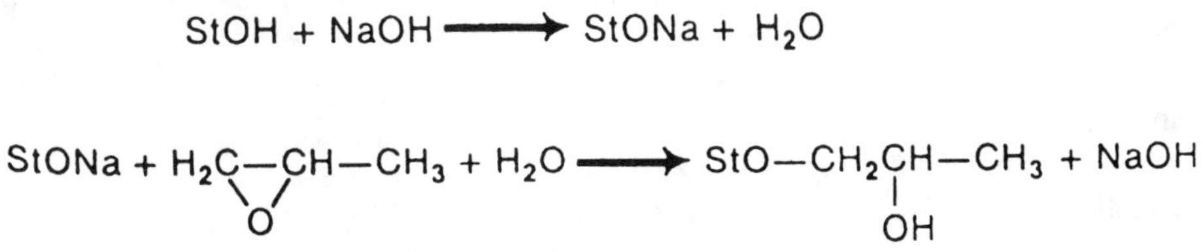

Figure 4–28 Hydroxypropylation of Starch

Table 4–10 Properties and Applications of Modified Starches

Process	Function/Property	Application
Acid conversion	Viscosity lowering	Gum candies, formulated liquid foods
Oxidation	Stabilization; adhesion gelling; clarification	Formulated foods, batters, gum confectionery
Dextrins	Binding; coating; encapsulation; high solubility	Confectionery, baking (gloss), flavorings, spices, oils, fish pastes
Cross-linking	Thickening; stabilization; suspension; texturizing	Pie fillings, breads, frozen bakery products, puddings, infant foods, soups, gravies, salad dressings
Esterification	Stabilization; thickening; clarification; when combined with cross-linking, alkali sensitive	Candies, emulsions, products gelatinized at lower temperatures
Etherification	Stabilization; low-temperature storage	Soups, puddings, frozen foods
Dual modification	Combinations of properties	Bakery, soups and sauces, salad dressings, frozen foods

Source: Reprinted with permission from O.B. Wurzburg, Modified Starches, in *Food Polysaccharides and Their Applications,* A.M. Stephen ed., p. 93, 1995. By courtesy of Marcel Dekker, Inc.

---

and recombination with free hydroxyl groups at other locations. In this manner new randomly branched structures or dextrins are formed; this reaction happens in the more highly converted products known as yellow dextrins. The dextrins have film-forming properties and are used for coating and as binders.

The properties and applications of modified starches are summarized in Table 4–10 (Wurzburg 1995). The application of modified starches as functional food ingredients has been described by Luallen (1985).

Glycogen

This animal reserve polysaccharide consists of a highly branched system of glucose units, joined by α-1→4 linkages with branching through α-1→6 linkages. It gives a red-brown color with iodine and is chemically very similar to starch. The outer branches of the molecule (Figure 4–29) consist of six or seven glucose residues; the branches that are formed by attachment to the 6-positions contain an average of three glucose residues.

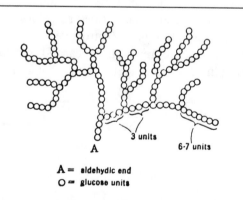

A = aldehydic end
O = glucose units

Figure 4–29 Schematic Representation of the Structure of Glycogen

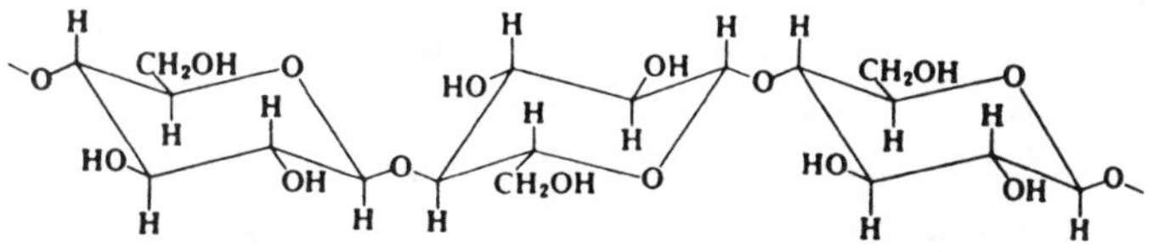

Figure 4–30 Section of a Cellulose Molecule

Cellulose

Cellulose is a polymer of β-glucose with β-1→4 linkages between glucose units. It functions as structural material in plant tissues in the form of a mixture of homologous polymers and is usually accompanied by varying amounts of other polysaccharides and lignins. The cellulose molecule (Figure 4–30) is elongated and rigid, even when in solution. The hydroxyl groups that protrude from the chain may readily form hydrogen bonds, resulting in a certain amount of crystallinity. The crystallinity of cellulose occurs in limited areas. The areas of crystallinity are more dense and more resistant to enzymes and chemical reagents than the noncrystalline areas. Crystalline areas absorb water poorly. A high degree of crystallinity results in an increased elastic modulus and greater tensile strength of cellulose fibers and should lead to greater toughness of a cellulose-containing food. Dehydrated carrots have been shown to increase in crystallinity with time, and digestibility of the cellulose decreases with this change. The amorphous regions of cellulose absorb water and swell. Heating of cellulose can result in a limited decrease of hydrogen bonding, leading thus to greater swelling because of decrease in crystalline content.

The amorphous gel regions of cellulose can become progressively more crystalline when moisture is removed from a food. Drying of cellulose-containing foods, such as vegetables, may lead to increased toughness, decreased plasticity, and swelling power.

Hydrolysis of cellulose leads to cellobiose and finally to glucose. The nature of the 1→4 linkage has been established by X-ray diffraction studies and by the fact that the bond is attacked only by β-glucosidases. The number of glucose units or degree of polymerization of cellulose is variable and can be as high as a DP of 10,000, which therefore has a molecular weight of 1,620,000.

The crystalline nature of cellulose fibers can be easily demonstrated by examination in the polarizing microscope. X-ray diffraction has demonstrated that the unit cell of cellulose crystals consists of two cellobiose units. According to Gortner and Gortner (1950), three different kinds of forces hold the lattice structure together. Along the *b* axis, the glucose units are held by β-1→4 glucosidic bonds; along the *c* axis, relatively weak van der Waals forces result in a distance between atomic centers of about 0.31 nm. Along the *a* axis, stronger hydrogen bond forces result in distances between oxygen atoms of only 0.25 nm.

Hemicelluloses and Pentosans

Hemicelluloses and pentosans are noncellulosic, nonstarchy complex polysaccharides that occur in many plant tissues.

Hemicellulose refers to the water-insoluble, non-starchy polysaccharides; pentosan refers to water-soluble, nonstarchy polysaccharides (D'Appolonia et al., 1971).

Hemicelluloses are not precursors of cellulose and have no part in cellulose biosynthesis but are independently produced in plants as structural components of the cell wall. Hemicelluloses are classified according to the sugars present. Xylans are polymers of xylose, mannans of mannose, and galactans of galactose. Most hemicelluloses are heteropolysaccharides, which usually contain two to four different sugar units. The sugars most frequently found in cereal hemicelluloses and pentosans are D-xylose and L-arabinose. Other hexoses and their derivatives include D-galactose, D-glucose, D-glucuronic acid, and 4-O-methyl-D-glucuronic acid. The basic structure of a wheat flour water-soluble pentosan is illustrated in Figure 4–31 (D'Appolonia et al. 1971).

The hemicellulose of wheat bran constitutes about 43 percent of the carbohydrates. It can be obtained by alkali extraction of wheat bran and contains 59 percent L-arabi-nose, 38.5 percent D-xylose, and 9 percent D-glucuronic acid. This compound is a highly branched araboxylan with a degree of polymerization of about 300. Graded acid hydrolysis of wheat bran hemicellulose preferentially removes L-arabinose and leaves an insoluble acidic polysaccharide comprised of seven to eight D-xylopyranose units joined by 1→4 linkages. One D-glucoronic acid unit is attached via a 1→2 linkage as a branch. The repeating unit is illustrated in Figure 4–32. Wheat endosperm contains about 2.4 percent hemicellulose. This mucilaginous component yields the following sugars on acid hydrolysis: 59 percent D-xylose, 39 percent L-arabinose, and 2 percent D-glucose. The molecule is highly branched.

Water-soluble pentosans occur in wheat flour at a level of 2 to 3 percent. They contain mainly arabinose and xylose. The structure consists of a straight chain of anhydro-D-xylopyranosyl residues linked beta 1→4 with branches consisting of anhydro L-arabinofuranosyl units attached at the 2- or 3-position of some of the anhydro xylose units.

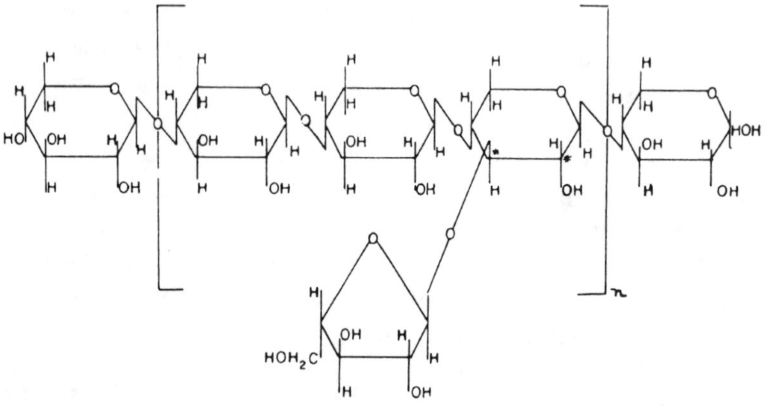

Figure 4–31 Structure of a Water-Soluble Wheat Flour Pentosan. (*n* indicates a finite number of polymer units; * indicates positions at which branching occurs). *Source:* From B.L. D'Appolonia et al., Carbohydrates, in *Wheat: Chemistry and Technology,* Y. Pomeranz, ed., 1971, American Association of Cereal Chemists, Inc.

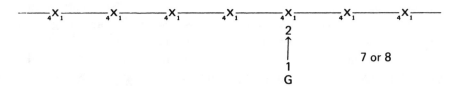

Figure 4–32 Repeating Unit of Insoluble Hemicellulose of Wheat Bran. *X* represents D-xylopyranose acid, *G* represents D-glucuronic acid. Subscripts refer to carbon atoms at which adjacent sugars are joined. *Source:* From B.L. D'Appolonia et al., Carbohydrates, in *Wheat: Chemistry and Technology,* Y. Pomeranz, ed., 1971, American Association of Cereal Chemists, Inc.

The water-soluble pentosans are highly branched, highly viscous, and gel forming. Because of these properties, it is thought that the pentosans may contribute to the structure of bread dough. Hoseny (1984) has described the functional properties of pentosans in baked foods. One of the more significant properties is due to the water-soluble pentosans, which form very viscous aqueous solutions. These solutions are subject to oxidative gelation with certain oxidizing agents. The cross-linking of protein and polysaccharide chains creates high molecular weight compounds that increase the viscosity and thereby change the rheological properties of dough.

Lignin

Although lignin is not a polysaccharide, it is included in this chapter because it is a component of dietary fiber and an important constituent of plant tissues. Lignin is present in mature plant cells and provides mechanical support, conducts solutes, and provides resistance to microbial degradation (Dreher 1987). Lignin is always associated in the cell wall with cellulose and hemicelluloses, both in close physical contact but also joined by covalent bonds. Lignins are defined as polymeric natural products resulting from enzyme-initiated dehydrogenative polymerization of three primary precursors: *trans*-coniferyl, *trans*-sinapyl, and *trans*-p-coumaryl alcohol (Figure 4–33). Lignin occurs in plant cell walls as well as in wood, with the latter having higher molecular weights. Lignin obtained from different sources differs in the relative amounts of the three constituents as well as in molecular weight. The polymeric units have molecular weights between 1,000 and 4,000. The polymeric

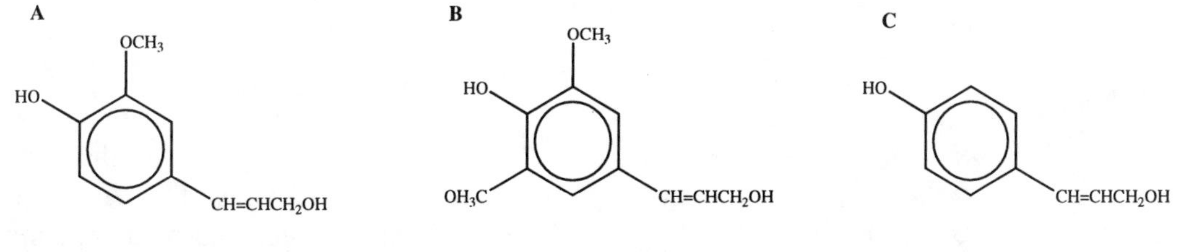

Figure 4–33 Monomeric Components of Lignin: (A) *trans*-coniferyl alcohol, (B) *trans*-sinapyl alcohol, (C) *trans*-p-coumaryl alcohol.

units contain numerous hydroxylic and ether functions, which provide opportunities for internal hydrogen bonds. These properties lend a good deal of rigidity to lignin molecules. One of the problems in the study of lignin composition is that separation from the cell wall causes rupturing of lignin-polysaccharide bonds and a reduction in molecular weight so that isolated lignin is never the same as the *in situ* lignin (Sarkanen and Ludwig 1971).

Cyclodextrins

When starch is treated with a glycosyl-transferase enzyme (CGTase), cyclic polymers are formed that contain six, seven, or eight glucose units. These are known as α-, β-, and γ-cyclodextrins, respectively. The structure of β-cyclodextrin is shown in Figure 4–34. These ring structures have a hollow cavity that is relatively hydrophobic in nature because hydrogen atoms and glycosidic oxygen atoms are directed to the interior. The outer surfaces of the ring are hydrophilic because polar hydroxyl groups are located on the outer edges. The hydrophobic nature of the cavity allows molecules of suitable size to be complexed by hydrophobic interaction. These stable complexes may alter the physical and chemical properties of the guest molecule. For example, vitamin molecules could be complexed by cyclodextrin to prevent degradation. Other possible applications have been described by Pszczola (1988). A disadvantage of this method is that the complexes may become insoluble. This can be overcome by derivatization of the cyclodextrin, for instance, by selective methylation of the C(2) and C(3) hydroxyl groups (Szejtli 1984).

Polydextrose

Polydextrose is a randomly bonded condensation polymer of glucose. It is synthesized in the presence of minor amounts of sorbitol and citric acid. The polymer contains all possible types of linkages between glucose monomers, resulting in a highly branched complex structure (Figure 4–35). Because of the material's unusual structure, it is not readily broken down in the human intestinal tract and therefore supplies only 1 calorie per gram. It is described as a *bulking agent* and can be used in low-calorie diets. It provides no sweetness. When polydextrose use is combined with artificial sweeteners, a reduction in calories of 50 percent or more can be achieved (Smiles 1982).

Pectic Substances

Pectic substances are located in the middle lamellae of plant cell walls; they function in the movement of water and as a cementing material for the cellulose network. Pectic substances can be linked to cellulose fibers

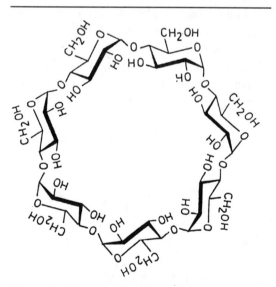

Figure 4–34 Structure of β-Cyclodextrin

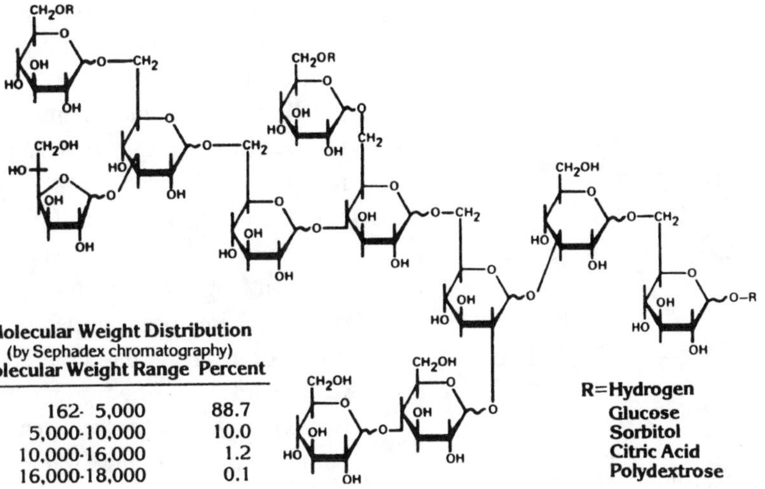

Molecular Weight Distribution
(by Sephadex chromatography)

Molecular Weight Range	Percent
162- 5,000	88.7
5,000-10,000	10.0
10,000-16,000	1.2
16,000-18,000	0.1

R=Hydrogen
Glucose
Sorbitol
Citric Acid
Polydextrose

Figure 4–35 Hypothetical Structure of Polydextrose Repeating Unit

and also by glucosidic bonds to xyloglucan chains that, in turn, can be covalently attached to cellulose. When pectic substances are heated in acidified aqueous medium, they are hydrolyzed to form pectin. A similar reaction, which leads to the formation of soluble pectin, occurs during the ripening of fruit. The level of pectin found in some plant tissues is listed in Table 4–11. The structure of pectin consists mostly of repeating units of D-galacturonic acid, which are joined by α 1-4 linkages (Okenfull 1991) (Figure 4–36). The carboxylic acid groups are in part present as esters of methanol. This structure is a homopolymer of 1-4 α-D-galactopyranosyluronic acid units. In addition, pectins contain an α-D-galacturonan, which is a heteropolymer formed from repeating units of 1-2 α-L-rhammosyl-(1-L) α-D galactosyluronic acid. This type of structure makes pectin a block copolymer, which means that it contains blocks of different composition. The main blocks are branched galacturonan chains interrupted and bent by rhamnose units. There are many rhamnose units, and these may carry side chains. The branched blocks

alternate with unbranched blocks containing few rhamnose units. The rhamnose in the branched blocks are joined to arabinan and galactan chains or arabinogalactan chains, which form 1-4 linkages to the rhamnose. In these side chains a number of neutral sugars may be present, mostly consisting of D-galactopyranose and L-arabinofuranose, making up 10 to 15 percent of the weight of pectin. The rhamnogalacturonan areas with

Table 4–11 Pectin Content of Some Plant Tissues

Plant Material	Pectin (%)
Potato	2.5
Tomato	3
Apple	5–7
Apple pomace	15–20
Carrot	10
Sunflower heads	25
Sugar beet pulp	15–20
Citrus albedo	30–35

Source: From R.L. Whistler, Pectin and Gums, in *Symposium on Foods: Carbohydrates and Their Roles,* H.W. Schultz et al., eds., 1969, AVI Publishing Co.

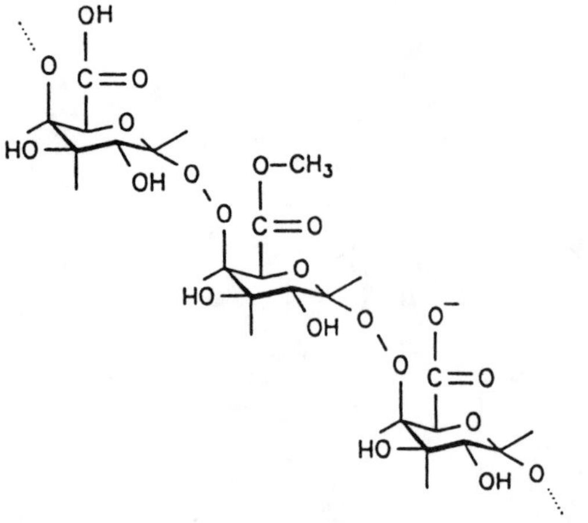

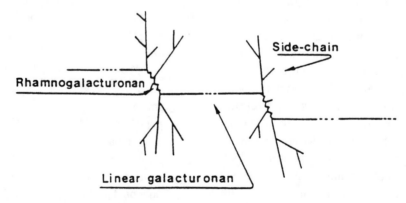

Figure 4–36 A Repeating Segment of Galacturonic Acid Units in Pectin. *Source:* Reprinted with permission from D.G. Oakenfull, The Chemistry of High Methoyxl Pectins, in *The Chemistry and Technology of Pectin*, R.H. Walter, ed., p. 87, © 1991, Academic Press.

Pectins from different sources, although similar in general structure, may differ in details. Different sugars may occupy some of the positions on the side chains. The length of the side chains may vary between about 8 and 20 residues, and some pectins may have acylation on the uronide residue (Whistler 1969; Thakur et al. 1997).

Chain length and degree of methylation are especially important in determining the properties of pectins, particularly gel formation. Completely esterified pectins would have 16 percent methoxyl content but do not occur in nature. The usual range is 9 to 12 percent ester methoxyl, although some pectins may have a very low methoxyl content. When the methyl ester group is removed by alkaline hydrolysis or enzyme action, a number of intermediates named pectinic acids are formed. When all methyl groups are removed, the product becomes insoluble and is called pectic acid. Pectin is thus only a generic name for a range of products with differing composition, which can be classified as pectinic acids.

their side chains have been described as "hairy" regions and the linear areas as "smooth" regions (Axelos and Thibault 1991). This type of structure is shown in schematic representation in Figure 4–37.

Pectins are widely used because of their unique ability to form gels in the presence of

Figure 4–37 Schematic Representation of Pectin Structure Showing "Hairy" Regions (Rhamnogalacturonan and Side Chains) and Smooth Regions (Linear Galacturonan). *Source:* Reprinted with permission from M.A.V. Axelos and J.F. Thibault, The Chemistry of Low Methoxyl Pectin, in *The Chemistry and Technology of Pectin*, R.M. Walter, ed., p. 109, © 1991, Academic Press.

calcium ions, sugar, and acid. In a gel, a three-dimensional network is formed that binds a relatively large volume of water. This sort of network requires some specific properties of the molecules that form the network. They should not be linear but branched and should form interchain associations based on ionic, hydrogen, and hydrophobic bonds. The properties of pectin gels depend on the degree of polymerization, the nature of the side chains, degree of methylation, composition of the side chains, and cross-linking of the molecules (Sterling 1963).

Two types of pectin are recognized, low and high methoxyl, and these form different kinds of gels. Low-methoxyl (LM) pectin has a degree of methylation of 25 to 50 percent and forms calcium gels; high-methoxyl (HM) pectin has 50 to 80 percent methylation and forms acid gels. In LM pectin gels, calcium ions act as a bridge between neighboring pectin molecules. HM pectin gels require the presence of at least 55 percent by weight of sugar and a pH below 3.6. HM pectin gels are formed by noncovalent forces, hydrogen and hydrophobic, that are thought to arise from stabilization of the junctions between molecules by the sugar, which acts as a dehydrating agent.

Pectins are evaluated for industrial use by *pectin grades*. Pectin grade is the number of parts of sugar required to gel one part of pectin to acceptable firmness. Usual conditions are pH 3.2 to 3.5, sugar 65 to 70 percent, and pectin 0.2 to 1.5 percent. Commercial grades vary from 100 to 500. Several types of pectin exist in the trade. Rapid-set pectin has a degree of methoxylation of over 70 percent. This type forms gels with sugar and acid at an optimum pH of 3.0 to 3.4. Gel strength depends on molecular weight; the higher the molecular weight, the firmer the gel. Gel strength is not influenced by degree of meth-oxylation. Slow-set pectin has a degree of methoxylation of 50 to 70 percent and forms gels with sugar and acid at an optimum pH of 2.8 to 3.2 and at a lower temperature than the rapid-set pectin. LM pectins have methoxylation levels of below 50 percent and do not form gels with sugar and acid; they gel with calcium ions. In these products, gel strength depends on the degree of methoxylation and not especially on molecular weight.

Gums

This large group of polysaccharides and their derivatives is characterized by its ability to give highly viscous solutions at low concentrations. Gums are widely used in the food industry as gelling, stabilizing, and suspending agents. Compounds in this group come from different sources and may include naturally occurring compounds as well as their derivatives, such as exudate gums, seaweed gums, seed gums, microbial gums, and starch and cellulose derivatives.

Table 4–12 lists the source and molecular structure of many of the gums as well as other polysaccharides (Stephen 1995). These polysaccharides are extensively used in the food industry as stabilizers, thickeners, emulsifiers, and gel formers as indicated in Table 4–13 (Stephen 1995).

All these materials have hydrophilic molecules, which can combine with water to form viscous solutions or gels. The nature of the molecules influences the properties of the various gums. Linear polysaccharide molecules occupy more space and are more viscous than highly branched molecules of the same molecular weight. The branched compounds form gels more easily and are more stable because extensive interaction along the chains is not possible. The linear neutral

Table 4–12 Sources and Molecular Structure of Food Polysaccharides

Polysaccharide	Main Sources	Molecular Structure[a]
Starch (amylose)	Cereal grains, tubers	Essentially linear (1→4)-α-D-glucan
Starch (amylopectin)	Cereal grains, tubers	Clusters of short (1→4)-α-D-Glc chains attached by α-linkages to 0-6 of other chains
Modified starches	Corn kernels	Cross-linked starch molecules; some C-6 oxidized; acetates
Maltodextrins	Corn and potato starches	Acid- and enzyme-catalyzed hydrolysates, Mw < 4000
Carboxymethylcellulose	Cotton cellulose	HO_2CCH_2-groups at 0-6 of linear (1→4)-β-D-glucan
Galactomannans	Seeds of guar, locust bean, tara	α-D-Galp groups at 0-6 of (1→4)-β-D-mannan chains
Carrageenans	Red seaweeds (*Gracilaria, Gigartina, Eucheuma* spp.)	Sulfated D-galactans, units of (1→3)-β-D-Gal and (1→4)-3,6-anhydro-α-D-Gal alternating; pyruvate and Me groups
Agars	Red seaweeds (*Gelidium* spp.)	As for carrageenans, anhydrosugar units L
Gum arabic	Stem exudate of *Acacia senegal*	Acidic L-arabino-, (1→3)- and (1→6)-β-D-galactan, highly branched with peripheral L-Rhap attached to D-GlcA. Minor component a glycoprotein
Gum tragacanth	*Astragalus* spp.	Modified, acidic arabinogalactan, and a modified pectin
Pectins	Citrus, apple, and other fruits	Linear and branched (1→4)-α-D-galacturonan (partly Me esterifed and acetylated); chains include (1→2)-L-Rhap, and branches D-Galp, L-Araf, D-Xylp, D-GlcA
Alginates	Brown seaweeds (*Macrocystis, Ascophyllum, Laminaria, Ecklonia* spp.)	Linear (1→4)-β-D-mannuronan and -α-L-guluronan
Xanthan gum	*Xanthomonas campestris*	Cellulosic structure, D-Manp (two) and GlcA-containing side chains, acetylated and pyruvylated on Man

[a] Abbreviations for usual forms of the sugar units: D-glucopyranose, D-Glcp; D-galactopyranose, D-Galp; D-glucuronic acid, D-GlcA; D-galacturonic acid, D-GalA; D-mannopyranose, D-Manp; D-mannuronic acid, D-ManA; L-arabinofuranose, L-Araf, D-xylopyranose, D-Xylp; L-rhamnopyranose, L-Rhap; L-fucopyranose, L-Fucp; D-fructofuranose, D-Fruf, L-guluronic acid, L-GulA.

Source: Reprinted with permission from A.M. Stephen, *Food Polysaccharides and Their Applications*, 1995. By courtesy of Marcel Dekker, Inc.

Table 4–13 Function and Food Applications of Hydrocolloids

Hydrocolloid	Function	Application
Guar and locust bean gums	Stabilizer, water retention	Dairy, ice cream, desserts, bakery
Carrageenans	Stabilizer, thickener, gelation	Ice cream, flan, desserts, meat products, dressing, instant puddings
Agars	Gelation	Dairy, confectionery, meat products
Gum arabic	Stabilizer, thickener, emulsifer, encapsulating agent	Confectionery, bakery, beverages, sauces
Gum tragacanth	Stabilizer, thickener, emulsifer	Dairy, dressings, sauces, confectionery
Pectins	Gelation, thickener, stabilizer	Jams, preserves, beverages, bakery, confectionery, dairy
Alginates	Stabilizer, gelation	Ice cream, instant puddings, beverages
Xanthan gum	Stabilizer, thickener	Dressings, beverages, dairy
Carboxymethylcellulose	Stabilizer, thickener, water retention	Ice cream, batters, syrups, cake mixes, meat products

Source: Reprinted with permission from A.M. Stephen, *Food Polysaccharides and Their Applications*, 1995. By courtesy of Marcel Dekker, Inc.

polysaccharides readily form coherent films when dry, and they are good coating agents. Solutions are not tacky. Solutions of branched polysaccharides are tacky because of extensive entangling of the side chains and because the dried solutions do not form films readily. The dried material can be more easily redissolved than can the dried linear compounds.

Neutral polysaccharides are only slightly affected by change in pH, and salts at low concentrations also have little effect. High salt concentration may result in removal of the bound water and precipitation of the polysaccharide. Some polysaccharides have long straight chains with many short branches. Such compounds, for example, locust bean gum and guar gum, combine many properties of the linear and the branched polysaccharides. Some gums have molecules containing many carboxyl groups

along the chains; examples are pectin and alginate. These molecules are precipitated below pH 3 when free carboxyl groups are formed. At higher pH values, alkali metal salts of these compounds are highly ionized, and the charges keep the molecules in extended form and extensively hydrated. This results in stable solutions. Divalent cations such as calcium may form salt bridges between neighboring molecules, resulting in gel formation and—if much calcium is present—precipitation.

Examples of polysaccharides with strong acid groups are furcellaran and carrageenan. Both are seaweed extractives with sulfuric acid ester groups. Because the ionization of sulfuric acid groups is not reduced much at low pH, such gums are stable in solutions of low pH values.

Gums can be chemically modified by introduction of small amounts of neutral or ionic substituent groups. Substitution or derivatization to a degree of substitution (DS) of 0.01 to 0.04 is often sufficient to completely alter the properties of a gum. The effect of derivatization is much less dramatic with charged molecules than with neutral ones.

Introduction of neutral substituents along the chains of linear polysaccharides results in increased viscosity and solution stability. Some of the commonly introduced groups are methyl, ethyl, and hydroxymethyl. Acid groups can be carboxyl, introduced by oxidation, or sulfate and phosphate groups. Introduction of strongly ionized acid groups may make the polysaccharides mucilaginous.

Gum Arabic

Gum arabic is a dried exudate from acacia trees. It is a neutral or slightly acidic salt of a complex polysaccharide containing calcium, magnesium, and potassium anions. The molecule exists in a stiff coil with many side chains and a molecular weight of about 300,000. The molecule is made up of four sugars, L-arabinose, L-rhamnose, D-galactose, and D-glucuronic acid. It is one of the few gums that require high concentration to give increased viscosity and is used as crystallization inhibitor and emulsifier. Gum arabic forms coacervates with gelatin and many other proteins.

Locust Bean Gum

This gum is obtained from the carob bean, which is cultivated exclusively around the Mediterranean. The commercial gum contains 88 percent of D-galacto-D-mannoglycan, 4 percent of pentoglycan, 6 percent protein, 1 percent cellulose, and 1 percent ash. The molecular weight is about 310,000,

and the molecule is a linear chain of D-mannopyranosyl units linked 1→4. Every fourth or fifth D-mannopyranosyl unit is substituted on carbon 6 with a D-galactopyranosyl residue. Locust bean gum forms tough, pliable films.

Guar Gum

Guar gum is obtained from the seed of the guar plant and was only introduced in 1954. It is a straight chain of D-galacto-D-mannoglycan with many single galactose branches. The D-mannopyranose units are joined by β-1→4 bonds, and the single D-galactopyranose units are attached by α-1→6 linkages. The branches occur at every second mannose unit. The compound has a molecular weight of 220,000 and forms viscous solutions at low concentration. At concentrations of 2 to 3 percent, gels are formed. Guar gum shows no incompatibility with proteins or other polysaccharides. Guar gum forms tough, pliable films.

Agar

Agar is extracted from algae of the class Rhodophyceae. It is soluble in boiling water but is insoluble in cold water. The gels are heat-resistant, and agar is widely used as an emulsifying, gelling, and stabilizing agent in foods. The gel formation properties of agar are unique. It shows hysteresis in that gelation takes place at temperatures far below the gel-melting temperature. It is also the most potent gel-former known, as gelation becomes perceptible at 0.04 percent concentration. Molecular weight determinations have given varying results. Osmotic pressure measurements indicate values from 5,000 to 30,000; other methods, as high as 110,000. Agar is a mixture of at least two polysaccha-

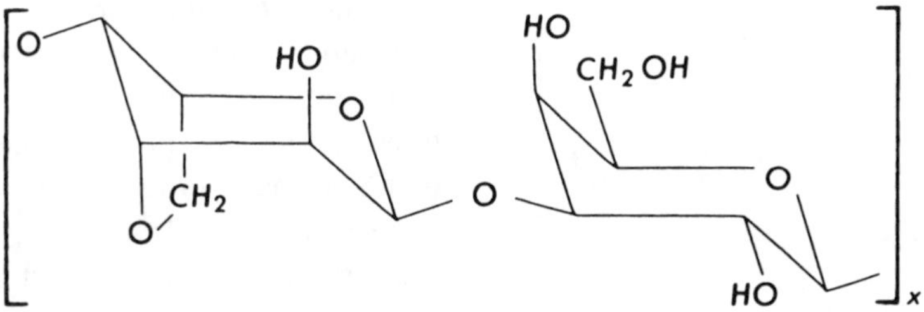

Figure 4–38 Structure of Agarose

rides (Glicksman 1969): agarose, a neutral polysaccharide with little or no ester sulfate groups, and agaropectin with 5 to 10 percent sulfate groups. The ratio of the two polymers can vary widely. Agarose consists of a linear chain of agarobiose disaccharide units. The structure, as shown in Figure 4–38, indicates alternating 1→4 linked, 3,6-anhydro-L-galactose units and 1→3 linked D-galactose units. Agaropectin is a sulfated molecule composed of agarose and ester sulfate, D-glucuronic acid, and small amounts of pyruvic acid. In neutral solutions, agar is compatible with proteins and other polysaccharides. At pH 3, mixing of warm agar and gelatin dispersions causes flocculation. Some gums, such as alginate and starch, decrease the strength of agar gels. Locust bean gum can improve rupture strain of agar gels several times.

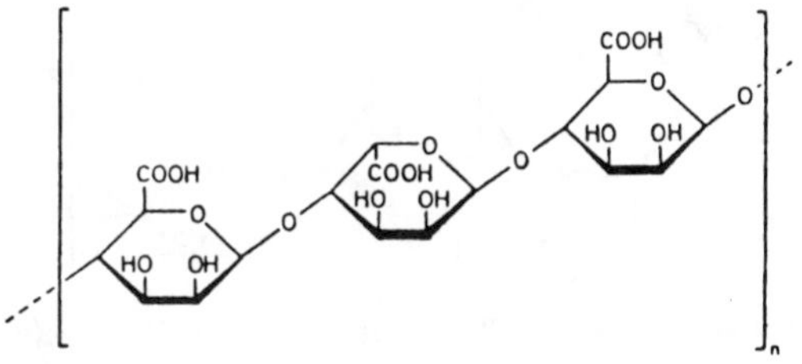

Figure 4–39 Structure of Alginic Acid

Algin

This gum is obtained from the giant kelp *Macrocystis pyrifera*. Algin is a generic designation of the derivatives of alginic acid. Alginic acid is a mixed polymer of anhydro-1→4-β-D-mannuronic acid and L-guluronic acid (Figure 4–39). The most common form is sodium alginate. Algin has thickening, suspending, emulsifying, stabilizing, gel-forming, and film-forming properties and is soluble in hot or cold water. When no divalent cations are present, solutions have long flow properties. Increasing amounts of calcium ions increase viscosity and result in short flow properties. Algin solutions do not gel on cooling or coagulate on heating, and the viscosity is little affected by pH in the range of 4 to 10. Algin can form gels with calcium, acid, or both.

Carrageenan

Extracted from Irish moss (*Chondrus crispus*), a red seaweed, carrageenan consists of salts or sulfate esters with a ratio of sulfate to hexose units of close to unity. Three fractions of carrageenan have been isolated, named κ-, λ-, and ι-carrageenan. The idealized structure of κ-carrageenan (Figure 4–40) is made up of 1→3 linked galactose-4-sulfate units and 1→4 linked 3,6-anhydro-D-

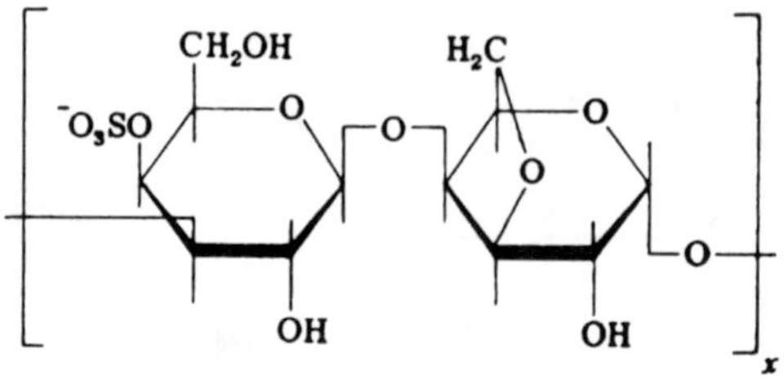

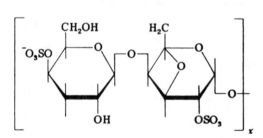

Figure 4–40 Idealized Structure of κ-Carrageenan

galactose units. Actually, up to 20 to 25 percent of the 3,6-anhydro-D-galactose units can be sulfated at carbon 2 and some of the 3,6-anhydro-D-galactose may occur as galactose-6-sulfate. The 6-sulfate group can be removed by heating with lime to yield 3,6-anhydro residues, and this treatment results in greatly increased gel strength. The major portion of λ-carrageenan consists of 1→3 linked galactose 2-sulfate and 1→4 linked galactose 2,6-disulfate (Figure 4–41); about 30 percent of the 1,3 galactose units are not sulfated. The 6-sulfate group can be removed with lime treatment but does not result in gel formation. Iota-carrageenan consists mainly of 1→3 linked galactose 4-sulfate and 1→4 linked 3,6-anhydro-D-galactose 2-sulfate (Figure 4–42). A certain

amount of 6-sulfate groups present can be changed to 3,6-anhydro groups by alkali treatment. The comparative properties of the three types of carrageenan have been listed by Glicksman (1969). Molecular weights of carrageenan vary from 100,000 to 800,000. Carrageenan can form thermally reversible gels whose strengths and gelation temperatures are dependent on the cations potassium and ammonium. The mechanism has been visualized as a zipper arrangement between aligned sections of linear polymer sulfates, with the potassium ions locked between alternating sulfate residues. Other monovalent cations, such as sodium, are not effective, probably because of larger ionic diameter. At low concentrations, carrageenan can alter the degree of agglomeration of caseinate particles in milk. It is a highly effective suspending agent and is used to suspend cacao particles in chocolate milk at concentrations as low as 0.03 percent. Carrageenan is often used in combination with starch. The two compounds form complexes that have useful properties in foods. The complexes permit a lowering of the starch content by as much as 50 percent (Descamps et al. 1986). An example of mixed gels combining carrageenan and whey protein has been described by Mleko et al. (1997). Optimal gelation occurred at pH 6 to 7. The shear stress value of whey protein isolate at 3 per-

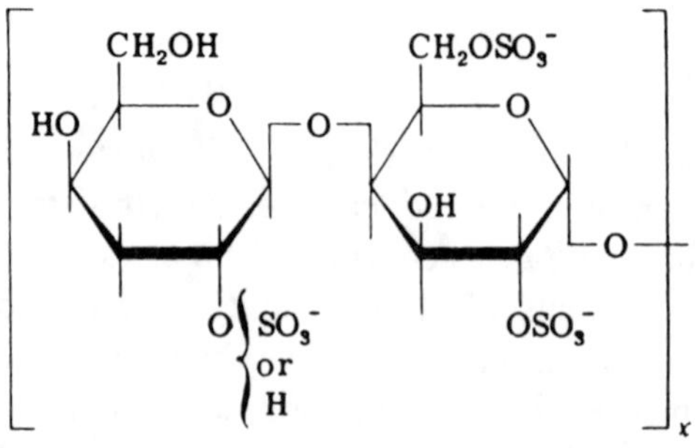

Figure 4–41 Idealized Structure of λ-Carrageenan

Figure 4–42 Idealized Structure of ι-Carrageenan

cent concentration was significantly enhanced by the presence of 0.5 percent κ-carrageenan.

Modified Celluloses

Modified celluloses are compounds of D-glucoglycans with various possible substituents, such as ethyl, methyl, hydroxyethyl, hydroxymethyl, hydroxypropyl, and carboxymethyl groups. Cellulose theoretically contains three OH-groups that could be derivatized, but the crystallinity of cellulose makes reaction difficult. The first step in the modification is preparation of alkali cellulose with strong alkali, followed by reaction with, for example, methyl chloride. Methyl cellulose is soluble in cold water at 1.3 to 2.6 DS. For the methoxyl derivative, solubility starts at a DS of 0.4, and for ethoxyl, at 0.9. Methyl cellulose is insoluble in hot water but soluble in cold water. Wetting the dry material with hot water ensures rapid dissolution in cold water. Because methyl cellulose is nonionic, it is not sensitive to divalent cations. Methyl cellulose can be used to make water-soluble sheets and bags that have sealing properties.

DIETARY FIBER

Originally, the fiber content of food was known as crude fiber and defined as the residue remaining after acid and alkaline extraction of a defatted sample. During the 1970s, the physiological effect of dietary fiber began to attract attention (Ink and Hurt 1987) and the need for better methods for the determination of fiber became apparent. Dietary fiber can be defined as a complex group of plant substances that are resistant to mammalian digestive enzymes. Because the definition is based on physiological properties rather than common chemical properties, the analysis of dietary fiber is not simple. Included in the definition of dietary fiber are cellulose, hemicellulose, lignin, cell wall components such as cutin, minerals, and soluble polysaccharides such as pectin. A method for determining total dietary fiber (TDF) that is based on enzymatic digestion has been accepted by the Association of Official Analytical Chemists (AOAC 1984) and is recognized for labeling food products. To determine the calorie content of a food, the TDF can be subtracted from the total carbohydrate content.

Table 4–14 Difference Between Crude Fiber (CF) and Total Dietary Fiber (TDF) of Some Plant Materials g/100g

Plant Material	CF	TDF	Ratio
Cellulose	72.5	94.0	1:1.3
Pea hulls	36.3	51.8	1:1.4
Corn bran	19.0	88.6	1:4.7
Distiller's dried grains	10.9	45.9	1:4.2
White wheat bran	8.7	36.4	1:4.2
Citrus pulp	14.4	24.8	1:1.7

Source: Reprinted with permission from M.L. Dreher, *Handbook of Dietary Fiber: An Applied Approach,* p. 58, 1987. By courtesy of Marcel Dekker, Inc.

Table 4–15 Total Dietary Fiber (TDF) and Components as Determined by the Southgate Method (g/100g Dry Weight)

Fiber Source	Noncellulose Polysaccharides			Cellulose	Lignin	TDF
	Hexose	Pentose	Uronic Acid			
Wheat bran	6.9	20.9	1.5	7.6	2.9	39.8
Rye biscuit	7.9	8.0	0.5	2.5	0.9	19.8
Dried apple	1.3	1.8	2.7	3.2	1.0	10.0
Citrus pectin	7.6	7.0	77.3	1.6	–	93.5
Potato powder	11.8	1.3	0.8	3.6	–	17.6
Soya flour	3.3	3.8	1.6	2.1	0.3	11.1

Source: Reprinted with permission from M.L. Dreher, *Handbook of Dietary Fiber: An Applied Approach*, p. 66, 1987. By courtesy of Marcel Dekker, Inc.

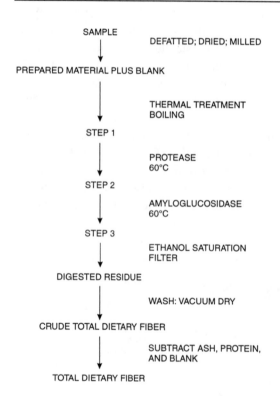

Figure 4–43 Association of Official Analytical Chemists (AOAC) Method for the Determination of Total Dietary Fiber. *Source:* Reprinted with permission from AOAC Collaborative Study, *Total Dietary Fiber Method,* © 1984, Association of Official Analytical Chemists.

Earlier literature refers to crude fiber, which consists of part of the cellulose and lignin only. This method is now obsolete. The dietary fiber content of foods is usually from 2 to 16 times greater than the crude fiber content. Examples of the difference between the two measurements are given in Table 4–14. One of the first alternative methods was the acid detergent fiber (ADF) method developed by Van Soest (1963). In this procedure, hemicellulose is completely extracted, and the residue contains lignin and cellulose. Thereafter, neutral dietary fiber (NDF) methods were developed. These methods measure cellulose, hemicellulose, lignin, cutin, minerals, and protein but do not include soluble polysaccharides such as pectins. One of these methods, enzyme-modified neutral detergent fiber (ENDF), has been approved for the determination of insoluble dietary fiber. Chemical methods of determining TDF are known as Southgate type methods (Southgate 1981). This procedure measures cellulose, lignin, and soluble and insoluble noncellulose polysaccharides (NCP) in terms of hexose, pentose, and uronic acid units. Examples of the determination of TDF by the Southgate

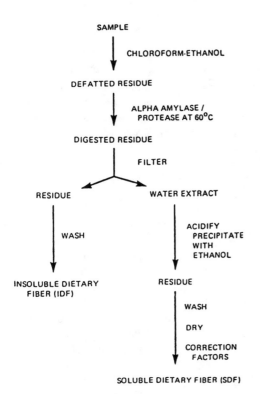

SAMPLE

CHLOROFORM-ETHANOL

DEFATTED RESIDUE

ALPHA AMYLASE / PROTEASE AT 60°C

DIGESTED RESIDUE

FILTER

RESIDUE — WATER EXTRACT

WASH

ACIDIFY PRECIPITATE WITH ETHANOL

INSOLUBLE DIETARY FIBER (IDF) — RESIDUE

WASH

DRY

CORRECTION FACTORS

SOLUBLE DIETARY FIBER (SDF)

Figure 4–44 Furda Method for the Determination of Insoluble and Soluble Dietary Fiber. *Source:* Reprinted with permission from I. Furda, Simultaneous Analysis of Soluble and Insoluble Dietary Fiber, in *The Analysis of Dietary Fiber in Food,* W.P.I. James and O. Theander, eds., 1981. By courtesy of Marcel Dekker, Inc.

Table 4–16 Total Dietary Fiber (TDF) Content of Some Foods as Determined by the AOAC Method in g/100g (Dry Basis)

Food	TDF
All-bran cereal	28.1
Whole wheat flour	11.8
Lettuce	26.0
Potatoes, cooked	11.6
Tomatoes	29.6
Carrots	23.9
Mushrooms	19.2
Green peas	24.6
Apples	14.7
Raspberries	53.5
Strawberries	24.2

Source: Reprinted with permission from M.L. Dreher, *Handbook of Dietary Fiber: An Applied Approach,* p. 59, 1987. By courtesy of Marcel Dekker, Inc.

Table 4–17 Composition of American Association of Cereal Chemists (AACC)–Certified Food-Grade Wheat Bran

Component	%
Acid detergent fiber	11.9
Neutral detergent fiber	40.2
Total dietary fiber	42.4
Protein	14.3
Lipid	5.2
Ash	5.1
Moisture	10.4
Lignin	3.2
Pectin	3.0
Cutin	0.0
Total starch	17.4
Total sugar	7.0
Pentosan	22.1
Phytic acid	3.4

Source: Reprinted with permission from M.L. Dreher, *Handbook of Dietary Fiber: An Applied Approach,* p. 82, 1987. By courtesy of Marcel Dekker, Inc.

method are given in Table 4–15. Finally, enzymatic gravimetric methods were developed and adopted to determine TDF in food. In these methods the defatted sample is treated with enzymes to degrade proteins and starch. Starch removal is an essential step in these procedures. The various steps evolved in the AOAC method for the determination of total dietary fiber are shown in Figure 4–43. There is no separation of soluble and insoluble fiber. The method developed by Furda

Table 4–18 Overview of Dietary Fiber Methods

Method	Portion Removed During Analysis	Fiber Components Determined by Method
Crude fiber (CF)	80% lignin, 85% hemicellulose, and 20–60% cellulose	Remainder of the lignin, hemicellulose, and cellulose
Acid detergent fiber (ADF)	Solubilizes cellular components (starch, sugars, fat, nitrogen compounds, and some minerals) plus hemicellulose	Cell wall components, except hemicellulose, as one unit
Neutral detergent fiber (NDF) or enzyme-modified NDF	Solubilizes cellular components: soluble fiber	Cell wall components as one unit
NDF - ADF	—	Hemicellulose
72% sulfuric acid (Klason lignin)	Cellulose	Lignin, insoluble nitrogen compounds, cutin, silica
Permanganate oxidation	Lignin	Lignin (loss in weight)
Southgate-type methods (unavailable carbohydrate)	Solubilizes cellular components; hydrolysis starch	Individual chemical components (including soluble and insoluble polysaccharides) = total dietary fiber (TDF)
Enzymatic methods	Solubilizes cellular components; hydrolysis starch and protein	TDF (indigestible residue); isolation soluble and insoluble fractions
Fractionation methods	—	Isolation and determination of individual components

Source: Reprinted with permission from M.L. Dreher, *Handbook of Dietary Fiber: An Applied Approach,* p. 106, 1987. By courtesy of Marcel Dekker, Inc.

(1981) and shown in Figure 4–44 does provide for separation of soluble and insoluble fiber. The TDF content of some foods as determined by the AOAC method is given in Table 4–16. The American Association of Cereal Chemists (AACC) makes available a certified standard for reference purposes. The composition of the AACC wheat bran standard is listed in Table 4–17 and illustrates the complexity of what is now known as dietary fiber. An overview of different methods for fiber determinations is presented in Table 4–18 (Dreher 1987).

One of the beneficial effects of dietary fiber is its bulking capacity, and the water-holding capacity of the gums plays an important role in this effect.

Dietary fiber, as now defined, includes the following three major fractions:

1. *Structural polysaccharides*—associated with the plant cell wall, including cellulose, hemicellulose, and some pectins
2. *Structural nonpolysaccharides*—mainly lignins
3. *Nonstructural polysaccharides*—including the gums and mucilages (Schneeman 1986).

REFERENCES

Angyal, S.J. 1976. Conformational analysis in carbohydrate chemistry III. The ^{13}C NMR spectra of hexuloses. *Aust. J. Chem.* 29: 1249–1265.

Angyal, S.J. 1984. The composition of reducing sugars in solution. *Adv. Carbohydr. Chem. Biochem.* 42: 15–68.

Association of Official Analytical Chemists Collaborative Study. 1984. *Total dietary fiber method.* Washington, DC: Association of Official Analytical Chemists.

Axelos, M.A.V., and J.F. Thibault. 1991. The chemistry of low methoxyl pectin. In *The chemistry and technology of pectin,* ed. R.M. Walter. New York: Academic Press.

Blankers, I. 1995. Properties and applications of lactitol. *Food Technol.* 49: 66–68.

Buttrose, M.S. 1963. Influence of environment on the shell structure of starch granules. *Staerke* 15: 213–217.

Commerford, J.D. 1974. Corn sweetener industry. In *Symposium: Sweeteners,* ed. I.E. Inglett. Westport, CT: AVI Publishing Co.

D'Appolonia, B.L., et al. 1971. Carbohydrates. In *Wheat: Chemistry and technology,* ed. Y. Pomeranz. St. Paul, MN: American Association of Cereal Chemists, Inc.

deMan, J.M., D.W. Stanley, and V. Rasper. 1975. Composition of Ontario soybeans and soymilk. *Can. Inst. Food Sci. Technol. J.* 8: 1–8.

deMan, L., J.M. deMan, and R.I. Buzzell. 1987. Composition and properties of soymilk and tofu made from Ontario light hilum soybeans. *Can. Inst. Food Sci. Technol. J.* 20: 363–367.

Descamps, O., P. Langevin, and D.H. Combs. 1986. Physical effect of starch/carrageenan interactions in water and milk. *Food Technol.* 40, no 4: 81–88.

Dreher, M.L. 1987. *Handbook of dietary fiber: An applied approach.* New York: Marcel Dekker, Inc.

Dziezak, J.D. 1987. Crystalline fructose: A breakthrough in corn sweetener process technology. *Food Technol.* 41, no. 1: 66–67, 72.

Furda, I. 1981. Simultaneous analysis of soluble and insoluble dietary fiber. In *The analysis of dietary fiber in food,* ed. W.P.I. James and O. Theander. New York: Marcel Dekker, Inc.

Gidley, M.J., and S.M. Bociek. 1985. Molecular organization in starches: A ^{13}C CP/MAS NMR study. *J. Am. Chem. Soc.* 107: 7040–7044.

Glicksman, M. 1969. *Gum technology in the food industry.* New York: Academic Press.

Gortner, R.A., and W.A. Gortner. 1950. *Outlines of biochemistry.* New York: John Wiley & Sons.

Heume, M., and A. Rapaille. 1996. Versatility of maltitol in different forms as a sugar substitute. In *Advances in Sweeteners,* ed. T.H. Grenby. London: Blackie Academic and Professional.

Hoseny, R.C. 1984. Functional properties of pentosans in baked foods. *Food Technol.* 38, no 1: 114–117.

Hudson, C.S. 1907. Catalysis by acids and bases of the mutarotation of glucose. *J. Am. Chem. Soc.* 29: 1571–1574.

Ink, S.L., and H.D. Hurt. 1987. Nutritional implications of gums. *Food Technol.* 41, no. 1: 77–82.

Jenness, R., and S. Patton. 1959. *Principles of dairy chemistry.* New York: John Wiley & Sons.

Jurch, Jr., G.R., and J.H. Tatum. 1970. Degradation of D-glucose with acetic acid and methyl amine. *Carbohyd. Res.* 15: 233–239.

Kennedy, J.F., et al. 1985. Oligosaccharide component composition and storage properties of commercial low DE maltodextrins and their further modification by enzymatic treatment. *Starch* 37: 343–351.

Luallen, T.E. 1985. Starch as a functional ingredient. *Food Technol.* 39, no 1: 59–63.

Mleko, S., E.C.Y. Li-Chan, and S. Pikus. 1997. Interaction of κ carrageenan with whey proteins in gels formed at different pH. *Food. Res. Intern.* 30: 427–434.

Oates, C.G. 1997. Towards an understanding of starch granule structure and hydrolysis. *Trends Food Sci. Technol.* 8: 375–382.

Okenfull, D.G. 1991. The chemistry of high methoyxl pectins. In *The chemistry and technology of pectin,* ed. R.H. Walter. New York: Academic Press.

Olsen, H.S. 1995. Enzymic production of glucose syrups. In *Handbook of starch hydrolysis products and their derivatives,* ed. M.W. Kearsley and S.Z. Dziedzic. London: Chapman and Hall.

Pictet, A., and P. Stricker. 1924. Constitution and synthesis of isosacchrosan. *Helv. Chim. Acta* 7: 708–713.

Pszczola, D.E. 1988. Production and potential food applications of cyclodextrins. *Food Technol.* 42, no. 1: 96–100.

Roos, Y.H. 1993. Melting and glass transitions of low molecular weight carbohydrates. *Carbohydrate Res.* 238: 39–48.

Roos, Y.H., and M. Karel. 1991a. Amorphous state and delayed ice formation in sucrose solutions. *Int. Food Sci. Technol.* 26: 553–566.

Roos, Y.H., and M. Karel. 1991b. Phase transitions of amorphous sucrose and frozen sucrose solutions. *J. Food Sci.* 56: 266–267.

Roos, Y.H., and M. Karel. 1991c. Phase transitions of mixtures of amorphous polysaccharides and sugars. *Biotechnol. Prog.* 7: 49–53.

Roos, Y.H., and M. Karel. 1991d. Water and molecular weight effects on glass transitions in amorphous carbohydrates and carbohydrate solutions. *J. Food Sci.* 56: 1676–1681.

Roos, Y.H., and M. Karel. 1992. Crystallization of amorphous lactose. *J. Food Sci.* 57: 775–777.

Sarkanen, K.V., and C.H. Ludwig. 1971. *Lignins: Occurrence, formation, structure and reactions.* New York: Wiley-Interscience.

Schneeman, B.O. 1986. Dietary fiber: Physical and chemical properties, methods of analysis, and physiological effects. *Food Technol.* 40, no. 2: 104–110.

Shallenberger, R.S., and G.G. Birch. 1975. *Sugar chemistry.* Westport, CT: AVI Publishing Co.

Smiles, R.E. 1982. The functional applications of polydextrose. In *Chemistry of foods and beverages: Recent developments,* eds. G. Charalambous and G. Inglett. New York: Academic Press.

Smythe, B.M. 1971. Sucrose crystal growth. *Sugar Technol. Rev.* 1: 191–231.

Southgate, D.A.T. 1981. Use of the Southgate method for unavailable carbohydrate in the measurement of dietary fiber. In *The analysis of dietary fiber in food,* ed. W.P.I. James and O. Theander. New York: Marcel Dekker, Inc.

Spiegel, J.E. et al. 1994. Safety and benefits of fructooligosaccharides as food ingredients. *Food Technol.* 48, no. 1: 85–89.

Stephen, A.M. 1995. *Food polysaccharides and their applications.* New York: Marcel Dekker, Inc.

Sterling, C. 1963. Texture and cell wall polysaccharides in foods. In *Recent advances in food science,* Vol. 3, ed. J.M. Leitch and D.N. Rhodes. London: Butterworths.

Szejtli, J. 1984. Highly soluble β-cyclodextrin derivatives. *Starch* 36: 429–432.

Thakur, B.R., R.K. Singh, and A.K. Handa. 1997. Chemistry and uses of pectin: A review. *Crit. Rev. Food Sci. Nutr.* 37: 47–73.

Thompson, A., et al. 1954. Acid reversion products from D-glucose. *J. Am. Chem. Soc.* 76: 1309–1311.

Van Soest, P.J. 1963. Use of detergents in the analysis of fibrous seeds II. A rapid method for the determination of fiber and lignin. *J. Assoc. Off. Anal. Chem.* 46: 829–835.

Washüttl, J., P. Reiderer, and E. Bancher. 1973. A qualitative and quantitative study of sugar-alcohols in several foods: A research role. *J. Food Sci.* 38: 1262–1263.

Whistler, R.L. 1969. Pectin and gums. In *Symposium on foods: Carbohydrates and their roles,* ed. H.W. Schultz et al. Westport, CT: AVI Publishing Co.

Whistler, R.L., and E.F. Paschall. 1967. *Starch: Chemistry and technology. Vol. 2. Industrial aspects.* New York: Academic Press.

Wurzburg, O.B. 1995. Modified starches. In *Food polysaccharides and their applications,* ed. A.M. Stephen. New York: Marcel Dekker, Inc.

Ziesenitz, S.C. 1996. Basic structure and metabolism of isomalt. In *Advances in sweeteners,* ed. T.H. Grenby. London: Blackie Academic and Professional.

CHAPTER 5

Minerals

INTRODUCTION

In addition to the major components, all foods contain varying amounts of minerals. The mineral material may be present as inorganic or organic salts or may be combined with organic material, as the phosphorus is combined with phosphoproteins and metals are combined with enzymes. More than 60 elements may be present in foods. It is customary to divide the minerals into two groups, the major salt components and the trace elements. The major salt components include potassium, sodium, calcium, magnesium, chloride, sulfate, phosphate, and bicarbonate. Trace elements are all others and are usually present in amounts below 50 parts per million (ppm). The trace elements can be divided into the following three groups:

1. essential nutritive elements, which include Fe, Cu, I, Co, Mn, Zn, Cr, Ni, Si, F, Mo, and Se.
2. nonnutritive, nontoxic elements, including Al, B, and Sn
3. nonnutritive, toxic elements, including Hg, Pb, As, Cd, and Sb

The minerals in foods are usually determined by ashing or incineration. This destroys the organic compounds and leaves the minerals behind. However, determined in this way, the ash does not include the nitrogen contained in proteins and is in several other respects different from the real mineral content. Organic anions disappear during incineration, and metals are changed to their oxides. Carbonates in ash may be the result of decomposition of organic material. The phosphorus and sulfur of proteins and the phosphorus of lipids are also part of ash. Some of the trace elements and some salts may be lost by volatilization during the ashing. Sodium chloride will be lost from the ash if the incineration temperature is over 600°C. Clearly, when we compare data on mineral composition of foods, we must pay great attention to the methods of analysis used.

Some elements appear in plant and animal products at relatively constant levels, but in a number of cases an abundance of a certain element in the environment may result in a greatly increased level of that mineral in plant or animal products. Enrichment of elements in a biological chain may occur; note, for instance, the high mercury levels reported in some large predatory fish species such as swordfish and tuna.

MAJOR MINERALS

Some of the major mineral constituents, especially monovalent species, are present in foods as soluble salts and mostly in ionized form. This applies, for example, to the cations sodium and potassium and the anions chloride and sulfate. Some of the polyvalent ions, however, are usually present in the form of an equilibrium between ionic, dissolved nonionic, and colloidal species. Such equilibria exist, for instance, in milk and in meat. Metals are often present in the form of *chelates*. Chelates are metal complexes formed by coordinate covalent bonds between a ligand and a metal cation; the ligand in a chelate has two or more coordinate covalent bonds to the metal. The name *chelate is* derived from the claw-like manner in which the metal is held by the coordinate covalent bonds of the ligand. In the formation of a chelate, the ligand functions as a Lewis base, and the metal ion acts as a Lewis acid. The stability constant of a chelate is influenced by a number of factors. The chelate is more stable when the ligand is relatively more basic. The chelate's stability depends on the nature of the metal ion and is related to the electronegative character of the metal. The stability of a chelate normally decreases with decreasing pH. In a chelate the donor atoms can be N, O, P, S, and Cl; some common donor groups are $-NH_2$, $=C=O$, $=NH$, $-COOH$, and $-OH-O-PO(OH)_2$. Many metal ions, especially the transition metals, can serve as acceptors to form chelates with these donor groups. Formation of chelates can involve ring systems with four, five, or six members. Some examples of four- and five-membered ring structures are given in Figure 5–1. An example of a six-membered chelate ring system is chlorophyll. Other examples of food components that can be considered metal chelates are hemoglobin and myoglobin, vitamin B_{12}, and calcium caseinate (Pfeilsticker 1970). It has also been proposed that the gelation of certain polysaccharides, such as alginates and pec-

tates, with metal ions occurs through chelation involving both hydroxyl and carboxyl groups (Schweiger 1966). A requirement for the formation of chelates by these polysaccharides is that the OH groups be present in vicinal pairs.

Concerns about the role of sodium in human hypertension have drawn attention to the levels of sodium and potassium in foods and to measures intended to lower our sodium intake. The total daily intake by Americans of salt is 10 to 12 g, or 4 to 5 g of sodium. This is distributed as 3 g occurring naturally in food, 3 g added during food preparation and at the table, and 4 to 6 g added during commercial processing. This amount is far greater than the daily requirement, estimated at 0.5 g (Marsh 1983). Salt has an important effect on the flavor and acceptability of a variety of foods. In addition to lowering the level of added salt in food, researchers have suggested replacing salt with a mixture of sodium chloride and potassium chloride (Maurer 1983; Dunaif and Khoo 1986). It has been suggested that calcium also plays an important role in regulating blood pressure.

Interactions with Other Food Components

The behavior of minerals is often influenced by the presence of other food constituents. The recent interest in the beneficial effect of dietary fiber has led to studies of the role fiber plays in the absorption of minerals. It has been shown (Toma and Curtis 1986) that mineral absorption is decreased by fiber. A study of the behavior of iron, zinc, and calcium showed that interactions occur with phytate, which is present in fiber. Phytates can form insoluble complexes with iron and zinc and may interfere with the

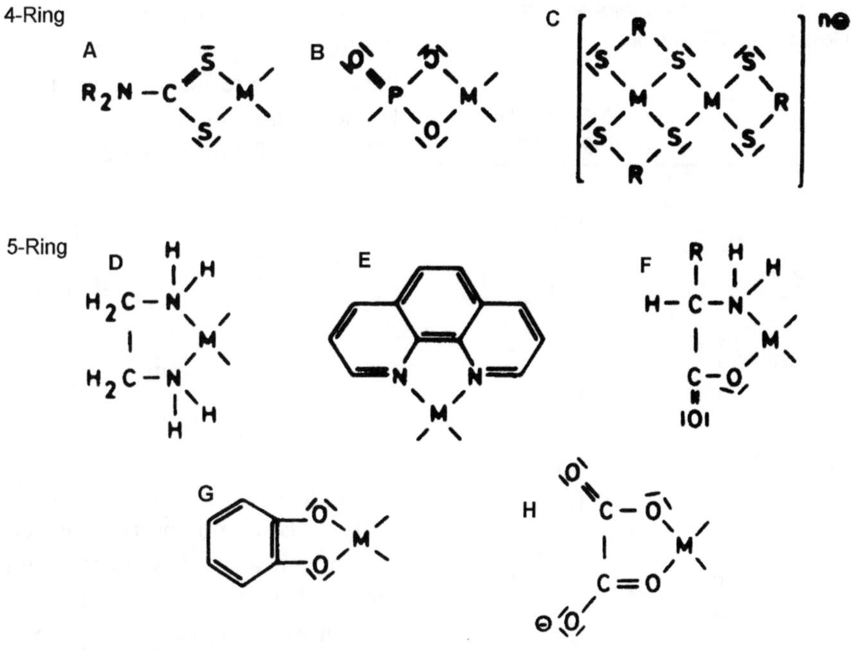

Figure 5–1 Examples of Metal Chelates. Only the relevant portions of the molecules are shown. The chelate formers are: (A) thiocarbamate, (B) phosphate, (C) thioacid, (D) diamine, (E) *o*-phenantrolin, (F) α-aminoacid, (G) *o*-diphenol, (H) oxalic acid. *Source*: From K. Pfeilsticker, Food Components as Metal Chelates, *Food Sci. Technol.*, Vol. 3, pp. 45–51, 1970.

absorption of calcium by causing formation of fiber-bound calcium in the intestines.

Iron bioavailability may be increased in the presence of meat (Politz and Clydesdale 1988). This is the so-called meat factor. The exact mechanism of this effect is not known, but it has been suggested that amino acids or polypeptides that result from digestion are able to chelate nonheme iron. These complexes would facilitate the absorption of iron. In nitrite-cured meats some factors promote iron bioavailability (the meat factor), particularly heme iron and ascorbic acid or erythorbic acid. Negative factors may in-clude nitrite and nitrosated heme (Lee and Greger 1983).

Minerals in Milk

The normal levels of the major mineral constituents of cow's milk are listed in Table 5–1. These are average values; there is a considerable natural variation in the levels of these constituents. A number of factors influence the variations in salt composition, such as feed, season, breed and individuality of the cow, stage of lactation, and udder infections. In all but the last case, the variations in individual mineral constituents do not affect the milk's osmotic pressure. The ash content of milk is relatively constant at about 0.7 percent. An important difference between milk and blood plasma is the rela-

Table 5–1 Average Values for Major Mineral Content of Cow's Milk (Skim Milk)

Constituent	Normal Level (mg/100 mL)
Sodium	50
Potassium	145
Calcium	120
Magnesium	13
Phosphorus (total)	95
Phosphorus (inorganic)	75
Chloride	100
Sulfate	10
Carbonate (as CO_2)	20
Citrate (as citric acid)	175

tive levels of sodium and potassium. Blood plasma contains 330 mg/100 mL of sodium and only 20 mg/100 mL of potassium. In contrast, the potassium level in milk is about three times as high as that of sodium. Some of the mineral salts of milk are present at levels exceeding their solubility and therefore occur in the colloidal form. Colloidal particles in milk contain calcium, magnesium, phosphate, and citrate. These colloidal particles precipitate with the curd when milk is coagulated with rennin. Dialysis and ultrafiltration are other methods used to obtain a serum free from these colloidal particles. In milk the salts of the weak acids (phosphates, citrates, and carbonates) are distributed among the various possible ionic forms. As indicated by Jenness and Patton (1959), the ratios of the ionic species can be calculated by using the Henderson-Hasselbach equation,

$$pH = pK_\alpha + \log \frac{[\text{salt}]}{[\text{acid}]}$$

The values for the dissociation constants of the three acids are listed in Table 5–2. When these values are substituted in the Henderson-Hasselbach equation for a sample of milk at pH 6.6, the following ratios will be obtained:

$$\frac{\text{Citrate}^-}{\text{Citric acid}} = 3,000 \qquad \frac{\text{Citrate}^=}{\text{Citrate}^-} = 72$$

$$\frac{\text{Citrate}^\equiv}{\text{Citrate}^=} = 16$$

From these ratios we can conclude that in milk at pH 6.6 no appreciable free citric acid or monocitrate ion is present and that tricitrate and dicitrate are the predominant ions, present in a ratio of about 16 to 1. For phosphates, the following ratios are obtained:

$$\frac{H_2PO_4^-}{H_3PO_4} = 43,600 \qquad \frac{HPO_4^=}{H_2PO_4^-} = 0.30$$

$$\frac{PO_4^\equiv}{HPO_4^=} = 0.000002$$

This indicates that mono- and diphosphate ions are the predominant species. For carbonates the ratios are as follows:

$$\frac{HCO_3^-}{H_2CO_3} = 1.7$$

$$\frac{CO_3^=}{HCO_3^-} = 0.0002$$

Table 5–2 Dissociation Constants of Weak Acids

Acid	pK_1	pK_2	pK_3
Citric	3.08	4.74	5.40
Phosphoric	1.96	7.12	10.32
Carbonic	6.37	10.25	—

The predominant forms are bicarbonates and the free acid.

Note that milk contains considerably more cations than anions; Jenness and Patton (1959) have suggested that this can be explained by assuming the formation of complex ions of calcium and magnesium with the weak acids. In the case of citrate (symbol ©$^-$) the following equilibria exist:

$$H©^= \rightleftharpoons ©^\equiv + H^+$$

$$©^\equiv + Ca^{++} \rightleftharpoons Ca\ ©^-$$

$$Ca©^- + H^+ \rightleftharpoons CaH\ ©$$

$$2Ca©^- + Ca^{++} \rightleftharpoons Ca_3\ ©_2$$

Soluble complex ions such as Ca ©$^-$ can account for a considerable portion of the calcium and magnesium in milk, and analogous complex ions can be formed with phosphate and possibly with bicarbonate.

The equilibria described here are represented schematically in Figure 5–2, and the levels of total and soluble calcium and phosphorus are listed in Table 5–3. The mineral equilibria in milk have been extensively studied because the ratio of ionic and total calcium exerts a profound effect on the stability of the caseinate particles in milk. Pro-

cessing conditions such as heating and evaporation change the salt equilibria and therefore the protein stability. When milk is heated, calcium and phosphate change from the soluble to the colloidal phase. Changes in pH result in profound changes of all of the salt equilibria in milk. Decreasing the pH results in changing calcium and phosphate from the colloidal to the soluble form. At pH 5.2, all of the calcium and phosphate of milk becomes soluble. An equilibrium change results from the removal of CO_2 as milk leaves the cow's udder. This loss of CO_2 by stirring or heating results in an increased pH. Concentration of milk results in a dual effect. The reduction in volume leads to a change of calcium and phosphate to the colloidal phase, but this also liberates hydrogen ions, which tend to dissolve some of the colloidal calcium phosphate. The net result depends on initial salt balance of the milk and the nature of the heat treatment.

The stability of the caseinate particles in milk can be measured by a test such as the heat stability test, rennet coagulation test, or alcohol stability test. Addition of various phosphates—especially polyphosphates, which are effective calcium complexing agents—can increase the caseinate stability of milk. Addition of calcium ions has the opposite effect and decreases the stability of milk. Calcium is bound by polyphosphates in the form of a chelate, as shown in Figure 5–3.

Minerals in Meat

The major mineral constituents of meat are listed in Table 5–4. Sodium, potassium, and phosphorus are present in relatively high amounts. Muscle tissue contains much more potassium than sodium. Meat also contains considerably more magnesium than calcium. Table 5–4 also provides information

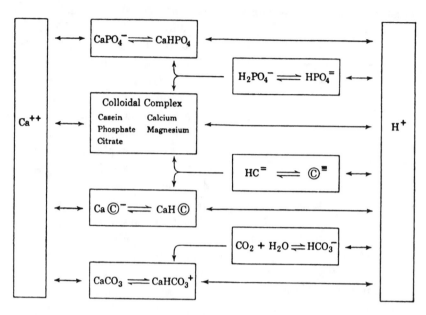

Figure 5–2 Equilibrium Among Milk Salts. *Source*: Reprinted with permission from R. Jenness and S. Patton, *Principles of Dairy Chemistry*, © 1959, John Wiley & Sons.

about the distribution of these minerals between the soluble and nonsoluble forms. The nonsoluble minerals are associated with the proteins. Since the minerals are mainly associated with the nonfatty portion of meat, the leaner meats usually have a higher mineral or ash content. When liquid is lost from meat (drip loss), the major element lost is sodium and, to a lesser extent, calcium,

phosphorus, and potassium. Muscle tissue consists of about 40 percent intracellular fluid, 20 percent extracellular fluid, and 40 percent solids. The potassium is found almost entirely in the intracellular fluid, as are magnesium, phosphate, and sulfate. Sodium is mainly present in the extracellular

Table 5–3 Total and Soluble Calcium and Phosphorus Content of Milk

Constituent	mg/100 mL
Total calcium	112.5
Soluble calcium	35.2
Ionic calcium	27.0
Total phosphorus	69.6
Soluble phosphorus	33.3

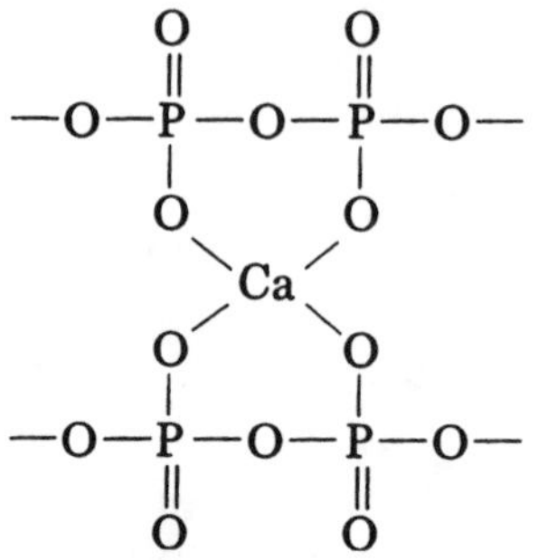

Figure 5–3 Calcium Chelate of a Polyphosphate

Table 5–4 Mineral Constituents in Meat (Beef)

Constituent	mg/100 g
Total calcium	8.6
Soluble calcium	3.8
Total magnesium	24.4
Soluble magnesium	17.7
Total citrate	8.2
Soluble citrate	6.6
Total inorganic phosphorus	233.0
Soluble inorganic phosphorus	95.2
Sodium	168
Potassium	244
Chloride	48

fluid in association with chloride and bicarbonate. During cooking, sodium may be lost, but the other minerals are well retained. Processing does not usually reduce the mineral content of meat. Many processed meats are cured in a brine that contains mostly sodium chloride. As a result, the sodium content of cured meats may be increased.

Ionic equilibria play an important role in the water-binding capacity of meat (Hamm 1971). The normal pH of rigor or post-rigor muscle (pH 5.5) is close to the isoelectric point of actomyosin. At this point the net charge on the protein is at a minimum. By addition of an acid or base, a cleavage of salt cross-linkages occurs, which increases the electrostatic repulsion (Figure 5–4), loosens the protein network, and thus permits more water to be taken up. Addition of neutral salts such as sodium chloride to meat increases water-holding capacity and swelling. The swelling effect has been attributed mainly to the chloride ion. The existence of intra- and extracellular fluid components has been described by Merkel (1971) and may explain the effect of salts such as sodium chloride. The proteins inside the cell membrane are nondiffusible, whereas the inorganic ions may move across this semipermeable membrane. If a solution of the sodium salt of a

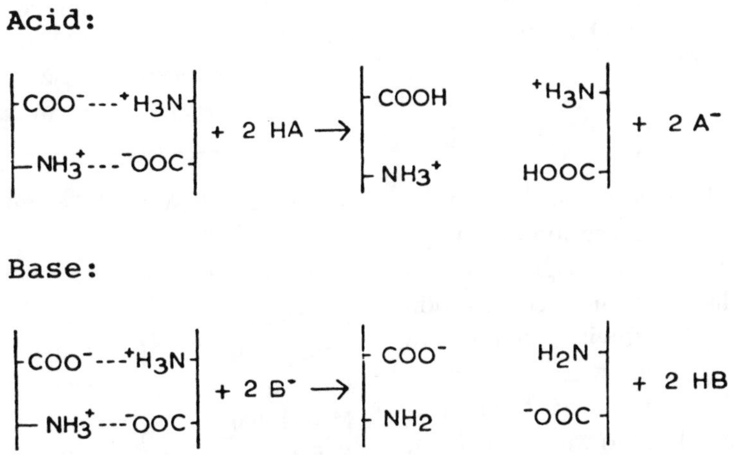

Acid:

Base:

Figure 5–4 Schematic Representation of the Addition of Acid (HA) or Base (B⁻) to an Isoelectric Protein. The isoelectric protein has equal numbers of positive and negative charges. The acid HA donates protons, the base B⁻ accepts protons. *Source*: Reprinted with permission from R. Hamm, *Colloid Chemistry of Meat*, © 1972, Paul Parey (in German).

protein is on one side of the membrane and sodium chloride on the other side, diffusion will occur until equilibrium has been reached. This can be represented as follows:

3 Na⁺	3 Na⁺		4 Na⁺	2 Na⁺
3 Pr⁻	3 Cl⁻		3 Pr⁻	2 Cl⁻
			1 Cl⁻	
At start			At equilibrium	

At equilibrium the product of the concentrations of diffusable ions on the left side of the membrane must be equal to the product on the right side, shown as follows:

$$[Na^+]_L \, [Cl^-]_L = [Na^+]_R \, [Cl^-]_R$$

In addition, the sum of the cations on one side must equal the sum of anions on the other side and vice versa:

$$[Na^+]_L = [Pr^-]_L + [Cl^-]_L \text{ and } [Na^+]_R = [Cl^-]_R$$

This is called the Gibbs-Donnan equilibrium and provides an insight into the reasons for the higher concentration of sodium ions in the intracellular fluid.

Struvite

Occasionally, phosphates can form undesirable crystals in foods. The most common example is struvite, a magnesium-ammonium phosphate of the composition $Mg \cdot (NH_4)PO_4 \cdot 6H_2O$. Struvite crystals are easily mistaken by consumers for broken pieces of glass. Most reports of struvite formation have been related to canned seafood, but occasionally the presence of struvite in other foods has been reported. It is assumed that in canned seafood, the struvite is formed from the magnesium of sea water and ammonia generated by the effect of heat on the fish or shellfish muscle protein.

Minerals in Plant Products

Plants generally have a higher content of potassium than of sodium. The major minerals in wheat are listed in Table 5–5 and include potassium, phosphorus, calcium, magnesium, and sulfur (Schrenk 1964). Sodium in wheat is present at a level of only about 80 ppm and is considered a trace element in this case. The minerals in a wheat kernel are not uniformly distributed; rather, they are concentrated in the areas close to the bran coat and in the bran itself. The various fractions resulting from the milling process have quite different ash contents. The ash content of flour is considered to be related to quality, and the degree of extraction of wheat in milling can be judged from the ash content of the flour. Wheat flour with high ash content is darker in color; generally, the lower the ash content, the whiter the flour. This general principle applies, but the ash content of wheat may vary within wide limits and is influenced by rainfall, soil conditions, fertilizers, and other factors. The distribution of mineral components in the various parts of the wheat kernel is shown in Table 5–6.

Table 5–5 Major Mineral Element Components in Wheat Grain

Element	Average (%)	Range (%)
Potassium	0.40	0.20–0.60
Phosphorus	0.40	0.15–0.55
Calcium	0.05	0.03–0.12
Magnesium	0.15	0.08–0.30
Sulfur	0.20	0.12–0.30

Source: Reprinted with permission from W.G. Schrenk, *Minerals in Wheat Grain*, Technical Bulletin 136, © 1964, Kansas State University Agricultural Experimental Station.

High-grade patent flour, which is pure endosperm, has an ash content of 0.30 to 0.35 percent, whereas whole wheat meal may have an ash content from 1.35 to 1.80 percent.

The ash content of soybeans is relatively high, close to 5 percent. The ash and major mineral levels in soybeans are listed in Table 5–7. Potassium and phosphorus are the elements present in greatest abundance. About 70 to 80 percent of the phosphorus in soybeans is present in the form of phytic acid, the phosphoric acid ester of inositol (Figure 5–5). Phytin is the calcium-magnesium-potassium salt of inositol hexaphosphoric acid or phytic acid. The phytates are important because of their effect on protein solubility and because they may interfere with absorption of calcium from the diet. Phytic acid is present in many foods of plant origin.

A major study of the mineral composition of fruits was conducted by Zook and Lehmann (1968). Some of their findings for the major minerals in fruits are listed in Table 5–8. Fruits are generally not as rich in minerals as vegetables are. Apples have the lowest mineral content of the fruits analyzed. The mineral levels of all fruits show great variation depending on growing region.

The rate of senescence of fruits and vegetables is influenced by the calcium content of the tissue (Poovaiah 1986.) When fruits and vegetables are treated with calcium solutions, the quality and storage life of the products can be extended.

TRACE ELEMENTS

Because trace metals are ubiquitous in our environment, they are found in all of the foods we eat. In general, the abundance of trace elements in foods is related to their abundance in the environment, although this relationship is not absolute, as has been indicated by Warren (1972b). Table 5–9 presents the order of abundance of some trace elements in soil, sea water, vegetables, and humans and the order of our intake. Trace elements may be present in foods as a result of uptake from soil or feeds or from contamination during and subsequent to processing

Table 5–6 Mineral Components in Endosperm and Bran Fractions of Red Winter Wheat

	P (%)	K (%)	Na (%)	Ca (%)	Mg (%)	Mn (ppm)	Fe (ppm)	Cu (ppm)
Total endosperm	0.10	0.13	0.0029	0.017	0.016	2.4	13	8
Total bran	0.38	0.35	0.0067	0.032	0.11	32	31	11
Wheat kernel								
Center section	0.35	0.34	0.0051	0.025	0.086	29	40	7
Germ end	0.55	0.52	0.0036	0.051	0.13	77	81	8
Brush end	0.41	0.41	0.0057	0.036	0.13	44	46	12
Entire kernel	0.44	0.42	0.0064	0.037	0.11	49	54	8

Source: From V.H. Morris et al., Studies on the Composition of the Wheat Kernel. II. Distribution of Certain Inorganic Elements in Center Sections, *Cereal Chem.*, Vol. 22, pp. 361–372, 1945.

Table 5–7 Mineral Content of Soybeans (Dry Basis)

Mineral	No. of Analyses	Range (%)	Mean (%)
Ash	—	3.30–6.35	4.60
Potassium	29	0.81–2.39	1.83
Calcium	9	0.19–0.30	0.24
Magnesium	7	0.24–0.34	0.31
Phosphorus	37	0.50–1.08	0.78
Sulfur	6	0.10–0.45	0.24
Chlorine	2	0.03–0.04	0.03
Sodium	6	0.14–0.61	0.24

Source: Reprinted with permission from A.K. Smith and S.J. Circle, *Soybeans: Chemistry and Technology*, © 1972, AVI Publishing Co.

of foods. For example, the level of some trace elements in milk depends on the level in the feed; for other trace elements, increases in levels in the feed are not reflected in increased levels in the milk. Crustacea and mollusks accumulate metal ions from the ambient sea water. As a result, concentrations of 8,000 ppm of copper and 28,000 ppm of zinc have been recorded (Meranger and Somers 1968). Contamination of food products with metal can occur as a result of pickup of metals from equipment or from packaging materials, especially tin cans. The nickel found in milk comes almost

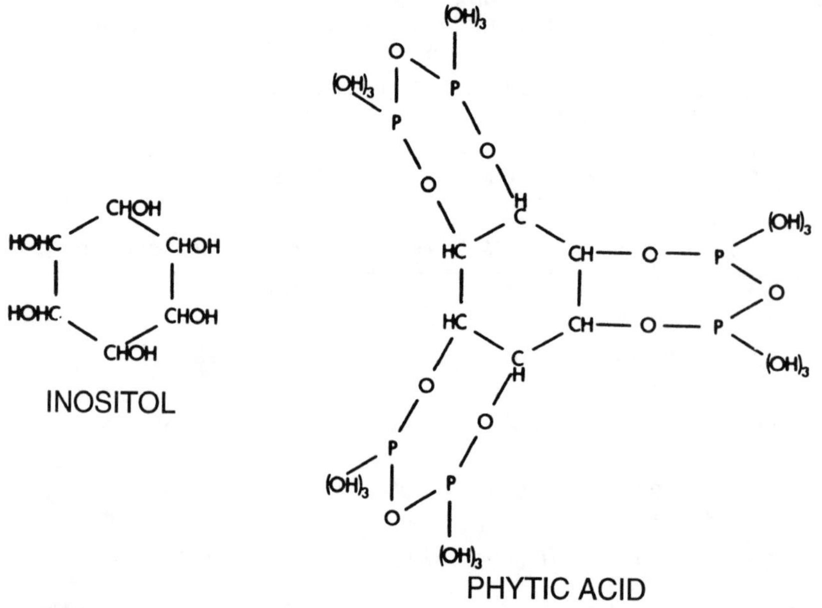

Figure 5–5 Inositol and Phytic Acid

Table 5–8 Mineral Content of Some Fruits

Fruit	Minerals (mg/100 g)				
	N	*Ca*	*Mg*	*P*	*K*
Orange (California navel)	162	23.7	10.2	15.8	175
Apple (McIntosh)	30	2.4	3.6	5.4	96
Grape (Thompson)	121	6.2	5.8	12.8	200
Cherry (Bing)	194	9.6	16.2	13.3	250
Pear (Bartlett)	63	4.8	6.5	9.3	129
Banana (Ecuador)	168	2.7	25.4	16.4	373
Pineapple (Puerto Rico)	71	2.2	3.9	3.0	142

Source: From E.G. Zook and J. Lehmann, Mineral Composition of Fruits, *J. Am. Dietetic Assoc.*, Vol. 52, pp. 225–231, 1968.

exclusively from stainless steel in processing equipment. Milk coming from the udder has no detectable nickel content. On the other hand, nutritionists are concerned about the low iron intake levels for large numbers of the population; this low intake can in part be explained by the disappearance of iron equipment and utensils from processing and food preparation.

Originally, nine of the trace elements were considered to be essential to humans: cobalt, copper, fluorine, iodine, iron, manganese, molybdenum, selenium, and zinc. Recently, chromium, silicon, and nickel have been added to this list (Reilly 1996). These are mostly metals; some are metalloids. In addition to essential trace elements, several trace elements have no known essentiality and

Table 5–9 Order of Abundance of Some Trace Elements in Various Media

Element	Soil	Sea Water	Vegetables	Man	Man's Intake
Iron	1	1	1	1	1
Manganese	2	4	3	5	3
Nickel	4	7	6	6	5
Zinc	3	2	2	2	2
Copper	5	3	4	3	4
Cobalt	7	8	8	8	8
Lead	6	5	5	4	6
Molybdenum	8	6	7	7	7
Cadmium	9	?	9	9	9
Mercury	?	9	?	10	?

Source: From H.V. Warren, Geology and Medicine, *Western Miner*, pp. 34–37, 1972.

some are toxic (such as lead, mercury, and cadmium). These toxic trace elements, which are classified as contaminants, are dealt with in Chapter 11.

Trace elements get into foods by different pathways. The most important source is from the soil, by absorption of elements in aqueous solution through the roots. Another, minor, source is foliar penetration. This is usually associated with industrial air pollution and vehicle emissions. Other possible sources are fertilizers, agricultural chemicals, and sewage sludge. Sewage sludge is a good source of nitrogen and phosphate but may contain high levels of trace minerals, many of these originating from industrial activities such as electroplating. Trace minerals may also originate from food processing and handling equipment, food packaging materials, and food additives.

Cobalt

Cobalt is an integral part of the only metal containing vitamin B_{12}. The level of cobalt in foods varies widely, from as little as 0.01 ppm in corn and cereals to 1 ppm in some legumes. The human requirement is very small and deficiencies do not occur.

Copper

Copper is present in foods as part of several copper-containing enzymes, including the polyphenolases. Copper is a very powerful prooxidant and catalyzes the oxidation of unsaturated fats and oils as well as ascorbic acid. The normal daily diet contains from 2 to 5 mg of copper, more than ample to cover the daily requirement of 0.6 to 2 mg.

Iron

Iron is a component of the heme pigments and of some enzymes. In spite of the fact that some foods have high iron levels, much of the population has frequently been found to be deficient in this element. Animal food products may have high levels that are well absorbed; liver may contain several thousand ppm of iron. The iron from other foods such as vegetables and eggs is more poorly absorbed. In the case of eggs the uptake is poor because the ferric iron is closely bound to the phosphate of the yolk phosphoproteins. Iron is used as a food additive to enrich flour and cereal products. The form of iron used significantly determines how well it will be taken up by the body. Ferrous sulfate is very well absorbed, but will easily discolor or oxidize the food to which it is added. Elemental iron is also well absorbed and is less likely to change the food. For these reasons, it is the preferred form of iron for the enrichment of flour.

Zinc

Zinc is the second most important of the essential trace elements for humans. It is a constituent of some enzymes, such as carbonic anhydrase. Zinc is sufficiently abundant that deficiencies of zinc are unknown. The highest levels of zinc are found in shellfish, which may contain 400 ppm. The level of zinc in cereal grains is 30 to 40 ppm. When acid foods such as fruit juices are stored in galvanized containers, sufficient zinc may be dissolved to cause zinc poisoning. The zinc in meat is tightly bound to the myofibrils and has been speculated to influence meat's water-binding capacity (Hamm 1972).

Manganese

Manganese is present in a wide range of foods but is not easily absorbed. This metal is associated with the activation of a number of enzymes. In wheat, a manganese content of 49 ppm has been reported (Schrenk 1964). This is mostly concentrated in the germ and bran; the level in the endosperm is only 2.4 ppm. Information on the manganese content of seafoods has been supplied by Meranger and Somers (1968). Values range from a low of 1.1 ppm in salmon to a high of 42 ppm in oyster.

Molybdenum

Molybdenum plays a role in several enzyme reactions. Some of the molybdenum-containing enzymes are aldehyde oxidase, sulfite oxidase, xanthine dehydrogenase, and xanthine oxidase. This metal is found in cereal grains and legumes; leafy vegetables, especially those rich in chlorophyll; animal organs; and in relatively small amounts, less than 0.1 ppm, in fruits. The molybdenum content of foods is subject to large variations.

Selenium

Selenium has recently been found to protect against liver necrosis. It usually occurs bound to organic molecules. Different selenium compounds have greater or lesser protective effect. The most active form of selenium is selenite, which is also the least stable chemically. Many selenium compounds are volatile and can be lost by cooking or processing. Kiermeier and Wigand (1969) found about a 5 percent loss of selenium as a result of drying of skim milk. The variation in selenium content of milk is wide

and undoubtedly associated with the selenium content of the soil. The same authors report figures for selenium in milk in various parts of the world ranging from 5 to 1,270 $\mu g/kg$. The selenium in milk is virtually all bound to the proteins. Morris and Levander (1970) determined the selenium content of a wide variety of foods. Most fruits and vegetables contain less than 0.01 $\mu g/g$. Grain products range from 0.025 to 0.66 $\mu g/g$, dried skim milk from 0.095 to 0.24 $\mu g/g$, meat from 0.1 to 1.9 $\mu g/g$, and seafood from 0.4 to 0.7 $\mu g/g$.

Fluorine

Fluorine is a constituent of skeletal bone and helps reduce the incidence of dental caries. The fluorine content of drinking water is usually below 0.2 mg/L but in some locations may be as high as 5 mg/L. The optimal concentration for dental health is 1 mg/L. The fluoride content of vegetables is low, with the exception of spinach, which contains 280 $\mu g/100$ g. Milk contains 20 $\mu g/100$ g and beef about 100 $\mu g/100$ g. Fish foods may contain up to 700 $\mu g/100$ g and tea about 100 $\mu g/g$.

Iodine

Iodine is not present in sufficient amounts in the diet in several areas of the world; an iodine deficiency results in goiter. The addition of iodine to table salt has been extremely effective in reducing the incidence of goiter. The iodine content of most foods is in the area of a few mg/100 g and is subject to great local variations. Fish and shellfish have higher levels. Saltwater fish have levels of about 50 to 150 mg/100 g and shellfish may have levels as high as 400 mg/100 g.

Nickel

Foods with a relatively high nickel content include nuts, legumes, cocoa products, shellfish, and hydrogenated fats. The source of nickel in the latter results from the use of nickel catalyst in the hydrogenation process. Animal products are generally low in nickel, plant products high (Table 5–10). The intake of nickel from the diet depends, therefore, on the origin and amounts of various foods consumed. Dietary nickel intake has been estimated to be in the range of 150 to 700 µg/day (Nielsen 1988), and the suggested dietary nickel requirement is about 35 µg/day.

Finished hydrogenated vegetable oils contain less than 1 mg/kg nickel. Treatment of the finished oil with citric or phosphoric acid followed by bleaching should result in nickel levels of less than 0.2 mg/kg.

Chromium

Recent well-controlled studies (Anderson 1988) have found that dietary intake of chromium is in the order of 50 µg/day. Refining and processing of foods may lead to loss of chromium. As an example, in the milling of flour, recovery of chromium in white flour is only 35 to 44 percent of that of the parent wheat (Zook et al. 1970). On the other hand, the widespread use of stainless steel equipment in food processing results in leaching of chromium into the food products (Offenbacher and Pi-Sunyer 1983). No foods are known to contain higher-than-average levels of chromium. The average daily intake of chromium from various food groups is shown in Table 5–11. It has been suggested that the dietary intake of chromium in most normal individuals is suboptimal and can lead to nutritional problems (Anderson 1988).

Silicon

Silicon is ubiquitous in the environment and present in many foods. Foods of animal origin are relatively low in silicon; foods of plant origin are relatively high. Good plant sources are unrefined grains, cereal products, and root crops. The dietary intake of silicon is poorly known but appears to be in the range of 20 to 50 µg/day. Although silicon is now regarded as an essential mineral for humans, a minimum requirement has not been established.

Additional Information on Trace Elements

The variations in trace elements in vegetables may be considerable (Warren 1972a) and may depend to a large extent on the nature of the soil in which the vegetables are grown. Table 5–12 illustrates the extent of the variability in the content of copper, zinc, lead, and molybdenum of a number of vege-

Table 5–10 Nickel Content of Some Foods

Food	Nickel Content (µg/g Fresh Weight)
Cashew nuts	5.1
Peanuts	1.6
Cocoa powder	9.8
Bittersweet chocolate	2.6
Milk chocolate	1.2
Red kidney beans	0.45
Peas, frozen	0.35
Spinach	0.39
Shortening	0.59–2.78

Source: Reprinted with permission from F.H. Nielsen, The Ultratrace Elements, in *Trace Minerals in Foods*, K.T. Smith, ed., p. 385, 1988, by courtesy of Marcel Dekker, Inc.

Table 5–11 Chromium Intake from Various Food Groups

Food Group	Average Daily Intake (μg)	Comments
Cereal products	3.7	55% from wheat
Meat	5.2	55% from pork
		25% from beef
Fish and seafood	0.6	
Fruits, vegetables, nuts	6.8	70% from fruits and berries
Dairy products, eggs, margarine	6.2	85% from milk
Beverages, confectionery, sugar, and condiments	6.6	45% from beer, wine, and soft drinks
Total	29.1	

Source: Reprinted with permission from R.A. Anderson, Chromium, in *Trace Minerals in Foods*, K.T. Smith, ed., p. 238, 1988, by courtesy of Marcel Dekker, Inc.

tables. The range of concentrations of these metals frequently covers one order of magnitude and occasionally as much as two orders of magnitude. Unusually high concentrations of certain metals may be associated with the incidence of diseases such as multiple sclerosis and cancer in humans.

Aluminum, which has been assumed to be nonnutritious and nontoxic, has come under increasing scrutiny. Its presence has been suggested to be involved in several serious conditions, including Alzheimer's disease (Greger 1985). Since aluminum is widely used in utensils and packaging materials, there is great interest in the aluminum content of foods. Several aluminum salts are used as food additives, for example, sodium aluminum phosphate as a leavening agent and aluminum sulfate for pH control. The estimated average daily intake of aluminum is 26.5 mg, with 70 percent coming from grain products (Greger 1985).

Fruits contain relatively high levels of organic acids, which may combine with metal ions. It is now generally agreed that these compounds may form chelates of the general formula $M_yH_pL_m(OH)_x$, where M and L represent the metal and the ligand, respectively. According to Pollard and Timberlake (1971), cupric ions form strong complexes with acids containing α-hydroxyl groups. The major fruit acids, citric, malic, and tartaric, are multidendate ligands capable of forming polynuclear chelates. Cupric and ferric ions form stronger complexes than ferrous ions. The strongest complexes are formed by citrate, followed by malate and then tartrate.

METAL UPTAKE IN CANNED FOODS

Canned foods may take up metals from the container, tin and iron from the tin plate, and tin and lead from the solder. There are several types of internal can corrosion. *Rapid detinning is* one of the most serious problems of can corrosion. With most acid foods, when canned in the absence of oxygen, tin forms the anode of the tin-iron couple. The tin under these conditions goes into solution

Table 5–12 Extreme Variation in the Content of Copper, Zinc, Lead, and Molybdenum in Some Vegetables

	"Normal" Content in ppm Wet Weight	Minimum as Fraction of "Normal"	Maximum as Multiple of "Normal"	Extreme Range
Copper				
Lettuce	0.74	1/15	8	1–120
Cabbage	0.26	1/6	2.5	1–15
Potato	0.92	1/9	4	1–36
Bean (except broad)	0.56	2/5	2.5	1–22
Carrot	0.52	1/9	2.5	1–22
Beet	0.78	1/9	2.5	1–20
Zinc				
Lettuce	4.9	1/6	15	1–90
Cabbage	1.9	1/2	6	1–12
Potato	2.9	1/2	5	1–10
Bean (except broad)	3.6	1/2	2	1–4
Carrot	3.4	1/2	8	1–48
Beet	4.1	1/4	12	1–16
Lead				
Lettuce	0.25	1/10	30	1–300
Cabbage	0.10	1/8	2.5	1–20
Potato	0.40	1/10	15	1–150
Bean (except broad)	0.24	1/5	4	1–20
Carrot	0.22	1/3	9	1–27
Beet	0.20	1/6	11	1–66
Molybdenum				
Lettuce	0.06	1/8	12	1–96
Cabbage	0.20	1/30	8	1–240
Potato	0.15	1/16	7.5	1–120
Bean (except broad)	0.48	1/30	7	1–210
Carrot	0.22	1/4	3.5	1–14
Beet	0.04	1/30	10	1–300

Source: From H.V. Warren, Variations in the Trace Element Contents of Some Vegetables, *J. Roy. Coll. Gen. Practit.*, Vol. 22, pp. 56–60, 1972.

at an extremely slow rate and can provide product protection for two years or longer. There are, however, conditions where iron forms the anode, and in the presence of depolarizing or oxidizing agents the dissolution of tin is greatly accelerated. The food is protected until most of the tin is dissolved; thereafter, hydrogen is produced and the can swells and becomes a *springer*. Some foods are more likely to involve rapid detinning, including spinach, green beans, tomato products, potatoes, carrots, vegetable soups, and certain fruit juices such as prune and grapefruit juice.

Another corrosion problem of cans is sulfide staining. This may happen when the food contains the sulfur-containing amino acids cysteine, cystine, or methionine. When the food is heated or aged, reduction may result in the formation of sulfide ions, which can then react with tin and iron to form SnS and FeS. The compound SnS is the major component of the sulfide stain. This type of corrosion may occur with foods such as pork, fish, and peas (Seiler 1968). Corrosion of tin cans depends on the nature of the canned food as well as on the type of tin plate used. Formerly, hot dipped tin plate was used, but this has been mostly replaced by electrolytically coated plate. It has been shown (McKirahan et al. 1959) that the size of the crystals in the tin coating has an important effect on corrosion resistance. Tin plate with small tin crystals easily develops hydrogen swell, whereas tin plate containing large crystals is quite resistant. Seiler (1968) found that the orientation of the different crystal planes also significantly affected the ease of forming sulfide stains.

The influence of processing techniques for grapefruit juice on the rate of can corrosion was studied by Bakal and Mannheim (1966). They found that the dissolved tin content can serve as a corrosion indicator. In Israel the maximum prescribed limit for tin content of canned food is 250 ppm. Deaeration of the juice significantly lowers tin dissolution. In a study of the in-can shelf life of tomato paste, Vander Merwe and Knock (1968) found that, depending on maturity and variety, 1 g of tomato paste stored at 22°C could corrode tin at rates ranging from 9×10^{-6} g/month to 68×10^{-6} g/month. The useful shelf life could vary from 24 months to as few as 3 months. Up to 95 percent of the variation could be related to effects of maturity and variety and the associated differences in contents of water-insoluble solids and nitrate.

Severe detinning has often been observed with applesauce packed in plain cans with enameled ends. This is usually characterized by detinning at the headspace interface. Stevenson and Wilson (1968) found that steam flow closure reduced the detinning problem, but the best results were obtained by complete removal of oxygen through nitrogen closure. Detinning by canned spinach was studied by Lambeth et al. (1969) and was found to be significantly related to the oxalic acid content of the fresh leaves and the pH of the canned product. High-oxalate spinach caused detinning in excess of 60 percent after 9 months' storage.

In some cases the dissolution of tin into a food may have a beneficial effect on food

Table 5–13 Iron and Tin Content of Fruit Juices

Product	Iron (ppm)	Tin (ppm)
Fresh orange juice	0.5	7.5
Bottled orange juice	2.5	25
Bottled orange juice	2.0	50
Bottled pineapple juice	15.0	50
Canned orange juice	2.5	60
Canned orange juice	0.5	115
Canned orange juice	2.5	120
Canned pineapple juice	17.5	135

Source: From W.J. Price and J.T.H. Roos, Analysis of Fruit Juice by Atomic Absorption Spectrophotometry. I. The Determination of Iron and Tin in Canned Juice, *J. Sci. Food Agric.*, Vol. 20, pp. 427–439, 1969.

color, with iron having the opposite effect. This is the case for canned wax beans (Van Buren and Downing 1969). Stannous ions were effective in preserving the light color of the beans, whereas small amounts of iron resulted in considerable darkening. A black discoloration has sometimes been observed in canned all-green asparagus after opening of the can. This has been attributed (Lueck 1970) to the formation of a black, water-insoluble coordination compound of iron and rutin. The iron is dissolved from the can, and the rutin is extracted from the asparagus during the sterilization. Rutin is a flavonol, the 3-rutinoside of quercetin. The black discoloration occurs only after the iron has been oxidized to the ferric state. Tin forms a yellow, water-soluble complex with rutin, which does not present a color problem. The uptake of iron and tin from canned foods is a common occurrence, as is demonstrated by Price and Roos (1969), who studied the presence of iron and tin in fruit juice (Table 5–13).

REFERENCES

Anderson, R.A. 1988. Chromium. In *Trace minerals in foods*, ed. K.T. Smith. New York: Marcel Dekker.

Bakal, A., and H.C. Mannheim. 1966. The influence of processing variants of grapefruit juice on the rate of can corrosion and product quality. *Israel J. Technol.* 4: 262–267.

Dunaif, G.D., and C.-S. Khoo. 1986. Developing low and reduced-sodium products: An industrial perspective. *Food Technol.* 40, no. 12: 105–107.

Greger, J.L. 1985. Aluminum content of the American diet. *Food Technol.* 39, no. 5: 73–80.

Hamm, R. 1971. Interactions between phosphates and meat proteins. In *Phosphates in food processing*, ed. J.M. deMan and P. Melnychyn. Westport, CT: AVI Publishing Co.

Hamm, R. 1972. *Colloid chemistry of meat* (in German). Berlin: Paul Parey.

Jenness, R., and S. Patton. 1959. *Principles of dairy chemistry*. New York: John Wiley & Sons.

Kiermeier, F., and W. Wigand. 1969. Selenium content of milk and milk powder (in German). *Z. Lebensm. Unters. Forsch.* 139: 205–211.

Lambeth, V.N., et al. 1969. Detinning by canned spinach as related to oxalic acid, nitrates and mineral composition. *Food Technol.* 23, no. 6: 132–134.

Lee, K., and J.L. Greger. 1983. Bioavailability and chemistry of iron from nitrite-cured meats. *Food Technol.* 37, no. 10: 139–144.

Lueck, R.H. 1970. Black discoloration in canned asparagus. Interrelations of iron, tin, oxygen, and rutin. *Agr. Food Chem.* 18: 607–612.

Marsh, A.C. 1983. Processes and formulations that affect the sodium content of foods. *Food Technol.* 37, no. 7: 45–49.

Maurer, A.J. 1983. Reduced sodium usage in poultry muscle foods. *Food Technol.* 37, no. 7: 60–65.

McKirahan, R.D., et al. 1959. Application of differentially coated tin plate for food containers. *Food Technol.* 13: 228–232.

Meranger, J.C., and E. Somers. 1968. Determination of the heavy metal content of seafoods by atomic absorption spectrophotometry. *Bull. Environ. Contamination Toxicol.* 3: 360–365.

Merkel, R.A. 1971. Inorganic constituents. In *The science of meat and meat products*, ed. J.F. Price and B.S. Schweigert. San Francisco: W.H. Freeman and Co.

Morris, V.C., and O.A. Levander. 1970. Selenium content of foods. *J. Nutr.* 100: 1383–1388.

Nielsen, F.H. 1988. The ultratrace elements. In *Trace minerals in foods*, ed. K.T. Smith. New York: Marcel Dekker.

Offenbacher, E.G., and F.X. Pi-Sunyer. 1983. Temperature and pH effects on the release of chromium from stainless steel into water and fruit juices. *J. Agr. Food Chem.* 31: 89–92.

Pfeilsticker, K. 1970. Food components as metal chelates. *Food Sci. Technol.* 3: 45–51.

Politz, M.L., and F.M. Clydesdale. 1988. Effect of enzymatic digestion, pH and molecular weight on the iron solubilizing properties of chicken muscle. *J. Food Sci.* 52: 1081–1085, 1090.

Pollard, A., and C.F. Timberlake. 1971. Fruit juices. In *The biochemistry of fruits and their products*, Vol. 2, ed. A.C. Hulme. New York: Academic Press.

Poovaiah, B.W. 1986. Role of calcium in prolonging storage life of fruits and vegetables. *Food Technol.* 40, no. 5: 86–89.

Price, W.J., and J.T.H. Roos. 1969. Analysis of fruit juice by atomic absorption spectrophotometry I. The determination of iron and tin in canned juice. *J. Sci. Food Agric.* 20: 427–439.

Reilly, C. 1996. *Selenium in food and health*. London: Blackie Academic and Professional.

Schrenk, W.G. 1964. *Minerals in wheat grain*. Technical Bulletin 136. Manhattan, KS: Kansas State University Agricultural Experimental Station.

Schweiger, R.G. 1966. Metal chelates of pectate and comparison with alginate. *Kolloid Z.* 208: 28–31.

Seiler, B.C. 1968. The mechanism of sulfide staining of tin foodpacks. *Food Technol.* 22: 1425–1429.

Stevenson, C.A., and C.H. Wilson. 1968. Nitrogen enclosure of canned applesauce. *Food Technol.* 33: 1143–1145.

Toma, R.B., and D.J. Curtis. 1986. Dietary fiber: Effect on mineral bioavailability. *Food Technol.* 40, no. 2: 111–116.

Van Buren, J.P., and D.L. Downing. 1969. Can characteristics, metal additives, and chelating agents: Effect on the color of canned wax beans. *Food Technol.* 23: 800–802.

Vander Merwe, H.B., and G.G. Knock. 1968. In-can shelf life of tomato paste as affected by tomato variety and maturity. *J. Food Technol.* 3: 249–262.

Warren, H.V. 1972a. Variations in the trace element contents of some vegetables. *J. Roy. Coll. Gen. Practit.* 22: 56–60.

Warren, H.V. 1972b. Geology and medicine. *Western Miner*, Sept., 34–37.

Zook, E.G. et al. 1970. Nutrient composition of selected wheat and wheat products. *Cereal Chem.* 47: 720–727.

Zook, E.G., and J. Lehmann. 1968. Mineral composition of fruits. *J. Am. Dietetic Assoc.* 52: 225–231.

CHAPTER 6

Color

INTRODUCTION

Color is important to many foods, both those that are unprocessed and those that are manufactured. Together with flavor and texture, color plays an important role in food acceptability. In addition, color may provide an indication of chemical changes in a food, such as browning and caramelization. For a few clear liquid foods, such as oils and beverages, color is mainly a matter of transmission of light. Other foods are opaque—they derive their color mostly from reflection.

Color is the general name for all sensations arising from the activity of the retina of the eye. When light reaches the retina, the eye's neural mechanism responds, signaling color among other things. Light is the radiant energy in the wavelength range of about 400 to 800 nm. According to this definition, color (like flavor and texture) cannot be studied without considering the human sensory system. The color perceived when the eye views an illuminated object is related to the following three factors: the spectral composition of the light source, the chemical and physical characteristics of the object, and the spectral sensitivity properties of the eye. To evaluate the properties of the object, we must standardize the other two factors. Fortunately, the characteristics of different people's eyes

for viewing colors are fairly uniform; it is not too difficult to replace the eye by some instrumental sensor or photocell that can provide consistent results. There are several systems of color classification; the most important is the CIE system (Commission International de l'Eclairage—International Commission on Illumination). Other systems used to describe food color are the Munsell, Hunter, and Lovibond systems.

When the reflectance of different colored objects is determined by means of spectrophotometry, curves of the type shown in Figure 6–1 are obtained. White materials reflect equally over the whole visible wavelength range, at a high level. Gray and black materials also reflect equally over this range but to a lower degree. Red materials reflect in the higher wavelength range and absorb the other wavelengths. Blue materials reflect in the low-wavelength range and absorb the high-wavelength light.

CIE SYSTEM

The spectral energy distribution of CIE light sources A and C is shown in Figure 6–2. CIE illuminant A is an incandescent light operated at 2854°K, and illuminant C is the same light modified by filters to result in a

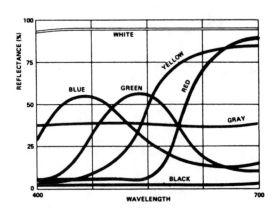

Figure 6–1 Spectrophotometric Curves of Colored Objects. *Source:* From Hunter Associates Lab., Inc.

spectral composition that approximates that of normal daylight. Figure 6–2 also shows the luminosity curve of the standard observer as specified by CIE. This curve indicates

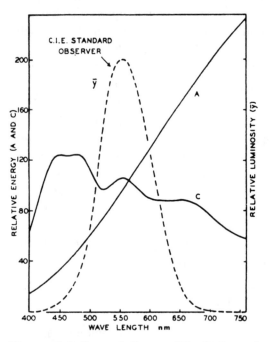

Figure 6–2 Spectral Energy Distribution of Light Sources *A* and *C*, the CIE, and Relative Luminosity Function γ for the CIE Standard Observer

how the eyes of normal observers respond to the various spectral light types in the visible portion of the spectrum. By breaking down the spectrum, complex light types are reduced to their component spectral light types. Each spectral light type is completely determined by its wavelength. In some light sources, a great deal of radiant energy is concentrated in a single spectral light type. An example of this is the sodium lamp shown in Figure 6–3, which produces monochromatic light. Other light sources, such as incandescent lamps, give off a continuous spectrum. A fluorescent lamp gives off a continuous spectrum on which is superimposed a line spectrum of the primary radiation produced by the gas discharge (Figure 6–3).

In the description of light sources, reference is sometimes made to the *black body*. This is a radiating surface inside a hollow space, and the light source's radiation comes out through a small opening. The radiation is independent of the type of material the light source is made of. When the temperature is very high, about 6000°K the maximum of the energy distribution will fall about in the middle of the visible spectrum. Such energy distribution corresponds with that of daylight on a cloudy day. At lower temperatures, the maximum of the energy distribution shifts to longer wavelengths. At 3000° K, the spectral energy distribution is similar to that of an incandescent lamp; at this temperature the energy at 380 nm is only one-sixteenth of that at 780 nm, and most of the energy is concentrated at higher wavelengths (Figure 6–3). The uneven spectral distribution of incandescent light makes red objects look attractive and blue ones unattractive. This is called color rendition. The human eye has the ability to adjust for this effect.

The CIE system is a trichromatic system; its basis is the fact that any color can be

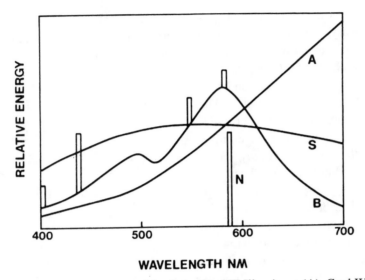

WAVELENGTH NM

Figure 6–3 Spectral Energy Distribution of Sunlight (S), CIE Illuminant (A), Cool White Fluorescent Lamp (B), and Sodium Light (N)

matched by a suitable mixture of three primary colors. The three primary colors, or *primaries*, are red, green, and blue. Any possible color can be represented as a point in a triangle. The triangle in Figure 6–4 shows how colors can be designated as a ratio of the three primaries. If the red, green, and blue values of a given light type are represented by a, b, and c, then the ratios of each to the total light are given by $a/(a + b + c)$, $b/(a + b + c)$, and $c/(a + b + c)$, respectively. Since the sum of these is one, then only two have to be known to know all three. Color, therefore, is determined by two, not three, of these mutually dependent quantities. In Figure 6–4, a color point is represented by P. By determining the distance of P from the right angle, the quantities $a/(a + b + c)$ and $b/(a + b + c)$ are found. The quantity $c/(a + b + c)$ is then found, by first extending the horizontal dotted line through P until it crosses the hypotenuse at Q and by then constructing another right angle triangle with Q at the top. All combinations

of a, b, and c will be points inside the triangle.

The relative amounts of the three primaries required to match a given color are called the

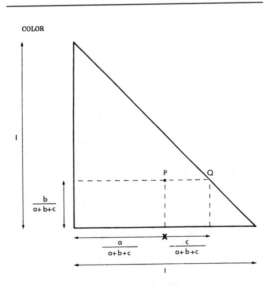

Figure 6–4 Representation of a Color as a Point in a Color Triangle

tristimulus values of the color. The CIE primaries are imaginary, because there are no real primaries that can be combined to match the highly saturated hues of the spectrum.

In the CIE system the red, green, and blue primaries are indicated by X, Y, and Z. The amount of each primary at any particular wavelength is given by the values $\bar{x}$, $\bar{y}$, and $\bar{z}$. These are called the *distribution coefficients* or the red, green, and blue factors. They represent the tristimulus values for each chosen wavelength. The distribution coefficients for the visible spectrum are presented in Figure 6–5. The values of $\bar{y}$ correspond with the luminosity curve of the standard observer (Figure 6–2). The distribution coefficients are dimensionless because they are the numbers by which radiation energy at each wavelength must be multiplied to arrive at the X,

Y, and Z content. The amounts of X, Y, and Z primaries required to produce a given color are calculated as follows:

$$X = \int_{380}^{780} \bar{x} \, IRdh$$

$$XY = \int_{380}^{780} \bar{y} \, IRdh$$

$$XZ = \int_{380}^{780} \bar{z} \, IRdh$$

where

I = spectral energy distribution of illuminant

R = spectral reflectance of sample

dh = small wavelength interval

$\bar{x}, \bar{y}, \bar{z}$ = red, green, and blue factors

The ratios of the primaries can be expressed as

$$x = \frac{X}{X + Y + Z}$$

$$y = \frac{Y}{X + Y + Z}$$

$$z = \frac{Z}{X + Y + Z}$$

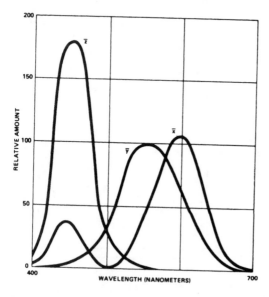

Figure 6–5 Distribution Coefficients $\bar{x}$, $\bar{y}$, and $\bar{z}$ for the Visible Spectrum. *Source:* From Hunter Associates Lab., Inc.

The quantities x and y are called the chromaticity coordinates and can be calculated for each wavelength from

$$x = \overline{x}/(\overline{x} + \overline{y} + \overline{z})$$

$$y = \overline{y}/(\overline{x} + \overline{y} + \overline{z})$$

$$z = 1 - (x + y)$$

A plot of *x* versus *y* results in the CIE chromaticity diagram (Figure 6–6). When the chromaticities of all of the spectral colors are placed in this graph, they form a line called the locus. Within this locus and the line connecting the ends, represented by 400 and 700 nm, every point represents a color that can be made by mixing the three primaries. The point at which exactly equal amounts of each

of the primaries are present is called the equal point and is white. This white point represents the chromaticity coordinates of illuminant *C*. The red primary is located at *x* = 1 and *y* = 0; the green primary at *x* = 0 and *y* = 1; and the blue primary at *x* = 0 and *y* = 0. The line connecting the ends of the locus represents purples, which are nonspectral colors resulting from mixing various amounts of red and blue. All points within the locus represent real colors. All points outside the locus are unreal, including the imaginary primaries *X*, *Y*, and *Z*. At the red end of the locus, there is only one point to represent the wavelength interval of 700 to 780 nm. This

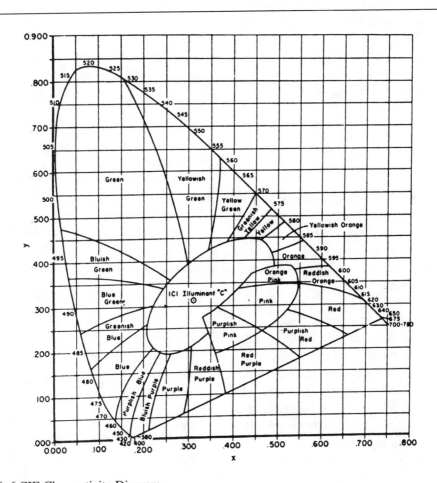

Figure 6–6 CIE Chromaticity Diagram

means that all colors in this range can be simply matched by adjustment of luminosity. In the range of 540 to 700 nm, the spectrum locus is almost straight; mixtures of two spectral light types along this line segment will closely match intervening colors with little loss of purity. In contrast, the spectrum locus below 540 nm is curved, indicating that a combination of two spectral lights along this portion of the locus results in colors of decreased purity.

A pure spectral color is gradually diluted with white when moving from a point on the spectrum locus to the white point P. Such a straight line with purity decreasing from 100 to 0 percent is known as a line of constant dominant wavelength. Each color, except the purples, has a dominant wavelength. The position of a color on the line connecting the locus and P is called excitation purity (p_e) and is calculated as follows:

$$P_e = \frac{x - x_w}{x_p - x_w} = \frac{y - y_w}{y_p - y_w}$$

where

x and y are the chromaticity coordinates of a color

x_w and y_w are the chromaticity coordinates of the achromatic source

x_p and y_p are the chromaticity coordinates of the pure spectral color

Achromatic colors are white, black, and gray. Black and gray differ from white only in their relative reflection of incident light. The purples are nonspectral chromatic colors. All other colors are chromatic; for example, brown is a yellow of low lightness and low saturation. It has a dominant wavelength in the yellow or orange range.

A color can be specified in terms of the tristimulus value Y and the chromaticity coor-

dinates x and y. The Y value is a measure of luminous reflectance or transmittance and is expressed in percent simply as Y/1000.

Another method of expressing color is in terms of luminance, dominant wavelength, and excitation purity. These latter are roughly equivalent to the three recognizable attributes of color: lightness, hue, and saturation. Lightness is associated with the relative luminous flux, reflected or transmitted. Hue is associated with the sense of redness, yellowness, blueness, and so forth. Saturation is associated with the strength of hue or the relative admixture with white. The combination of hue and saturation can be described as chromaticity.

Complementary colors (Table 6–1) are obtained when a straight line is drawn through the equal energy point P. When this is done for the ends of the spectrum locus, the wavelength complementary to the 700 to 780 point is at 492.5 nm, and for the 380 to 410 point is at 567 nm. All of the wavelengths between 492.5 and 567 nm are complementary to purple. The purples can be described in terms of dominant wavelength by using the wavelength complementary to each purple, and purity can be expressed in a manner similar to that of spectral colors.

Table 6–1 Complementary Colors

Wavelength (nm)	Color	Complementary Color
400	Violet	
450	Blue	Yellow
500	Green	Orange
550	Yellow	Red
600	Orange	Violet
650	Red	Blue
700		Green

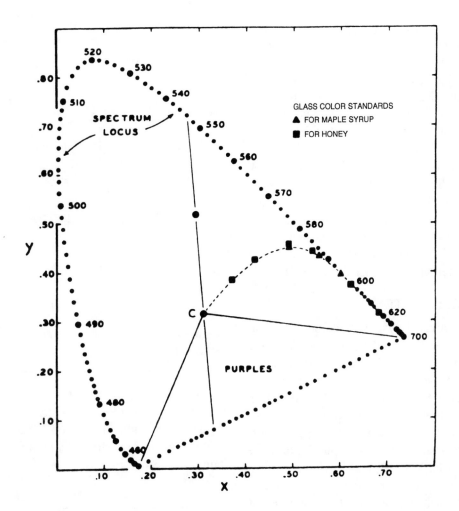

Figure 6–7 CIE Chromaticity Diagram with Color Points for Maple Syrup and Honey Glass Color Standards

An example of the application of the CIE system for color description is shown in Figure 6–7. The curved, dotted line originating from *C* represents the locus of the chromaticity coordinates of caramel and glycerol solutions. The chromaticity coordinates of maple syrup and honey follow the same locus. Three triangles on this curve represent the chromaticity coordinates of U.S. Department of Agriculture (USDA) glass color standards for maple syrup. These are described as light amber, medium amber, and dark amber. The six squares are chromaticity coordinates of honey, designated by USDA as water white, extra white, white, extra light amber, light amber, and amber. Such specifications are useful in describing color standards for a variety of products. In the case of the light amber standard for maple syrup, the following values apply: *x* = 0.486, *y* = 0.447, and *T* = 38.9 percent. In this way, x and y provide a specification for chromaticity and T for luminous transmittance or lightness. This is easily expressed as the mixture of primaries under illuminant C as follows: 48.6 percent of red primary, 44.7 percent of green primary, and 6.7 percent of blue primary. The light transmittance is 38.9 percent.

The importance of the light source and other conditions that affect viewing of samples cannot be overemphasized. Many substances are metameric; that is, they may have equal transmittance or reflectance at a certain wavelength but possess noticeably different colors when viewed under illuminant C.

MUNSELL SYSTEM

In the Munsell system of color classification, all colors are described by the three attributes of hue, value, and chroma. This can be envisaged as a three-dimensional system (Figure 6–8). The hue scale is based on ten hues which are distributed on the circumference of the hue circle. There are five hues: red, yellow, green, blue, and purple; they are written as R, Y, G, B, and P. There are also five intermediate hues, YR, GY, BG, PB, and RP. Each of the ten hues is at the midpoint of a scale from 1 to 10. The value scale is a lightness scale ranging from 0 (black) to 10 (white). This scale is distributed on a line perpendicular to the plane of the hue circle and intersecting its center. Chroma is a measure of the difference of a color from a gray of same lightness. It is a measure of purity. The chroma scale is of irregular length, and begins with 0 for the central gray. The scale extends outward in steps to the limit of purity obtainable by available pigments. The shape of the complete Munsell color space is indicated in Figure 6–9. The description of a color in the Munsell system is given as H, V/C. For example, a color indicated as 5R

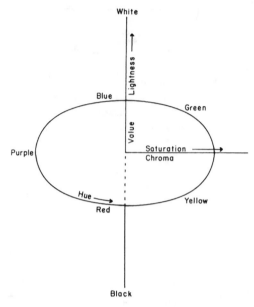

Figure 6–8 The Munsell System of Color Classification

2.8/3.7 means a color with a red hue of 5R, a value of 2.8, and a chroma of 3.7. All colors that can be made with available pigments are laid down as color chips in the Munsell book of color.

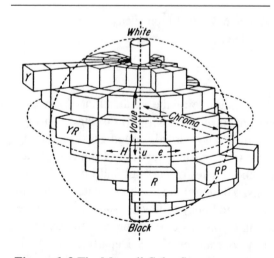

Figure 6–9 The Munsell Color Space

HUNTER SYSTEM

The CIE system of color measurement is based on the principle of color sensing by the human eye. This accepts that the eyes contain three light-sensitive receptors—the red, green, and blue receptors. One problem with this system is that the X, Y, and Z values have no relationship to color as perceived, though a color is completely defined. To overcome this problem, other color systems have been suggested. One of these, widely used for food colorimetry, is the Hunter L, a, b, system. The so-called uniform-color, opponent-colors color scales are based on the opponent-colors theory of color vision. In this theory, it is assumed that there is an intermediate signal-switching stage between the light receptors in the retina and the optic nerve, which transmits color signals to the brain. In this switching mechanism, red responses are compared with green and result in a red-to-green color dimension. The green response is compared with blue to give a yellow-to-blue color dimension. These two color dimensions are

represented by the symbols a and b. The third color dimension is lightness L, which is non-linear and usually indicated as the square or cube root of Y. This system can be represented by the color space shown in Figure 6–10. The L, a, b, color solid is similar to the Munsell color space. The lightness scale is common to both. The chromatic spacing is different. In the Munsell system, there are the polar hue and chroma coordinates, whereas in the L, a, b, color space, chromaticity is defined by rectangular a and b coordinates. CIE values can be converted to color values by the equations shown in Table 6–2 into L, a, b, values and vice versa (MacKinney and Little 1962; Clydesdale and Francis 1970). This is not the case with Munsell values. These are obtained from visual comparison with color chips (called Munsell renotations) or from instrumental measurements (called Munsell renotations), and conversion is difficult and tedious.

The Hunter tristimulus data, L (value), a (redness or greenness), and b (yellowness or blueness), can be converted to a single color

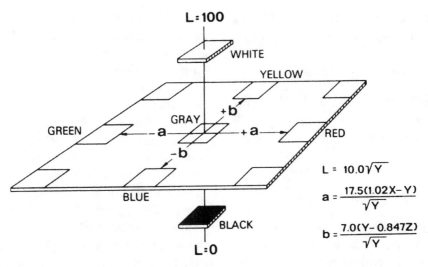

$$L = 10.0\sqrt{Y}$$

$$a = \frac{17.5(1.02X - Y)}{\sqrt{Y}}$$

$$b = \frac{7.0(Y - 0.847Z)}{\sqrt{Y}}$$

Figure 6–10 The Hunter L, a, b Color Space. *Source:* From Hunter Associates Lab., Inc.

Table 6–2 Mathematical Relationship Between Color Scales

To Convert	To L, a, b	To X%, Y, Z%	To Y, x, y
From X%, Y, Z%	$L = 10\sqrt{Y}$ $a = \dfrac{17.5(X\% - Y)}{\sqrt{Y}}$ $b = \dfrac{7.0(Y - Z\%)}{\sqrt{Y}}$		$Y = Y\%$ $x = \dfrac{X}{X + Y + Z}$ $y = \dfrac{Y}{X + Y + Z}$
From L, a, b		$Y = 0.01L^2$ $X\% = 0.01L^2 + \dfrac{aL}{175}$ $Z\% = 0.01L^2 - \dfrac{bL}{70}$	$Y = 0.01L^2$ $x = \dfrac{a + 1.75L}{5.645L + a - 3.012b}$ $y = \dfrac{1.786L}{5.645L + a - 3.012b}$
From Y, x, y	$L = 10\sqrt{Y}$ $a = 17.5\sqrt{Y}\,\dfrac{1.02x}{y} - 1$ $b = 5.929\sqrt{Y}\,\dfrac{2.181y + x - 1}{y}$	$X\% = 1.02 \times \dfrac{Y}{y}$ $Z\% = .847\,[1 - (x + y)]\,\dfrac{Y}{y}$	

Source: From Hunter Associates Lab., Inc.

function called color difference (ΔE) by using the following relationship:

$$\Delta E = (\Delta L)^2 + (\Delta a)^2 + (\Delta b)^2$$

The color difference is a measure of the distance in color space between two colors. It does not indicate the direction in which the colors differ.

LOVIBOND SYSTEM

The Lovibond system is widely used for the determination of the color of vegetable oils. The method involves the visual comparison of light transmitted through a glass cuvette filled with oil at one side of an inspection field; at the other side, colored glass filters are placed between the light source and the observer. When the colors on each side of the field are matched, the nominal value of the filters is used to define the color of the oil. Four series of filters are used—red, yellow, blue, and gray filters. The gray filters are used to compensate for intensity when measuring samples with intense chroma (color purity) and are used in the light path going through the sample. The red, yellow, and blue filters of increasing intensity are placed in the light path until a match with the sample is obtained. Vegetable oil colors are usually expressed in terms of red and yellow; a typical example of the Lovibond color of an oil would be R1.7 Y17. The visual determination of oil color by the Lovibond method is widely used in industry and is an official method of the American Oil Chemists' Society. Visual methods of this type are subject to a number of errors, and the results obtained are highly variable. A study has been reported (Maes et al., 1997) to calculate CIE and Lovibond color values of oils based on their visible light transmission spectra as measured by a spectrophotometer. A computer software has been developed that can easily convert light transmission spectra into CIE and Lovibond color indexes.

GLOSS

In addition to color, there is another important aspect of appearance, namely gloss. Gloss can be characterized as the reflecting property of a material. Reflection of light can be diffused or undiffused (specular). In specular reflection, the surface of the object acts as a mirror, and the light is reflected in a highly directional manner. Surfaces can range from a perfect mirror with completely specular reflection to a surface reflecting in a completely diffuse manner. In the latter, the light from an incident beam is scattered in all directions and the surface is called matte.

FOOD COLORANTS

The colors of foods are the result of natural pigments or of added colorants. The natural pigments are a group of substances present in animal and vegetable products. The added colorants are regulated as food additives, but some of the synthetic colors, especially carotenoids, are considered "nature identical"

and therefore are not subject to stringent toxicological evaluation as are other additives (Dziezak 1987).

The naturally occurring pigments embrace those already present in foods as well as those that are formed on heating, storage, or processing. With few exceptions, these pigments can be divided into the following four groups:

1. tetrapyrrole compounds: chlorophylls, hemes, and bilins
2. isoprenoid derivatives: carotenoids
3. benzopyran derivatives: anthocyanins and flavonoids
4. artefacts: melanoidins, caramels

The chlorophylls are characteristic of green vegetables and leaves. The heme pigments are found in meat and fish. The carotenoids are a large group of compounds that are widely distributed in animal and vegetable products; they are found in fish and crustaceans, vegetables and fruits, eggs, dairy products, and cereals. Anthocyanins and flavonoids are found in root vegetables and fruits such as berries and grapes. Caramels and melanoidins are found in syrups and cereal products, especially if these products have been subjected to heat treatment.

Tetrapyrrole Pigments

The basic unit from which the tetrapyrrole pigments are derived is pyrrole.

The basic structure of the heme pigments consists of four pyrrole units joined together into a porphyrin ring as shown in Figure 6–11.

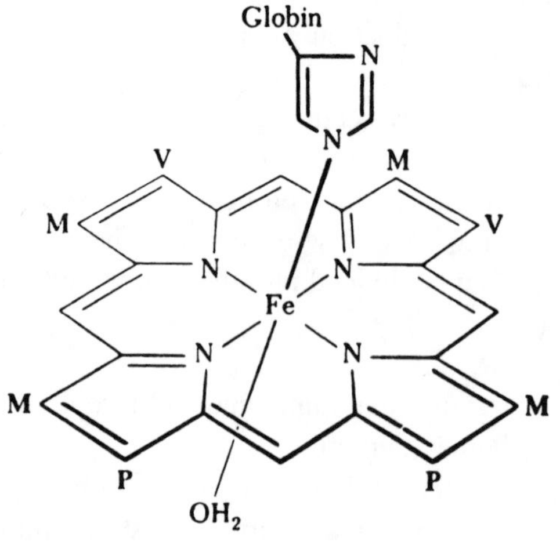

Figure 6–11 Schematic Representation of the Heme Complex of Myoglobin. M = methyl, P = propyl, V = vinyl. *Source*: From C.E. Bodwell and P.E. McClain, Proteins, in *The Sciences of Meat Products*, 2nd ed., J.E. Price and B.S. Schweigert, eds., 1971, W.H. Freeman & Co.

In the heme pigments, the nitrogen atoms are linked to a central iron atom. The color of meat is the result of the presence of two pigments, myoglobin and hemoglobin. Both pigments have globin as the protein portion, and the heme group is composed of the porphyrin ring system and the central iron atom. In myoglobin, the protein portion has a molecular weight of about 17,000. In hemoglobin, this is about 67,000—equivalent to four times the size of the myoglobin protein. The central iron in Figure 6–11 has six coordination bonds; each bond represents an electron pair accepted by the iron from five nitrogen atoms, four from the porphyrin ring and one from a histidyl residue of the globin. The sixth bond is available for joining with any atom that has an electron pair to donate. The ease with which an electron pair is donated determines the nature of the bond formed and the color of the complex. Other

factors playing a role in color formation are the oxidation state of the iron atom and the physical state of the globin.

In fresh meat and in the presence of oxygen, there is a dynamic system of three pigments, oxymyoglobin, myoglobin, and metmyoglobin. The reversible reaction with oxygen is

$$Mb + O_2 \rightleftharpoons MbO_2$$

In both pigments, the iron is in the ferrous form; upon oxidation to the ferric state, the compound becomes metmyoglobin. The bright red color of fresh meat is due to the presence of oxymyoglobin; discoloration to brown occurs in two stages, as follows:

$$MbO_2 \rightleftharpoons Mb \rightleftharpoons MetMb$$

MbO$_2$	Mb	MetMb
Red	Purplish red	Brownish

Oxymyoglobin represents a ferrous covalent complex of myoglobin and oxygen. The absorption spectra of the three pigments are shown in Figure 6–12 (Bodwell and McClain 1971). Myoglobin forms an ionic complex with water in the absence of strong electron pair donors that can form covalent complexes. It shows a diffuse absorption band in the green area of the spectrum at about 555 nm and has a purple color. In metmyoglobin, the major absorption peak is shifted toward the blue portion of the spectrum at about 505 nm with a smaller peak at 627 nm. The compound appears brown.

As indicated above, oxymyoglobin and myoglobin exist in a state of equilibrium with oxygen; therefore, the ratio of the pigments is dependent on oxygen pressure. The oxidized form of myoglobin, the metmyoglobin, cannot bind oxygen. In meat, there is a slow and continuous oxidation of the heme pigments to the metmyoglobin state. Reducing substances in the tissue reduce the metmyoglobin to the ferrous form. The oxygen pressure, which is so important for the state of the equilibrium, is greatly affected by packaging materials used for meats. The maximum rate of conversion to metmyoglobin occurs at partial pressures of 1 to 20 nm of mercury, depending on pigment, pH, and temperature (Fox 1966). When a packaging film with low oxygen permeability is used, the oxygen pressure drops to the point where oxidation is favored. To prevent this, Landrock and Wallace (1955) established that oxygen permeability of the packaging film must be at least 5 liters of oxygen/square meter/day/atm.

Fresh meat open to the air displays the bright red color of oxymyoglobin on the surface. In the interior, the myoglobin is in the reduced state and the meat has a dark purple color. As long as reducing substances are present in the meat, the myoglobin will remain in the reduced form; when they are used up, the brown color of metmyoglobin will predominate. According to Solberg (1970), there is a thin layer a few nanometers below the bright red surface and just before the myoglobin region, where a definite brown color is visible. This is the area where the oxygen partial pressure is about 1.4 nm and the brown pigment dominates. The growth of bacteria at the meat surface may reduce the partial oxygen pressure to

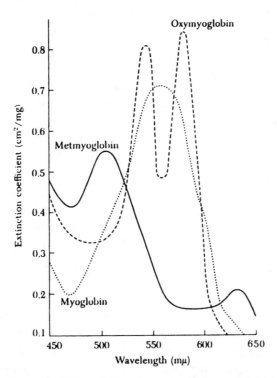

Figure 6–12 Absorption Spectra of Myoglobin, Oxymyoglobin, and Metmyoglobin. *Source*: From C.E. Bodwell and P.E. McClain, Proteins, in *The Sciences of Meat Products*, 2nd ed., J.E. Price and B.S. Schweigert, eds., 1971, W.H. Freeman & Co.

below the critical level of 4 nm. Microorganisms entering the logarithmic growth phase may change the surface color to that of the purplish-red myoglobin (Solberg 1968).

In the presence of sulfhydryl as a reducing agent, myoglobin may form a green pigment, called sulfmyoglobin. The pigment is green because of a strong absorption band in the red region of the spectrum at 616 nm. In the presence of other reducing agents, such as ascorbate, cholemyoglobin is formed. In this pigment, the porphyrin ring is oxidized. The conversion into sulfmyoglobin is reversible; cholemyoglobin formation is irreversible, and this compound is rapidly oxidized to yield globin, iron, and tetrapyrrole. According to Fox (1966), this reaction may happen in the pH range of 5 to 7.

Heating of meat results in the formation of a number of pigments. The globin is denatured. In addition, the iron is oxidized to the ferric state. The pigment of cooked meat is brown and called hemichrome. In the presence of reducing substances such as those that occur in the interior of cooked meat, the iron may be reduced to the ferrous form; the resulting pigment is pink hemochrome.

In the curing of meat, the heme reacts with nitrite of the curing mixture. The nitrite-heme complex is called nitrosomyoglobin, which has a red color but is not particularly stable. On heating the more stable nitrosohemochrome, the major cured meat pigment is formed, and the globin portion of the molecule is denatured. This requires a temperature of 65°C. This molecule has been called nitrosomyoglobin and nitrosylmyoglobin, but Möhler (1974) has pointed out that the only correct name is nitric oxide myoglobin. The first reaction of nitrite with myoglobin is oxidation of the ferrous iron to the ferric form and formation of MetMb. At the same time, nitrate is formed according to the following reaction (Möhler 1974):

$$4MbO_2 + 4NO_2^- + 2H_2O \rightarrow$$
$$4MetMbOH + 4NO_3^- + O_2$$

During the formation of the curing pigment, the nitrite content is gradually lowered; there are no definite theories to account for this loss.

The reactions of the heme pigments in meat and meat products have been summarized in the scheme presented in Figure 6–13 (Fox 1966). Bilin-type structures are formed when the porphyrin ring system is broken.

Chlorophylls

The chlorophylls are green pigments responsible for the color of leafy vegetables and some fruits. In green leaves, the chlorophyll is broken down during senescence and the green color tends to disappear. In many fruits, chlorophyll is present in the unripe state and gradually disappears as the yellow and red carotenoids take over during ripening. In plants, chlorophyll is isolated in the chloroplastids. These are microscopic particles consisting of even smaller units, called grana, which are usually less than one micrometer in size and at the limit of resolution of the light microscope. The grana are highly structured and contain laminae between which the chlorophyll molecules are positioned.

The chlorophylls are tetrapyrrole pigments in which the porphyrin ring is in the dihydro form and the central metal atom is magnesium. There are two chlorophylls, a and b, which occur together in a ratio of about 1:25. Chlorophyll b differs from chlorophyll a in that the methyl group on carbon 3 is replaced with an aldehyde group. The structural for-

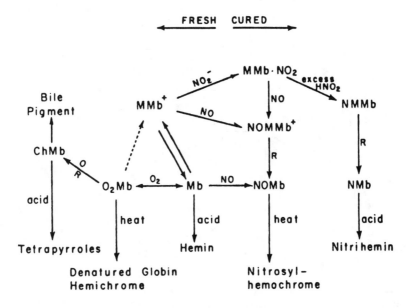

Figure 6–13 Heme Pigment Reactions in Meat and Meat Products. ChMb, cholemyoglobin (oxidized porphyrin ring); O_2Mb, oxymyoglobin (Fe^{+2}); MMb metmyoglobin (Fe^{+3}); Mb, myoglobin (Fe^{+2}); $MMb \cdot NO_2$, metmyoglobin nitrate; NOMMb, nitrosylmetmyoglobin; NOMb, nitrosylmyoglobin; NMMb, nitrimetmyoglobin; NMb, nitrimyoglobin, the latter two being reaction products of nitrous acid and the heme portion of the molecule; R, reductants; O, strong oxidizing conditions. *Source:* From J.B. Fox, The Chemistry of Meat Pigments, *J. Agr. Food Chem.*, Vol. 14, no. 3, pp. 207–210, 1966, American Chemical Society.

mula of chlorophyll *a* is given in Figure 6–14. Chlorophyll is a diester of a dicarboxylic acid (chlorophyllin); one group is esterified with methanol, the other with phytyl alcohol. The magnesium is removed very easily by acids, giving pheophytins *a* and *b*. The action of acid is especially important for fruits that are naturally high in acid. However, it appears that the chlorophyll in plant tissues is bound to lipoproteins and is protected from the effect of acid. Heating coagulates the protein and lowers the protective effect. The color of the pheophytins is olive-brown. Chlorophyll is stable in alkaline medium. The phytol chain confers insolubility in water on the chlorophyll molecule. Upon hydrolysis of the phytol group, the water-sol-

uble methyl chlorophyllides are formed. This reaction can be catalyzed by the enzyme chlorophyllase. In the presence of copper or zinc ions, it is possible to replace the magnesium, and the resulting zinc or copper complexes are very stable. Removal of the phytol group and the magnesium results in pheophorbides. All of these reactions are summarized in the scheme presented in Figure 6–15.

In addition to those reactions described above, it appears that chlorophyll can be degraded by yet another pathway. Chichester and McFeeters (1971) reported on chlorophyll degradation in frozen beans, which they related to fat peroxidation. In this reaction, lipoxidase may play a role, and no

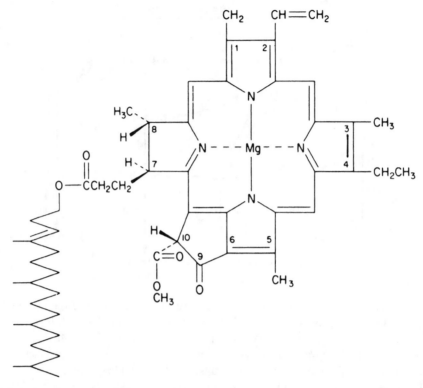

Figure 6–14 Structure of Chlorophyll *a*. (Chlorophyll *b* differs in having a formyl group at carbon 3). *Source*: Reprinted with permission from J.R. Whitaker, *Principles of Enzymology for the Food Sciences*, 1972, by courtesy of Marcel Dekker, Inc.

pheophytins, chlorophyllides, or pheophorbides are detected. The reaction requires oxygen and is inhibited by antioxidants.

Carotenoids

The naturally occurring carotenoids, with the exception of crocetin and bixin, are tetraterpenoids. They have a basic structure of eight isoprenoid residues arranged as if two 20-carbon units, formed by head-to-tail condensation of four isoprenoid units, had joined tail to tail. There are two possible ways of classifying the carotenoids. The first system recognizes two main classes, the car-

otenes, which are hydrocarbons, and the xanthophylls, which contain oxygen in the form of hydroxyl, methoxyl, carboxyl, keto, or epoxy groups. The second system divides the carotenoids into three types (Figure 6–16), acyclic, monocyclic, and bicyclic. Examples are lycopene (I)—acyclic, γ-carotene (II)—monocyclic, and α-carotene (III) and β-carotene (IV)—bicyclic.

The carotenoids take their name from the major pigments of carrot (*Daucus carota*). The color is the result of the presence of a system of conjugated double bonds. The greater the number of conjugated double bonds present in the molecule, the further the major absorption bands will be shifted to the

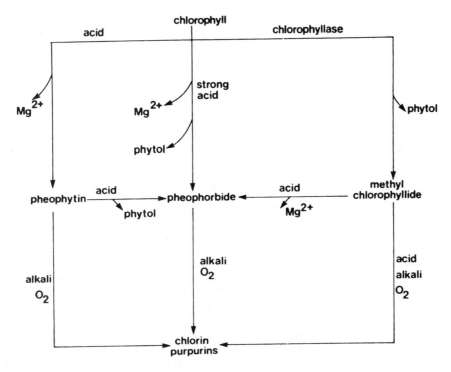

Figure 6–15 Reactions of Chlorophylls

region of longer wavelength; as a result, the hue will become more red. A minimum of seven conjugated double bonds are required before a perceptible yellow color appears. Each double bond may occur in either *cis* or *trans* configuration. The carotenoids in foods are usually of the all-*trans* type and only occasionally a mono-*cis* or di-*cis* compound occurs. The prefix *neo-* is used for stereoisomers with at least one *cis* double bond. The prefix *pro-* is for poly-*cis* carotenoids. The effect of the presence of *cis* double bonds on the absorption spectrum of β-carotene is shown in Figure 6–17. The configuration has an effect on color. The all-*trans* compounds have the deepest color; increasing numbers of *cis* bonds result in gradual lightening of the color. Factors that cause change of bonds from *trans* to *cis* are light, heat, and acid.

In the narrower sense, the carotenoids are the four compounds shown in Figure 6–16— α-, β-, and γ-carotene and lycopene—polyene hydrocarbons of overall composition $C_{40}H_{56}$. The relation between these and carotenoids with fewer than 40 carbon atoms is shown in Figure 6–18. The prefix *apo-* is used to designate a carotenoid that is derived from another one by loss of a structural element through degradation. It has been suggested that some of these smaller carotenoid molecules are formed in nature by oxidative degradation of C_{40} carotenoids (Grob 1963).

Several examples of this possible relationship are found in nature. One of the best known is the formation of retinin and vitamin A from β-carotene (Figure 6–19). Another obvious relationship is that of lycopene and bixin (Figure 6–20). Bixin is a food

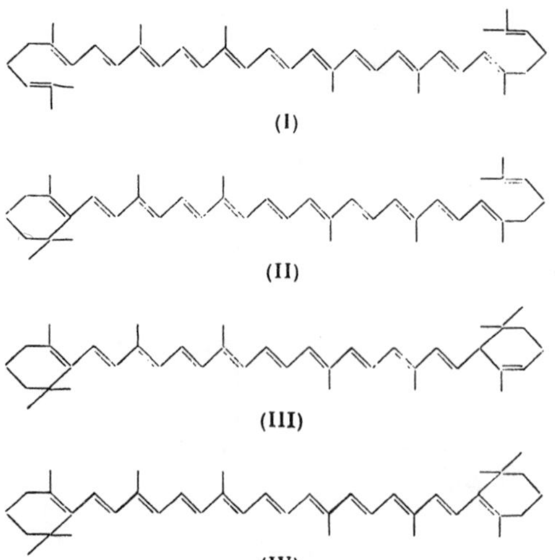

Figure 6–16 The Carotenoids: (I) Lycopene, (II) γ-carotene, (III) α-Carotene, and (IV) β-Carotene. *Source*: From E.C. Grob, The Biogenesis of Carotenes and Carotenoids, in *Carotenes and Carotenoids*, K. Lang, ed., 1963, Steinkopff Verlag.

color additive obtained from the seed coat of the fruit of a tropical brush, *Bixa orellana*. The pigment bixin is a dicarboxylic acid esterified with one methanol molecule. A pigment named crocin has been isolated from saffron. Crocin is a glycoside containing two molecules of gentiobiose. When these are removed, the dicarboxylic acid crocetin is formed (Figure 6–21). It has the same general structure as the aliphatic chain of the carotenes. Also obtained from saffron is the bitter compound picrocrocin. It is a glycoside and, after removal of the glucose, yields saffronal. It is possible to imagine a combination of two molecules of picrocrocin and one of crocin; this would yield protocrocin. Protocrocin, which is directly related to zeaxanthin, has been found in saffron (Grob 1963).

The structure of a number of important xanthophylls as they relate to the structure of β-carotene is given in Figure 6–22. Carotenoids may occur in foods as relatively simple mixtures of only a few compounds or as very complex mixtures of large numbers of carotenoids. The simplest mixtures usually exist in animal products because the animal organism has a limited ability to absorb and deposit carotenoids. Some of the most complex mixtures are found in citrus fruits.

Beta-carotene as determined in fruits and vegetables is used as a measure of the provitamin A content of foods. The column chromatographic procedure, which determines this content, does not separate α-carotene, β-carotene, and cryptoxanthin. Provitamin A values of some foods are given in Table 6–3. Carotenoids are not synthesized by animals, but they may change ingested carotenoids into animal carotenoids—as in, for example, salmon, eggs, and crustaceans. Usually carotenoid content of foods does not exceed 0.1 percent on a dry weight basis.

In ripening fruit, carotenoids increase at the same time chlorophylls decrease. The ratio of carotenes to xanthophylls also increases. Common carotenoids in fruits are α- and γ-carotene and lycopene. Fruit xanthophylls are usually present in esterified form. Oxygen, but not light, is required for carotenoid synthesis and the temperature range is critical. The relative amounts of different carotenoids are related to the characteristic color of some fruits. In the sequence of peach, apricot, and tomato, there is an increasing proportion of lycopene and increasing redness. Many peach varieties are devoid of lycopene. Apricots may have about 10 percent and tomatoes up to 90 percent. The lycopene content of tomatoes increases during ripening. As the chlorophyll breaks down during ripening, large amounts of carot-

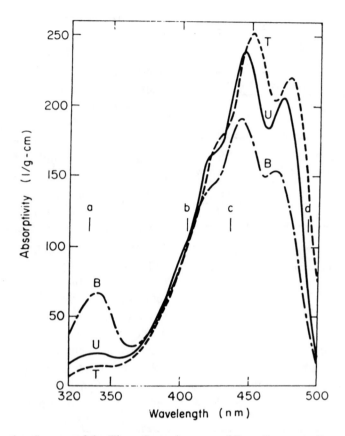

Figure 6–17 Absorption Spectra of the Three Stereoisomers of Beta Carotene. B = neo-β-carotene; U = neo-β-carotene-U; T = all-trans-β-carotene. *a, b, c,* and *d* indicate the location of the mercury arc lines 334.1, 404.7, 435.8 and 491.6 nm, respectively. *Source*: From F. Stitt et al., Spectrophotometric Determination of Beta Carotene Stereoisomers in Alfalfa, *J. Assoc. Off. Agric. Chem.* Vol. 34, pp. 460–471, 1951.

enoids are formed (Table 6–4). Color is an important attribute of citrus juice and is affected by variety, maturity, and processing methods. The carotenoid content of oranges is used as a measure of total color. Curl and Bailey (1956) showed that the 5,6-epoxides of fresh orange juice isomerize completely to 5,8-epoxides during storage of canned juice. This change amounts to the loss of one double bond from the conjugated double bond system and causes a shift in the wavelength of maximum absorption as well as a decrease in molar absorbance. In one year's storage at 70°F, an apparent carotenoid loss of 20 to 30 percent occurs.

Peaches contain violaxanthin, cryptoxanthin, β-carotene, and persicaxanthin as well as 25 other carotenoids, including neoxanthin. Apricots contain mainly β- and γ-carotene, lycopene, and little if any xanthophyll. Carrots have been found to have an average of 54 ppm of total carotene (Borenstein and Bunnell 1967), consisting mainly of α-, β, and ζ-carotene and some lycopene and xan-

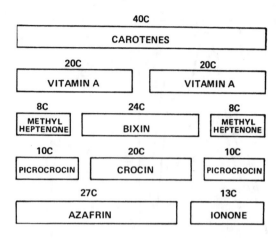

Figure 6–18 Relationship Between the Carotene and Carotenoids with Fewer than 40 Carbons

thophyll. Canning of carrots resulted in a 7 to 12 percent loss of provitamin A activity because of *cis-trans* isomerization of α- and β-carotene (Weckel et al. 1962). In dehy-

drated carrots, carotene oxidation and off-flavor development have been correlated (Falconer et al. 1964). Corn contains about one-third of the total carotenoids as carotenes and two-thirds xanthophylls. Compounds found in corn include zeaxanthin, cryptoxanthin, β-carotene, and lutein.

One of the highest known concentrations of carotenoids occurs in crude palm oil. It contains about 15 to 300 times more retinol equivalent than carrots, green leafy vegetables, and tomatoes. All of the carotenoids in crude palm oil are destroyed by the normal processing and refining operations. Recently, improved gentler processes have been developed that result in a "red palm oil" that retains most of the carotenoids. The composition of the carotenes in crude palm oil with a total carotene concentration of 673 mg/kg is shown in Table 6–5.

Milkfat contains carotenoids with seasonal variation (related to feed conditions) ranging from 2 to 13 ppm.

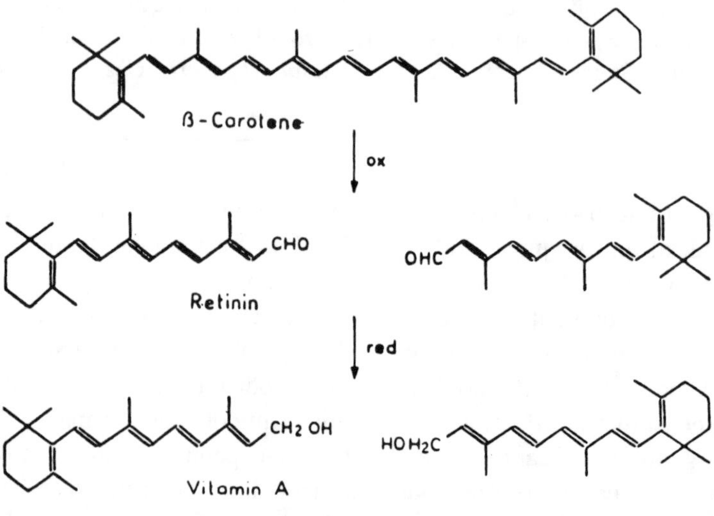

Figure 6–19 Formation of Retinin and Vitamin A from β-Carotene. *Source*: From E.C. Grob, The Biogenesis of Carotenes and Carotenoids, in *Carotenes and Carotenoids*, K. Lang, ed., 1963, Steinkopff Verlag.

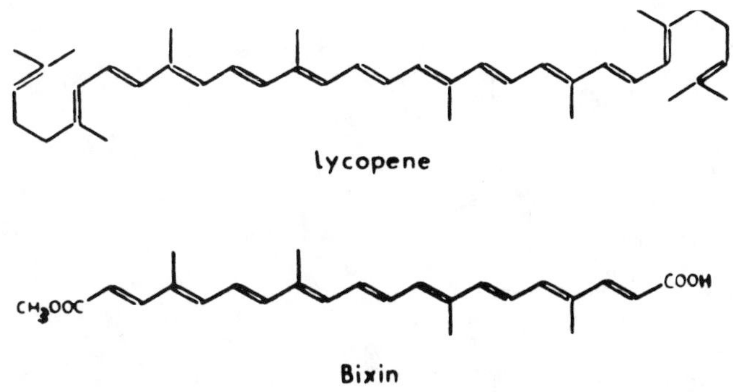

Figure 6–20 Relationship Between Lycopene and Bixin. *Source*: From E.C. Grob, The Biogenesis of Carotenes and Carotenoids, in *Carotenes and Carotenoids*, K. Lang, ed., 1963, Steinkopff Verlag.

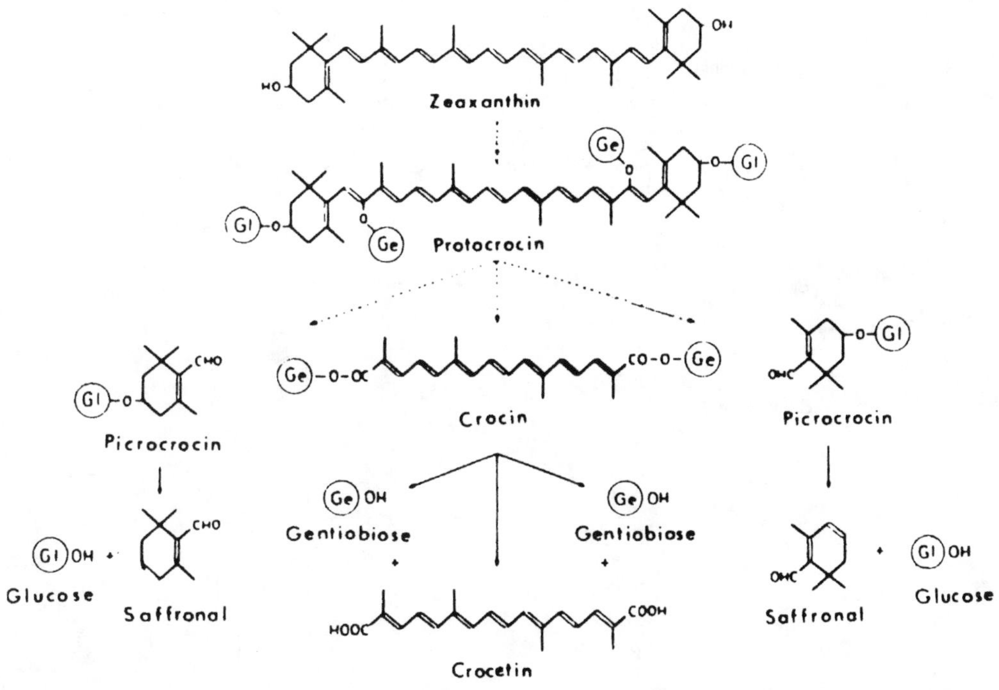

Figure 6–21 Relationship Between Crocin and Picrocrocin and the Carotenoids. *Source*: From E.C. Grob, The Biogenesis of Carotenes and Carotenoids, in *Carotenes and Carotenoids*, K. Lang, ed., 1963, Steinkopff Verlag.

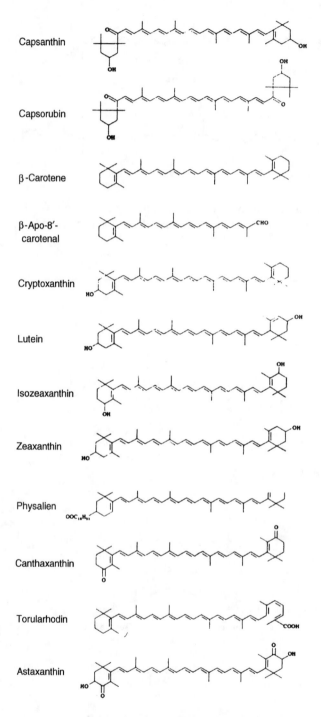

Figure 6–22 Structure of Some of the Important Carotenoids. *Source*: From B. Borenstein and R.H. Bunnell, Carotenoids: Properties, Occurrence, and Utilization in Foods, in *Advances in Food Research*, Vol. 15, C.O. Chichester et al., eds., 1967, Academic Press.

Table 6–3 Provitamin A Value of Some Fruits and Vegetables

Product	IU/100 g
Carrots, mature	20,000
Carrots, young	10,000
Spinach	13,000
Sweet potato	6,000
Broccoli	3,500
Apricots	2,000
Lettuce	2,000
Tomato	1,200
Asparagus	1,000
Bean, french	1,000
Cabbage	500
Peach	800
Brussels sprouts	700
Watermelon	550
Banana	400
Orange juice	200

Source: From B. Borenstein and R.H. Bunnell, Carotenoids: Properties, Occurrence, and Utilization in Foods, in *Advances in Food Research*, Vol. 15, C.O. Chichester et al., eds., 1967, Academic Press.

Egg yolk contains lutein, zeaxanthin, and cryptoxanthin. The total carotenoid content ranges from 3 to 89 ppm.

Crustaceans contain carotenoids bound to protein resulting in a blue or blue-gray color. When the animal is immersed in boiling water, the carotenoid-protein bond is broken and the orange-red color of the free carotenoid appears. Widely distributed in crustaceans is astaxanthin. Red fish contain astaxanthin, lutein, and taraxanthin.

Common unit operations of food processing are reported to have only minor effects on the carotenoids (Borenstein and Bunnell 1967). The carotenoid-protein complexes are generally more stable than the free carotenoids. Because carotenoids are highly unsaturated, oxygen and light are major factors in their breakdown. Blanching destroys enzymes that cause carotenoid destruction. Carotenoids in frozen or heat-sterilized foods are quite stable. The stability of carotenoids in dehydrated foods is poor, unless the food is packaged in inert gas. A notable exception is dried apricots, which keep their color well. Dehydrated carrots fade rapidly.

Several of the carotenoids are now commercially synthesized and used as food colors. A possible method of synthesis is described by Borenstein and Bunnell (1967). Beta-ionone is obtained from lemon grass oil and converted into a C14 aldehyde. The C14 aldehyde is changed to a C16 aldehyde, then to a C19 aldehyde. Two moles of the C19 aldehyde are condensed with acetylene dimagnesium bromide and, after a series of reactions, yield β-carotene.

Three synthetically produced carotenoids are used as food colorants, β-carotene, β-apo-8′-carotenal (apocarotenal), and canthaxanthin. Because of their high tinctorial power, they are used at levels of 1 to 25 ppm

Table 6–4 Development of Pigments in the Ripening Tomato

Pigment	Green (mg/100 g)	Half-ripe (mg/100 g)	Ripe (mg/100 g)
Lycopene	0.11	0.84	7.85
Carotene	0.16	0.43	0.73
Xanthophyll	0.02	0.03	0.06
Xanthophyll ester	0	0.02	0.10

in foods (Dziezak 1987). They are unstable in light but otherwise exhibit good stability in food applications. Although they are fat soluble, water-dispersible forms have been developed for use in a variety of foods. Beta-carotene imparts a light yellow to orange color, apocarotenal a light orange to reddish-orange, and canthaxanthin, orange-red to red. The application of these compounds in a variety of foods has been described by Counsell (1985). Natural carotenoid food colors are annatto, oleoresin of paprika, and unrefined palm oil.

Anthocyanins and Flavonoids

The anthocyanin pigments are present in the sap of plant cells; they take the form of glycosides and are responsible for the red, blue, and violet colors of many fruits and vegetables. When the sugar moiety is removed by hydrolysis, the aglucone remains and is called anthocyanidin. The sugar part usually consists of one or two molecules of glucose, galactose, and rhamnose. The basic structure consists of 2-phenyl-benzopyrylium or flavylium with a number of hydroxy and methoxy substituents. Most of the anthocyanidins are derived from 3,5,7-trihydroxy-flavylium chloride (Figure 6–23) and the sugar moiety is usually attached to the hydroxyl group on carbon 3. The anthocyanins are highly colored, and their names are derived from those of flowers. The structure of some of the more important anthocyanidins is shown in Figure 6–24, and the occurrence of anthocyanidins in some fruits and vegetables is listed in Table 6–6. Recent studies have indicated that some anthocyanins contain additional components such as organic acids and metals (Fe, Al, Mg).

Substitution of hydroxyl and methoxyl groups influences the color of the anthocyanins. This effect has been shown by Braverman (1963) (Figure 6–25). Increase in the number of hydroxyl groups tends to deepen the color to a more bluish shade. Increase in the number of methoxyl groups increases redness. The anthocyanins can occur in different forms. In solution, there is an equilibrium between the colored cation R^+ or oxonium salt and the colorless pseudobase ROH, which is dependent on pH.

$$R^+ + H_2O \rightleftharpoons ROH + H^+$$

As the pH is raised, more pseudobase is formed and the color becomes weaker. However, in addition to pH, other factors influence the color of anthocyanins, including metal chelation and combination with other flavonoids and tannins.

Anthocyanidins are highly colored in strongly acid medium. They have two absorption maxima—one in the visible spec-

Table 6–5 Composition of the Carotenes in Crude Palm Oil

Carotene	% of Total Carotenes
Phytoene	1.27
Cis-β-carotene	0.68
Phytofluene	0.06
β-carotene	56.02
α-carotene	35.06
ζ-carotene	0.69
γ-carotene	0.33
δ-carotene	0.83
Neurosporene	0.29
β-zeacarotene	0.74
α-zeacarotene	0.23
Lycopene	1.30

Source: Reprinted with permission from Choo Yuen May, Carotenoids from Palm Oil, *Palm Oil Developments*, Vol. 22, pp. 1–6, Palm Oil Research Institute of Malaysia.

trum at 500–550 nm, which is responsible for the color, and a second in the ultraviolet (UV) spectrum at 280 nm. The absorption maxima relate to color. For example, the relationship in 0.01 percent HCl in methanol is as follows: at 520 nm pelargonidin is scarlet, at 535 nm cyanidin is crimson, and at 546 nm delphinidin is blue-mauve (Macheix et al. 1990).

About 16 anthocyanidins have been identified in natural products, but only the following six of these occur frequently and in many different products: pelargonidin, cyanidin, delphinidin, peonidin, malvidin, and petunidin. The anthocyanin pigments of Red Delicious apples were found to contain mostly cyanidin-3-galactoside, cyanidin-3-arabinoside, and cyanidin-7-arabinoside (Sun and Francis 1968). Bing cherries contain primarily cyanidin-3-rutinoside, cyanidin-3-glucoside, and small amounts of the pigments cyanidin, peonidin, peonidin-3-glucoside, and peonidin-3-ruti-

noside (Lynn and Luh 1964). Cranberry anthocyanins were identified as cyanidin-3-monogalactoside, peonidin-3-monogalactoside, cyanidin monoarabinoside, and peonidin-3-monoarabinoside (Zapsalis and Francis 1965). Cabernet Sauvignon grapes contain four major anthocyanins: delphinidin-3-monoglucoside, petunidin-3-monoglucoside, malvidin-3-monoglucoside, and malvidin-3-monoglucoside acetylated with chlorogenic acid. One of the major pigments is petunidin (Somaatmadja and Powers 1963).

Anthocyanin pigments can easily be destroyed when fruits and vegetables are processed. High temperature, increased sugar level, pH, and ascorbic acid can affect the rate of destruction (Daravingas and Cain 1965). These authors studied the change in anthocyanin pigments during the processing and storage of raspberries. During storage, the absorption maximum of the pigments shifted, indicating a change in color. The

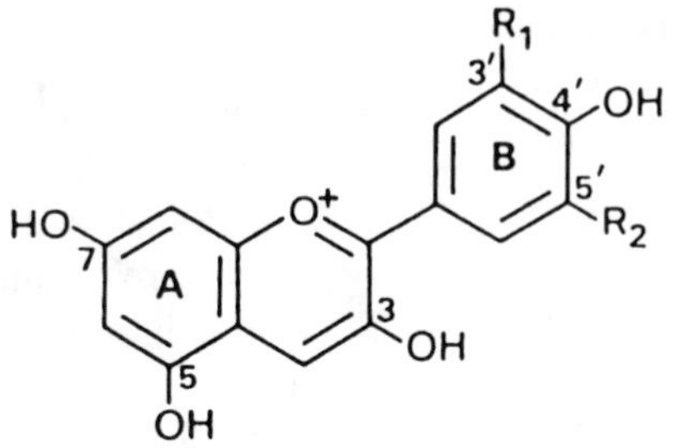

R_1 = H	R_2 = H	PELARGONIDIN
R_1 = OH	R_2 = H	CYANIDIN
R_1 = OH	R_2 = OH	DELPHINIDIN
R_1 = OCH$_3$	R_2 = H	PEONIDIN
R_1 = OCH$_3$	R_2 = OH	PETUNIDIN
R_1 = OCH$_3$	R_2 = OCH$_3$	MALVIDIN

Figure 6–23 Chemical Structure of Fruit Anthocyanidins

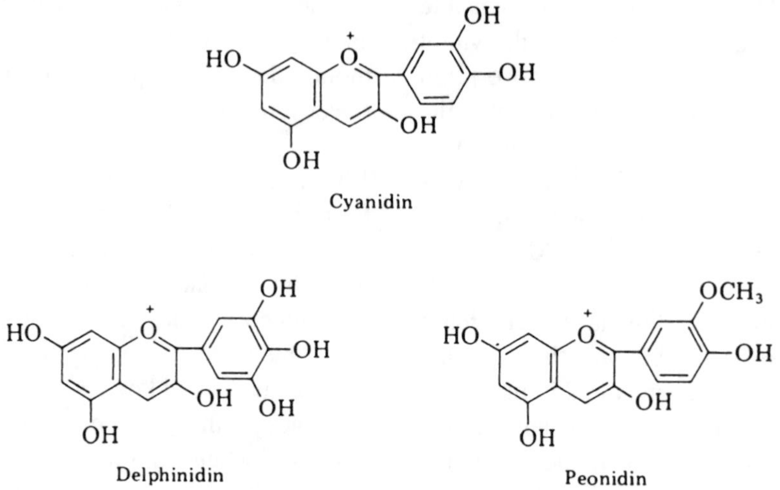

Cyanidin

Delphinidin

Peonidin

Figure 6–24 Structure of Some Important Anthocyanidins

Table 6–6 Anthocyanidins Occurring in Some Fruits and Vegetables

Fruit or Vegetable	Anthocyanidin
Apple	Cyanidin
Black currant	Cyanidin and delphinidin
Blueberry	Cyanidin, delphinidin, mal- vidin, petunidin, and peonidin
Cabbage (red)	Cyanidin
Cherry	Cyanidin and peonidin
Grape	Malvidin, peonidin, delphini- din, cyanidin, petunidin, and pelargonidin
Orange	Cyanidin and delphinidin
Peach	Cyanidin
Plum	Cyanidin and peonidin
Radish	Pelargonidin
Raspberry	Cyanidin
Strawberry	Pelargonidin and a little cyanidin

Source: From P. Markakis, Anthocyanins, in *Encyclopedia of Food Technology*, A.H. Johnson and M.S. Peterson, eds., 1974, AVI Publishing Co.

level of pigments was lowered by prolonged times and higher temperatures of storage. Higher concentration of the ingoing sugar syrup and the presence of oxygen resulted in greater pigment destruction.

The stability of anthocyanins is increased by acylation (Dougall et al. 1997). These acylated anthocyanins may occur naturally as in the case of an anthocyanin from the purple yam (Yoshida et al. 1991). This anthocyanin has one sinapic residue attached through a disaccharide and was found to be stable at pH 6.0 compared to other anthocyanins without acylation. Dougall et al. (1977) were able to produce stable anthocyanins by acylation of carrot anthocyanins in cell cultures. They found that a wide range of aromatic acids could be incorporated into the anthocyanin.

Anthocyanins can form purplish or slate-gray pigments with metals, which are called lakes. This can happen when canned foods take up tin from the container. Anthocyanins

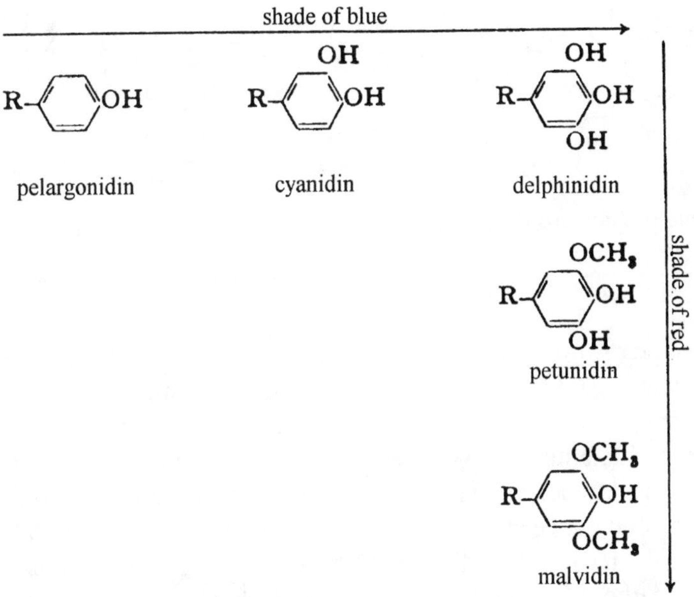

Figure 6–25 Effect of Substituents on the Color of Anthocyanidins. *Source*: Reprinted with permission from J.B.S. Braverman, *Introduction to the Biochemistry of Foods*, © 1963, Elsevier Publishing Co.

can be bleached by sulfur dioxide. According to Jurd (1964), this is a reversible process that does not involve hydrolysis of the glycosidic linkage, reduction of the pigment, or addition of bisulfite to a ketonic, chalcone derivative. The reactive species was found to be the anthocyanin carbonium ion (R^+), which reacts with a bisulfite ion to form a colorless chromen-2(or 4)-sulfonic acid (R–SO_3H), similar in structure and properties to an anthocyanin carbinol base (R–OH). This reaction is shown in Figure 6–26.

The colors of the anthocyanins at acid pH values correspond to those of the oxonium salts. In slightly alkaline solutions (pH 8 to 10), highly colored ionized anhydro bases are formed. At pH 12, these hydrolyze rapidly to fully ionized chalcones (Figure 6–27). Leuco bases are the reduced form of the anthocyanins. They are usually without much color but are widely distributed in fruits and vegetables. Under the influence of oxygen and acid hydrolysis, they may develop the characteristic color of the car-

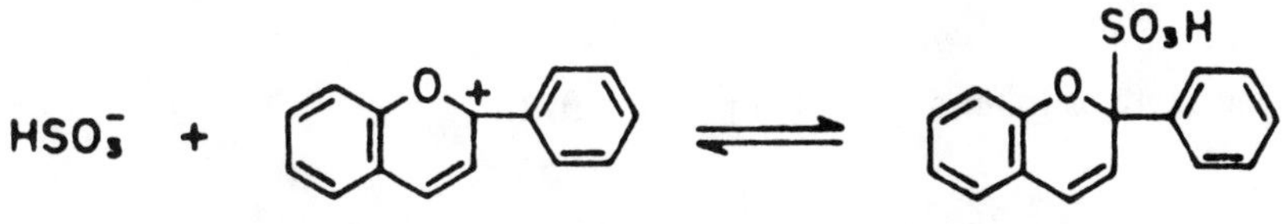

Figure 6–26 Reaction of Bisulfite with the Anthocyanin Carbonium Ion

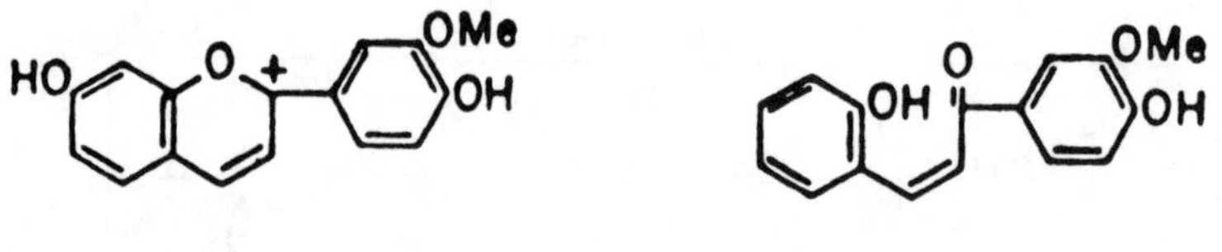

Figure 6–27 Structure of Anhydro Base (I) and Chalcone (II)

bonium ion. Canned pears, for example, may show "pinking"—a change from the leuco base to the anthocyanin.

The flavonoids or anthoxanthins are glycosides with a benzopyrone nucleus. The flavones have a double bond between carbons 2 and 3. The flavonols have an additional hydroxyl group at carbon 3, and the flavanones are saturated at carbons 2 and 3 (Figure 6–28). The flavonoids have low coloring power but may be involved in discolorations; for example, they can impart blue and green colors when combined with iron. Some of these compounds are also potential substrates for enzymic browning and can cause undesirable discoloration through this mechanism. The most ubiquitous flavonoid is quercetin, a 3,5,7,3′,4′-pentahydroxy flavone (Figure 6–29). Many flavonoids contain the sugar rutinose, a disaccharide of glucose

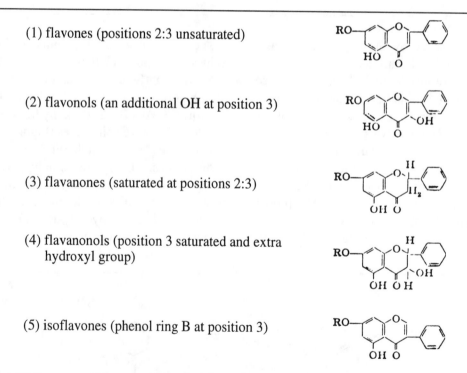

(1) flavones (positions 2:3 unsaturated)

(2) flavonols (an additional OH at position 3)

(3) flavanones (saturated at positions 2:3)

(4) flavanonols (position 3 saturated and extra hydroxyl group)

(5) isoflavones (phenol ring B at position 3)

Figure 6–28 Structure of Flavones, Flavonals, Flavanones, Flavanonols, and Isoflavones

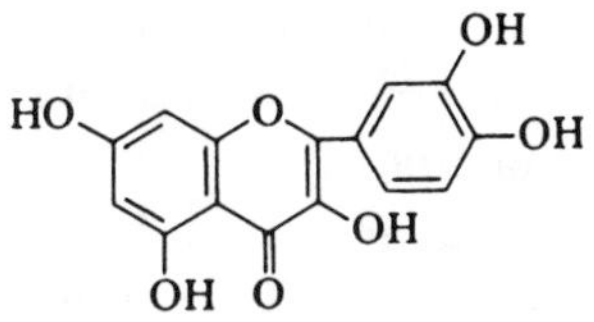

Figure 6–29 Structure of Quercetin

and rhamnose. Hesperidin is a flavanone occurring in citrus fruits and, at pH 12, the inner ring opens to form a chalcone in a similar way as shown for the anthocyanins. The chalcones are yellow to brown in color.

Tannins

Tannins are polyphenolic compounds present in many fruits. They are important as color compounds and also for their effect on taste as a factor in astringency (see Chapter 7). Tannins can be divided into two classes—hydrolyzable tannins and nonhydrolyzable or condensed tannins. The tannins are characterized by the presence of a large number of hydroxyl groups, which provide the ability to form reversible bonds with other macromolecules, polysaccharides, and proteins, as well as other substances such as alkaloids. This bond formation may occur during the development of the fruit or during the mechanical damage that takes place during processing.

Hydrolyzable tannins are composed of phenolic acids and sugars that can be broken down by acid, alkaline, or enzymic hydroly-

sis. They are polyesters based on gallic acid and/or hexahydroxydiphenic acid (Figure 6–30). The usual sugar is D-glucose and molecular weights are in the range of 500 to 2,800. Gallotannins release gallic acid on hydrolysis, and ellagitannins produce ellagic acid. Ellagic acid is the lactone form of hexahydroxydiphenic acid, which is the compound originally present in the tannin (Figure 6–30).

Nonhydrolyzable or condensed tannins are also named proanthocyanidins. These are polymers of flavan-3-ols, with the flavan bonds most commonly between C4 and C8 or C6 (Figure 6–23) (Macheix et al. 1990). Many plants contain tannins that are polymers of (+)-catechin or (–)-epicatechin. These are hydrogenated forms of flavonoids or anthocyanidins. Other monomers occupying places in condensed fruit tannins have trihydroxylation in the B-ring: (+)-gallocatechin and (–)-epigallocatechin. Oligomeric and polymeric procyanidins are formed by addition of more flavan-3-ol units and result in the formation of helical structures. These structures can form bonds with proteins.

Tannins are present in the skins of red grapes and play an important part in the flavor profile of red wine. Tannins in grapes are usually estimated in terms of the content of gallic acid (Amerine and Joslyn 1970).

Oxidation and polymerization of phenolic compounds as a result of enzymic activity of phenoloxidases or peroxidases may result in

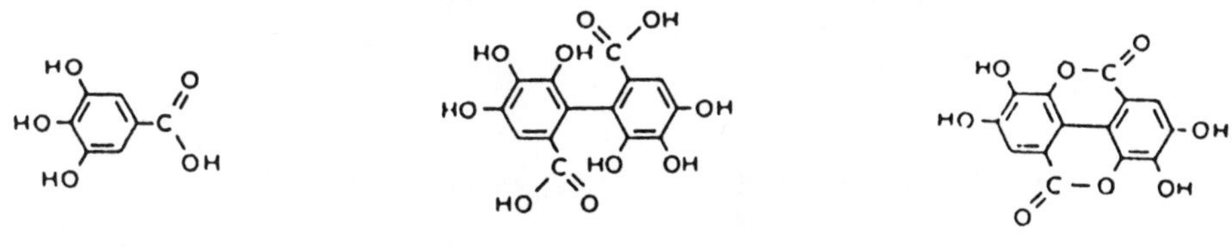

GALLIC ACID   **HEXAHYDROXYDIPHENIC ACID**   **ELLAGIC ACID**

Figure 6–30 Structure of Components of Hydrolyzable Tannins

the formation of brown pigments. This can take place during the growth of fruits (e.g., in dates) or during mechanical damage in processing.

Betalains

Table beets are a good source of red pigments; these have been increasingly used for food coloring. The red and yellow pigments obtained from beets are known as betalains and consist of the red betacyanins and the yellow betaxanthins (Von Elbe and Maing 1973). The structures of the betacyanins are shown in Figure 6–31. The major betacyanin is betanin, which accounts for 75 to 95 percent of the total pigments of beets. The remaining pigments contain isobetanin, prebetanin, and isoprebetanin. The latter two are sulfate monoesters of betanin and isobetanin,

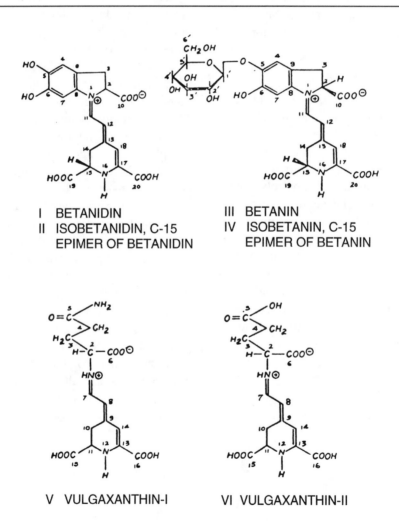

I   BETANIDIN
II   ISOBETANIDIN, C-15
     EPIMER OF BETANIDIN

III BETANIN
IV   ISOBETANIN, C-15
     EPIMER OF BETANIN

V   VULGAXANTHIN-I

VI VULGAXANTHIN-II

Figure 6–31 Structure of Naturally Occurring Betalains in Red Beets. *Source*: From J.H. Von Elbe and I.-Y. Maing, Betalains as Possible Food Colorants of Meat Substitutes, *Cereal Sci. Today*, Vol. 18, pp. 263–264, 316–317, 1973.

respectively. The major yellow pigments are vulgaxanthin I and vulgaxanthin II. Betanin is the glucoside of betanidin, and isobetanin is the C-15 epimer of betanin.

Betanidin has three carboxyl groups (pk_a = 3.4), two phenol groups (pH_a = 8.5), and asymmetric carbons at positions 2 and 15. The 15-position is easily isomerized under acid or basic conditions in the absence of oxygen to yield isobetanidin. Under alkaline conditions and in the presence of glutamine or glutamic acid, betanin can be converted to vulgaxanthin (Mabry 1970).

The color of betanin solutions is influenced by pH. In the range of 3.5 to 7.0, the spectrum shows a maximum of 537 nm (Figure 6–32). Below pH 3, the intensity of this maximum decreases and a slight increase in the region of 570 to 640 nm occurs and the color shifts toward violet. At pH values over

7, a shift of the maximum occurs to longer wavelength. At pH 9, the maximum is about 544 nm and the color shifts toward blue. Von Elbe et al. (1974) found that the color of betanin is most stable between pH 4.0 and 6.0. The thermostability is greatest between pH 4.0 and 5.0. Light and air have a degrading effect on betanin, and the effect is cumulative.

Caramel

Caramel color can be produced from a variety of carbohydrate sources, but usually corn sugar syrup is used. Corn starch is first hydrolyzed with acid to a DE of 8 to 9, followed by hydrolysis with bacterial α-amylase to a DE of 12 to 14, then with fungal amyloglucosidase up to a DE of 90 to 95. Several types of caramel are produced. The largest amount is

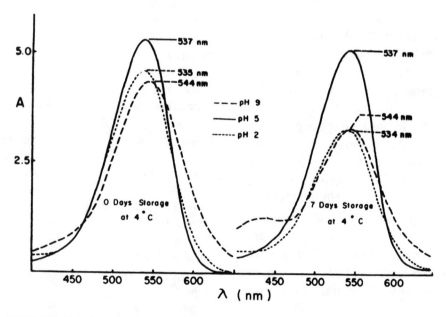

Figure 6–32 Visible Spectra of Betanin at pH Values of 2.0, 5.0, and 9.0. *Source*: From J.H. Von Elbe, I.-Y. Maing, and C.H. Amundson, Color Stability of Betanin, *Journal of Food Science*, Vol. 39, pp. 334–337, 1974, Institute of Food Technologists.

electropositive or positive caramel, which is made with ammonia. Electronegative or negative caramel is made with ammonium salts. A slightly electronegative caramel is soluble in alcohol and is used for coloring beverages (Greenshields 1973). The composition and coloring power of caramel depends on the type of raw materials and the process used. Both Maillard-type reactions and pure caramelizing reactions are thought to be involved, and the commercial product is extremely complex in composition. Caramels contain high and low molecular weight colored compounds, as well as a variety of volatile components.

Other Colorants

Synthetic colorants, used commercially, are also known as certified color additives. There are two types, FD&C dyes and FD&C lakes. FD&C indicates substances approved for use in food, drug, and cosmetic use by U.S. federal regulations. Dyes are water-soluble compounds that produce color in solution. They are manufactured in the form of powders, granules, pastes, and dispersions. They are used in foods at concentrations of less than 300 ppm (Institute of Food Technologists 1986). Lakes are made by combining dyes with alumina to form insoluble colorants, which have dye contents in the range of 20 to 25 percent (Pearce 1985). The lakes produce color in dispersion and can be used in oil-based foods when insufficient water is present for the solubilization of the dye. The list of approved water-soluble colorants has changed frequently; the current list is given in Chapter 11.

The uncertified color additives (Institute of Food Technologists 1986) include a number of natural extracts as well as inorganic substances such as titanium dioxide. Some of these can be used only with certain restrictions (Table 6–7). The consumer demand for more natural colorants has provided an impetus for examining many natural coloring substances. These have been described in detail by Francis (1987). The possibility of using plant tissue culture for the production of natural pigments has also been considered (Ilker 1987).

Table 6–7 Color Additives Not Requiring Certification

Colorant	Restriction
Annatto extract	—
Beta-apo-8′-carotenal	33 mg/kg
Beta-carotene	—
Beet powder	—
Canthaxanthin	66 mg/kg
Caramel	—
Carrot oil	—
Cochineal extract (carmine)	—
Ferrous gluconate	Ripe olives only
Fruit juice	—
Grape color extract	Nonbeverage foods only
Grape skin extract (enocianina)	Beverages
Paprika and its oleoresin	—
Riboflavin	—
Saffron	—
Titanium dioxide	1%
Turmeric and its oleoresin	—
Vegetable juice	—

REFERENCES

Amerine, M.A., and M.A. Joslyn. 1970. *Table wines. The technology of their production.* Berkeley, CA: University of California Press.

Bodwell, C.E., and P.E. McClain. 1971. Proteins. In *The sciences of meat products*, ed. J. Price and B.S. Schweigert. San Francisco: W.H. Freeman and Co.

Borenstein, B., and R.H. Bunnell. 1967. Carotenoids: Properties, occurrence and utilization in foods. In *Advances in food research*, Vol. 15, ed. C.O. Chichester, E.M. Mrak, and G.F. Stewart. New York: Academic Press.

Braverman, J.B.S. 1963. *Introduction to the biochemistry of foods.* New York: Elsevier Publishing Co.

Chichester, C.O., and R. McFeeters. 1971. Pigment degeneration during processing and storage. In *The biochemistry of fruits and their products*, ed. A.C. Hulme. New York: Academic Press.

Clydesdale, F.M., and F.J. Francis. 1970. Color scales. *Food Prod. Dev.* 3: 117–125.

Counsell, J.N. 1985. Uses of carotenoids in foods. *IFST Proceedings* 18: 156–162.

Curl, A.L., and G.F. Bailey. 1956. Carotenoids of aged canned Valencia orange juice. *J. Agr. Food Chem.* 4: 159–162.

Daravingas, G., and R.F. Cain. 1965. Changes in the anthocyanin pigments of raspberries during processing and storage. *J. Food Sci.* 30: 400–405.

Dougall, D.K., et al. 1997. Biosynthesis and stability of monoacylated anthocyanins. *Food Technol.* 51, no. 11: 69–71.

Dziezak, J.D. 1987. Applications of food colorants. *Food Technol.* 41, no. 4: 78–88.

Falconer, M.E., et al. 1964. Carotene oxidation and off-flavor development in dehydrated carrot. *J. Sci. Food Agr.* 15: 897–901.

Fox, J.B. 1966. The chemistry of meat pigments. *J. Agr. Food Chem.* 14: 207–210.

Francis, F.J. 1987. Lesser known food colorants. *Food Technol.* 41, no. 4: 62–68.

Greenshields, R.N. 1973. Caramel—Part 2. Manufacture, composition and properties. *Process Biochem.* 8, no. 4: 17–20.

Grob, E.C. 1963. The biogenesis of carotenes and carotenoids. In *Carotenes and carotenoids*, ed. K. Lang. Darmstadt, Germany: Steinkopff Verlag.

Ilker, R. 1987. In-vitro pigment production: An alternative to color synthesis. *Food Technol.* 41, no. 4: 70–72.

Institute of Food Technologists. 1986. Food colors: Scientific status summary. *Food Technol.* 40, no. 7: 49–56.

Jurd, L. 1964. Reactions involved in sulfite bleaching of anthocyanins. *J. Food Sci.* 29: 16–19.

Landrock, A.H., and G.A. Wallace. 1955. Discoloration of fresh red meat and its relationship to film oxygen permeability. *Food Technol.* 9: 194–196.

Lynn, D.Y.C., and B.S. Luh. 1964. Anthocyanin pigments in Bing cherries. *J. Food Sci.* 29: 735–743.

Mabry, T.J. 1970. Betalains, red-violet and yellow alkaloids of *Centrospermae*. In *Chemistry of the Alkaloids*, ed. S.W. Pelletier. New York: Van Nostrand Reinhold Co.

Macheix, J.J., et al. 1990. *Fruit phenolics.* Boca Raton, FL: CRC Press.

MacKinney, G., and A.C. Little. 1962. Color of foods. Westport, CT: AVI Publishing Co.

Maes, P.J.A., et al. 1997. Converting spectra into color indices. *Inform* 8: 1245–1252.

Möhler, K. 1974. Formation of curing pigments by chemical, biochemical or enzymatic reactions. In *Proceedings of the International Symposium on Nitrite in Meat Products.* Wageningen, The Netherlands: Center for Agricultural Publishing and Documentation.

Pearce, A. 1985. Current synthetic food colors. *IFST Proceedings* 18: 147–155.

Solberg, M. 1968. Factors affecting fresh meat color. *Proc. Meat Ind. Research Conference*, Chicago. March 21, 22.

Solberg, M. 1970. The chemistry of color stability in meat: A review. *Can. Inst. Food Technol. J.* 3: 55–62.

Somaatmadja, D., and J.J. Powers. 1963. Anthocyanins, IV: Anthocyanin pigments of Cabernet Sauvignon grapes. *J. Food Sci.* 28: 617–622.

Sun, B.H., and F.J. Francis. 1968. Apple anthocyanins: Identification of cyanidin-7-arabinoside. *J. Food Sci.* 32: 647–649.

Von Elbe, J.H., and I.-Y. Maing. 1973. Betalains as possible food colorants of meat substitutes. *Cereal Sci. Today* 18: 263–264, 316–317.

Von Elbe, J.H., I.-Y. Maing, and C.H. Amundson. 1974. Color stability of betanin. *J. Food Sci.* 39: 334–337.

Weckel, K.G., et al. 1962. Carotene components of frozen and processed carrots. *Food Technol.* 16, no. 8: 91–94.

Yoshida, K., et al. 1991. Unusually stable monoacylated anthocyanin from purple yam *Dioscorea alata*. *Tetrahedron Lett.* 32: 5579–5580.

Zapsalis, C., and F.J. Francis. 1965. Cranberry anthocyanins. *J. Food Sci.* 30: 396–399.

CHAPTER 7

Flavor

INTRODUCTION

Flavor has been defined by Hall (1968) as follows: "Flavor is the sensation produced by a material taken in the mouth, perceived principally by the senses of taste and smell, and also by the general pain, tactile and temperature receptors in the mouth. Flavor also denotes the sum of the characteristics of the material which produce that sensation."

This definition makes clear that flavor is a property of a material (a food) as well as of the receptor mechanism of the person ingesting the food. The study of flavor includes the composition of food compounds having taste or smell, as well as the interaction of these compounds with the receptors in the taste and smell sensory organs. Following an interaction, the organs produce signals that are carried to the central nervous system, thus creating what we understand as flavor. This process is probably less well understood than the processes occurring in other organs (O'Mahony 1984). Beidler (1957) has represented the taste process schematically (Figure 7–1).

Although flavor is composed mainly of taste and odor, other qualities contribute to the overall sensation. Texture has a very definite effect. Smoothness, roughness, granularity, and viscosity can all influence

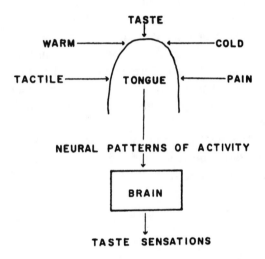

Figure 7–1 Schematic Representation of the Taste Process. *Source*: From L.M. Beidler, Facts and Theory on the Mechanism of Taste and Odor Perception, in *Chemistry of Natural Food Flavors*, 1957, Quartermaster Food and Container Institute for the Armed Forces.

flavor, as can hotness of spices, coolness of menthol, brothiness or fullness of certain amino acids, and the tastes described as metallic and alkaline.

TASTE

It is generally agreed that there are only four basic, or *true*, tastes: sweet, bitter, sour, and salty. The sensitivity to taste is located in taste buds of the tongue. The taste buds are grouped in papillae, which appear to be sensitive to more than one taste. There is undoubtedly a regional distribution of the four kinds of receptors at the tongue, creating areas of sensitivity—the sweet taste at the tip of the tongue, bitter at the back, sour at the edges, and salty at both edges and tip (Figure 7–2). The question of how the four types of receptors are able to respond this specifically has not been resolved. According to Teranishi et al. (1971), perception of the basic taste qualities results from a pattern of nerve activity coming from many taste cells; specific receptors for sweet, sour, bitter, and salty do not exist. It may be envisioned that a single taste cell possesses multiple receptor sites, each of which may have specificity.

The mechanism of the interaction between the taste substance and the taste receptor is not well understood. It has been suggested that the taste compounds interact with specific proteins in the receptor cells. Sweet- and bitter-sensitive proteins have been reported. Dastoli and Price (1966) isolated a protein from bovine tongue epithelium that showed the properties of a sweet taste receptor molecule. Dastoli et al. (1968) reported isolating a protein that had the properties of a bitter receptor.

We know that binding between stimulus and receptor is a weak one because no irreversible effects have been observed. A mechanism of taste stimulation with electrolytes has been proposed by Beidler (1957); it is shown in Figure 7–3. The time required for taste response to take place is in the order of 25 milliseconds. The taste molecule is weakly adsorbed, thereby creating a disturbance in the molecular geography of the surface and allowing an interchange of ions across the surface. This reaction is followed by an electrical depolarization that initiates a nerve impulse.

The taste receptor mechanism has been more fully described by Kurihara (1987). The process from chemical stimulation to transmitter release is schematically presented in Figure 7–4. The receptor membranes contain voltage-dependent calcium channels. Taste compounds contact the taste cells and depolarize the receptor membrane; this depolarization spreads to the synaptic area, activating the voltage-dependent calcium channels. Influx of calcium triggers the release of the transmitter norepinephrine.

The relationship between stimulus concentration and neural response is not a simple one. As the stimulus concentration increases, the response increases at a decreasing rate until a point is reached where further increase in stimulus concentration does not produce a further increase in response. Beidler (1954) proposed the following equa-

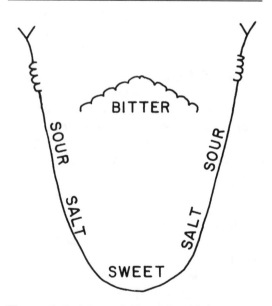

Figure 7–2 Areas of Taste Sensitivity of the Tongue

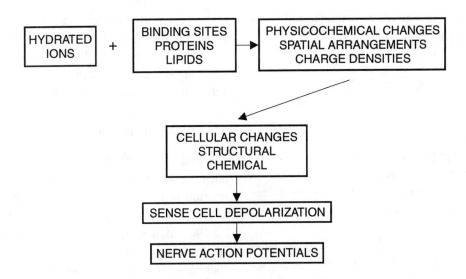

Figure 7–3 Mechanism of Taste Stimulation as Proposed by Beidler. *Source*: From L.M. Beidler, Facts and Theory on the Mechanism of Taste and Odor Perception, in *Chemistry of Natural Food Flavors*, 1957, Quartermaster Food and Container Institute for the Armed Forces.

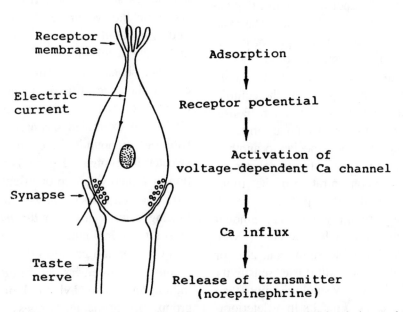

Figure 7–4 Diagram of a Taste Cell and the Mechanism of Chemical Stimulation and Transmitter Release. *Source*: Reprinted with permission from Y. Kawamura and M.R. Kare, *Umami: A Basic Tale*, © 1987, Marcel Dekker, Inc.

tion relating magnitude of response and stimulus concentration:

$$\frac{C}{R} = \frac{C}{R_s} + \frac{1}{KR_s}$$

where

C = stimulus concentration
R = response magnitude
R_s = maximum response
K = equilibrium constant for the stimulus-receptor reaction

K values reported by Beidler for many substances are in the range of 5 to 15.

It appears that the initial step in the stimulus-receptor reaction is the formation of a weak complex, as evidenced by the small values of K. The complex formation results in the initiation of the nerve impulse. Taste responses are relatively insensitive to changes in pH and temperature. Because of the decreasing rate of response, we know that the number of receptor sites is finite. The taste response is a function of the proportion of sites occupied by the stimulus compound.

According to Beidler (1957), the threshold value of a substance depends on the equilibrium constant and the maximum response. Since K and R_s both vary from one substance to another and from one species to another, the threshold also varies between substances and species. The concentration of the stimulus can be increased in steps just large enough to elicit an increase in response. This amount is called the just noticeable difference (JND).

There appear to be no significant age- or sex-related differences in taste sensitivity (Fisher 1971), but heavy smoking (more than 20 cigarettes per day) results in a deterioration in taste responsiveness with age.

Differences in taste perception between individuals seem to be common. Peryam

(1963) found that sweet and salt are usually well recognized. However, with sour and bitter taste some difficulty is experienced. Some tasters ascribe a bitter quality to citric acid and a sour quality to caffeine.

Chemical Structure and Taste

A first requirement for a substance to produce a taste is that it be water soluble. The relationship between the chemical structure of a compound and its taste is more easily established than that between structure and smell. In general, all acid substances are sour. Sodium chloride and other salts are salty, but as constituent atoms get bigger, a bitter taste develops. Potassium bromide is both salty and bitter, and potassium iodide is predominantly bitter. Sweetness is a property of sugars and related compounds but also of lead acetate, beryllium salts, and many other substances such as the artificial sweeteners saccharin and cyclamate. Bitterness is exhibited by alkaloids such as quinine, picric acid, and heavy metal salts.

Minor changes in chemical structure may change the taste of a compound from sweet to bitter or tasteless. For example, Beidler (1966) has examined saccharin and its substitution compounds. Saccharin is 500 times sweeter than sugar (Figure 7–5). Introduction of a methyl group or of chloride in the *para* position reduces the sweetness by half. Placing a nitro group in the *meta* position makes the compound very bitter. Introduction of an amino group in the *para* position retains the sweetness. Substitutions at the imino group by methyl, ethyl, or bromoethyl groups all result in tasteless compounds. However, introduction of sodium at this location yields sodium saccharin, which is very sweet.

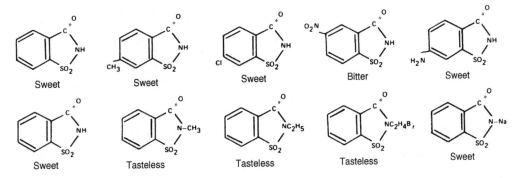

Figure 7–5 The Effect of Substitutions in Saccharin on Sweetness. *Source*: From L.M. Beidler, Chemical Excitation of Taste and Odor Receptors, in *Flavor Chemistry*, I. Hornstein, ed., 1966, American Chemistry Society.

The compound 5-nitro-*o*-toluidine is sweet. The positional isomers 3-nitro-*o*-toluidine and 3-nitro-*p*-toluidine are both tasteless (Figure 7–6). Teranishi et al. (1971) provided another example of change in taste resulting from the position of substituent group: 2-amino-4-nitro-propoxybenzene is 4,000 times sweeter than sugar, 2-nitro-4-amino-propoxybenzene is tasteless, and 2,4-dinitro-propoxybenzene is bitter (Figure 7–7). Dulcin (*p*-ethoxyphenylurea) is extremely sweet, the thiourea analog is bitter, and the *o*-ethoxyphenylurea is tasteless (Figure 7–8).

Just as positional isomers affect taste, so do different stereoisomers. There are eight amino acids that are practically tasteless. A group of three has varying tastes; except for glutamic acid, these are probably derived from sulfur-containing decomposition prod-

ucts. Seven amino acids have a bitter taste in the L form or a sweet taste in the D form, except for L-alanine, which has a sweet taste (Table 7–1). Solms et al. (1965) reported on the taste intensity, especially of aromatic amino acids. L-tryptophan is about half as bitter as caffeine; D-tryptophan is 35 times sweeter than sucrose and 1.7 times sweeter than calcium cyclamate. L-phenylalanine is about one-fourth as bitter as caffeine; the D form is about seven times sweeter than sucrose. L-tyrosine is about one-twentieth as bitter as caffeine, but D-tyrosine is still 5.5 times sweeter than sucrose.

Some researchers claim that differences exist between the L and D forms of some sugars. They propose that L-glucose is slightly salty and not sweet, whereas D-glucose is sweet. There is even a difference in taste

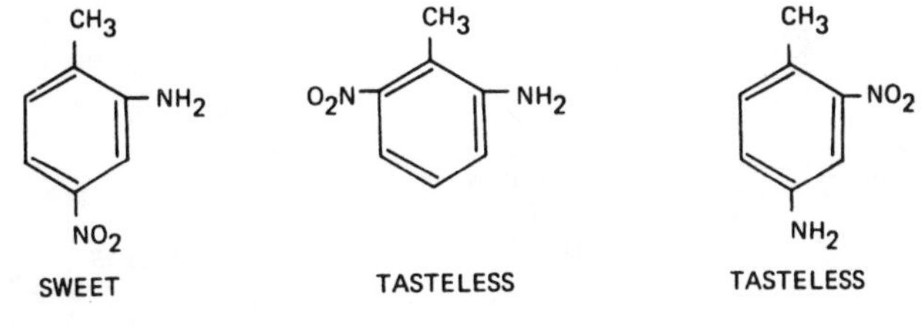

Figure 7–6 Taste of Nitrotoluidine Isomers

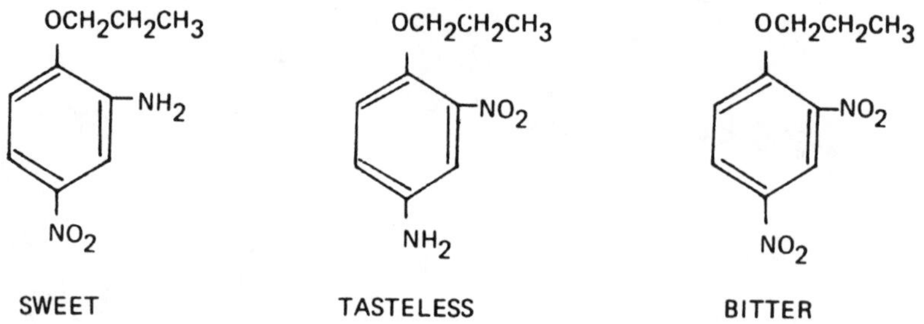

Figure 7–7 Taste of Substituted Propoxybenzenes

between the two anomers of D-mannose. The α form is sweet as sugar, and the β form is bitter as quinine.

Optical isomers of carvone have totally different flavors. The D+ form is characteristic of caraway; the L- form is characteristic of spearmint.

The ability to taste certain substances is genetically determined and has been studied with phenylthiourea. At low concentrations, about 25 percent of subjects tested do not taste this compound; for the other 75 percent, the taste is bitter. The inability to taste phenylthiourea is probably due to a recessive gene. The compounds by which tasters and nontasters can be differentiated all contain the following isothiocyanate group:

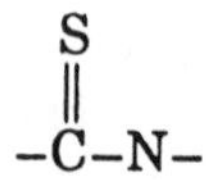

$$\begin{array}{c} S \\ \parallel \\ -C-N- \end{array}$$

These compounds—phenylthiourea, thiourea, and thiouracil—are illustrated in Figure

7–9. The corresponding compounds that contain the group,

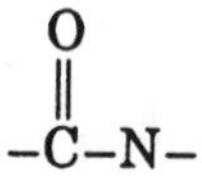

$$\begin{array}{c} O \\ \parallel \\ -C-N- \end{array}$$

phenylurea, urea, and uracil, do not show this phenomenon. Another compound containing the isothiocyanate group has been found in many species of the Cruciferae family; this family includes cabbage, turnips, and rapeseed and is well known for its goitrogenic effect. The compound is goitrin, 5-vinyloxazolidine-2-thione (Figure 7–10).

Sweet Taste

Many investigators have attempted to relate the chemical structure of sweet tasting compounds to the taste effect, and a series of theories have been proposed (Shallenberger 1971). Shallenberger and Acree (1967, 1969) pro-

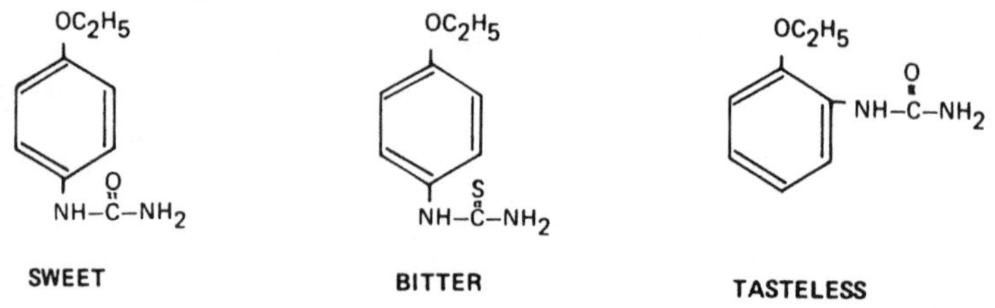

Figure 7–8 Taste of Substituted Ethoxybenzenes

Table 7–1 Difference in Taste Between the L- and D-Forms of Amino Acids

Amino Acid	Taste of L Isomer	Taste of D Isomer
Asparagine	Insipid	Sweet
Glutamic acid	Unique	Almost tasteless
Phenylalanine	Faintly bitter	Sweet, bitter aftertaste
Leucine	Flat, faintly bitter	Strikingly sweet
Valine	Slightly sweet, bitter	Strikingly sweet
Serine	Faintly sweet, stale aftertaste	Strikingly sweet
Histidine	Tasteless to bitter	Sweet
Isoleucine	Bitter	Sweet
Methionine	Flat	Sweet
Tryptophane	Bitter	Very sweet

posed a theory that can be considered a refinement of some of the ideas incorporated in previous theories. According to this theory, called the AH,B theory, all compounds that bring about a sweet taste response possess an electronegative atom A, such as oxygen or nitrogen. This atom also possesses a proton attached to it by a single covalent bond; therefore, AH can represent a hydroxyl group, an imine or amine group, or a methine group.

Within a distance of about 0.3 nm from the AH proton, there must be a second electronegative atom B, which again can be oxygen or nitrogen (Figure 7–11). Investigators have recognized that sugars that occur in a favored chair conformation yield a glycol unit conformation with the proton of one hydroxyl group at a distance of about 0.3 nm from the oxygen of the next hydroxyl group; this unit can be considered as an AH,B system. It was also found that the π bonding cloud of the benzene ring could serve as a B moiety. This explains the sweetness of benzyl alcohol and the sweetness of the *anti* isomer of anisaldehyde oxime, as well as the lack of sweetness of the *syn* isomer. The structure of these compounds is given in Figure 7–12. The AH,B system present in sweet compounds is, according to Shallenberger, able to react with a similar AH,B unit that exists at the taste bud receptor site through the formation of simultaneous hydrogen bonds. The relatively strong nature of such bonds could explain why the sense of sweetness is a lingering sensation. According to the AH,B theory, there should not be a difference in sweetness between the L and D isomers of sugars. Experiments by Shallenberger (1971) indicated that a panel could not distinguish among the sweet taste of the enantiomorphic forms of glucose, galactose, mannose, arabinose, xylose, rhamnose, and glucoheptulose. This suggests that the notion that L sugars are tasteless is a myth.

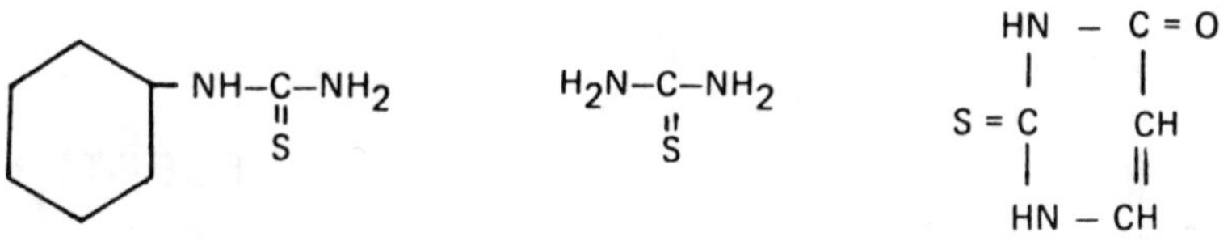

Phenylthiourea   **Thiourea**   **Thiouracil**

Figure 7–9 Compounds Containing the $-\overset{S}{\overset{\|}{C}}-N-$ Group by Which Tasters and Nontasters Can Be Differentiated

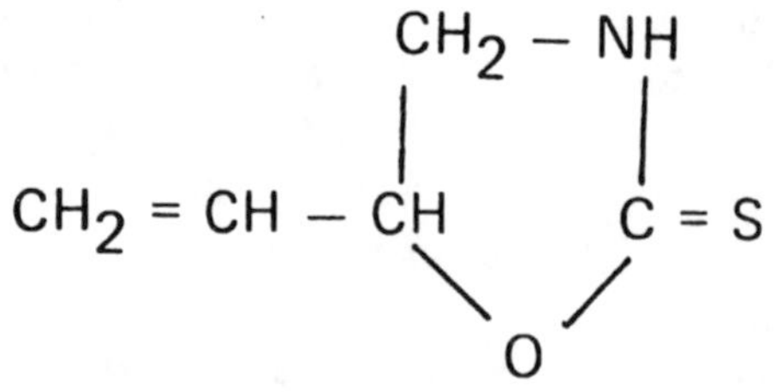

$$CH_2 = CH - CH \quad \overset{CH_2 - NH}{\underset{O}{\diagup}} \quad C = S$$

Figure 7–10 5-Vinyloxazolidine-2-thione

Spillane (1996) has pointed out that the AH,B theory appears to work quite well, although spatial, hydrophobic/hydrophilic, and electronic effects are also important. Shallenberger (1998) describes the initiation of sweetness as being due to a concerted intermolecular, antiparallel hydrogen-bonding interaction between the glycophore (Greek *glyks*, sweet; *phoros*, to carry) and receptor dipoles. The difficulty in explaining the sweetness of compounds with different chemical structures is also covered by Shallenberger (1998) and how this has resulted in alternative taste theories. The application of sweetness theory is shown to have important applications in the food industry.

Extensive experiments with a large number of sugars by Birch and Lee (1971) support Shallenberger's theory of sweetness and indicate that the fourth hydroxyl group of glucopyranosides is of unique importance in determining sweetness, possibly by donating the proton as the AH group. Ap-

parently the primary alcohol group is of little importance for sweetness. Substitution of acetyl or azide groups confers intense bitterness to sugars, whereas substitution of benzoyl groups causes tastelessness.

As the molecular weight of saccharides increases, their sweetness decreases. This is best explained by the decrease in solubility and increase in size of the molecule. Apparently, only one sugar residue in each oligosaccharide is involved in the interaction at the taste bud receptor site.

The relative sweetness of a number of sugars and other sweeteners has been reported by Solms (1971) and is given in Table 7–2. These figures apply to compounds tasted singly and do not necessarily apply to sugars in foods, except in a general sense. The relative sweetness of mixtures of sugars changes with the concentration of the components. Synergistic effects may increase the sweetness by as much as 20 to 30 percent in such mixtures (Stone and Oliver 1969).

Sour Taste

Although it is generally recognized that sour taste is a property of the hydrogen ion, there is no simple relationship between sourness and acid concentration. Acids have different tastes; the sourness as experienced in the mouth may depend on the nature of the acid group, pH, titratable acidity, buffering

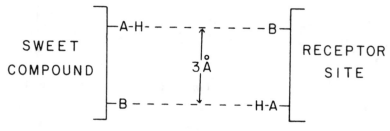

Figure 7–11 The AH,B Theory of Sweet Taste Perception

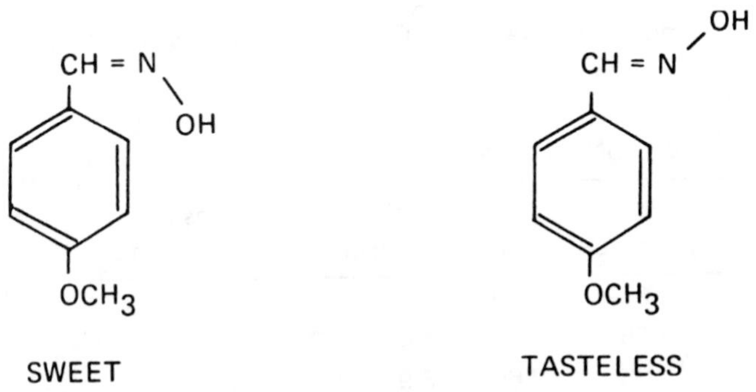

Figure 7–12 *Anti*-Anisaldehyde Oxime, Sweet; and *Syn*-Anisaldehyde Oxime, Tasteless

effects and the presence of other compounds, especially sugars. Organic acids have a greater taste effect than inorganic acids (such as hydrochloric acid) at the same pH. Information on a number of the most common acids found in foods and phosphoric acid (which is also used in soft drinks) has been collected by Solms (1971) and compared with hydrochloric acid. This information is presented in Table 7–3.

According to Beatty and Cragg (1935), relative sourness in unbuffered solutions of acids is not a function of molarity but is proportional to the amount of phosphate buffer required to bring the pH to 4.4. Ough (1963) determined relative sourness of four organic acids added to wine and also preference for these acids. Citric acid was judged the most sour, fumaric and tartaric about equal, and adipic least sour. The tastes of citric and tartaric acids were preferred over those of fumaric and adipic acids.

Pangborn (1963) determined the relative sourness of lactic, tartaric, acetic, and citric acid and found no relation between pH, total acidity, and relative sourness. It was also found that there may be considerable differences in taste effects between sugars and acids when they are tested in aqueous solutions and in actual food products.

Table 7–2 Relative Sweetness of Sugars and Other Sweeteners

Compound	Relative Sweetness
Sucrose	1
Lactose	0.27
Maltose	0.5
Sorbitol	0.5
Galactose	0.6
Glucose	0.5–0.7
Mannitol	0.7
Glycerol	0.8
Fructose	1.1–1.5
Cyclamate	30–80
Glycyrrhizin	50
Aspartyl-phenylalanine methylester	100–200
Stevioside	300
Naringin dihydrochalcone	300
Saccharin	500–700
Neohesperidin dihydrochalcone	1000–1500

Source: From J. Solms, Nonvolatile Compounds and the Flavor of Foods, in *Gustation and Olfaction*, G. Ohloff and A.F. Thomas, eds., 1971, Academic Press.

Table 7–3 Properties of Some Acids, Arranged in Order of Decreasing Acid Taste and with Tartaric Acid as Reference

Acid	Taste	Properties of 0.05N Solutions		Ionization Constant	Taste Sensation	Found In
		Total Acid g/L	pH			
Hydrochloric	+1.43	1.85	1.70	—	—	—
Tartaric	0	3.75	2.45	1.04×10^{-3}	Hard	Grape
Malic	−0.43	3.35	2.65	3.9×10^{-4}	Green	Apple, pear, prune, grape, cherry, apricot
Phosphoric	−1.14	1.65	2.25	7.52×10^{-3}	Intense	Orange, grapefruit
Acetic	−1.14	3.00	2.95	1.75×10^{-5}	Vinegar	—
Lactic	−1.14	4.50	2.60	1.26×10^{-4}	Sour, tart	—
Citric	−1.28	3.50	2.60	8.4×10^{-4}	Fresh	Berries, citrus, pineapple
Propionic	−1.85	3.70	2.90	1.34×10^{-5}	Sour, cheesy	—

Source: From J. Solms, Nonvolatile Compounds and the Flavor of Foods, in *Gustation and Olfaction*, G. Ohloff and A.F. Thomas, eds., 1971, Academic Press.

Buffering action appears to help determine the sourness of various acids; this may explain why weak organic acids taste more sour than mineral acids of the same pH. It is suggested that the buffering capacity of saliva may play a role, and foods contain many substances that could have a buffering capacity.

Wucherpfennig (1969) examined the sour taste in wine and found that alcohol may decrease the sourness of organic acids. He examined the relative sourness of 17 organic acids and found that the acids tasted at the same level of undissociated acid have greatly different intensities of sourness. Partially neutralized acids taste more sour than pure acids containing the same amount of undissociated acids. The change of malic into lactic acid during the malolactic fermentation of wines leads to a decrease in sourness, thus making the flavor of the wine milder.

Salty Taste

The salty taste is best exhibited by sodium chloride. It is sometimes claimed that the taste of salt by itself is unpleasant and that the main purpose of salt as a food component is to act as a flavor enhancer or flavor potentiator. The taste of salts depends on the nature of both cation and anion. As the molecular weight of either cation or anion—or both—increases, salts are likely to taste bitter. The lead and beryllium salts of acetic acid have a sweet taste. The taste of a number of salts is presented in Table 7–4.

The current trend of reducing sodium intake in the diet has resulted in the formulation of low-sodium or reduced-sodium foods. It has been shown (Gillette 1985) that sodium chloride enhances mouthfeel, sweetness, balance, and saltiness, and also masks

Table 7–4 Taste Sensations of Salts

Taste	Salts
Salty	LiCl, LiBr, LiI, NaNO$_3$, NaCl, NaBr, NaI, KNO$_3$, KCl
Salty and bitter	KBr, NH$_4$I
Bitter	CsCl, CsBr, KI, MgSO$_4$
Sweet	Lead acetate,[1] beryllium acetate[1]

[1]Extremely toxic

or decreases off-notes. Salt substitutes based on potassium chloride do not enhance mouthfeel or balance and increase bitter or metallic off-notes.

Bitter Taste

Bitter taste is characteristic of many foods and can be attributed to a great variety of inorganic and organic compounds. Many substances of plant origin are bitter. Although bitter taste by itself is usually considered to be unpleasant, it is a component of the taste of many foods, usually those foods that are sweet or sour. Inorganic salts can have a bitter taste (Table 7–4). Some amino acids may be bitter (Table 7–1). Bitter peptides may be formed during the partial enzymic hydrolysis of proteins—for example, during the ripening of cheese. Solms (1969) has given a list of peptides with different taste sensations (Table 7–5).

The compounds best known for their bitter taste belong to the alkaloids and glycosides. Alkaloids are basic nitrogen-containing organic compounds that are derived from pyridine, pyrrolidine, quinoline, isoquinoline, or purine. Quinine is often used as a standard for testing bitterness (Figure 7–13).

The bitterness of quinine hydrochloride is detectable in a solution as dilute as 0.00004 molar, or 0.0016 percent. If 5 mL of this solution is tasted, the amount of substance a person detects would be 0.08 mg (Moncrieff 1951). Our sensitivity to bitterness is more extreme than our sensitivity to other tastes; the order of sensitivity is from bitter to sour to salty and our least sensitivity is to sweet taste. Threshold values reported by Moncrieff are as follows: sour—0.007 percent HCl; salt—0.25 percent NaCl; and sweet—0.5 percent sucrose. If the artificial sweeteners such as saccharine are considered, the sweet sensitivity is second to bitter. Quinine is used as a component of some soft drinks to produce bitterness. Other alkaloids occurring as natural bitter constituents of foods are caffeine and theobromine (Figure 7–14), which are derivatives of purine. Another naturally occurring bitter substance is the glycoside naringin, which occurs in grapefruit and some other citrus fruits. Naringin in pure form is more bitter than quinine and can be detected in concen-

Table 7–5 Taste of Some Selected Peptides

Taste	Composition of Peptides
Flat	L-Lys-L-Glu, L-PhE-L-Phe, Gly-Gly-Gly-Gly
Sour	L-Ala-L-Asp, γ-L-Glu-L-Glu, Gly-L-Asp-L-Ser-Gly
Bitter	L-Leu-L-Leu, L-Arg-L-Pro, L-Val-L-Val-L-Val
Sweet	L-Asp-L-Phe-OMe, L-Asp-L-Met-OMe
Biting	γ-L-Glutamyl-S-(prop-1-enyl)-L-cystein

Source: From J. Solms, Nonvolatile Compounds and the Flavor of Foods, in *Gustation and Olfaction*, G. Ohloff and A.F. Thomas, eds., 1971, Academic Press.

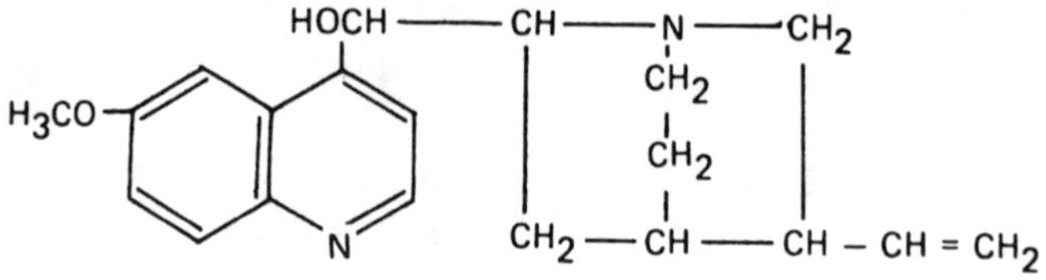

Figure 7–13 Structure of Quinine. This has an intensely bitter taste.

trations of less than 0.002 percent. Naringin (Figure 7–15) contains the sugar moiety rutinose (L-rhamnose-D-glucose), which can be removed by hydrolysis with boiling mineral acid. The aglucose is called naringenin, and it lacks the bitterness of naringin. Since naringin is only slightly soluble in water (0.05 percent at 20°C), it may crystallize out when grapefruit is subjected to below-freezing temperatures. Hesperidin (Figure 7–15) occurs widely in citrus fruits and is also a rutinose glycoside. It occurs in oranges and lemons. Dried orange peel may contain as much as 8 percent hesperidin. The aglycone of hesperidin is called hesperetin. The sugar moiety is attached to carbon 7. Horowitz and Gentili (1969) have studied the relationship between bitterness and the structure of 7-rhamnoglycosides of citrus fruits; they found that the structure of the disaccharide moiety plays an important role in bitterness. The point of attachment of rhamnose to glucose determines whether the substance will be bitter or tasteless. Thus, neohesperidin contains the disaccharide neohesporidose, which contains rham-

nose linked 1→2 to glucose; therefore, the sugar moiety is 2-*O*-α-L-rhamnopyranosyl-D-glucose. Glycosides containing this sugar, including neohesperidin, have a bitter taste. When the linkage between rhamnose and glucose is 1→6, the compound is tasteless as in hesperidin, where the sugar part, rutinose, is 6-*O*-α-L-rhamnopyranosyl-D-glucose.

Bitterness occurs as a defect in dairy products as a result of casein proteolysis by enzymes that produce bitter peptides. Bitter peptides are produced in cheese because of an undesirable pattern of hydrolysis of milk casein (Habibi-Najafi and Lee 1996). According to Ney (1979), bitterness in amino acids and peptides is related to hydrophobicity. Each amino acid has a hydrophobicity value (Δf), which is defined as the free energy of transfer of the side chains and is based on solubility properties (Table 7–6). The average hydrophobicity of a peptide, Q, is obtained as the sum of the Δf of component amino acids divided by the number of amino acid residues. Ney (1979) reported that bitterness is found only in peptides with molecular weights

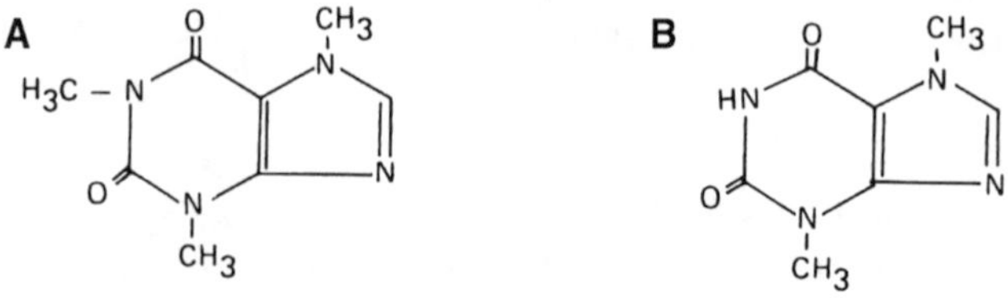

Figure 7–14 (A) Caffeine and (B) Theobromine

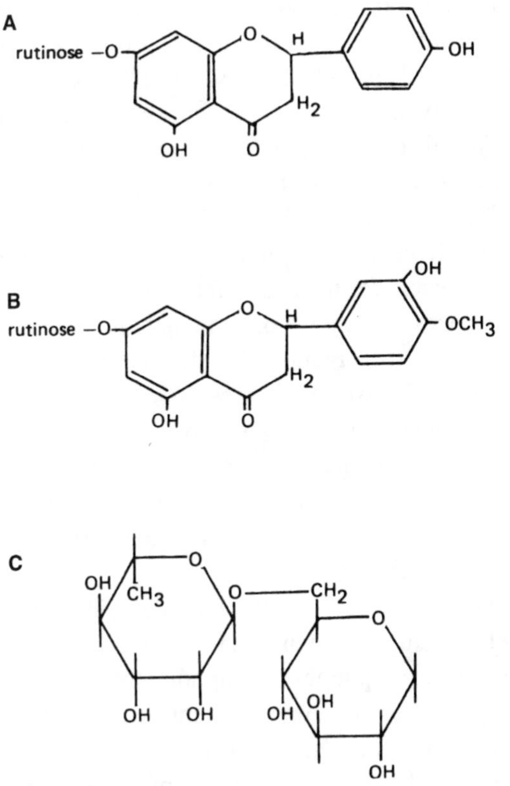

A

B

C

Figure 7–15 (A) Naringin; (B) Hesperidin; (C) Rutinose, 6-*O*-α-L-Rhamnopyranosyl-D-Glucopyranose

Table 7–6 Hydrophobicity Values (Δf) of the Side Chains of Amino Acids

Amino Acid	Abbrevia-tion	Δf (cal/mol)
Glycine	Gly	0
Serine	Ser	40
Threonine	Thr	440
Histidine	His	500
Aspartic acid	Asp	540
Glutamic acid	Glu	550
Arginine	Arg	730
Alanine	Ala	730
Methionine	Met	1,300
Lysine	Lys	1,500
Valine	Val	1,690
Leucine	Leu	2,420
Proline	Pro	2,620
Phenylalanine	Phe	2,650
Tyrosine	Tyr	2,870
Isoleucine	Ile	2,970
Tryptophan	Trp	3,000

Source: Reprinted with permission from K.H. Ney, Bitterness of Peptides: Amino Acid Composition and Chain Length, in *Food Taste Chemistry*, J.C. Boudreau, ed., ACS Symp. Ser. 115, © 1979, American Chemical Society.

below 6,000 Da when their Q value is greater than 1,400. These findings indicate the importance of molecular weight and hydrophobicity. In a more detailed study of the composition of bitter peptides, Kanehisa (1984) reported that at least six amino acids are required for strong bitterness. A bitter peptide requires the presence of a basic amino acid at the N-terminal position and a hydrophobic one at the C-terminal position. It appears that at least two hydrophobic amino acids are required in the C-terminal area of the peptide to produce intense bitterness. The high hydrophobicity of leucine and the number of leu-cine and possibly proline residues in the peptide probably play a role in the bitterness.

Other Aspects of Taste

The basic sensations—sweet, sour, salty, and bitter—account for the major part of the taste response. However, it is generally agreed that these basic tastes alone cannot completely describe taste. In addition to the four individual tastes, there are important interrelationships among them. One of the most important in foods is the interrelationship between sweet and sour. The sugar-acid ratio plays an important part in many foods, especially fruits. Kushman and Ballinger (1968) have demonstrated the change in sugar-acid ratio in ripening blueberries (Table 7–7). Sugar-acid ratios play an important role in the flavor quality of fruit juices and wines (Ough 1963). Alkaline taste has been attributed to the hydroxyl ion. Caustic compounds can be detected in solutions containing only 0.01 percent of the alkali. Probably the major effect of alkali is irritation of the general nerve endings in the mouth. Another effect that is difficult to describe is astringency. Borax is known for its ability to produce this effect, as are the tannins present in foods, especially those that occur in tea. Even if astringency is not considered a part of the taste sense, it must still be considered a feature of food flavor.

Another important taste sensation is coolness, which is a characteristic of menthol. The cooling effect of menthol is part of the mint flavor complex and is exhibited by only some of the possible isomeric forms. Only (–) and (+) menthol show the cooling effect, the former to a higher degree than the latter, but the isomers isomenthol, neomenthol, and neoisomenthol do not give a cooling effect (Figure 7–16) (Kulka 1967). Hotness is a property associated with spices and is also referred to as pungency. The compound primarily responsible for the hotness of black pepper is piperine (Figure 7–17). In red pepper or capsicum, nonvolatile amides are responsible for the heat effect. The heat effect of spices and their constituents can be measured by an organoleptic threshold method (Rogers 1966) and expressed in heat units. The pungent principle of capsicum is capsaicin. The structure of capsaicin is given in Figure 7–18. Capsaicin shows similarity to the compound zingerone, the pungent principle of ginger (Figure 7–19).

Govindarajan (1979) has described the relationship between pungency and chemical structure of pungent compounds. There are three groups of natural pungent compounds—the capsaicinoids, piperine, and the gingerols. These have some common structural aspects, including an aromatic ring and an alkyl side chain with a carbonyl function (Figures 7–18 and 7–19). Structural variations in these compounds affect the intensity of the pungent response. These structural variations include the length of the alkyl side chain, the position of the amide group near the polar aromatic end, the nature of the groupings at the alkyl end, and the unsaturation of the alkyl chain.

The metallic taste has been described by Moncrieff (1964). There are no receptor sites for this taste or for the alkaline and meaty tastes. However, according to Moncrieff, there is no doubt that the metallic taste is a real one. It is observable over a wide area of the surface of the tongue and mouth and, like irritation and pain, appears to be a modality of the common chemical sense. The metallic taste can be generated by salts of metals such

Table 7–7 Change in Sugar-Acid Ratio During Ripening of Blueberries*

	Unripe	Ripe	Overripe
Total sugar (%)	5.8	7.9	12.4
pH	2.83	3.91	3.76
Titr acidity (mEq/100 g)	23.9	12.9	7.5
Sugar-acid ratio	3.8	9.5	25.8

*The sugars are mainly glucose and fructose, and the acidity is expressed as citric acid.

Source: From L.J. Kushman and W.E. Ballinger, Acid and Sugar Changes During Ripening in Wolcott Blueberries, *Proc. Amer. Soc. Hort. Soc.*, Vol. 92, pp. 290–295, 1968.

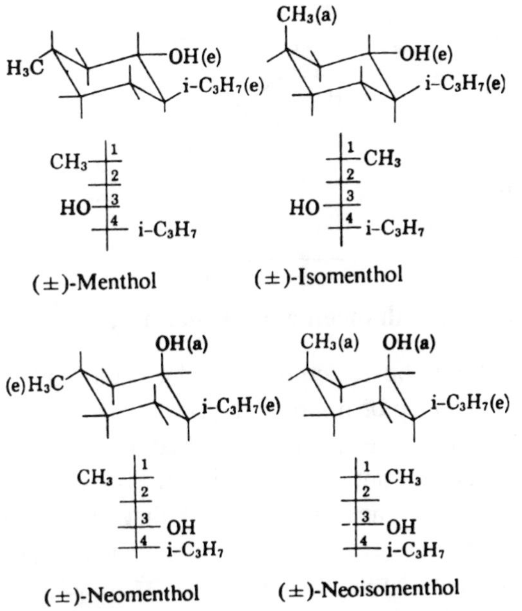

Figure 7–16 Isomeric Forms of Menthol

as mercury and silver (which are most potent) but normally by salts of iron, copper, and tin. The threshold concentration is in the order of 20 to 30 ppm of the metal ion. In canned foods, considerable metal uptake may occur and the threshold could be

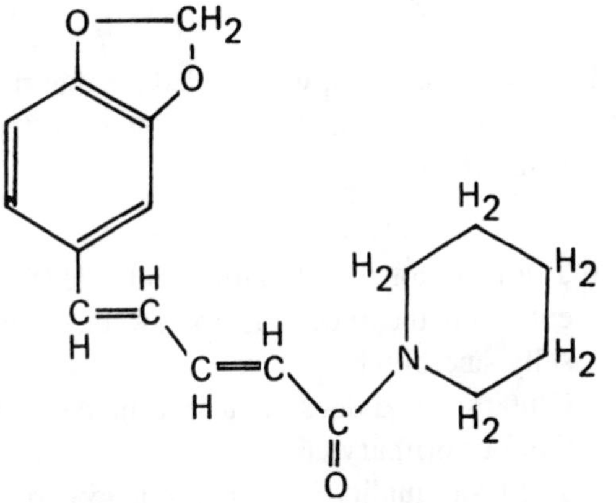

Figure 7–17 Piperine, Responsible for the Hotness of Pepper

exceeded in such cases. Moncrieff (1964) also mentions the possibility of metallic ion exchange between the food and the container. The threshold concentration of copper is increased by salt, sugar, citric acid, and alcohol. Tannin, on the other hand, lowers the threshold value and makes the copper taste more noticeable. The metallic taste is frequently observed as an aftertaste. The lead salt of saccharin gives an impression of intense sweetness, followed by a metallic aftertaste. Interestingly, the metallic taste is frequently associated with oxidized products. Tressler and Joslyn (1954) indicate that 20 ppm of copper is detectable by taste in orange juice. Copper is well known for its ability to catalyze oxidation reactions. Stark and Forss (1962) have isolated and identified oct-1-en-3-one as the compound responsible for the metallic flavor in dairy products.

Taste Inhibition and Modification

Some substances have the ability to modify our perception of taste qualities. Two such compounds are gymnemagenin, which is able to suppress the ability to taste sweetness, and the protein from miracle fruit, which changes the perception of sour to sweet. Both compounds are obtained from tropical plants.

The leaves of the tropical plant *Gymnema sylvestre,* when chewed, suppress the ability to taste sweetness. The effect lasts for hours, and sugar seems like sand in the mouth. The ability to taste other sweeteners such as saccharin is equally suppressed. There is also a decrease in the ability to taste bitterness. The active principle of leaves has been named gymnemic acid and has been found (Stöcklin et al. 1967) to consist of four components, designated as gymnemic acids, A_1, A_2, A_3, and A_4. These are D-glucuronides of acety-

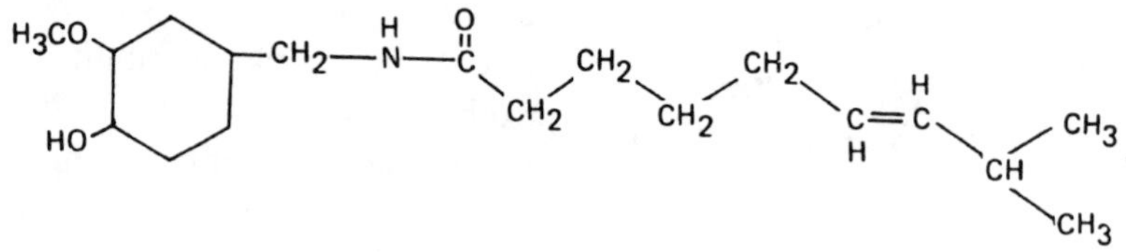

Figure 7–18 Capsaicin, the Pungent Principle of Red Pepper

lated gymnemagenins. The unacetylated gymnemagenin is a hexahydroxy pentacyclic triterpene; its structure is given in Figure 7–20.

The berries of a West African shrub (*Synsepalum dulcificum*) contain a substance that has the ability to make sour substances taste sweet. The berry, also known as miracle fruit, has been shown to contain a taste-modifying protein (Kurihara and Beidler 1968; 1969). The protein is a basic glycoprotein with a molecular weight of 44,000. It is suggested that the protein binds to the receptor membrane near the sweet receptor site. The low pH changes the conformation of the membrane so that the sugar part of the protein fits into the sweet receptor site. The taste-modifying protein was found to contain 6.7 percent of arabinose and xylose.

These taste-modifying substances provide an insight into the mechanism of the production of taste sensations and, therefore, are a valuable tool in the study of the interrelationship between taste and chemical structure.

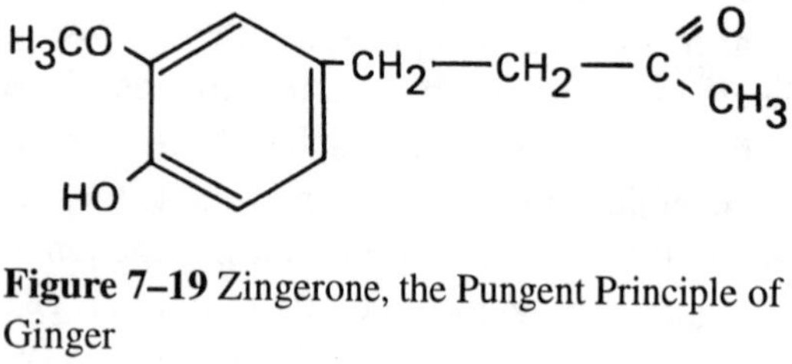

Figure 7–19 Zingerone, the Pungent Principle of Ginger

Flavor Enhancement—Umami

A number of compounds have the ability to enhance or improve the flavor of foods. It has often been suggested that these compounds do not have a particular taste of their own. Evidence now suggests that there is a basic taste response to amino acids, especially glutamic acid. This taste is sometimes described by the word *umami*, derived from the Japanese for deliciousness (Kawamura and Kare 1987). It is suggested that a primary taste has the following characteristics:

• The receptor site for a primary taste chemical is different from those of other primary tastes.
• The taste quality is different from others.
• The taste cannot be reproduced by a mixture of chemicals of different primary tastes.

From these criteria, we can deduce that the glutamic acid taste is a primary taste for the following reasons:

• The receptor for glutamic acid is different from the receptors for sweet, sour, salty, and bitter.
• Glutamic acid does not affect the taste of the four primary tastes.
• The taste quality of glutamic acid is different from that of the four primary tastes.

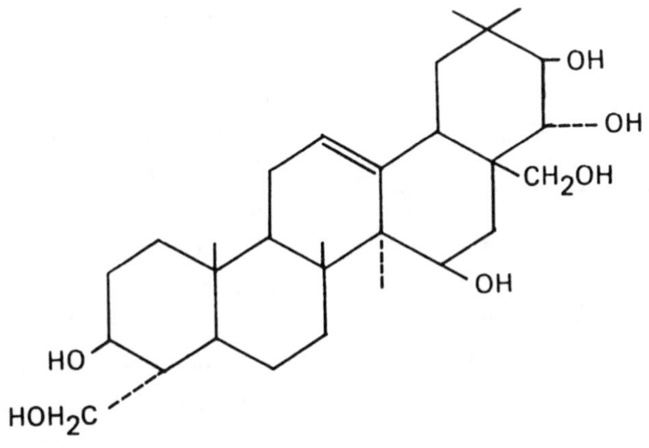

Figure 7–20 Structure of Gymnemagenin

- Umami cannot be reproduced by mixing any of the four primary tastes.

Monosodium glutamate has long been recognized as a flavor enhancer and is now being considered a primary taste, umami. The flavor potentiation capacity of monosodium glutamate in foods is not the result of an intensifying effect of the four primary tastes. Glutamate may exist in the L and D forms and as a racemic mixture. The L form is the naturally occurring isomer that has a flavor-enhancing property. The D form is inert. Although glutamic acid was first isolated in 1866, the flavor-enhancing properties of the sodium salt were not discovered until 1909 by the Japanese chemist Ikeda. Almost immediately, commercial production of the compound started and total production for the year 1954 was estimated at 13,000,000 pounds. The product as first described by Ikeda was made by neutralizing a hydrolysate of the seaweed *Laminaria japonica* with soda. Monosodium glutamate is now produced from wheat gluten, beet sugar waste, and soy protein and is used in the form of the pure crystallized compound. It can also be used in the form of protein hydrolysates derived from proteins that contain 16 percent or more of glutamic acid. Wheat gluten, casein, and soy flour are good sources of glutamic acid and are used to produce protein hydrolysates. The glutamic acid content of some proteins is listed in Table 7–8 (Hall 1948). The protein is hydrolyzed with hydrochloric acid, and the neutralized hydrolysate is used in liquid form or as a dry powder. Soy sauce, which is similar to these hydrolysates, is produced wholly or partially by enzymic hydrolysis. This results in the formation of ammonia from acid amides; soy sauce contains ammonium complexes of amino acids, including ammonium glutamate.

The flavor of glutamate is difficult to describe. It has sometimes been suggested that glutamate has a meaty or chickeny taste, but it is now generally agreed that glutamate flavor is unique and has no similarity to meat. Pure sodium glutamate is detectable in concentrations as low as 0.03 percent; at 0.05 percent the taste is very strong and does not increase at higher concentrations. The taste has been described (Crocker 1948) as a mixture of the four tastes. At about 2 threshold values of glutamate concentration, it could

Table 7–8 Glutamic Acid Content of Some Proteins

Protein Source	Glutamic Acid (%)
Wheat gluten	36.0
Corn gluten	24.5
Zein	36.0
Peanut flour	19.5
Cottonseed flour	17.6
Soybean flour	21.0
Casein	22.0
Rice	24.1
Egg albumin	16.0
Yeast	18.5

Source: From L.A. Hall, Protein Hydrolysates as a Source of Glutamate Flavors, in *Monosodium Glutamate—A Symposium,* 1948, Quartermaster Food and Container Institute for the Armed Forces.

be well matched by a solution containing 0.6 threshold of sweet, 0.7 of salty, 0.3 of sour, and 0.9 of bitter. In addition, glutamate is said to cause a tingling feeling and a marked persistency of taste sensation. This feeling is present in the whole of the mouth and provides a feeling of satisfaction or fullness. Apparently glutamate stimulates our tactile sense as well as our taste receptors. The presence of salt is required to produce the glutamate effect. Glutamate taste is most effective in the pH range of 6 to 8 and decreases at lower pH values. Sugar content also affects glutamate taste. The taste in a complex food, therefore, depends on a complex interaction of sweet, sour, and salty, as well as the added glutamate.

Monosodium glutamate improves the flavor of many food products and is therefore widely used in processed foods. Products benefiting from the addition of glutamate include meat and poultry, soups, vegetables, and seafood.

For many years glutamate was the only known flavor enhancer, but recently a number of compounds that act similarly have been discovered. The 5′-nucleotides, especially 5′-inosinate and 5′-guanylate, have enhancement properties and also show a synergistic effect in the presence of glutamate. This synergistic effect has been demonstrated by determining the threshold levels of the compounds alone and in mixtures. The data in Table 7–9 are quoted from Kuninaka (1966). The 5′-nucleotides were discovered many years ago in Japan as components of dried bonito (a kind of fish). However, they were not produced commercially and used as flavor enhancers until recently, when technical problems in their production were solved. The general structure of the nucleotides with flavor activity is presented in Figure 7–21. There are three types of inosinic acid, 2′-, 3′-, and 5′-isomers; only the 5′-isomer has flavor activity. Both riboside and 5′-phosphomonoester linkages are required for flavor activity, which is also the case for the OH group at the 6-position of the ring. Replacing the OH group with other groups, such as an amino group, sharply reduces flavor activity but this is not true for the group at the 2-position. Hydrogen at the 2-position corresponds with inosinate and an amino group with guanylate; both have comparable flavor activity, and the effect of the two compounds is additive.

The synergistic effect of umami substances is exceptional. The subjective taste intensity of a blend of monosodium glutamate and disodium 5′-inosinate was found to be 16 times stronger than that of the glutamate by itself at the same total concentration (Yamaguchi 1979).

Table 7–9 Threshold Levels of Flavor Enhancers Alone and in Mixtures in Aqueous Solution

	Threshold Level (%)		
Solvent	*Disodium 5′-Inosinate*	*Disodium 5′-Guanylate*	*Monosodium L-Glutamate*
Water	0.012	0.0035	0.03
0.1% glutamate	0.0001	0.00003	—
0.01% inosinate	—	—	0.002

Source: From A. Kuninaka, Recent Studies of 5′-Nucleotides as New Flavor Enhancers, in *Flavor Chemistry*, I. Hornstein, ed., 1966, American Chemical Society.

5′-nucleotides can be produced by degradation of ribonucleic acid. The problem is that most enzymes split the molecule at the 3′-phosphodiester linkages, resulting in nucleotides without flavor acitivity. Suitable enzymes were found in strains of *Penicillium* and *Streptomyces*. With the aid of these enzymes, the 5′-nucleotides can be manufactured industrially from yeast ribonucleic acid. Another process produces the nucleoside inosine by fermentation, followed by chemical phosphorylation to 5′-inosinic acid (Kuninaka 1966).

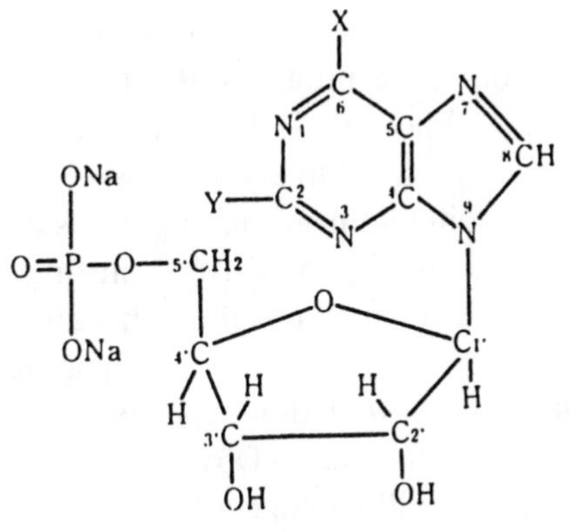

The search for other flavor enhancers has brought to light two new amino acids, tricholomic acid and ibotenic acid, obtained from fungi (Figure 7–22). These amino acids have flavor activities similar to that of monosodium glutamate. Apparently, the flavor enhancers can be divided into two groups; the first consists of 5′-inosinate and 5′-guanylate with the same kind of activity and an additive relationship. The other group consists of glutamate, tricholomic, and ibotenic acid, which are additive in action. Between the members of the two groups, the activity is synergistic.

A different type of flavor enhancer is maltol, which has the ability to enhance sweetness produced by sugars. Maltol is formed during roasting of malt, coffee, cacao, and grains. During the baking process, maltol is formed in the crust of bread. It is also found in many dairy products that have been heated, as a product of decomposition of the casein-lactose system. Maltol (Figure 7–23) is formed from di-, tri-, and tetrasaccharides including isomaltose, maltotretraose, and

X = OH   Y = H   IMP
          Y = NH$_2$ GMP
          Y = OH   XMP

Figure 7–21 Structure of Nucleotides with Flavor Activity

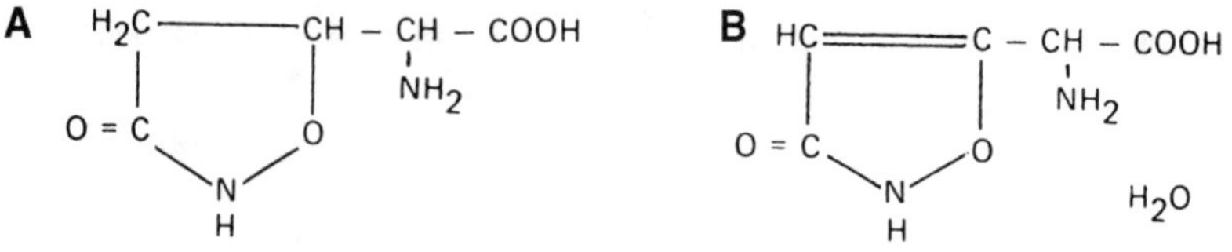

Figure 7–22 (A) Tricholomic and (B) Ibotenic Acid

panose but not from maltotriose. Formation of maltol is brought about by high temperatures and is catalyzed by metals such as iron, nickel, and calcium.

Maltol has antioxidant properties. It has been found to prolong storage life of coffee and roasted cereal products. Maltol is used as a flavor enhancer in chocolate and candies, ice cream, baked products, instant coffee and tea, liqueurs, and flavorings. It is used in concentrations of 50 to 250 ppm and is commercially produced by a fermentation process.

ODOR

The olfactory mechanism is both more complex and more sensitive than the process of gustation. There are thousands of odors, and the sensitivity of the smell organ is about 10,000 times greater than that of the taste organ. Our understanding of the odor receptor's mechanism is very limited, and there is no single, generally accepted theory accounting for the relationship between molecular structure and odor. The odorous substance arrives at the olfactory tissue in the nasal cavity, contained in a stream of air. This method of sensing requires that the odorous compound be volatile. Most odorous compounds are soluble in a variety of solvents, but it appears that solubility is less important than type of molecular arrangement, which confers both solubility and chemical reactiv-

ity (Moncrieff 1951). The number of volatile compounds occurring in foods is very high. Maarse (1991) has given the following numbers for some foods: beef (boiled, cooked)—486; beer—562; butter—257; coffee—790; grape—466; orange—203; tea—541; tomato—387; and wine (white)—644. Not all of these substances may be essential in determining the odor of a product. Usually, the relative amounts of a limited number of these volatile compounds are important in establishing the characteristic odor and flavor of a food product.

The sensitivity of the human olfactory organ is inferior to that of many animals. Dogs and rats can detect odorous compounds at threshold concentrations 100 times lower than man. When air is breathed in, only a small part of it is likely to flow over the olfactory epithelium in the upper nasal cavity. When a smell is perceived, sniffing may increase the amount reaching the olfactory tissue. When foods are eaten, the passage of breath during exhalation reaches the nasal cavity from the back. Döving (1967) has quoted the threshold concentrations of odorous substances listed in Table 7–10. Apparently, it is possible to change odor thresholds by a factor of 100 or more by stimulating the sympathetic nervous system so that more odor can reach the olfactory tissue. What is remarkable about the olfactory mechanism is not only that thousands of odors can be recognized, but that it is possible to store the

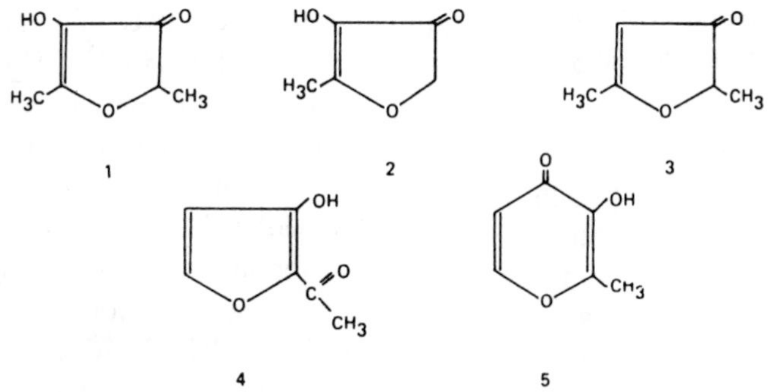

Figure 7–23 Some Furanones (1,2,3), Isomaltol (4), and Maltol (5)

information in the brain for retrieval after long periods of time. The ability to smell is affected by several conditions, such as colds, menstrual cycle, and drugs such as penicillin. Odors are usually the result of the presence of mixtures of several, sometimes many, different odorous compounds. The combined effect creates an impression that may be very different from that of the individual components. Many food flavors, natural as well as artificial, are of this compound nature.

Odor and Molecular Structure

M. Stoll wrote in 1957: "The whole subject of the relation between molecular structure and odor is very perplexing, as there is no doubt that there exist as many relationships of structure and odor as there are structures of odorous substances." In 1971 (referring to Stoll 1957), Teranishi wrote: "The relation between molecular structure and odor was perplexing then. It is now." We can observe a number of similarities between the chemical structure of compounds and their odors. However, the field of food flavors, as is the field of perfumery, is still very much an art, albeit one greatly supported by scientists' advancing ability to classify structures and identify the effect of certain molecular configurations. The odor potency of various compounds ranges widely. Table 7–11 indicates a range of about eight orders of magnitude (Teranishi 1971). This indicates that volatile flavor compounds may be present in greatly differing quantities, from traces to relatively large amounts.

The musks are a common illustration of compounds with different structures that all

Table 7–10 Odor Threshold Concentrations of Odorous Substances Perceived During Normal Inspiration

Compound	Threshold Concentration (Molecules/cc)
Allyl mercaptan	6×10^7
Sec. butyl mercaptan	1×10^8
Isopropyl mercaptan	1×10^8
Isobutyl mercaptan	4×10^8
Tert. butyl mercaptan	6×10^8
Thiophenol	8×10^8
Ethyl mercaptan	1×10^9
1,3-Xylen-4-ol	2×10^{12}
μ-Xylene	2×10^{12}
Acetone	6×10^{13}

Source: From K.B. Döving, Problems in the Physiology of Olfaction, in *Symposium on Foods: The Chemistry and Physiology of Flavors*, H.W. Schultz et al., eds., 1967, AVI Publishing.

Table 7–11 Odor Thresholds of Compounds Covering a Wide Range of Intensity

Odorant	Threshold (µg/L of Water)
Ethanol	100,000
Butyric acid	240
Nootkatone	170
Humulene	160
Myrcene	15
n-Amyl acetate	5
n-Decanal	0.0
α- and β-Sinensal	0.05
Methyl mercaptan	0.02
β-Ionone	0.007
2-methoxy-3-isobutylpyrazine	0.002

Source: From R. Teranishi, Odor and Molecular Structure, in *Gustation and Olfaction*, G. Ohloff and A.F. Thomas, eds., 1971, Academic Press.

give similar odors. These may include tricyclic compounds, macrocyclic ketones and lactones, steroids, nitrocyclohexanes, indanes, tetrahydronaphthalenes, and acetophenoses. Small changes in the structure of these molecules may significantly change in potency but will not affect quality, since all are musky. There are also some compounds that have similar structures and very different odors, such as nootkatone and related compounds (Teranishi 1971). Nootkatone is a flavor compound from grapefruit oil. This compound and 1,10-dihydronootkatone have a grapefruity flavor (Figure 7–24). Several other related compounds have a woody flavor. The odor character of stereoisomers may be quite different. The case of menthol has already been described. Only menthol isomers have peppermint aroma. The iso-, neo-, and neoisomenthols have an unpleasant musty flavor. Naves (1957) describes the

difference between the *cis-* and *trans-* forms of 3-hexenol ($CH_2OH–CH_2–CH=CH–CH_2CH_3$). The *cis*-isomer has a fresh green odor, whereas the *trans*-isomer has a scent reminiscent of chrysanthemum. The 2-*trans*-6-*cis* nonadienal smells of cucumber and is quite different from the smell of the 2-*trans*-6-*trans* isomer (nonadienal, $CHO–CH=CH–(CH_2)_2–CH=CH–CH_2–CH_3$). Lengthening of the carbon chain may affect odorous properties. The odor of saturated acids changes remarkably as chain length increases. The lower fatty acids, especially butyric, have very intense and unpleasant flavors, because an increased chain length changes flavor character (Table 7–12) and lessens intensity. The fatty acids with 16 or 18 carbon atoms have only a faint flavor.

Another example is given by Kulka (1967). Gamma-nonalactone has a strong coconut-like flavor; γ-undecalactone has a peach aroma. As the chain length is increased by one more carbon atom, the flavor character becomes peach-musk. The lactones are compounds of widely differing structure and odor quality and are found as components of many food flavors. Gamma- and δ-lactones with 10 to 16 carbon atoms have been reported (Juriens and Oele 1965) as flavor components of butter, contributing to the butter flavor in concentrations of only parts per million. The flavor character and chemical structure of some γ-lactones as reported by Teranishi (1971) are shown in Figure 7–25. One of these, the γ-lactone with a total chain length of 10 carbons, has peach flavor. The α-hydroxy-β-methyl-γ-carboxy-$\Delta^{\alpha-\beta}$-γ-hexeno-lactone occurs in protein hydrolysate and has very strong odor and flavor of beef bouillon. Gold and Wilson (1963) found that the volatile flavor compounds of celery contain a number of phthalides (phthalides are lactones of phthalic acid, lactones are internal esters of hydroxy acids). These include the following:

- 3-isobutyliden-3a,4-dihydrophthalide (Figure 7–26)
- 3-isovalidene-3a,4-dihydrophthalide
- 3-isobutylidene phthalide
- 3-isovalidene phthalide

These compounds exhibit celery-like odors at levels of 0.1 ppm in water. Pyrazines have been identified as the compounds giving the characteristic intense odor of green peppers (Seifert et al. 1970). A number of pyrazine derivatives were tested and, within this single class of compounds, odor potencies showed a range of eight orders of magnitude equal to that of the widely varying compounds listed in Table 7–11. The compounds examined by Seifert et al. (1970) are listed in Table 7–13. 2-methoxy-3-isobutylpyrazine appears to be the compound responsible for the green pepper odor. Removal of the methoxy- or alkyl-

groups reduces the odor potency by 10^5 to 10^6 times, as is the case with 2-methoxypyrazine, 2-iosbutylpyrazine, and 2,5-dimethylpyrazine. Thus, small changes in molecular structure may greatly affect flavor potency. The odors of isobutyl, propyl, and hexyl methoxypyrazines are similar to that of green peppers. The isopropyl compound is moderately similar to peppers and its odor is somewhat similar to raw potato. The ethyl compound is even more similar to raw potato and less to pepper. In fact, this compound can be isolated from potatoes. The methyl compound has an odor like roasted peanuts. The structure of some of the pyrazines is shown in Figure 7–27. Pyrazines have been identified as flavor components in a number of foods that are normally heated during processing. Rizzi (1967) demonstrated the presence of seven alkyl-substituted pyrazines in chocolate aroma. These were isolated by steam distillation, separated by gas-liquid chromatography, and identified by mass spectrometry. The components are methyl pyra-

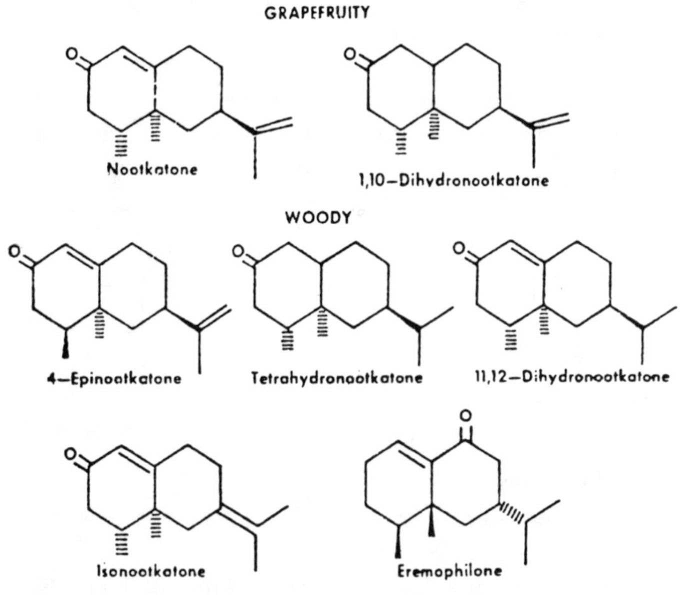

Figure 7–24 Odor Character of Nootkatone and Related Compounds

Table 7–12 Flavor Character of Some N-Carboxylic Acids

Acid	Flavor Character
Formic	Acid, pungent
Acetic	Acid, vinegary, pungent
Propionic	Acid, pungent, rancid, cheesy
Butyric	Acid, rancid
Hexanoic	Sweaty, goaty
Octanoic	Rancid
Decanoic	Waxy
Lauric	Tallowy
Myristic	Soapy, cardboard
Palmitic	Soapy

zine; 2,3-dimethylpyrazine; 2-ethyl-5-methyl-pyrazine; trimethylpyrazine; 2,5-dimethyl-3-ethylpyrazine; 2,6-dimethyl-3-ethylpyrazine; and tetramethylpyrazine. Other researchers (Flament et al. 1967; Marion et al. 1967) have isolated these and other pyrazines from the aroma components of cocoa. Pyrazines are also aroma constituents of coffee. Goldman et al. (1967) isolated and identified 24 pyrazines and pyridines and revealed the presence of possibly 10 more. Bondarovich et al. (1967) isolated and identified a large number of pyrazines from coffee aroma and drew

attention to the importance of pyrazines and dihydropyrazines to the flavor of roasted or otherwise cooked foods. These authors also drew attention to the instability of the dihydropyrazines. This instability not only makes their detection and isolation difficult, but may help explain why flavors such as that of roasted coffee rapidly change with time. Another roasted product from which pyrazines have been isolated is peanuts. Mason et al. (1966) found methylpyrazine; 2,5-dimethylpyrazine; trimethylpyrazine; methylethylpyrazine; and dimethylethylpyrazine in the flavor of roasted peanuts. The pyrazines appear to be present in unprocessed as well as in heated foods.

Another group of compounds that have been related to the aroma of heated foods is the furanones. Teranishi (1971) summarized the findings on several of the furanones (see Figure 7–23). The 4-hydroxy-2,5-dimethyl-3-dihydrofuranone (1) has a caramel or burnt pineapple odor. The 4-hydroxy-5-methyl-3-dihydrofuranone (2) has a roasted chicory root odor. Both compounds may contribute to beef broth flavor. The 2,5-dimethyl-3-dihydrofuranone (3) has the odor of freshly baked bread. Isomaltol (4) and maltol (5) are products of the caramelization and pyrolysis of carbohydrates.

R = C_5H_{11} (coconut)

R = C_6H_{13} (peach)

R = C_7H_{15} (peach)

R = C_8H_{17} (peach-musk)

Beef bouillon

Figure 7–25 Flavor Character of Some Lactones. *Source*: From R. Teranishi, Odor and Molecular Structure, in *Gustation and Olfaction*, G. Ohloff and A.F. Thomas, eds., 1971, Academic Press.

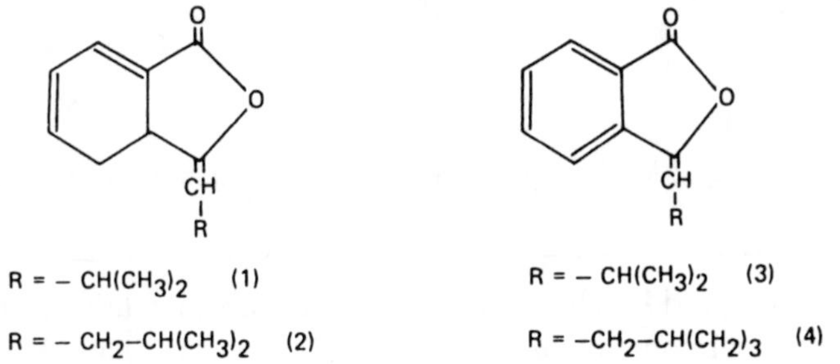

$R = - CH(CH_3)_2$ (1)

$R = - CH_2-CH(CH_3)_2$ (2)

$R = - CH(CH_3)_2$ (3)

$R = -CH_2-CH(CH_2)_3$ (4)

Figure 7–26 Phthalides of Celery Volatiles

Theories of Olfaction

When an odoriferous compound, or *odorivector,* arrives at the olfactory organ, a reaction takes place between the odor molecules and the chemoreceptors; this reaction produces a neural pulse, which eventually reaches the brain. The exact nature of the interaction between odorivector and chemoreceptor is not well known. The number of olfactory receptors in the smell organs is in the order of 100 million, and Moncrieff (1951) has calculated that the number of molecules at the threshold concentration of one of the powerful mercaptans in a sniff (about 20 mL) of air would be 1×10^{10} molecules. Obviously, only a fraction of these would interact with the receptors, but undoubtedly numerous interactions are required to produce a neural response. Dravnieks (1966) has indicated that according to information theory, 13 types of sensors are needed to distinguish 10,000 odors on a yes-or-no basis, but more than 20 might be required to respond rapidly and without error. Many attempts have been made to classify odors into a relatively small number of groups of related odors. These so-called primary odors have been used in olfaction theories to explain odor quality. One theory, the stereochemical site theory (Amoore et al. 1964; Amoore 1967), is based on molecular size and shape. Amoore compared the various odor qualities that have been used to characterize odors and concluded that seven primary odors would suffice to cover them all: camphoraceous, pungent, ethereal, floral,

Table 7–13 Odor Threshold of Pyrazine and Derivatives

Compound	Odor Threshold (Parts per 10^{12} Parts of Water)
2-methoxy-3-hexylpyrazine	1
2-methoxy-3-isobutylpyrazine	2
2-methoxy-3-propylpyrazine	6
2-methoxy-3-isopropylpyrazine	2
2-methoxy-3-ethylpyrazine	400
2-methoxy-3-methylpyrazine	4000
2-methoxypyrazine	700,000
2-isobutylpyrazine	400,000
2-5-dimethylpyrazine	1,800,000
pyrazine	175,000,000

Source: From R.M. Seifert et al., Synthesis of Some 2-Methoxy-3-Alkylpyrazines with Strong Bell Pepper–Like Odors, *J. Agr. Food Chem.*, Vol. 18, pp. 246–249, 1970, American Chemical Society.

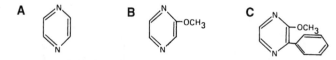

Figure 7–27 (A) Pyrazine, (B) 2-Methoxypyrazine, and (C) 2-Methoxy-3-Hexylpyrazine

pepperminty, musky, and putrid. Table 7–14 lists some of the chemical compounds that can be used to demonstrate these primary odors. The theory is based on the assumption that all odorous compounds have a distinctive molecular shape and size that fit into a socket on the receptor site. This would be similar to the "lock-and-key" concept of enzyme action. Five of the receptor sites would accept the flavor compound according to shape and size, and two (pungent and putrid) on the basis of electronic status (Figure 7–28). The site-fitting concept as initially proposed was inadequate because it assessed only one-half of the molecule; subsequent refinements considered all aspects of molecular surface in a "shadow-matching" technique (Amoore 1967). It was also suggested that there may be more than seven primaries. The primary odors may have to be split into subgroups and others added as new primaries. Molecular model silhouettes as developed for five primary odors are reproduced in Figure 7–29.

A membrane-puncturing theory has been proposed by Davis (Dravnieks 1967). According to this theory, the odorous substance molecules are adsorbed across the interface of the thin lipid membrane, which forms part of the cylindrical wall of the neuron in the chemoreceptor and the aqueous phase that surrounds the neuron. Adsorbed molecules orient themselves with the hydrophilic end toward the aqueous phase. When the adsorbed molecules are desorbed, they move into the aqueous phase, leaving a defect. Ions may adsorb into this puncture and cause a neural response. This theory could be considered a thermodynamic form of the profile functional group concept, since the free energy of adsorption of the odor substance at the interface is related to shape, size, functional groups and their distribution, and position. The adsorption is a dynamic process with a free energy of adsorption of about 1 to 8 kcal/mole for different substances. Davies prepared a plot of molecular cross-sectional area versus free energy of adsorption and obtained a diagram (Figure 7–30) in which groups of related odors occupy distinct areas.

The suggestion that odorous character is related to vibrational specificity of odor molecules has led to the vibrational theory of olfaction (Wright 1957). Vibrational energy levels can be derived from the infrared or Raman spectra. The spectral area of greatest interest is that below 700 cm^{-1}, which is related to vibrations of chains and flexing or twisting of bonds between groups of atoms in the molecule. Wright and others have demonstrated that correlations exist between spectral properties and odor quality in a number of cases, but inconsistencies in other cases have yet to be explained.

Obviously, none of the many theories of olfaction proposed so far have been entirely satisfactory. It might be better to speak of hypotheses rather than of theories. Most of these theories deal with the explanation of odor quality and do not account for the quantitative aspects of the mechanism of olfaction. The classification of odor and the

Table 7–14 Primary Odors for Humans and Compounds Eliciting These Odors

Primary Odor	Odor Compounds
Camphoraceous	Borneol, *tert*-butyl alcohol *d*-camphor, cineol, pentamethyl ethyl alcohol
Pungent	Allyl alcohol, cyanogen, formaldehyde, formic acid, methylisothiocyanate
Ethereal	Acetylene, carbon tetrachloride, chloroform, ethylene dichloride, propyl alcohol
Floral	Benzyl acetate, geraniol, α-ionone, phenylethyl alcohol, terpineol
Pepperminty	*tert*-butylcarbinol, cyclohexanone, menthone, piperitol, 1,1,3-trimethyl-cyclo-5-hexanone
Musky	Androstan-3α-ol (strong), cyclohexadecanone, ethylene cebacate, 17-methylandrostan-3α-ol, pentadecanolactone
Putrid	Amylmercaptan, cadaverine, hydrogen sulfide, indole (when concentrated, floral when dilute), skatole

Source: From J.E. Amoore et al., The Stereochemical Theory of Odor, *Sci. Am.*, Vol. 210, No. 2, pp. 42–49, 1964.

correlation of chemical structure and odor remain difficult to resolve.

Odor Description

An odor can be described by the combination of threshold value and odor quality. The threshold value, the lowest concentration that creates an odor impression, can be considered the intensity factor, whereas the odor quality describes the character of the aroma. As has been mentioned under olfactory theories, attempts at reducing the number of characteristic odor qualities to a small number have not been successful. In many cases, the aroma and flavor of a food can be related to the presence of one or a few compounds that create an impression of a particular food when smelled alone. Such compounds have been named *contributory flavor compounds* by Jennings and Sevenants (1964). Some such compounds are the pyrazines, which give the odor quality of green bell peppers; nootkatone for grapefruit; esters for fruits;

and nona-2-*trans*-6-*cis*-dienal for cucumbers (Forss et al. 1962). In a great number of other cases, there are no easily recognizable contributory flavor compounds, but the flavor seems to be the integrated impression of a large number of compounds.

Determining the threshold value is difficult because subthreshold levels of one compound may affect the threshold levels of another. Also, the flavor quality of a compound may be different at threshold level and at suprathreshold levels. The total range of perception can be divided into units that represent the smallest additional amount that can be perceived. This amount is called just noticeable difference (JND). The whole intensity scale of odor perception covers about 25 JNDs; this is similar to the number of JNDs that comprise the scale of taste intensity. Flavor thresholds for some compounds depend on the medium in which the compound is dispersed or dissolved. Patton (1964) found large differences in the threshold values of saturated fatty acids dissolved in water and in oil.

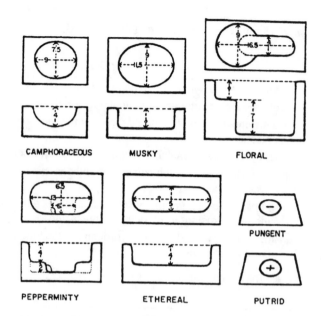

Figure 7–28 Olfactory Receptor Sites According to the Stereochemical Theory of Odor

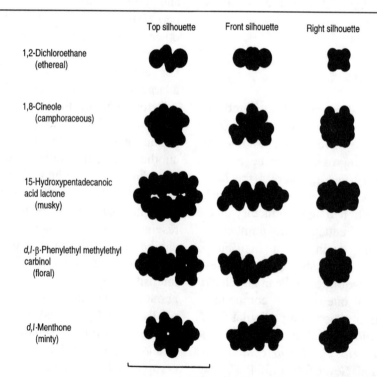

Figure 7–29 Molecular Model Silhouettes of Five Standard Odorants. *Source*: From J. Amoore, Stereochemical Theory of Olfaction, in *Symposium on Foods: The Chemistry and Physiology of Flavors*, H.W. Schultz et al., eds., 1967, AVI Publishing Co.

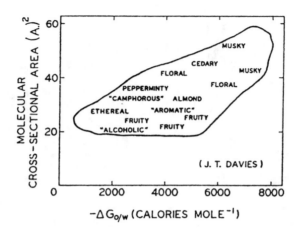

Figure 7–30 Plot of Molecular Cross-Sectional Area Versus Free Energy of Adsorption for Davies' Theory of Olfaction

DESCRIPTION OF FOOD FLAVORS

The flavor impression of a food is influenced by compounds that affect both taste and odor. The analysis and identification of many volatile flavor compounds in a large variety of food products have been assisted by the development of powerful analytical techniques. Gas-liquid chromatography was widely used in the early 1950s when commercial instruments became available. Introduction of the flame ionization detector increased sensitivity by a factor of 100 and, together with mass spectrometers, gave a method for rapid identification of many components in complex mixtures. These methods have been described by Teranishi et al. (1971). As a result, a great deal of information on volatile flavor components has been obtained in recent years for a variety of food products. The combination of gas chromatography and mass spectrometry can provide identification and quantitation of flavor compounds. However, when the flavor consists of many compounds, sometimes several hun-

dred, it is impossible to evaluate a flavor from this information alone. It is then possible to use pattern recognition techniques to further describe the flavor. The pattern recognition method involves the application of computer analysis of complex mixtures of compounds. Computer multivariate analysis has been used for the detection of adulteration of orange juice (Page 1986) and Spanish sherries (Maarse et al. 1987).

Flavors are often described by using the human senses on the basis of widely recognized taste and smell sensations. A proposed wine aroma description system is shown in Figure 7–31 (Noble et al. 1987). Such systems attempt to provide an orderly and reliable basis for comparison of flavor descriptions by different tasters.

The aroma is divided into first-, second-, and third-tier terms, with the first-tier terms in the center. Examination of the descriptors in the aroma wheel shows that they can be divided into two types, flavors and off-flavors. Thus, it would be more useful to divide the flavor wheel into two tables—one for fla-

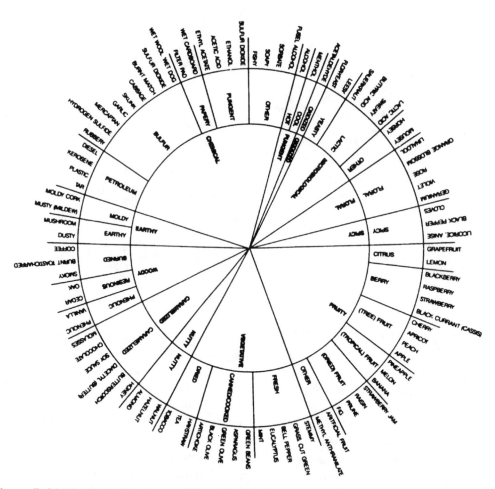

Figure 7–31 Modified Wine Aroma Wheel for the Description of Wine Aroma. *Source*: From A.C. Noble et al., Modification of a Standardized System of Wine Aroma Terminology, *Am. J. Enol. Vitic.*, Vol. 38, pp. 143–146, 1987, American Society of Enology and Viticulture.

vors and one for off-flavors, as shown in Tables 7–15 and 7–16.

The difficulty in relating chemical composition and structure to the aroma of a food that contains a multitude of flavor compounds is evident from the work of Meyboom and Jongenotter (1981). They studied the flavor of straight-chain, unsaturated aldehydes as a function of double-bond position and geometry. Some of their results are presented in Table 7–17. Flavors of unsaturated aldehydes of different chain length and geometry may vary from bitter almond to lemon and cucumber when tasted separately.

A method of flavor description, developed by researchers at A.D. Little Inc. (Sjöström 1972), has been named the flavor profile method. The flavor profile method uses the recognition, description, and comparison of aroma and flavor by a trained panel of four to

Table 7–15 Aroma Description of Wine as Listed in the Aroma Wheel, Listing Only the Flavor Contribution

First Tier	Second Tier	Third Tier	First Tier	Second Tier	Third Tier
Floral	Floral	Geranium		Canned/cooked	Green beans
		Violet			Asparagus
		Rose			Green olive
		Orange blossom			Black olive
		Linalool			Artichoke
Spicy	Spicy	Licorice anise		Dried	Hay/straw
		Black pepper			Tea
		Cloves			Tobacco
Fruity	Citrus	Grapefruit	Nutty	Nutty	Walnut
		Lemon			Hazelnut
	Berry	Blackberry			Almond
		Raspberry	Caramelized	Caramelized	Honey
		Strawberry			Butterscotch
		Black currant			Diacetyl (butter)
	Tree fruit	Cherry			Soy sauce
		Apricot			Chocolate
		Peach			Molasses
		Apple	Woody	Phenolic	Phenolic
	Tropical fruit	Pineapple			Vanilla
		Melon		Resinous	Cedar
		Banana			Oak
	Dried fruit	Strawberry jam		Burned	Smoky
		Raisin			Burnt toast/charred
		Prune			Coffee
		Fig			
	Other	Artificial fruit			
		Methyl anthranilate			
Vegetative	Fresh	Stemmy			
		Grass, cut green			
		Bell pepper			
		Eucalyptus			
		Mint			

six people. Through training, the panel members are made familiar with the terminology used in describing flavor qualities. In addition to describing flavor quality, intensity values are assigned to each of the quality aspects. The intensity scale is threshold, slight, moderate, and strong, and these are represented by the symbols)(, 1, 2, 3. With the exception of threshold value, the units are ranges and can be more precisely defined by the use of reference standards. In the panel work, the evaluation of aroma is conducted

Table 7–16 Aroma Description of Wine as Listed in the Aroma Wheel, Listing Only the Off-Flavors

First Tier	Second Tier	Third Tier
Earthy	Moldy	Moldy cork
		Musty (mildew)
	Earthy	Mushroom
		Dusty
Chemical	Petroleum	Diesel
		Kerosene
		Plastic
		Tar
	Sulfur	Wet wool, wet dog
		Sulfur dioxide
		Burnt match
		Cabbage
		Skunk
		Garlic
		Mercaptan
		Hydrogen sulfide
		Rubbery
	Papery	Wet cardboard
		Filterpad
	Pungent	Sulfur dioxide
		Ethanol
		Acetic acid
		Ethyl acetate
	Other	Fusel alcohol
		Sorbate
		Soapy
		Fishy
Pungent	Cool	Menthol
	Hot	Alcohol
Oxidized	Oxidized	Acetaldehyde
Microbiological	Yeasty	Leesy
		Flor yeast
	Lactic	Lactic acid
		Sweaty
		Butyric acid
		Sauerkraut
	Other	Mousey
		Horsey

Table 7–17 Flavor Description of Unsaturated Aldehydes Dissolved in Paraffin Oil

Aldehyde	Flavor Description
trans-3-hexenal	Green, odor of pine tree needles
cis-3-hexenal	Green beans, tomato green
trans-2-heptenal	Bitter almonds
cis-6-heptenal	Green, melon
trans-2-octenal	Nutty
trans-5-octenal	Cucumber
cis-5-octenal	Cucumber
trans-2-nonenal	Starch, glue
trans-7-nonenal	Melon

Source: From P.W. Meyboom and G.A. Jongenotter, Flavor Perceptibility of Straight Chain, Unsaturated Aldehydes as a Function of Double Bond Position and Geometry, *J. Am. Oil Chem. Soc.*, Vol. 58, pp. 680–682, 1981.

first because odor notes can be overpowered when the food is eaten. This is followed by flavor analysis, called "flavor by mouth," a specialists' description of what a consumer would experience eating the food. Flavor analysis includes such factors as taste, aroma, feeling, and aftertaste. A sample flavor profile of margarine is given in Table 7–18.

ASTRINGENCY

The sensation of astringency is considered to be related more to touch than to taste. Astringency causes a drying and puckering over the whole surface of the mouth and tongue. This sensation is caused by interaction of astringent compounds with proteins and glycoproteins in the mouth. Astringent compounds are present in fruits and bever-

Table 7–18 Flavor Profile of Margarine

Aroma		Flavor by Mouth	
Amplitude	2	Amplitude	2½
Sweet cream	½	Sweet cream	1½
Oil	)(	Oil	½
Sour	½	Salt	1½
Vanillin sweet	)(	Butter mouthfeel	2
		Sour	1

Note:)(= threshold; 1 = slight; 2 = moderate; 3 = strong.

Source: Reprinted with permission from L.B. Sjöström, *The Flavor Profile*, © 1972, A.D. Little, Inc.

ages derived from fruit (such as juice, wine, and cider), in tea and cocoa, and in beverages matured in oak casks. Astringency is caused by tannins, either those present in the food or extracted from the wood of oak barrels. The astringent reaction involves a bonding to proteins in the mouth, followed by a physiological response. The astringent reaction has been found to occur between salivary proteins that are rich in proline (Luck et al. 1994). These proline-rich proteins (PRPs) have a high affinity for polyphenols. The effect of the structure of PRP is twofold: (1) proline causes the protein to have an open and flexible structure, and (2) the proline residue itself plays an important role in recognizing the polyphenols involved in the complex formation. The complex formation between PRP and polyphenol has been represented by Luck et al. (1994) in pictorial form (Figure 7–32). The reaction is mediated by hydrophobic effects and hydrogen bonding on protein sites close to prolyl residues in the PRP. The resulting cross-linking, aggregation, and precipitation of the PRP causes the sensation of astringency.

Some anthocyanins are both bitter and astringent. Bitter compounds such as quinine and caffeine compete with the tannins in complexing with buccal proteins and thereby lower the astringent response. Astringency is caused by higher molecular weight tannins, whereas the lower molecular weight tannins up to tetramers are associated with bitterness (Macheix et al. 1990).

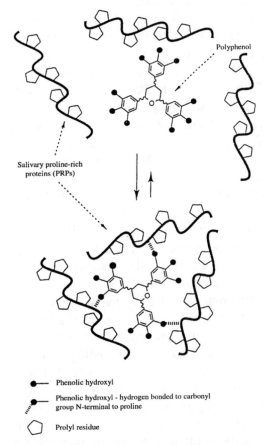

● — Phenolic hydroxyl
∽ Phenolic hydroxyl - hydrogen bonded to carbonyl group N-terminal to proline
⬠ Prolyl residue

Figure 7–32 Complex Formation Between Proline-Rich Proteins and Polyphenols *Source*: Reprinted with permission from G. Luck et al., The Cup That Cheers: Polyphenols and the Astringency of Tea, Lecture Paper No. 0030, © 1994, Society of Chemical Industry.

FLAVOR AND OFF-FLAVOR

It is impossible to deal with the subject of flavor without considering off-flavors. In many cases the same chemical compounds are involved in both flavors and off-flavors. The only distinction appears to be whether a flavor is judged to be pleasant or unpleasant. This amounts to a personal judgment, although many unpleasant flavors (or off-flavors) are universally found to be unpleasant. A distinction is sometimes made between off-flavors—defined as unpleasant odors or flavors imparted to food through internal deteriorative change—and taints—defined as unpleasant odors or flavors imparted to food through external sources (Saxby 1996). Off-flavors in animal products, meat and milk, may be caused by transfer of substances from feed. Off-flavors in otherwise sound foods can be caused by heat, oxidation, light, or enzymic action. The perception of taste and flavor can be defined for a given group of people by the International Standards Organization (ISO) 5492 standard (ISO 1992) as follows: The odor or taste threshold is the lowest concentration of a compound detectable by a certain proportion (usually 50 percent) of a given group of people. A graphic representation of this relationship has been given by Saxby (1996). The graph in Figure 7–33 relates the percentage of people within a given group to the ability to detect a substance at varying concentrations. Of the population, 50 percent can detect the compound at the concentration of one unit. At a concentration of the compound 10 times greater than the mean threshold, about 10 percent of the population is still not able to detect it. At the other end of the spectrum, 5 percent of the population can still detect the compound at a concentration 10 times less than the

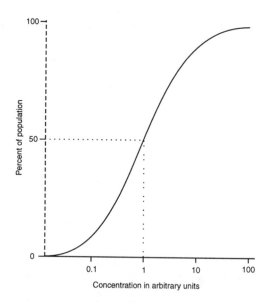

Figure 7–33 Variation of Taste Threshold within a Given Population. *Source*: Reprinted from M.J. Saxby, *Food Taints and Off-Flavors*, p. 43, © 1996, Aspen Publishers, Inc.

mean threshold. These findings have important consequences for the presence of compounds causing off-flavors. Even very low levels of a chemical that produces off-flavors may cause a significant number of people to complain.

Certain flavor compounds may appear quite pleasant in one case and extremely unpleasant in another. Many examples of this can be cited. One of the well-known cases is that of short-chain free fatty acids in certain dairy products. Many cheese flavors contain volatile fatty acids as flavor contributors (Day 1967). Yet, the same fatty acids in very low concentrations in milk and other dairy products cause a very unpleasant, rancid off-flavor. Forss (1969) has drawn attention to the compound non-2-enal. During studies of dairy product off-flavor, this compound was isolated as a component of the oxidation off-flavor and was found to have an odor reminiscent of cucumbers. The same compound was isolated from cucumbers, and the cucumber-like flavor was assigned to the molecular structure of a 2-*trans*-enal with 9 or 10 carbon atoms. Further unsaturation and conjugation to give a 2,4-dienal produces flavors reminiscent of cardboard or linoleum. Lactones were isolated by Keeney and Patton (1956) and Tharp and Patton (1960) and were considered to be the cause of stale off-flavors in certain dairy products. The same lactones, including δ-decalactone and δ-dodecalactone, were subsequently recognized as contributors to the pleasant aroma of butter (Day 1966). Dimethylsulfide is a component of the agreeable aroma of meat and fish but has also been found to cause an off-flavor in canned salmon (Tarr 1966). Acetaldehyde occurs naturally in many foods, especially fruits, and is reported to be essential for imparting the taste of freshness (Byrne and Sherman 1984). The same compound is responsible for a very unpleasant oxidized flavor in wine. Sinki (1988) has discussed the problems involved in creating a universally acceptable taste, and has stated that most individual flavor chemicals are either repugnant or painful outside their proper formulations. This complex interaction between flavor chemicals, and between flavors and the individual, makes the creation of a flavorful product both a science and an art, according to Sinki. The subject of pleasantness and unpleasantness of flavors is the basis of a chapter in *Odour Description and Odour Classification* by Harper et al. (1968) and is the main subject of Moncrieff's *Odour Preferences* (1966).

FLAVOR OF SOME FOODS

As indicated previously, the two main factors affecting flavor are taste and odor. In a general way, food flavors can be divided into two groups. The first consists of foods whose flavor cannot be attributed to one or a few outstanding flavor notes; their flavor is the result of the complex interaction of a variety of taste and odor components. Examples include bread, meat, and cheese. The second group consists of foods in which the flavor can be related to one or a few easily recognized components (contributory flavor compounds). Examples include certain fruits, vegetables, and spices. Another way of differentiating food flavors is by considering one group in which the flavor compounds are naturally present and another group in which the flavor compounds are produced by processing methods.

Bread

The flavor of white bread is formed mainly from the fermentation and baking processes. Freshly baked bread has a delightful aroma that is rapidly lost on cooling and storage. It has been suggested that this loss of flavor is the result of disappearance of volatile flavor components. However, it is well known that the aroma may be at least partially regenerated by simply heating the bread. Schoch (1965) suggested that volatile flavor compounds may become locked in by the linear fraction of wheat starch. The change in texture upon aging may be a contributory factor in the loss of flavor. During fermentation, a number of alcohols are formed, including ethanol, *n*-propanol, isoamyl and amyl alcohol, isobutyl alcohol, and β-phenol alcohol. The importance of the alcohols to bread flavor is a matter of controversy. Much of the alcohols are lost to the oven air during baking. A large number of organic acids are also formed (Johnson et al. 1966). These include many of the odd and even carbon number saturated aliphatic acids, from formic to capric, as well as lactic, succinic, pyruvic, hydrocinnamic, benzilic, itaconic, and levulinic acid. A large number of carbonyl compounds has been identified in bread, and these are believed to be important flavor components. Johnson et al. (1966) list the carbonyl compounds isolated by various workers from bread; this list includes 14 aldehydes and 6 ketones. In white bread made with glucose, the prevalent carbonyl compound is hydroxymethylfurfural (Linko et al. 1962). The formation of the crust and browning during baking appear to be primary contributors to bread flavor. The browning is mainly the result of a Maillard-type browning reaction rather than caramelization. This accounts for the presence of the carbonyl compounds, especially furfural, hydroxymethylfurfural, and other aldehydes. In the Maillard reaction, the amino acids are transformed into aldehydes with one less carbon atom. Specific aldehydes can thus be formed in bread crust if the necessary amino acids are present. The formation of aldehydes in bread crust is accompanied by a lowering of the amino acid content compared to that in the crumb. Johnson et al. (1966) have listed the aldehydes that can be formed from amino acids in bread crust as a result of the Strecker degradation (Table 7–19).

Grosch and Schieberle (1991) reported the aroma of wheat bread to include ethanol, 2-methylpropanal, 3-methylbutanal, 2,3-butanedione, and 3-methylbutanol. These compounds contribute significantly to bread aroma, whereas other compounds are of minor importance.

Meat

Meat is another food in which the flavor is developed by heating from precursors present

Table 7–19 Aldehydes That Can Be Formed from Amino Acids in Bread Crust as a Result of the Strecker Degradation

Amino Acid	Aldehyde
Alanine	Acetaldehyde
Glycine	Formaldehyde
Isoleucine	2-Methylbutanal
Leucine	Isovaleraldehyde
Methionine	Methional
Phenylalanine	Phenylacetaldehyde
Threonine	2-Hydroxypropanal
Serine	Glyoxal

Source: From J.A. Johnson et al., Chemistry of Bread Flavor, in *Flavor Chemistry*, I. Hornstein, ed., 1966, American Chemical Society.

in the meat; this occurs in a Maillard-type browning reaction. The overall flavor impression is the result of the presence of a large number of nonvolatile compounds and the volatiles produced during heating. The contribution of nonvolatile compounds in meat flavor has been summarized by Solms (1971). Meat extracts contain a large number of amino acids, peptides, nucleotides, acids, and sugars. The presence of relatively large amounts of inosine-5′-monophosphate has been the reason for considering this compound as a basic flavor component. In combination with other compounds, this nucleotide would be responsible for the meaty taste. Living muscle contains adenosine-5′-triphosphate; this is converted after slaughter into adenosine-5′-monophosphate, which is deaminated to form inosine-5′-monophosphate (Jones 1969). The volatile compounds produced on heating can be accounted for by reactions involving amino acids and sugars present in meat extract. Lean beef, pork, and lamb are surprisingly similar in flavor; this reflects the similarity in composition of extracts in terms of amino acid and sugar components. The fats of these different species may account for some of the normal differences in flavor. In the volatile fractions of meat aroma, hydrogen sulfide and methyl mercaptan have been found; these may be important contributors to meat flavor. Other volatiles that have been isolated include a variety of carbonyls such as acetaldehyde, propionaldehyde, 2-methylpropanal, 3-methylbutanal, acetone, 2-butanone, *n*-hexanal, and 3-methyl-2-butanone (Moody 1983).

Fish

Fish contains sugars and amino acids that may be involved in Maillard-type reactions during heat processing (canning). Proline is a prominent amino acid in fish and may contribute to sweetness. The sugars ribose, glucose, and glucose-6-phosphate are flavor contributors, as is 5'-inosinic acid, which contributes a meaty flavor note. Volatile sulfur compounds contribute to the flavor of fish; hydrogen sulfide, methylmercaptan, and dimethylsulfide may contribute to the aroma of fish. Tarr (1966) described an off-flavor problem in canned salmon that is related to dimethylsulfide. The salmon was found to feed on zooplankton containing large amounts of dimethyl-2-carboxyethyl sulfonium chloride. This compound became part of the liver and flesh of the salmon and in canning degraded to dimethylsulfide according to the following equation:

$$(CH_3)_2-SH-CH_2-CH_2-COOH \rightarrow$$
$$(CH_3)_2S + CH_3-CH_2-COOH$$

The flavor of cooked, fresh fish is caused by the presence of sugars, including glucose and fructose, giving a sweet impression as well as a umami component arising from the synergism between inosine monophosphate and free amino acids. The fresh flavor of fish is rapidly lost by bacterial spoilage. In fresh fish, a small amount of free ammonia, which has a pH level of below 7, exists in protonated form. As spoilage increases, the pH rises and ammonia is released. The main source of ammonia is trimethylamine, produced as a degradation product of trimethylamineoxide.

The taste-producing properties of hypoxanthine and histidine in fish have been described by Konosu (1979). 5'-inosinate accumulates in fish muscle as a postmortem degradation product of ATP. The inosinate slowly degrades into hypoxanthine, which has a strong bitter taste. Some kinds of fish, such as tuna and mackerel, contain very high levels of free histidine, which has been postulated to contribute to the flavor of these fish.

Milk

The flavor of normal fresh milk is probably produced by the cow's metabolism and is comprised of free fatty acids, carbonyl compounds, alkanols, and sulfur compounds. Free fatty acids may result from the action of milk lipase or bacterial lipase. Other decomposition products of lipids may be produced by the action of heat. In addition to lipids, proteins and lactose may be precursors of flavor compounds in milk (Badings 1991). Sulfur compounds that can be formed by heat from β-lactoglobulin include dimethyl sulfide, hydrogen sulfide, dimethyl disulfide, and methanethiol. Some of these sulfur compounds are also produced from methionine when milk is exposed to light. Heterocyclic compounds are produced by nonenzymatic browning reactions. Bitter peptides can be formed by milk or bacterial proteinases.

The basic taste of milk is very bland, slightly sweet, and salty. Processing conditions influence flavor profiles. The extent of heat treatment determines the type of flavor produced. Low heat treatment produces traces of hydrogen sulfide. Ultra-high temperature treatment results in a slight fruity, ketone-like flavor. Sterilization results in strong ketone-like and caramelization/sterilization flavors. Sterilization flavors of milk are caused by the presence of 2-alkanones and heterocyclic compounds resulting from the Maillard reaction. Because of the bland flavor of milk, it is relatively easy for off-flavors to take over.

Cheese

The flavor of cheese largely results from the fermentation process that is common to most varieties of cheese. The microorganisms used as cultures in the manufacture of cheese act on many of the milk components and produce a large variety of metabolites. Depending on the type of culture used and the duration of the ripening process, the cheese may vary in flavor from mild to extremely powerful. Casein, the main protein in cheese, is hydrolyzed in a pattern and at a rate that is characteristic for each type of cheese. Proteolytic enzymes produce a range of peptides of specific composition that are related to the specificity of the enzymes present. Under certain conditions bitter peptides may be formed, which produce an off-flavor. Continued hydrolysis yields amino acids. The range of peptides and amino acids provides a "brothy" taste background to the aroma of cheese. Some of these compounds may function as flavor enhancers. Breakdown of the lipids is essential for the produc-

tion of cheese aroma since cheese made from skim milk never develops the full aroma of normal cheese. The lipases elaborated by the culture organisms hydrolyze the triglycerides to form fatty acids and partial glycerides. The particular flavor of some Italian cheeses can be enhanced by adding enzymes during the cheese-making process that cause preferential hydrolysis of short-chain fatty acids. Apparently, a variety of minor components are important in producing the characteristic flavor of cheese. Carbonyls, esters, and sulfur compounds are included in this group. The relative importance of many of these constituents is still uncertain. Sulfur compounds found in cheese include hydrogen sulfide, dimethylsulfide, methional, and methyl mercaptan. All of these compounds are derived from sulfur-containing amino acids. The flavor of blue cheese is mainly the result of the presence of a number of methyl ketones with odd carbon numbers ranging in chain length from 3 to 15 carbons (Day 1967). The most important of these are 2-heptanone and 2-nonanone. The methyl ketones are formed by β-oxidation of fatty acids by the spores of *P. roqueforti*.

Fruits

The flavor of many fruits appears to be a combination of a delicate balance of sweet and sour taste and the odor of a number of volatile compounds. The characteristic flavor of citrus products is largely due to essential oils contained in the peel. The essential oil of citrus fruits contains a group of terpenes and sesquiterpenes and a group of oxygenated compounds. Only the latter are important as contributors to the citrus flavor. The volatile oil of orange juice was found to be 91.6 mg per kg, of which 88.4 was hydrocarbons (Kefford 1959). The volatile water-soluble

constituents of orange juice consist mainly of acetaldehyde, ethanol, methanol, and acetic acid. The hydrocarbons include mainly D-limonene, β-myrcene, and a compound of composition $C_{15}H_{24}$. The esters include iso-valerate, methyl alphaethyl-*n*-caproate, citronellyl acetate, and terpinyl acetate. In the group of carbonyls, the following compounds were identified: *n*-hexanal, *n*-octanal, *n*-decanal, and citronella; and in the group of alcohols, linalool, α-terpineol, *n*-hexane-1-ol, *n*-octan-1-ol, *n*-decan-1-ol, and 3-hexen-1-ol were identified. The flavor deterioration of canned orange juice during storage results in stale off-flavors. This is due to reactions of the nonvolatile water-soluble constituents. As in the case of citrus fruits, no single compound is completely responsible for any single fruit aroma. However, some organoleptically important compounds characteristic for particular fruits have been found. These include amyl esters in banana aroma, citral in lemon, and lactones in peaches. The major flavor component of Bartlett pears was identified by Jennings and Sevenants (1964) as ethyl *trans*-2-*cis*-4-decadienoate.

Vegetables

Vegetables contain an extensive array of volatile flavor compounds, either in original form or produced by enzyme action from precursors. Maarse (1991) has reviewed these in detail. Onion and garlic have distinctive and pungent aromas that result mostly from the presence of sulfur-containing compounds. A large number of flavor compounds in vegetables are formed after cooking or frying. In raw onions, an important compound is thio-propanal *s*-oxide—the lachrymatory factor. The distinctive odor of freshly cut onions involves two main compounds, propyl methane-thiosulfonate and propyl propanethiosulfonate. Raw garlic contains virtually exclusively sulfur compounds: four thiols, three sulfides, seven disulfides, three trisulfides, and six dialkylthiosulfinates.

Tea

The flavor of black tea is the result of a number of compounds formed during the processing of green tea leaves. The processing involves withering, fermentation, and firing. Bokuchava and Skobeleva (1969) indicate that the formation of the aroma occurs mainly during firing. Aromatic compounds isolated and identified from black tea include acrolein, *n*-butyric aldehyde, ethanol, *n*-butanol, isobutanol, hexanal, pentanal, 2-hexanol, 3-hexen-1-ol, benzaldehyde, linalool, terpeneol, methylsalicylate, benzyl alcohol, β-phenylethanol, isobutyric aldehyde, geraniol, and acetophenone. The flavor substances of tea can be divided into the following four fractions: a carbonyl-free neutral fraction including a number of alcohols, a carbonyl fraction, a carboxylic acid fraction, and a phenolic fraction. A compilation (Maarse 1991) identifies a total of 467 flavor constituents in tea. The distinctive flavor of tea is due to its content of lactones, aldehydes, alcohols, acids, and pyridines.

Coffee

The flavor of coffee is developed during the roasting of the green coffee bean. Gas-liquid chromatography can be used to demonstrate (Figure 7–34) the development of volatile constituents in increasing amounts as intensity of roasting increases (Gianturco 1967). The total number of volatile compounds that have been isolated is in the hun-

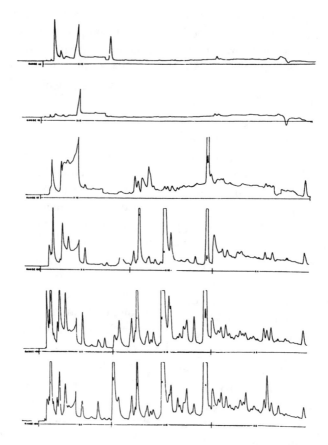

Figure 7–34 Development of Volatile Constituents During Roasting of Coffee. From top to bottom: green coffee after 2, 6, 8, 11, and 15 minutes of roasting. The gas chromatograms show increasing concentrations of volatile compounds. *Source*: From M.A. Gianturco, Coffee Flavor, in *Symposium on Foods: The Chemistry and Physiology of Flavors*, H.W. Schultz et al., eds., 1967, AVI Publishing Co.

dreds, and many of these have been identified. To determine the flavor contribution of each of these is a Herculean task. Many compounds result from the pyrolytic decomposition of carbohydrates into units of 2, 3, 4, or 5 carbons. Other compounds of carbohydrate origin are 16 furanic compounds, cyclic diketones, and maltol. Roasting of the proteins of the coffee bean can yield low molecular weight products such as amino acids, ammonia, amines, hydrogen sulfide, methyl mercaptan, dimethylsulfide, and dimethyl disulfide. A series of furanic and pyrrolic compounds identified include the following: furan, furfural, acetylfuran, 5-methylfuran, 5-methylfurfural, 5-methyl-2-acetylfuran and pyrrole, 2-pyrrolaldehyde, 2-acetylpyrrole, N-methylpyrrole, N-methyl-2-pyrrolaldehyde, and N-methyl-2-acetylpyrrole. Differences in the aroma of different coffees can be related to quantitative differences in some of the compounds isolated by gas chromatography, and ratios and amounts of these compounds may be different. Pyrazines, furanes, pyrroles, and thiophen derivatives are particularly abundant in coffee aroma. Furfuryl-methyl-sulfide and its homologs are important contributors to the aroma of coffee. The structures of some of the important aroma contributors are presented in Figure 7–35. The compounds identified in coffee aroma are listed and differentiated on the basis of functional groups in Table 7–20. It is, of course, impossible to compare the aroma of different coffees on the basis of one or a few of the flavor constituents. Computer-generated histograms can be used for comparisons after selection of important regions of gas-liquid chromatograms by using mathematical treatments. Biggers et al. (1969) differenti-

ated the beverage quality of two varieties of coffee (arabica and robusta) on the basis of contributions of flavor compounds.

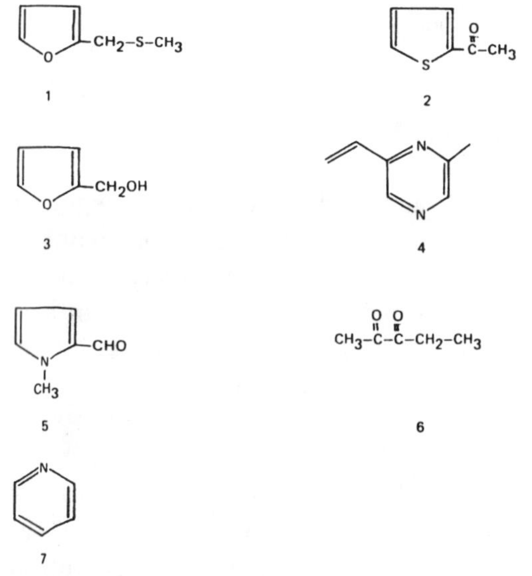

Recent studies have identified 655 compounds in the flavor of coffee, the principal ones being furans, pyrazines, pyrroles, and ketones (Maarse 1991). The distinctiveness of coffee flavor is related to the fact that it contains a large percentage of thiophenes, furans, pyrroles, as well as oxazoles, thiazoles, and phenols.

Alcoholic Beverages

In distilled beverages, one of the major flavor compounds is acetaldehyde. Acetaldehyde represents about 90 percent of the total aldehydes present in beverages like whiskey, cognac, and rum. Together with other short-chain aliphatic aldehydes, it produces a pungent odor and sharp flavor, which is masked by other flavor components in cognac, fruit brandies, rum, and whiskey. In vodka the presence of acetaldehyde may result in an off-flavor. Propanol and 2-methylpropanol, as well as unsaturated aldehydes, are also present in distilled beverages. The aldehydes are very reactive and can form acetals by reacting with ethanol. This reaction results in a smoother flavor profile. Another important flavor compound in distilled beverages is the diketone, 2,3-butanedione (diacetyl), which is a product of fermentation. Depending on fermentation and distillation conditions, the level of diacetyl varies widely in different beverages.

Fusel alcohols, which are present in most distilled beverages, influence flavor. They are formed during fermentation from amino acids through decarboxylation and deamination, and include 1-propanol, 2-methylpropanol, 2-methylbutanol, 3-methylbutanol, and 2-phenylethanol.

Figure 7–35 Structure of Some Important Constituents of the Aroma of Coffee. (1) Furfuryl-methyl-sulfide, (2) 2-acetylthiophene, (3) 2-furfurylalcohol, (4) 2-methyl-6-vinyl-pyrazine, (5) n-methyl-pyrrole-2-aldehyde, (6) acetylpropionyl, (7) pyridine.

Table 7–20 Volatile Compounds in Roasted Coffee Aroma

Compound Type	Functional Group								
	None	—OH	—O—	$-C{\scriptstyle\nwarrow}^{O}_{OH}$	C=O	—CO—CO—	$-C{\scriptstyle\nwarrow}^{O}_{OR}$	—S—	Other
Aliphatic	17	19	—	13	30	10	16	9	25
Isocyclic	3	1	—	—	6	6	—	—	—
Benzenic	20	—	6	4	5	1	5	2	16
Furanic	15	1	4	3	13	4	11	7	3
Thiophenic	6	1	—	2	6	2	3	20	—
Pyrrolic	10	—	—	8	5	1	—	—	—
Pyrazinic	27	—	—	—	—	—	—	—	—
Other	5	2	8	—	7	—	1	5	13
Total number	103	24	18	30	72	24	36	43	57

Distilled beverages also contain fatty acids—from acetic acid (which is one of the major fatty acids) to long-chain unsaturated fatty acids.

Maturation in oak barrels has a major effect on flavor of distilled beverages. Maturing fresh distillates in oak barrels can transform a raw-tasting product into a mellow, well-rounded beverage. The reactions that take place during maturation involve reactions between components of the distillate and reactions between distillate components and compounds present in the oak wood. The alcoholic solution in the barrel extracts lignin from the oak to form an alcohol-soluble ethanol-lignin. Alcoholysis converts this to coniferic alcohol and then by oxidation to coniferaldehyde. Similarly, sinapic alcohol is converted to sinapaldehyde. These aldehydes then produce syringaldehyde and vanillin. The latter compound is important in the flavor of cognac and whiskey. A similar process occurs in the aging of wines in oak barrels to produce the distinctive smoothness of oak-aged wines.

Spices and Herbs

Spices and herbs are natural vegetable products used for adding flavor and aroma to foods. They are usually highly flavored themselves and are used in small quantities. There is no clear distinction between spices and herbs, other than the general rule that spices are produced from tropical plants and herbs from plants grown in cooler climates. Spices and herbs provide aroma because of the presence of aromatic constituents; in addition, spices often provide pungency or hotness. The flavor and pungency of spices can be provided by the dried or ground products themselves, by their essential oils (produced by steam distillation), or by their oleoresins (produced by extraction with solvents). Essential oils contain only volatile compounds; oleoresins also include nonvolatile fats or oils.

Spices and herbs differ in the nature of their volatile constituents (Boelens 1991). Spices contain higher levels of phenylpropanoids (Figure 7–36) such as eugenol, dillapiol, and cinnamaldehyde. Herbs have higher levels of *para*-menthanoids, such as menthol, carvone, thymol, carvacrol, and cuminaldehyde.

Numerous volatile compounds have been identified in the essential oils of spices. Maarse (1991) has reported the number of hydrocarbons, alcohols, aldehydes, ketones, esters, phenols, acids, and others (Table 7–21).

Ginger contains about 2 percent of volatile oil, composed mostly of sesquiterpene hydrocarbons. Other constituents are oxygenated sesquiterpenes, monoterpene hydrocarbons, and oxygenated monoterpenes. The pungent component of ginger is gingerol, which is a series of compounds consisting of zingerone-forming condensation products with saturated straight-chain aldehydes of chain lengths 6, 8, and 10. Fresh ginger has a lemony flavor resulting from the presence of citral and terpineol compounds. The lemony character may be lost because of flashing off during drying.

Pepper aroma and flavor are determined by the composition of the steam volatile oil (Purseglove et al. 1991). The steam volatiles consist of monoterpene hydrocarbons and smaller amounts of sesquiterpene hydrocarbons. The major pungent compound in pepper is piperine. Also contributing to pungency are five minor alkaloids, whose structure is shown in Figure 7–37. Nutmeg oil, which is obtained by steam distillation, contains the following major components: monoterpene hydrocarbons, oxygenated monoterpenes, and aromatic ethers. The monoterpene hydrocarbons contain alpha- and beta-pinene and sabinene. The aromatic ether fraction has as major constituent myristicin; this fraction is thought to play a major role in the flavor of nutmeg.

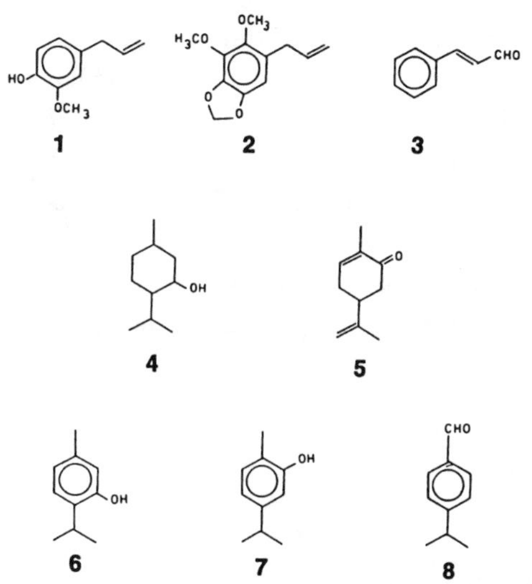

Figure 7–36 Volatile Constituents of Spices and Herbs: (1) Eugenol, (2) dillapiol, (3) cinnamaldehyde, (4) menthol, (5) carvone, (6) thymol, (7) carvacrol, (8) cuminaldehyde

Table 7–21 Number of Volatile Components in the Essential Oils of Some Spices

Spice	Number
Cinnamon	113
Cloves	95
Ginger	146
Nutmeg	80
Pepper	122
Vanilla	190

Source: Reprinted with permission from H. Maarse, *Volatile Compounds in Foods and Beverages*, p. 420, 1991, by courtesy of Marcel Dekker, Inc.

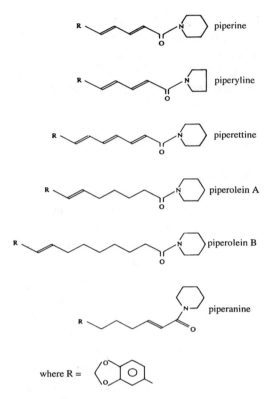

where R =

Figure 7–37 Alkaloids Contributing to the Pungency of Pepper. *Source*: Reprinted with permission from J.W. Purseglove et al., *Spices*, Vol. 1 and 2, p. 52, © 1991, Blackwell Science Ltd.

The many chilies are members of the species *Capsicum*. They include peppers used as a vegetable, paprika, and the various pungent forms used as a spice. The composition of the pungent *Capsicum* fruits varies widely and is influenced by the species, cultivars, growing conditions, stage of maturity at harvest, and postharvest processing. The bell peppers possess no pungency, and paprika is mainly used for its coloring power. The main pungent principle of hot chilies is capsaicin (see Figure 7–18). In addition, Purseglove et al. (1991) have reported a number of analogs and homologs of capsaicin that contribute to the pungency of chilies.

Vanilla

Vanilla is obtained from dried and cured vanilla beans. These can be used directly, in the form of an alcoholic extract, or as oleoresin. The major flavor compound is vanillin, which is present in the beans at a level of 1.3 to 3.8 percent (Maarse 1991). The extracts contain resins that contribute to the taste and serve in the fixation of flavor. The precursor of vanillin is probably lignin, of which the cured beans contain from 2.1 to 3.9 percent. Numerous other compounds are present at very low concentrations. These include *p*-hydroxybenzaldehyde and *p*-hydroxylbenzyl methyl ether. The composition of vanilla is influenced by the geographic origin of the beans.

REFERENCES

Amoore, J. 1967. Stereochemical theory of olfaction. In *Symposium on foods: The chemistry and physiology of flavors*, ed. H.W. Schultz et al. Westport, CT: AVI Publishing Co.

Amoore, J., et al. 1964. The stereochemical theory of odor. *Sci. Am.* 210, no. 2: 42–49.

Badings, H.T. 1991. Milk. In *Volatile compounds in foods and beverages*. New York: Marcel Dekker.

Beatty, R.M., and L.H. Cragg. 1935. The sourness of acids. *J. Am. Chem. Soc.* 57: 2347–2351.

Beidler, L.M. 1954. A theory of taste stimulation. *J. Gen. Physiol.* 38: 133–139.

Beidler, L.M. 1957. Facts and theory on the mechanism of taste and odor perception. In *Chemistry of natural food flavors*. Chicago: Quartermaster Food and Container Institute for the Armed Forces.

Beidler, L.M. 1966. Chemical excitation of taste and odor receptors. In *Flavor Chemistry*, ed. I. Hornstein. Washington, DC: American Chemical Society.

Biggers, R.E., et al. 1969. Differentiation between *Coffea arabica* and *Coffea robusta* by computer evaluation of gas chromatographic profiles: Comparison of numerically derived quality predictions with organoleptic evaluations. *J. Chrom. Sci.* 7: 453–472.

Birch, G.G., and C. Lee. 1971. Chemical basis of sweetness in model sugars. In *Sweetness and sweeteners*, ed. G.G. Birch. London: Applied Science Publishers, Ltd.

Boelens, M.H. 1991. Spices and condiments. II. In *Volatile compounds in foods and beverages*, ed. H. Maarse. New York: Marcel Dekker.

Bokuchava, M.A., and N.I. Skobeleva. 1969. The chemistry and biochemistry of tea and tea manufacture. In *Advances in food research*, Vol. 17, ed. E.M. Mrak and G.F. Stewart. New York: Academic Press.

Bondarovich, H.A., et al. 1967. Volatile constituents of coffee: Pyrazines and other compounds. *J. Agr. Food Chem.* 15: 1093–1099.

Byrne, B., and G. Sherman. 1984. Stability of dry acetaldehyde systems. *Food Technol.* 38, no. 7: 57–61.

Crocker, E.C. 1948. Meat flavor and observations on the taste of glutamate and other amino acids. In *Monosodium glutamate—A symposium*. Chicago: Quartermaster Food and Container Institute for the Armed Forces.

Dastoli, F.R., et al. 1968. Bitter sensitive protein from porcine taste buds. *Nature* 218: 884–885.

Dastoli, F.R., and S. Price. 1966. Sweet sensitive protein from bovine taste buds: Isolation and assay. *Science* 154: 905–907.

Day, E.A. 1966. Role of milk lipids in flavors of dairy products. In *Flavor chemistry*, ed. I. Hornstein. Washington, DC: American Chemical Society.

Day, E.A. 1967. Cheese flavor. In *Symposium on foods: The chemistry and physiology of flavors*, ed. H.W. Schultz et al. Westport, CT: AVI Publishing Co.

Döving, K.B. 1967. Problems in the physiology of olfaction. In *Symposium on foods: The chemistry and physiology of flavors*, ed. H.W. Schultz et al. Westport, CT: AVI Publishing Co.

Dravnieks, A. 1966. Current status of odor theories. In *Flavor Chemistry*, ed. I. Hornstein. Washington, DC: American Chemical Society.

Dravnieks, A. 1967. Theories of olfaction. In *Symposium on foods: The chemistry and physiology of flavors*, ed. H.W. Schultz et al. Westport, CT: AVI Publishing Co.

Fisher, R. 1971. Gustatory, behavioral and pharmacological manifestations of chemoreception in man. In *Gustation and olfaction*, ed. G. Ohloff and A.F. Thomas. New York: Academic Press.

Flament, I., et al. 1967. Research on flavor: Cocoa aroma III. *Helv. Chim. Acta* 50: 2233–2243 (French).

Forss, D.A. 1969. Role of lipids in flavors. *J. Agr. Food Chem.* 17: 681–685.

Forss, D.A., et al. 1962. The flavor of cucumbers. *J. Food Sci.* 27: 90–93.

Gianturco, M.A. 1967. Coffee flavor. In *Symposium on foods: The chemistry and physiology of flavors*, ed. H.W. Schultz et al. Westport, CT: AVI Publishing Co.

Gillette, M. 1985. Flavor effects of sodium chloride. *Food Technol.* 39, no. 6: 47–52, 56.

Gold, H.J., and C.W. Wilson. 1963. The volatile flavor substances of celery. *J. Food Sci.* 28: 484–488.

Goldman, I.M., et al. 1967. Research on flavor. Coffee aroma II. Pyrazines and pyridines. *Helv. Chim. Acta* 50: 694–705 (French).

Govindarajan, V.S. 1979. Pungency: The stimuli and their evaluation. In *Food taste chemistry*, ed. J.C. Boudreau. Washington, DC: American Chemical Society.

Grosch, W., and P. Schieberle. 1991. Bread. In *Volatile compounds in foods and beverages*. New York: Marcel Dekker.

Habibi-Najafi, M.B., and B.H. Lee. 1996. Bitterness in cheese: A review. *Crit. Rev. Food Sci. Nutr.* 36: 397–411.

Hall, L.A. 1948. Protein hydrolysates as a source of glutamate flavors. In *Monosodium glutamate—A symposium*. Chicago: Quartermaster Food and Container Institute for the Armed Forces.

Hall, R.L. 1968. Food flavors: Benefits and problems. *Food Technol.* 22: 1388–1392.

Harper, R., et al. 1968. *Odour description and odour classification*. London: J.A. Churchill, Ltd.

Horowitz, R.M., and B. Gentili. 1969. Taste and structure in phenolic glycosides. *J. Agr. Food Chem.* 17: 696–700.

International Standards Organization. 1992. *Glossary of terms relating to sensory analysis*. ISO Standard 5492.

Jennings, W.G., and M.R. Sevenants. 1964. Volatile esters of Bartlett pear. III. *J. Food Sci.* 29: 158–163.

Johnson, J.A., et al. 1966. Chemistry of bread flavor. In *Flavor chemistry*, ed. I. Hornstein. Washington, DC: American Chemical Society.

Jones, N.R. 1969. Meat and fish flavors: Significance of ribomononucleotides and their metabolites. *J. Agr. Food Chem.* 17: 712–716.

Juriens, G., and J.M. Oele. 1965. Determination of hydroxyacid triglycerides and lactones in butter. *J. Am. Oil Chem. Soc.* 42: 857–861.

Kanehisa, H. 1984. Studies of bitter peptides from casein hydrolyzates. VI. Synthesis and bitter taste of BPIC (Val-Tyr-Pro-Phe-Pro-Gly-Ile-Asn-His) and its analog and fragments. *Bull. Chem. Soc. Jpn.* 57: 301–308.

Kawamura, Y., and M.R. Kare. 1987. *Umami: A basic taste*. New York: Marcel Dekker.

Keeney, P.G., and S. Patton. 1956. The coconut-like flavor defect of milk fat. I. Isolation of the flavor compound from butter oil and its identification as δ-decalactone. *J. Dairy Sci.* 39: 1104–1113.

Kefford, J.F. 1959. The chemical constituents of citrus fruits. In *Advances in food research*, Vol. 9, eds. E.M. Mrak and G.F. Stewart. New York: Academic Press.

Konosu, S. 1979. The taste of fish and shell fish. In *Food taste chemistry*, ed. J.C. Boudreau. Washington, DC: American Chemical Society.

Kulka, K. 1967. Aspects of functional groups and flavor. *J. Agr. Food Chem.* 15: 48–57.

Kuninaka, A. 1966. Recent studies of 5′-nucleotides as new flavor enhancers. In *Flavor Chemistry*, ed. I. Hornstein. Washington, DC: American Chemical Society.

Kurihara, K. 1987. Recent progress in the taste receptor mechanism. In *Umami: A basic taste*, ed. Y. Kawamura and M.R. Kare. New York: Marcel Dekker.

Kurihara, K., and L.M. Beidler. 1968. Taste-modifying protein from miracle fruit. *Science* 161: 1241–1243.

Kurihara, K., and L.M. Beidler. 1969. Mechanism of the action of taste-modifying protein. *Nature* 222: 1176–1179.

Kushman, L.J., and W.E. Ballinger. 1968. Acid and sugar changes during ripening in Wolcott blueberries. *Proc. Amer. Soc. Hort. Sci.* 92: 290–295.

Linko, Y., et al. 1962. The origin and fate of certain carbonyl compounds in white bread. *Cereal Chem.* 29: 468–476.

Luck, G., et al. 1994. The cup that cheers: Polyphenols and the astringency of tea. Lecture paper No. 0030. London: Society of Chemical Industry.

Maarse, H. 1991. *Volatile compounds in foods and beverages*. New York: Marcel Dekker.

Maarse, H., et al. 1987. Characterization of Spanish medium sherries. In *Flavor science and technology*, ed. M. Martens et al. New York: John Wiley & Sons.

Macheix, J-J., et al. 1990. *Fruit phenolics*. Boca Raton, FL: CRC Press.

Marion, J.P., et al. 1967. The composition of cocoa aroma. *Helv. Chim. Acta* 50: 1509–1522 (French).

Mason, M.E., et al. 1966. Flavor components of roasted peanuts: Some low molecular weight pyrazines and a pyrrole. *J. Agr. Food Chem.* 14: 454–460.

Meyboom, P.W., and G.A. Jongenotter. 1981. Flavor perceptibility of straight chain, unsaturated aldehydes as a function of double bond position and geometry. *J. Am. Oil Chem. Soc.* 58: 680–682.

Moncrieff, R.W. 1951. *The chemical senses*. London: Leonard Hill, Ltd.

Moncrieff, R.W. 1964. The metallic taste. *Perf. Ess. Oil Rec.* 55: 205–207.

Moncrieff, R.W. 1966. *Odour preferences*. London: Leonard Hill, Ltd.

Moody, W.G. 1983. Beef flavor—A review. *Food Technol.* 37, no. 5: 227–232, 238.

Naves, Y.R. 1957. The relationship between the stereochemistry and odorous properties of organic substances. In *Molecular structure and organoleptic quality*. London: Society of Chemical Industry.

Ney, K.H. 1979. Bitterness of peptides: Amino acid composition and chain length. In *Food taste chemistry*, ed. J.C. Boudreau. Washington, DC: American Chemical Society.

Noble, A.C., et al. 1987. Modification of a standardized system of wine aroma terminology. *Am. J. Enol. Vitic.* 38: 143–146.

O'Mahony, M.A.P. 1984. How we perceive flavor. *Nutr. Today* 19, no. 3: 6–15.

Ough, C.S. 1963. Sensory examination of four organic acids added to wine. *J. Food Sci.* 28: 101–106.

Page, S.W. 1986. Pattern recognition methods for the determination of food composition. *Food Technol.* 40, no. 11: 104–109.

Pangborn, R.M. 1963. Relative taste intensities of selected sugars and organic acids. *J. Food Sci.* 28: 726–733.

Patton, S. 1964. Flavor thresholds of volatile fatty acids. *J. Food Sci.* 29: 679–680.

Peryam, D.R. 1963. Variability of taste perception. *J. Food Sci.* 28: 734–740.

Purseglove, J.W., et al. 1991. *Spices*. Vol. 1 and 2. New York: Longman Scientific and Technical.

Rizzi, G.P. 1967. The occurrence of simple alkylpyrazines in cocoa butter. *J. Agr. Food Chem.* 15: 549–551.

Rogers, J.A. 1966. Advances in spice flavor and oleoresin chemistry. In *Flavor chemistry*, ed. I. Hornstein. Washington, DC: American Chemical Society.

Saxby, M.J. 1996. *Food taints and off-flavors*. London: Blackie Academic and Professional.

Schoch, T.J. 1965. Starch in bakery products. *Baker's Dig.* 39, no. 2: 48–57.

Seifert, R.M., et al. 1970. Synthesis of some 2-methoxy-3-alkylpyrazines with strong bell pepper-like odors. *J. Agr. Food Chem.* 18: 246–249.

Shallenberger, R.S. 1971. Molecular structure and taste. In *Gustation and olfaction*, ed. G. Ohloff and A.G. Thomas. New York: Academic Press.

Shallenberger, R.S. 1998. Sweetness theory and its application in the food industry. *Food Technol.* 52: 72–76.

Shallenberger, R.S., and T.E. Acree. 1967. Molecular theory of sweet taste. *Nature* 216: 480–482.

Shallenberger, R.S., and T.E. Acree. 1969. Molecular structure and sweet taste. *J. Agr. Food Chem.* 17: 701–703.

Sinki, G.S. 1988. Finding the universally acceptable taste. *Food Technol.* 42, no. 7: 90–93.

Sjöström, L.B. 1972. *The flavor profile*. Cambridge, MA: A.D. Little, Inc.

Solms, J. 1969. The taste of amino acids, peptides and proteins. *J. Agr. Food Chem.* 17: 686–688.

Solms, J. 1971. Nonvolatile compounds and the flavor of foods. In *Gustation and olfaction*, ed. G. Ohloff and A.F. Thomas. New York: Academic Press.

Solms, J., et al. 1965. The taste of L and D amino acids. *Experientia* 21: 692–694.

Spillane, W.J. 1996. Molecular structure and sweet taste. In *Advances in sweeteners*, ed. T.H. Grenby. London: Blackie Academic and Professional.

Stark, W., and D.A. Forss. 1962. A compound responsible for metallic flavor in dairy products. I. Isolation and identification. *J. Dairy Res.* 29: 173–180.

Stöcklin, W., et al. 1967. Gymnemic acid, the antisaccharic principle of *Gymnema sylvestre* R. Br. Isolation and identification. *Helv. Chim. Acta* 50: 474–490 (German).

Stoll, M. 1957. Facts old and new concerning relationships between molecular structure and odour. In *Molecular structure and organoleptic quality*. London: Society of Chemical Industry.

Stone, H., and S.M. Oliver. 1969. Measurement of the relative sweetness of selected sweeteners and sweetener mixture. *J. Food Sci.* 34: 215–222.

Tarr, H.L.A. 1966. Flavor of fresh foods. In *Flavor chemistry*, ed. I. Hornstein. Washington, DC: American Chemical Society.

Teranishi, R., 1971. Odor and molecular structure. In *Gustation and olfaction*, ed. G. Ohloff and A.F. Thomas. New York: Academic Press.

Teranishi, R., et al. 1971. *Flavor research—Principles and techniques*. New York: Marcel Dekker.

Tharp, B.W., and S. Patton. 1960. Coconut-like flavor defect of milk fat. IV. Demonstration of δ-dodecalactone in the steam distillate from milk fat. *J. Dairy Sci.* 43: 475–479.

Tressler, D.K., and M.A. Joslyn. 1954. *Fruit and vegetable juice production*. Westport, CT: AVI Publishing Co.

Wright, R.H. 1957. Odor and molecular vibration. In *Molecular structure and organoleptic quality*. London: Society of Chemical Industry.

Wucherpfennig, K. 1969. Acids: A quality determining factor in wine. *Dtsch. Wein Ztg.* 30: 836–840.

Yamaguchi, S. 1979. The umami taste. In *Food taste chemistry*, ed. J.C. Boudreau. Washington, DC: American Chemical Society.

CHAPTER 8

Texture

INTRODUCTION

Food texture can be defined as the way in which the various constituents and structural elements are arranged and combined into a micro- and macrostructure and the external manifestations of this structure in terms of flow and deformation.

Most of our foods are complex physicochemical structures and, as a result, the physical properties cover a wide range—from fluid, Newtonian materials to the most complex disperse systems with semisolid character. There is a direct relationship between the chemical composition of a food, its physical structure, and the resulting physical or mechanical properties; this relationship is presented in Figure 8–1. Food texture can be evaluated by mechanical tests (instrumental methods) or by sensory analysis. In the latter case, we use the human sense organs as analytical tools. A proper understanding of textural properties often requires study of the physical structure. This is most often accomplished by light and electron microscopy, as well as by several other physical methods. X-ray diffraction analysis provides information about crystalline structure, differential scanning calorimetry provides information about melting and solidification and other phase transitions, and particle size analysis

and sedimentation methods provide information about particle size distribution and particle shape.

In the study of food texture, attention is given to two interdependent areas: the flow and deformation properties and the macro- and microstructure. The study of food texture is important for three reasons:

1. to evaluate the resistance of products against mechanical action, such as in mechanical harvesting of fruits and vegetables
2. to determine the flow properties of products during processing, handling, and storage
3. to establish the mechanical behavior of a food when consumed

There is sometimes a tendency to restrict texture to the third area. The other two are equally important, although the first area is generally considered to belong in the domain of agricultural engineering.

Because most foods are complex disperse systems, there are great difficulties in establishing objective criteria for texture measurement. It is also difficult in many cases to relate results obtained by instrumental techniques of measurement to the type of response obtained by sensory panel tests.

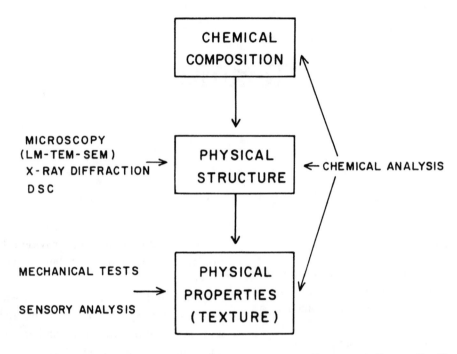

Figure 8–1 Interrelationships in Texture Studies. *Source*: From P. Sherman, A Texture Profile of Food-stuffs Based upon Well-Defined Rheological Properties, *J. Food Sci.*, Vol. 34, pp. 458–462, 1969.

The terms for the textural properties of foods have a long history. Many of the terms are accepted but are often poorly defined descriptive terms. Following are some examples of such terms:

- *Consistency* denotes those aspects of texture that relate to flow and deformation. It can be said to encompass all of the rheological properties of a product.
- *Hardness* has been defined as resistance to deformation.
- *Firmness* is essentially identical to hardness but is occasionally used to describe the property of a substance able to resist deformation under its own weight.
- *Brittleness* is the property of fracturing before significant flow has occurred.

- *Stickiness* is a surface property related to the adhesion between material and adjoining surface. When the two surfaces are of identical material, we use the term *cohesion*.

A variety of other words and expressions are used to describe textural characteristics, such as body, crisp, greasy, brittle, tender, juicy, mealy, flaky, crunchy, and so forth. Many of these terms have been discussed by Szczesniak (1963) and Sherman (1969); most have no objective physical meaning and cannot be expressed in units of measurement that are universally applicable. Kokini (1985) has attempted to relate some of these ill-defined terms to the physical properties involved in their evaluation. Through the years, many types of instruments have been developed for measuring certain aspects of food texture. Unfortunately, the instruments are often based on empirical procedures, and results cannot be compared with those obtained with other instruments. Recently, instruments have been developed that are more widely applicable and are based on sound physical and engineering principles.

TEXTURE PROFILE

Texture is an important aspect of food quality, sometimes even more important than flavor and color. Szczesniak and Kleyn (1963) conducted a consumer-awareness study of texture and found that texture significantly influences people's image of food. Texture was most important in bland foods and foods that are crunchy or crisp. The characteristics most often referred to were hardness, cohesiveness, and moisture content. Several attempts have been made to develop a classification system for textural characteristics. Szczesniak (1963) divided textural characteristics into three main classes, as follows:

1. mechanical characteristics
2. geometrical characteristics
3. other characteristics, related mainly to moisture and fat content

Mechanical characteristics include five basic parameters.

1. *Hardness*—the force necessary to attain a given deformation.
2. *Cohesiveness*—the strength of the internal bonds making up the body of the product.
3. *Viscosity*—the rate of flow per unit force.

4. *Elasticity*—the rate at which a deformed material reverts to its undeformed condition after the deforming force is removed.
5. *Adhesiveness*—the work necessary to overcome the attractive forces between the surface of the food and the surface of other materials with which the food comes in contact (e.g., tongue, teeth, and palate).

In addition, there are in this class the three following secondary parameters:

1. *Brittleness*—the force with which the material fractures. This is related to hardness and cohesiveness. In brittle materials, cohesiveness is low, and hardness can be either low or high. Brittle materials often create sound effects when masticated (e.g., toast, carrots, celery).
2. *Chewiness*—the energy required to masticate a solid food product to a state ready for swallowing. It is related to hardness, cohesiveness, and elasticity.
3. *Gumminess*—the energy required to disintegrate a semisolid food to a state ready for swallowing. It is related to hardness and cohesiveness.

Geometrical characteristics include two general groups: those related to size and shape of the particles, and those related to shape and orientation. Names for geometrical characteristics include smooth, cellular, fibrous, and so on. The group of other characteristics in this system is related to moisture and fat content and includes qualities such as moist, oily, and greasy. A summary of this system is given in Table 8–1.

Based on the Szczesniak system of textural characteristics, Brandt et al. (1963) devel-

Table 8–1 Classification of Textural Characteristics

MECHANICAL CHARACTERISTICS

Primary Parameters	Secondary Parameters	Popular Terms
Hardness		Soft → Firm → Hard
Cohesiveness	Brittleness	Crumbly → Crunchy → Brittle
	Chewiness	Tender → Chewy → Tough
	Gumminess	Short → Mealy → Pasty → Gummy
Viscosity		Thin → Viscous
Elasticity		Plastic → Elastic
Adhesiveness		Sticky → Tacky → Gooey

GEOMETRICAL CHARACTERISTICS

Class	Examples
Particle size and shape	Gritty, Grainy, Coarse, etc.
Particle shape and orientation	Fibrous, Cellular, Crystalline, etc.

OTHER CHARACTERISTICS

Primary Parameters	Secondary Parameters	Popular Terms
Moisture content		Dry → Moist → Wet → Watery
Fat content	Oiliness	Oily
	Greasiness	Greasy

Source: From A.S. Szczesniak, Classification of Textural Characteristics, *J. Food Sci.*, Vol. 28, pp. 385–389, 1963.

oped a method for profiling texture so that a sensory evaluation could be given that would assess the entire texture of a food. The texture profile method was based on the earlier development of the flavor profile (Cairncross and Sjöström 1950).

The Szczesniak system was critically examined by Sherman (1969), who proposed some modifications. In the improved system, no distinction is drawn among analytical, geometrical, and mechanical attributes. Instead, the only criterion is whether a charac-

teristic is a fundamental property or derived by a combination of two or more attributes in unknown proportions. The Sherman system contains three groups of characteristics (Figure 8–2). The primary category includes analytical characteristics from which all other attributes are derived. The basic rheological parameters, elasticity, viscosity, and adhesion form the secondary category; the remaining attributes form the tertiary category since they are a complex mixture of these secondary parameters. This system is

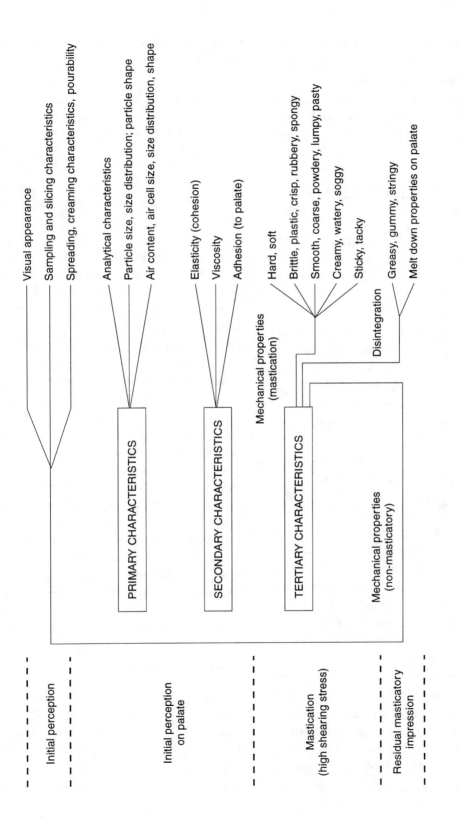

Figure 8–2 The Modified Texture Profile. *Source:* From P. Sherman, A Texture Profile of Foodstuffs Based upon Well-Defined Rheological Properties, *J. Food Sci.*, Vol. 34, pp. 458–462, 1969.

interesting because it attempts to relate sensory responses with mechanical strain-time tests. Sensory panel responses associated with masticatory tertiary characteristics of the Sherman texture profile for solid, semisolid, and liquid foods are given in Figure 8–3.

OBJECTIVE MEASUREMENT OF TEXTURE

The objective measurement of texture belongs in the area of rheology, which is the science of flow and deformation of matter. Determining the rheological properties of a food does not necessarily mean that the complete texture of the product is determined. However, knowledge of some of the rheological properties of a food may give important clues as to its acceptability and may be important in determining the nature and design of processing methods and equipment.

Food rheology is mainly concerned with forces and deformations. In addition, time is an important factor; many rheological phenomena are time-dependent. Temperature is another important variable. Many products show important changes in rheological behavior as a result of changes in temperature. In addition to flow and deformation of cohesive bodies, food rheology includes such phenomena as the breakup or rupture of solid materials and surface phenomena such as stickiness (adhesion).

Deformation may be of one or both of two types, irreversible deformation, called flow, and reversible deformation, called elasticity. The energy used in irreversible deformation is dissipated as heat, and the body is permanently deformed. The energy used in reversible deformation is recovered upon release of the deforming stress, when the body regains its original shape.

Force and Stress

When a force acts externally on a body, several different cases may be distinguished: tension, compression, and shear. Bending involves tension and compression, torque involves shear, and hydrostatic compression involves all three. All other cases may involve one of these three factors or a combination of them. In addition, the weight or inertia of a body may constitute a force leading to deformation. Generally, however, the externally applied forces are of much greater magnitude and the effect of weight is usually neglected. The forces acting on a body can be expressed in grams or in pounds. Stress is the intensity factor of force and is expressed as force per unit area; it is similar to pressure. There are several types of stress: compressive stress (with the stress components directed at right angles toward the plane on which they act); tensile stress (in which the stress components are directed away from the plane on which they act); and shearing stress (in which the stress components act tangentially to the plane on which they act). A uniaxial stress is usually designated by the symbol σ, a shearing stress by τ. Shear stress is expressed in dynes/cm^2 when using the metric system of measurement; in the SI system it is expressed in N/m^2 or pascal (P).

Deformation and Strain

When the dimensions of a body change, we speak of deformation. Deformation can be linear, as in a tensile test when a body of original length L is subjected to a tensile stress. The linear deformation ΔL can then be expressed as strain $\varepsilon = \Delta L/L$. Strain can be

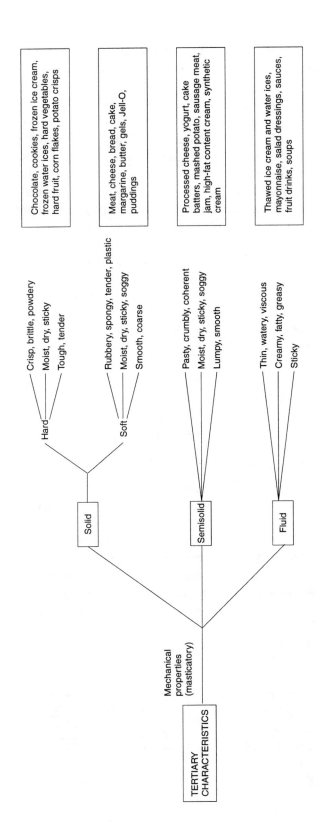

Figure 8–3 Panel Responses Associated with Masticatory Tertiary Characteristics of the Modified Texture Profile

expressed as a ratio or percent; inches per inch or centimeters per centimeter. In addition to linear deformations, there are other types of deformation, such as in a hydrostatic test where there will be a volumetric strain $\Delta V/V$.

For certain materials the deformation resulting from an applied force can be very large; this indicates the material is a liquid. In such cases, we deal with rate of deformation, or shear rate; $d\gamma/dt$ or $\dot{\gamma}$. This is the velocity difference per unit thickness of the liquid. $\dot{\gamma}$ is expressed in units of s^{-1}.

Viscosity

Consider a liquid contained between two parallel plates, each of area A cm^2 (Figure 8–4). The plates are h cm apart and a force of P dynes is applied on the upper plate. This shearing stress causes it to move with respect to the lower plate with a velocity of v cm s^{-1}. The shearing stress τ acts throughout the liquid contained between the plates and can be defined as the shearing force P divided by the area A, or P/A dynes/cm^2. The deformation can be expressed as the mean rate of shear $\dot{\gamma}$ or velocity gradient and is equal to the velocity difference divided by the distance between the plates $\dot{\gamma} = v/h$, expressed in units of s^{-1}.

The relationship between shearing stress and rate of shear can be used to define the flow properties of materials. In the simplest case, the shearing stress is directly proportional to the mean rate of shear $\tau = \eta\dot{\gamma}$ (Figure 8–5). The proportionality constant η is called the viscosity coefficient, or *dynamic viscosity,* or simply the viscosity of the liquid. The metric unit of viscosity is the dyne.s cm^{-2}, or Poise (P). The commonly used unit is 100 times smaller and called centiPoise (cP). In the SI system, η is expressed in N.s/m^2. or

Pa.s. Therefore, 1 Pa.s = 10 P = 1000 cP. Some instruments measure kinematic viscosity, which is equal to dynamic viscosity × density and is expressed in units of Stokes. The viscosity of water at room temperature is about 1 cP. Mohsenin (1970) has listed the viscosities of some foods; these, as well as their SI equivalents, are given in Table 8–2.

Materials that exhibit a direct proportionality between shearing stress and rate of shear are called Newtonian materials. These include water and aqueous solutions, simple organic liquids, and dilute suspensions and emulsions. Most foods are non-Newtonian in character, and their shearing stress–rate-of-shear curves are either not straight or do not go through the origin, or both. This introduces a considerable difficulty, because their flow behavior cannot be expressed by a single value, as is the case for Newtonian liquids.

The ratio of shearing stress and rate of shear in such materials is not a constant value, so the value is designated *apparent viscosity.* To be useful, a reported value for apparent viscosity of a non-Newtonian material should be given together with the value of rate of shear or shearing stress used in the determination. The relationship of shearing stress and rate of shear of non-Newtonian materials such as the dilatant and pseudo-plastic bodies of Figure 8–5 can be represented by a power law as follows:

$$\tau = A\dot{\gamma}^n$$

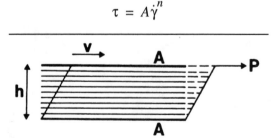

Figure 8–4 Flow Between Parallel Plates

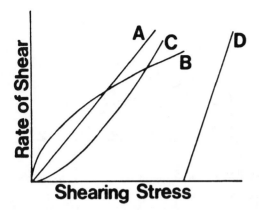

Figure 8–5 Shearing Stress-Rate of Shear Diagrams. (A) Newtonian liquid, viscous flow, (B) dilatant flow, (C) pseudoplastic flow, (D) plastic flow.

where A and n are constants. A is the consistency index or apparent viscosity and n is the flow behavior index. The exponent is $n = 1$ for Newtonian liquids; for dilatant materials, it is greater than 1; and for pseudoplastic materials, it is less than 1. In its logarithmic form,

$$\log \tau = \log A + n \log \dot{\gamma}$$

A plot of $\log \tau$ versus $\log \dot{\gamma}$ will yield a straight line with a slope of n.

For non-Newtonian materials that have a yield stress, the Casson or Hershel-Bulkley models can be used. The Casson model is represented by the equation,

$$\sqrt{\tau} = \sqrt{\tau_0} + A \sqrt{\dot{\gamma}}$$

where $\tau_0 =$ yield stress.

This model has been found useful for several food products, especially chocolate (Kleinert 1976).

The Hershel-Bulkley model describes material with a yield stress and a linear relationship between log shear stress and log shear rate:

$$\tau = \tau_0 + A\dot{\gamma}^n$$

Table 8–2 Viscosity Coefficients of Some Foods

Product	Temperature (°C)	Viscosity	
		(cP)	(Pa.s)
Water	0	1.79	0.00179
Water	20	1.00	0.00100
Skim milk	25	1.37	0.00137
Milk, whole	0	4.28	0.00428
Milk, whole	20	2.12	0.00212
Cream (20% fat)	4	6.20	0.00620
Cream (30% fat)	4	13.78	0.01378
Soybean oil	30	40.6	0.0406
Sucrose solution (60%)	21	60.2	0.0602
Olive oil	30	84.0	0.0840
Cottonseed oil	16	91.0	0.0910
Molasses	21	6600.0	6.600

Source: Reprinted with permission from N.N. Mohsenin, *Physical Properties of Plant and Animal Materials, Vol. 1, Structure, Physical Characteristics and Mechanical Properties*, © 1970, Gordon and Breach Science Publisher.

The value of *n* indicates how close the linear plot of shear stress and shear rate is to being a straight line.

Principles of Measurement

For Newtonian fluids, it is sufficient to measure the ratio of shearing stress and rate of shear from which the viscosity can be calculated. This can be done in a viscometer, which can be one of various types, including capillary, rotational, falling ball, and so on. For non-Newtonian materials, such as the dilatant, pseudoplastic, and plastic bodies shown in Figure 8–5, the problem is more difficult. With non-Newtonian materials, several methods of measurement involve the ratio of shear stress and rate of shear, the relationship of stress to time under constant strain (relaxation), and the relationship of strain to time under constant stress (creep). In relaxation measurements, a material is subjected to a sudden deformation ε_o, which is held constant. In many materials, the stress will decay with time according to the curve of Figure 8–6. The point at which the stress has decayed to σ/e, or 36.7 percent of the original value of σ_o, is called the relaxation time. When the strain is removed at time T, the stress returns to zero. In a creep experiment, a material is subjected to the instantaneous application of a constant load or stress and the strain measured as a function of time. The resulting creep curve has the shape indicated in Figure 8–7. At time zero, the applied load results in a strain ε_o, which increases with time. When the load is removed at time T, the strain immediately decreases, as indicated by the vertical straight portion of the curve at T; the strain continues to decrease thereafter with time. In many materials, the value of ε never reaches zero, and we know, therefore, a permanent deformation ε_p has

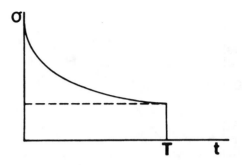

Figure 8–6 Relaxation Curve (Relationship of Stress to Time under Constant Strain)

resulted. The ratio of strain to applied stress in a creep experiment is a function of time and is called the creep compliance (J). Creep experiments are sometimes plotted as graphs relating J to time.

DIFFERENT TYPES OF BODIES

The Elastic Body

For certain solid bodies, the relationship between stress and strain is represented by a straight line through the origin (Figure 8–8)

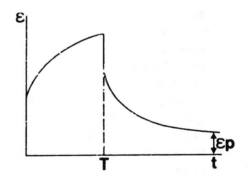

Figure 8–7 Creep Curve (Relationship of Strain to Time under Constant Stress)

up to the so-called limit of elasticity, according to the law of Hooke, $\sigma = E\varepsilon$. The proportionality factor E for uniaxial stress is called *modulus of elasticity*, or Young's modulus. For a shear stress, the modulus is G, or Coulomb modulus. Note that a modulus is the ratio of stress to strain, $E = \sigma/\varepsilon$. The behavior of a Hookean body is further exemplified by the stress-time and strain-time curves of Figure 8–9. When a Hookean body is subjected to a constant strain ε_o, the stress σ will remain constant with time and will return to zero when the strain is removed at time T. The strain ε will follow the same pattern when a constant stress is applied and released at time T.

The Retarded Elastic Body

In bodies showing retarded elasticity, the deformation is a function of time as well as stress. Such a stress-strain curve is shown in Figure 8–10. The upward part of the curve represents increasing values of stress; when the stress is reduced, the corresponding strains are greater on the downward part of the curve. When the stress reaches 0, the strain has a finite value, which will slowly return to zero. There is no permanent deformation. The corresponding relaxation (stress-time) and creep (strain-time) curves

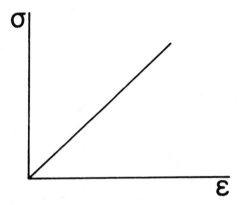

Figure 8–8 Stress-Strain Curve for a Perfectly Elastic Body

for this type of body are given in Figure 8–11.

The Viscous Body

A viscous or Newtonian liquid is one showing a direct proportionality between stress and rate of shear, as indicated by curve A in Figure 8–5.

The Viscoelastic Body

Certain bodies combine the properties of both viscous and elastic materials. The elas-

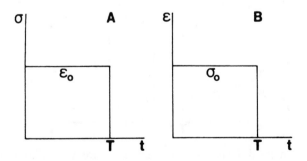

Figure 8–9 (A) Stress-Time and (B) Strain-Time Curves of a Hookean Body

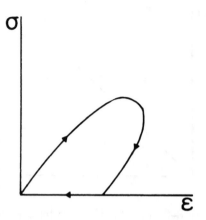

Figure 8–10 Stress-Strain Curve of a Retarded Elastic Body

tic component can be partially retarded elasticity. Viscoelastic bodies may flow slowly and nonreversibly under the influence of a small stress. Under larger stresses the elastic component becomes apparent. The relaxation curve of viscoelastic materials has the shape indicated in Figure 8–12A. The curve has the tendency to approach the time axis. The creep curve indicates that the strain increases for as long as the stress is applied (Figure 8–12B). The magnitude of the permanent deformation of the body increases with the applied stress and with the length of application.

Mechanical models can be used to visualize the behavior of different bodies. Thus, a spring denotes a Hookean body, and a dashpot denotes a purely viscous body or Newtonian fluid. These elements can be combined in a variety of ways to represent the rheological behavior of complex substances. Two basic viscoelastic models are the Voigt-Kelvin and the Maxwell bodies. The Voigt-Kelvin model employs a spring and dashpot in parallel, the Maxwell model a spring and dashpot in series (Figure 8–13). In the Voigt-Kelvin body, the stress is the sum of two components where one is proportional to the strain and the other to the rate of shear. Because the elements are in parallel, they must move together. In the Maxwell model the deformation is composed of two parts— one purely viscous, the other purely elastic. Although both the Voigt-Kelvin and Maxwell bodies represent viscoelasticity, they react differently in relaxation and creep experiments. When a constant load is applied in a creep test to a Voigt-Kelvin model, a final steady-state deformation is obtained because the compressed spring element resists further movement. The Maxwell model will give continuing flow under these conditions because the viscous element is not limited by the spring element. When the load is removed, the Voigt-Kelvin model recovers

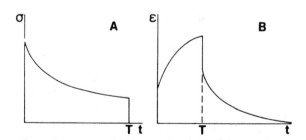

Figure 8–11 (A) Stress-Time and (B) Strain-Time Curves of a Retarded Elastic Body

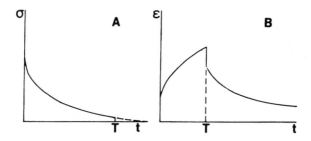

Figure 8–12 (A) Stress-Time and (B) Strain-Time Curves of a Viscoelastic Body

completely, but not instantaneously. The Maxwell body does not recover completely but, rather, instantly. The Voigt-Kelvin body, therefore, shows no stress relaxation but the Maxwell body does. A variety of models can be constructed to represent the rheological behavior of viscoelastic materials. By placing a number of Kelvin models in series, a so-called generalized Kelvin model is obtained. Similarly, a generalized Maxwell model is obtained by placing a number of Maxwell models in parallel. The combination of a Kelvin and a Maxwell model in series (Figure 8–13C) is called a Burgers model.

For ideal viscoelastic materials, the initial elastic deformation at the time the load is applied should equal the instantaneous elastic deformation when the load is removed (Figure 8–14). For most food products, this is not the case. As is shown by the example of butter in Figure 8–14, the initial deformation is greater than the elastic recovery at time *t*. This may result from the fact that these foods are plastic as well as viscoelastic, which means they have a yield value. There-

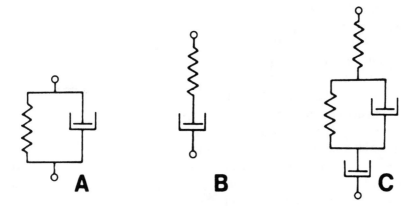

Figure 8–13 (A) Voigt-Kelvin, (B) Maxwell, and (C) Burgers Models

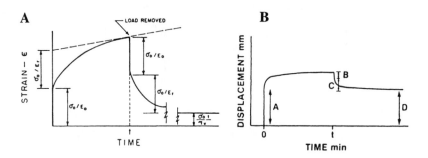

Figure 8–14 (A) Creep Curve for an Ideal Viscoelastic Body and (B) Creep Curve for Butter

fore, the initial deformation consists of both an instantaneous elastic deformation and a permanent deformation (viscous flow component). It has also been found (deMan et al. 1985) that the magnitude of the instantaneous elastic recovery in fat products is time dependent and decreases as the time of application of the load increases. It appears that the fat crystal network gradually collapses as the load remains on the sample.

The Plastic Body

A plastic material is defined as one that does not undergo a permanent deformation until a certain yield stress has been exceeded. A perfectly plastic body showing no elasticity would have the stress-strain behavior depicted in Figure 8–15. Under influence of a small stress, no deformation occurs; when the stress is increased, the material will suddenly start to flow at applied stress σ_o (the yield stress). The material will then continue to flow at the same stress until this is removed; the material retains its total deformation. In reality, few bodies are perfectly plastic; rather, they are plasto-elastic or plasto-viscoelastic. The mechanical model used to represent a plastic body, also called a St. Venant body, is a friction element. The

model is analogous to a block of solid material that rests on a flat horizontal surface. The block will not move when a force is applied to it until the force exceeds the friction existing between block and surface. The models for ideal plastic and plasto-elastic bodies are shown in Figure 8–16A and 8–16B.

A more common body is the plasto-viscoelastic, or Bingham body. Its mechanical model is shown in Figure 8–16C. When a stress is applied that is below the yield stress, the Bingham body reacts as an elastic body. At stress values beyond the yield stress, there are two components, one of which is constant and is represented by the friction ele-

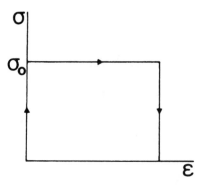

Figure 8–15 Stress-Strain Curve of an Ideal Plastic Body

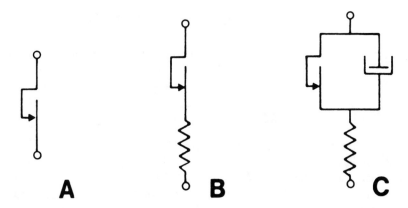

Figure 8–16 Mechanical Models for a Plastic Body. (A) St. Venant body, (B) plasto-elastic body, and (C) plasto-viscoelastic or Bingham body.

ment, and the other, which is proportional to the shear rate and represents the viscous flow element. In a creep experiment with stress not exceeding the yield stress, the creep curve would be similar to the one for a Hookean body (Figure 8–9B). When the shear stress is greater than the yield stress, the strain increases with time, similar to the behavior of a Maxwell body (Figure 8–17). Upon removal of the stress at time T, the strain decreases instantaneously and remains constant thereafter. The decrease represents the elastic component; the plastic deformation is permanent. The relationship of rate of shear and shear stress of a Bingham body would have the form shown in Figure 8–18A. When flow occurs, the relationship between shearing stress and rate of shear is given by

$$\sigma - \sigma_o = UD$$

where

σ_o = yield stress
U = proportionality constant
D = mean rate of shear

The constant U can be named plastic viscosity and its reciprocal $1/U$ is referred to as mobility.

In reality, plastic materials are more likely to have a curve similar to the one in Figure 8–18B. The yield stress or yield value can be taken at three different points—the lower yield value at the point where the curve starts on the stress axis; the upper yield value

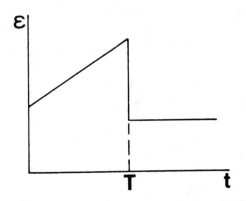

Figure 8–17 Creep Curve of a Bingham Body Subjected to a Stress Greater Than the Yield Stress

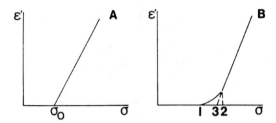

Figure 8–18 Rate-of-Shear–Shear Stress Diagrams of Bingham Bodies. (A) Ideal case, and (B) practical case. The yield values are as follows: lower yield value (1), upper yield value (2), and Bingham yield value (3).

where the curve becomes straight; and the Bingham yield value, which is found by extrapolating the straight portion of the curve to the stress axis.

The Thixotropic Body

Thixotropy can be defined as an isothermal, reversible, sol-gel transformation and is a behavior common to many foods. Thixotropy is an effect brought about by mechanical action, and it results in a lowered apparent viscosity. When the body is allowed sufficient time, the apparent viscosity will return to its original value. Such behavior would result in a shear stress–rate-of-shear diagram, as given in Figure 8–19. Increasing shear rate results in increased shear stress up to a maximum; after the maximum is reached, decreasing shear rates will result in substantially lower shear stress.

Dynamic Behavior

Viscoelastic materials are often characterized by their dynamic behavior. Because vis-coelastic materials are subject to structural breakdown when subjected to large strains, it is useful to analyze them by small amplitude sinusoidal strain. The relationship of stress and strain under these conditions can be evaluated from Figure 8–20 (Bell 1989). The applied stress is alternating at a selected frequency and is expressed in cycles s^{-1}, or ω in radians s^{-1}. The response of a purely elastic material will show a stress and strain response that is in phase, the phase angle $\delta = 0°$. A purely viscous material will show the stress being out of phase by $90°$, and a viscoelastic material shows intermediate behavior, with δ between $0°$ and $90°$. The viscoelastic dynamic response is composed of an in-phase component ($\sin \omega t$) and an out-of-phase component ($\cos \omega t$). The energy used for the viscous component is lost as heat; that used for the elastic component is retained as stored energy. This results in two moduli, the storage modulus (G') and the loss modulus (G''). The ratio of the two moduli is known as tan δ and is given by tan $\delta = G''/G'$.

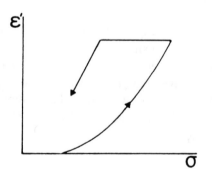

Figure 8–19 Shear Stress–Rate-of-Shear Diagram of a Thixotropic Body. *Source:* From J.M. deMan amd F.W. Wood, Hardness of Butter. II. Influence of Setting, *J. Dairy Sci.* Vol. 42, pp. 56–61, 1959.

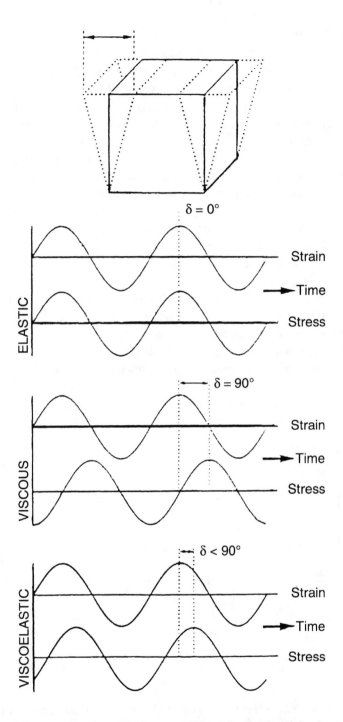

δ = 0°

Strain

Time

Stress

ELASTIC

δ = 90°

Strain

Time

Stress

VISCOUS

δ < 90°

Strain

Time

Stress

VISCOELASTIC

Figure 8–20 Dynamic (Oscillation) Measurement of Viscoelastic Materials. As an oscillating strain is applied, the resulting stress values are recorded. δ is the phase angle and its value indicates whether the material is viscous, elastic, or viscoelastic. *Source*: Reprinted from A.E. Bell, Gel Structure and Food Biopolymers, in *Water and Food Quality*, T.M. Hardman, ed., p. 253, © 1989, Aspen Publishers, Inc.

APPLICATION TO FOODS

Many of the rheological properties of complex biological materials are time-dependent, and Mohsenin (1970) has suggested that many foods can be regarded as viscoelastic materials. Many foods are disperse systems of interacting nonspherical particles and show thixotropic behavior. Such particles may interact to form a three-dimensional network that imparts rigidity to the system. The interaction may be the result of ionic forces in aqueous systems or of hydrophobic or van der Waals interactions in systems that contain fat crystals in liquid oil (e.g., butter, margarine, and shortening). Mechanical action, such as agitation, kneading, or working results in disruption of the network structure and a corresponding loss in hardness. When the system is then left undisturbed, the bonds between particles will reform and hardness will increase with time until maximum hardness is reached. The nature of thixotropy was demonstrated with butter by deMan and Wood (1959). Hardness of freshly worked butter was determined over a period of three weeks (Figure 8–21). The same butter was frozen and removed from frozen storage after three weeks. No thixotropic change had occurred with the frozen sample. The freezing had completely immobilized the crystal particles. Thixotropy is important in many food products; great care must be exercised that measurements are not influenced by thixotropic changes.

The viscosity of Newtonian liquids can be measured simply, by one-point determinations with viscometers, such as rotational, capillary, or falling ball viscometers. For non-Newtonian materials, measurement of

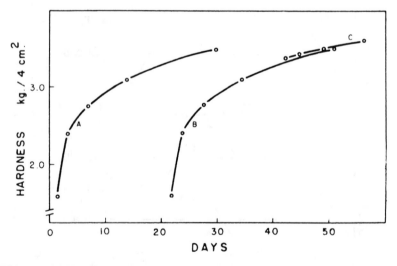

Figure 8–21 Thixotropic Hardness Change in Butter. (A) Freshly worked butter left undisturbed for four weeks at 5°C. (B) The same butter stored at –20°C for three weeks then left at 5°C. (C) The same butter left at 5°C for three weeks, then frozen for three weeks and again placed at 5°C. *Source*: From J.M. deMan and F.W. Wood, Hardness of Butter. II. Influence of Setting, *J. Dairy Sci.*, Vol. 42, pp. 56–61, 1959.

rheological properties is more difficult because single-point determinations (i.e., at one single shearing stress) will yield no useful information. We can visualize the rate of shear dependence of Newtonian fluids by considering a diagram of two fluids, as shown in Figure 8–22 (Sherman 1973). The behavior of these fluids is represented by two straight lines parallel to the shear-rate axis. With non-Newtonian fluids, a situation as shown in Figure 8–23 may arise. The fluids 3 and 4 have curves that intersect. Below this point of intersection, fluid 4 will appear more viscous; beyond the intersection, fluid 3 will appear more viscous. Fluids 5 and 6 do not intersect and the problem does not arise. In spite of the possibility of such problems, many practical applications of rheological measurements of non-Newtonian fluids are carried out at only one rate of shear. Note that results obtained in this way should be interpreted with caution. Shoemaker et al.

(1987) have given an overview of the application of rheological techniques for foods.

Probably the most widely used type of viscometer in the food industry is the Brookfield rotational viscometer. An example of this instrument's application to a non-Newtonian food product is given in the work of Saravacos and Moyer (1967) on fruit purees. Viscometer scale readings were plotted against rotational speed on a logarithmic scale, and the slope of the straight line obtained was taken as the exponent n in the following equation for pseudoplastic materials:

$$\tau = K\dot{\gamma}^{n}$$

where

τ = shearing stress (dyne/cm^2)
K = constant
$\dot{\gamma}$ = shear rate (s^{-1})

The instrument readings were converted into shear stress by using an oil of known viscos-

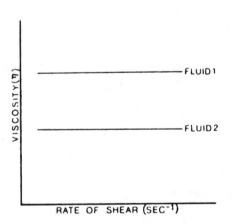

Figure 8–22 Rate of Shear Dependence of the Viscosity of Two Newtonian Fluids. *Source*: From P. Sherman, Structure and Textural Properties of Foods, in *Texture Measurement of Foods*, A. Kramer and A.S. Szczesniak, eds., 1973, D. Reidel Publishing Co.

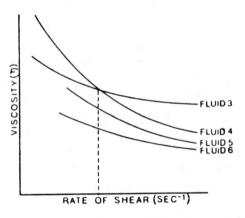

Figure 8–23 Rate of Shear Dependence of the Apparent Viscosity of Several Non-Newtonian Fluids. *Source*: From P. Sherman, Structure and Textural Properties of Foods, in *Texture Measurment of Foods*, A. Kramer and A.S. Szczesniak, eds., 1973, D. Reidel Publishing Co.

ity. The shear rate at a given rotational speed N was calculated from

$$\dot{\gamma} = 4\pi N/n$$

When shear stress τ was plotted against shear rate γ on a double logarithmic scale, the intercept of the straight line on the τ axis at $\gamma = 1\ s^{-1}$ was taken as the value of the constant K. The apparent viscosity μ_{app} at a given shear rate was then calculated from the equation

$$\mu_{app} = K\dot{\gamma}^{n-1}$$

Apparent viscosities of fruit purees determined in this manner are shown in Figure 8–24.

Factors have been reported in the literature (Johnston and Brower 1966) for the conversion of Brookfield viscometer scale readings to yield value or viscosity. Saravacos (1968) has also used capillary viscometers for rheological measurements of fruit purees.

For products not sufficiently fluid to be studied with viscometers, a variety of texture-measuring devices is available. These range from simple penetrometers such as the Magness-Taylor fruit pressure tester to complex universal testing machines such as the Instron. All these instruments either apply a known and constant stress and measure deformation or cause a constant deformation and measure stress. Some of the more sophisticated instruments can do both. In the Instron Universal Testing Machine, the crosshead moves at a speed that can be selected by changing gears. The drive is by rotating screws, and the force measurement is done with load cells. Mohsenin (1970) and coworkers have developed a type of universal testing machine in which the movement is achieved by air pressure. The Kramer shear

press uses a hydraulic system for movement of the crosshead.

Texture-measuring instruments can be classified according to their use of penetration, compression, shear, or flow.

Penetrometers come in a variety of types. One of the most widely used is the Precision penetrometer, which is used for measuring consistency of fats. The procedure and cone dimensions are standardized and described in the Official and Tentative Methods of the American Oil Chemists' Society. According to this method, the results are expressed in mm/10 of penetration depth. Haighton (1959) proposed the following formula for the conversion of depth of penetration into yield value:

$$C = KW/p^{1.6}$$

where
C = yield value
K = constant depending on the angle of the cone
p = penetration depth
W = weight of cone

Vasic and deMan (1968) suggested conversion of the depth of penetration readings into hardness by using the formula

$$H = G/A$$

where
H = hardness
G = total weight of cone assembly
A = area of impression

The advantage of this conversion is that changes in hardness are more uniform than changes in penetration depth. With the latter, a difference of an equal number of units at

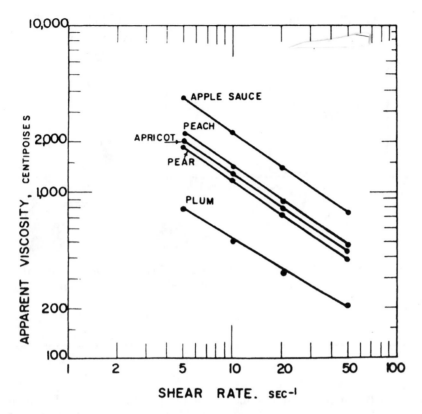

Figure 8–24 Apparent Viscosities of Fruit Purees Determined at 86°C. *Source*: From G.D. Saravacos and J.C. Moyer, Heating Rates of Fruit Products in an Agitated Kettle, *Food Technol.*, Vol. 21, pp. 372–376, 1967.

the tip of the cone and higher up on the cone is not at all comparable.

Many penetrometers use punches of various shapes and sizes as penetrating bodies. Little was known about the relationship between shape and size and penetrating force until Bourne's (1966) work. He postulated that when a punch penetrates a food, both compression and shear occur. Shear, in this case, is defined as the movement of interfaces in opposite directions. Bourne suggested that compression is proportional to the area under the punch and to the compressive strength of the food and also that the shear force is proportional to the perimeter of the punch and to the shear strength of the food (Figure 8–25). The following equation was suggested:

$$F = K_c A + K_s P + C$$

where

F = measured force
K_c = compression coefficient of tested food
K_s = shear coefficient of tested food
A = area of punch
P = perimeter of punch
C = constant

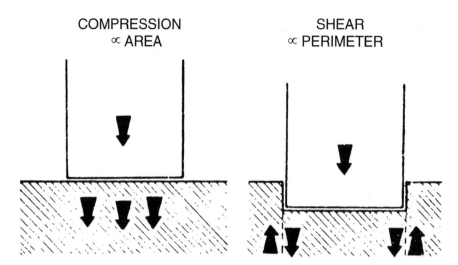

Figure 8–25 Compression and Shear Components in Penetration Tests. *Source*: From M.C. Bourne, Measure of Shear and Compression Components of Puncture Tests, *J. Food Sci.*, Vol. 31, pp. 282–291, 1966.

The relationship between penetration force and cross-sectional area of cylindrical punches has been established by Kàmel and deMan (1975).

Bourne did show that, for a variety of foods, the relationships between punch area and force and between punch perimeter and force were represented by straight lines. DeMan (1969) later showed that for certain products, such as butter and margarine, the penetrating force was dependent only on area and was not influenced by perimeter. deMan suggested that in such products flow is the only factor affecting force readings. It appears that useful conclusions can be drawn regarding the textural characteristics of a food by using penetration tests.

A variation on the penetration method is the back extrusion technique, where the sample is contained in a cylinder and the penetrating body leaves only a small annular gap for the product to flow. The application of the back extrusion method to non-Newtonian fluids has been described by Steffe and Osorio (1987).

Many instruments combine shear and compression testing. One of the most widely used is the Kramer shear press. Based on the principle of the shear cell used in the pea tenderometer, the shear press was designed to be a versatile and widely applicable instrument for texture measurement of a variety of products. The shear press is essentially a hydraulically driven piston, to which the standard 10-blade shear cell or a variety of other specialized devices can be attached. Force measurement is achieved either by a direct reading proving ring or by an electronic recording device. The results obtained with the shear press are influenced by the weight of the sample and the speed of the crosshead. These factors have been exhaustively studied by Szczesniak et al. (1970). The relationship between maximum force values and sample weight was found to be different for different foods. Products fitted into three categories—those having a constant force-to-weight ratio (e.g., white bread, sponge cake); those having a continuously decreasing force-to-weight ratio (e.g., raw apples, cooked white beans); and those giving a constant force, independent of sample weight beyond a certain fill level (e.g., canned beets, canned and frozen peas). This is demonstrated by the curves of Figure 8–26. Some of the attachments to the shear press are the succulometer cell, the single-blade meat shear cell, and the compression cell.

Based on the Szczesniak classification of textural characteristics, a new instrument was developed in the General Foods Research Laboratories; it is called the General Foods Texturometer. This device is an improved version of the MIT denture tenderometer (Proctor et al. 1956). From the reciprocating motion of a deforming body on the sample, which is contained in a tray provided with strain gages, a force record called a texture profile curve (Figure 8–27) is obtained. From this texturometer curve, a variety of rheological parameters can be obtained. Hardness is measured from the height of the first peak. Cohesiveness is expressed as the ratio of the areas under the second and first peaks. Elasticity is measured as the difference between

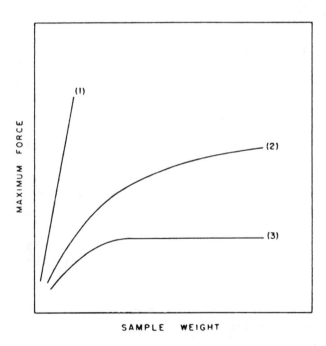

Figure 8–26 Effect of Sample Weight on Maximum Force Registered with the Shear Press and Using the 10-Blade Standard Cell. (1) White bread and sponge cake, (2) raw apples and cooked white beans, (3) canned beets and peas and frozen peas. *Source*: From A.S. Szczesniak, Instrumental Methods of Texture Measurements, in *Texture Measurement of Foods*, A. Kramer and A.S. Szczesniak, eds., 1973, D. Reidel Publishing Co.

distance *B*, measured from initial sample contact to sample contact on the second "chew," and the same distance (distance B) measured with a completely inelastic material such as clay. Adhesiveness is measured as the area of the negative peak A_3 beneath the baseline. In addition, other parameters can be derived from the curve such as brittleness, chewiness, and gumminess.

TEXTURAL PROPERTIES OF SOME FOODS

Meat Texture

Meat texture is usually described in terms of tenderness or the lack of it—toughness. This obviously is related to the ease with which a piece of meat can be cut with a knife or with the teeth. The oldest and most widely used device for measuring meat tenderness is the Warner-Bratzler shear device (Bratzler 1932). In this device, a cylindrical core of cooked meat is subjected to the shearing action of a steel blade and the maximum force is indicated by a springloaded mechanism. A considerable improvement was the shear apparatus described by Voisey and Hansen (1967). In this apparatus, the shearing force is sensed by a strain gage transducer and a complete shear-force time curve is recorded on a strip chart. The Warner-Bratzler shear method has several disadvantages. It is very difficult to obtain uniform meat cores. Cores from different positions in one cut of meat may vary in tenderness, and cooking method may affect tenderness.

Meat tenderness has been measured with the shear press. This can be done with the 10-blade universal cell or with the single-

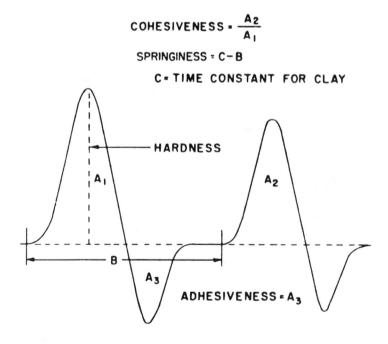

Figure 8–27 Typical Texturometer Curve

blade meat shear attachment. There is no standard procedure for measuring meat tenderness with the shear press; sample size, sample preparation, and rate of shear are factors that may affect the results.

A pressure method for measuring meat tenderness has been described by Sperring et al. (1959). A sample of raw meat is contained in a cylinder that has a small hole in its bottom. A hydraulic press forces a plunger into the cylinder, and the pressure required to squeeze the meat through the hole is taken as a measure of tenderness.

A portable rotating knife tenderometer has been described by Bjorksten et al. (1967). A rotating blunt knife is forced into the meat sample, and a tracing of the area traversed by a recording pen is used as a measure of tenderness.

A meat grinder technique for measuring meat tenderness was reported by Miyada and Tappel (1956); in this method, power consumption of the meat grinder motor was used as a measure of meat tenderness. The electronic recording food grinder described by Voisey and deMan (1970) measures the torque exerted on a strain gage transducer. This apparatus has been used successfully for measuring meat tenderness.

Other methods used for meat tenderness evaluation have included measurement of sarcomere length (Howard and Judge 1968) and determination of the amount of connective tissue present.

Stoner et al. (1974) have proposed a mechanical model for postmortem striated muscle; it is shown in Figure 8–28. The model is a combination of the Voigt model with a four-element viscoelastic model. The former includes a contractile element (CE), which is the force generator. The element SE is a spring that is passively elongated by the shortening of the CE and thus develops an internal force. The parallel elastic component (PE) contributes to the resting tension of the muscle. The combination of elements PE, CE, and SE represents the purely elastic properties of the muscle as the fourth component of a four-element model (of which E_2, η_3, and η_2 are the other three components).

Dough

The rheological properties of dough are important in determining the baking quality of flour. For many years the Farinograph was used to measure the physical properties of dough. The Farinograph is a dough mixer hooked up to a dynamometer for recording

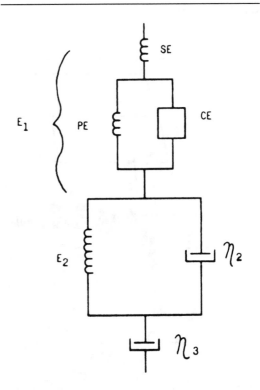

Figure 8–28 Mechanical Model for Postmortem Striated Muscle. *Source:* From C.W. Brabender Instruments, Inc., South Hackensack, New Jersey.

torque. The instrument can measure water absorption of flour. A typical Farinograph curve is presented in Figure 8–29. The amount of water required to bring the middle of the curve to 500 units, added from a buret, is a measure of the water absorption of the dough. The measurement from zero to the point where the top of the curve first intersects the 500 line on the chart is called arrival time; the measurement from zero to maximum consistency is called peak time; the point where the top of the curve leaves the 500 line is called departure time; the difference between departure and arrival time is stability. The elasticity of dough is measured with the extensigraph, which records the force required to stretch a piece of dough of standard dimensions.

The peculiar viscoelastic properties of wheat dough are the result of the presence of a three-dimensional network of gluten proteins. The network is formed by thiol-disulfide exchange reactions among gluten proteins. Peptide disulfides can interfere in a thiol-disulfide exchange system by reacting with a protein (PR)-thiol to liberate a peptide (R)-thiol and form a mixed disulfide, as follows:

$$PR\text{--}SH + R\text{--}SS\text{--}R \rightarrow R\text{--}H + PR\text{--}SS\text{--}R$$

Disulfide bonds between proteins have an energy of 49 kcal/mole and are not broken at room temperature except as the result of a chemical reaction. The effects of oxidizing agents on the rheological properties of dough

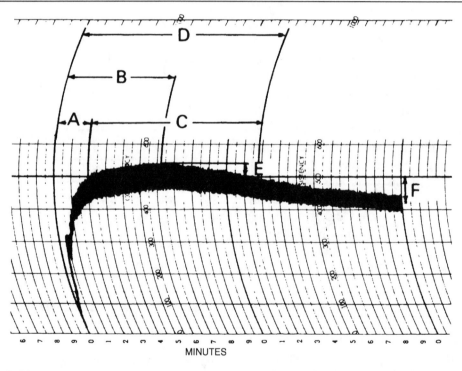

Figure 8–29 Typical Farinograph Curve. (A) Arrival time, (B) peak time, (C) stability, (D) departure, (E) mixing tolerance index, (F) 20-minute drop. *Source*: From H.P. Wehrli and Y. Pomeranz, The Role of Chemical Bonds in Dough, *Baker's Digest*, Vol. 43, no. 6, pp. 22–26, 1969.

may be quantitatively explained as the breaking of disulfide cross-links; their reformation may be explained as exchange reactions with sulfhydryl groups (Wehrli and Pomeranz 1969). The baking quality of wheat is strongly influenced by protein content and the disulfide/sulfhydryl ratio. A schematic diagram of the bonds within and between polypeptide chains in dough is given in Figure 8–30.

Fats

Consistency of fats is commonly determined with the cone penetrometer, as specified in the Official and Tentative Methods of the American Oil Chemists' Society (Method Cc 16-60). Other methods that have frequently been employed involve extrusion; they include the extrusion attachment to the shear press (Vasic and deMan 1967), an extrusion rheometer used with the Instron universal testing machine (Scherr and Wittnauer 1967), and the FIRA-NIRD extruder (Prentice 1954).

Other devices used for fat consistency measurements include wire-cutting instruments (sectilometers), penetration of a probe when the sample is contained in a small cup, and compression of cylindrical samples between two parallel plates. The compression method reveals detailed information about plastic fats (deMan et al., 1991) such as elasticity, viscous flow, and degree of brittleness. These characteristics are important in shortenings destined for cakes and puff pastries. Compression curves of a variety of shortenings are displayed in Figure 8–31. Temperature treatment of a fat has profound effect on its texture. deMan and deMan (1996) studied the effect of crystallization temperature and tempering temperature on the texture of palm oil and hydrogenated fats using the compression method and found that lowering the crystallization temperature from 10 to 0°C resulted in softer texture, especially for palm oil. Increasing the tempering temperature from 25° to 30°C also resulted in softer texture, especially for hydrogenated fats.

The hardness or consistency of fats is the result of the presence of a three-dimensional network of fat crystals. All fat products such

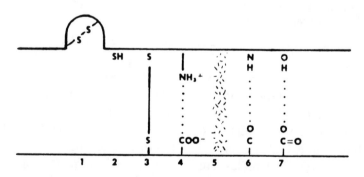

Figure 8–30 Schematic Diagram of Bonds Within and Between Polypeptide Chains in Dough. Solid lines represent covalent bonds, dotted lines other bonds. (1) Intramolecular disulfide bond, (2) free sulfhydryl group, (3) intermolecular disulfide bond, (4) ionic bond, (5) van der Waals bond, (6) interpeptide hydrogen bond, (7) side chain hydrogen bond. *Source:* From A.H. Bloksma, Rheology of Wheat Flour Dough, *J. Texture Studies*, Vol. 3, pp. 3–17, 1972.

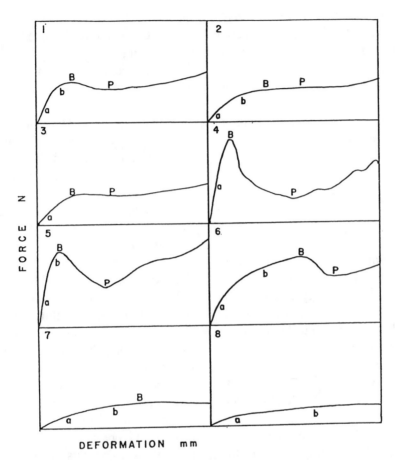

Figure 8–31 Examples of Compression Curves of Shortenings. (1) and (2) soy-palm; (3) soy-canola-palm; (4) soy only; (5) tallow-lard; (6) lard; (7) palm-vegetable; (8), palm-palm kernel. a = elastic non-recoverable deformation; b = viscous flow; B = breaking force; P = plateau force; distance between B and P is an indication of brittleness.

as margarine, shortening, and butter are mixtures of solid fat in crystallized form and liquid oil. Because the individual glycerides in fats have a wide range of melting points, the ratio of solid to liquid fat is highly temperature dependent. The crystal particles are linked by weak van der Waals forces. These bonds are easily broken by mechanical action during processing, and the consistency may be greatly influenced by such mechanical forces. After a rest period, some

of these bonds are reformed and the reversible sol-gel transformation taking place is called thixotropy. There appear to be two types of bonds in fats—those that are reformed after mechanical action, and those that do not reform. The latter result in a portion of the hardness loss that is irreversible. The nature of these bonds has not been established with certainty, but it is assumed to mainly involve van der Waals forces. The hardness loss of fats as a result of working is called work softening and can be expressed as follows:

$$WS = \frac{H_o - H_w}{H_o} \times 100\%$$

where H_o and H_w are the hardness before and after working.

The work softening is influenced not only by the nature of the mechanical treatment but also by temperature conditions and the size and quantity of fat crystals.

Tanaka et al. (1971) have used a two-element mechanical model (Figure 8–32) to represent fats as viscoplastic materials. The model consists of a dashpot representing the viscous element in parallel with a friction element that represents the yield value.

The theory of bond formation between the crystal particles in plastic fats needs revision. Recently, it has been proposed that a process of "sintering"—the formation of solid bridges between fat crystals—occurs during postcrystallization hardening (Heertje et al., 1987; Johansson and Bergenstahl 1995). The use of the word *sintering* is unfortunate since it describes the fusion of small particles into a solid block; this is not the case, however, because the fat crystals in fats can be suspended in a solvent such as isobutanol (Chawla and deMan 1990) without showing any sign of fusion into larger aggregates. At the solid fat level found in plastic fats (15 to 35 percent), the crystals are tightly packed together and exist in a state of entanglement (deMan 1997). Entanglement of crystals is a more realistic description of the structure formation in fats. The sintering process described above can be considered a form of entanglement.

Many interrelated factors influence the texture of plastic fats. Fatty acid and glyceride composition are basic factors in establishing the properties of a fat. These factors, in turn, are related to solid fat content, crystal size and shape, and polymorphic behavior. Once the crystal network is formed, mechanical treatment and temperature history may influence the texture.

The network systems in plastic fats differ from those in protein or carbohydrate systems. Fat crystals are embedded in liquid oil and the crystals have no ionized groups. Therefore, the interactive forces in fat crystal networks are low. The minimum concentration of solid particles in a fat to provide a yield value is in the range of 10 to 15 percent.

Fruits and Vegetables

Much of the texture work with fruits and vegetables has been done with the Kramer shear press. The shear press was developed because of the tenderometer's limitations and has been widely used for measuring tender-

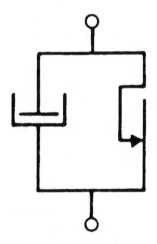

Figure 8–32 Mechanical Model for Foods as Viscoplastic Materials. *Source*: From M. Tanaka, et al., Measurement of Textural Properties of Foods with a Constant Speed Cone Penetrometer, *J. Texture Studies*, Vol. 2, pp. 301–315, 1971.

ness of peas for processing (Kramer 1961). The shear press is also used, for example, in the quality method suggested for raw and canned sweet corn (Kramer and Cooler 1962). This procedure determines shear force with the standard shear cell and amount of juice pressed out with the juice extraction cell. It is possible to relate quality of the corn to these parameters. In addition to the shear press, the Instron universal testing machine and others based on the same principle are popular for fruit and vegetable products. A special testing machine has been developed by Mohsenin (1970). This machine uses an air motor for movement of the crosshead but is otherwise similar to other universal testing machines. A mechanically driven test system has been developed by Voisey (1971). Voisey (1970) has also described a number of test cells that are simpler in design than the standard shear cell of the shear press.

The texture of fruit and vegetable products is related to the cellular structure of these materials. Reeve and Brown (1968a,b) studied the development of cellular structure in the green bean pod as it relates to texture and eating quality. Sterling (1968) studied the effect of solutes and pH on the structure and firmness of cooked carrot. Sterling also related histological changes such as cellular separation and collapse to the texture of the product. In fruits and vegetables the relationship between physical structure and physical properties is probably more evident than in many other products.

Morrow and Mohsenin (1966) have studied the physical properties of a variety of vegetables; they assumed these products to behave as viscoelastic materials and to behave according to the three-element model represented in Figure 8–33A. Such viscoelastic materials are characterized by the strain-time and stress-time relationships, as given in Figure 8–33B,C.

Starch

The texture of starch suspensions is determined by the source of the starch, the chemical and/or physical modification of the starch granule, and the cooking conditions of the starch (Kruger and Murray 1976). The texture of starch suspensions is measured by means of the viscoamylograph. The viscosity is recorded while the temperature of the suspension is raised from 30° to 95°C, held at 95°C for 30 minutes, lowered to 25°C, and held at that temperature for 30 minutes The viscosity of a 5 percent suspension of waxy corn is shown in Figure 8–34. Initially the viscosity is low, but it increases rapidly at the gelatinization temperature of about 73°C. As the granules swell, they become weaker and start to disintegrate causing the viscosity to drop. When the temperature is lowered to 25°C, there is another increase in viscosity caused by the interaction of the broken and deformed granules. This phenomenon is demonstrated by the width and irregularity of the recorded line, which is indicative of the cohesiveness of the starch particles.

Modification of the starch has a profound effect on the texture of the suspensions. Introduction of as little as 1 cross-bond per 100,000 glucose units slows the breakdown of the swollen granules during and after cooking (Figure 8–35). This results in a higher final viscosity. Increasing the cross-bonding to 1, 3, or 6 cross-bonds per 10,000 glucose units will result in no breakdown during the heating cycle (Figure 8–36). As the cross-bonding increases, the granule is strengthened and does not swell much during heating, but viscosity is decreased. Most food starches used at pH values of 4 to 8

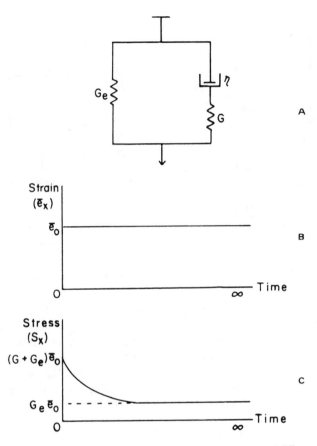

Figure 8–33 Mechanical Model Proposed by Morrow and Mohsenin. (A) Viscoelastic foods, (B) the strain time, (C) stress time, characteristics of the system.

have 2 to 3 cross-bonds per 10,000 glucose units.

Waxy corn starch contains only amylopectin; corn starch contains both amylopectin and amylose. This results in a different viscosity profile (Figure 8–37). Corn starch shows a lower peak viscosity and less breakdown during heating. After cooling, the viscosity continues to increase, probably because the amylose interlinks with the amylopectin. On further storage at 25°C, the slurry sets to a firm gel. Tapioca starch is intermediate between corn and waxy corn starch (Figure 8–37). This is explained by the

fact that tapioca amylose molecules are larger than those in corn starch.

Starches can be substituted by nonionic or ionic groups. The latter can be made anionic by introduction of phosphate or succinate groups. These have lower gelatinization temperature, higher peak viscosity, and higher final cold viscosity than nonionic starches (Figure 8–38).

MICROSTRUCTURE

With only a few exceptions, food products are non-Newtonian and possess a variety of

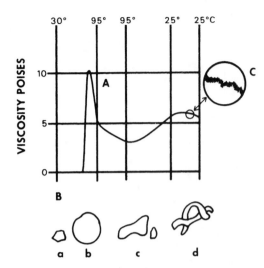

Figure 8–34 Viscosity and Granule Appearance in the Viscoamylograph Test of a 5% Suspension of Waxy Corn in Water. A = viscosity curve, B = granule shape and size, C = magnified portion of curve to indicate cohesiveness, a = unswollen granule, b = swollen granule, c = collapsed granule, d = entwined collapsed granules. *Source*: Reprinted from L.H. Kruger and R. Murray, Starch Texture, in *Rheology and Texture in Food Quality*, J.M. deMan, P.W. Voisey, V.F. Raspar, and D.W. Stanley, eds., © 1976, Aspen Publishers, Inc.

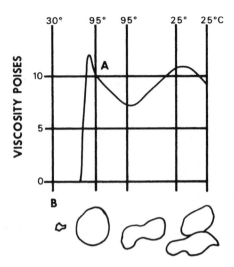

Figure 8–35 Viscosity and Granule Appearance in the Viscoamylograph Test of a 5% Suspension in Water of Waxy Corn with 1 Cross-Bond per 100,000 Glucose Units. A = viscosity curve, B = granule appearance. *Source*: Reprinted from L.H. Kruger and R. Murray, Starch Texture, in *Rheology and Texture in Food Quality*, J.M. deMan, P.W. Voisey, V.F. Raspar, and D.W. Stanley, eds., © 1976, Aspen Publishers, Inc.

internal structures. Cellular and fibrous structures are found in fruits and vegetables; fibrous structures are found in meat; and many manufactured foods contain protein, carbohydrate, or fat crystal networks.

Many of these food systems are dispersions that belong in the realm of colloids. Colloids are defined as heterogeneous or dispersed systems that contain at least two phases—the dispersed phase and the continuous phase. Colloids are characterized by their ability to exist in either the sol or the gel form. In the former, the dispersed particles exist as independent entities; in the latter, they associate to form network structures that may entrap large volumes of the continuous phase. The isothermal reversible sol-gel transformation exhibited by many foods is called thixotropy. Disperse systems can be classified on the basis of particle size. Coarse dispersions have particle size greater than 0.5 μm. They can be seen in the light microscope, can be filtered over a paper filter, and will sediment rapidly. Colloidal dispersions have particles in the range of 0.5 μm to 1 nm. These particles remain in suspension by Brownian movement and can run through a paper filter but cannot run through a membrane filter. Particles smaller than these are

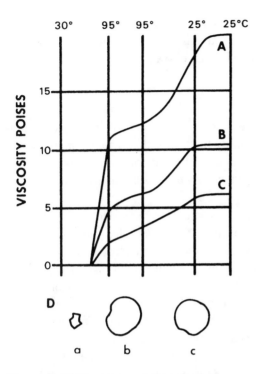

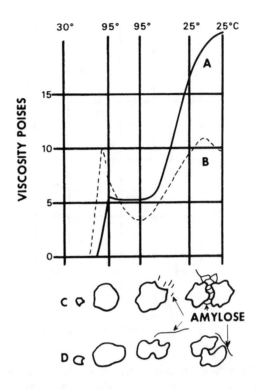

Figure 8–36 Viscosity and Granule Appearance in the Viscoamylograph of 5% Suspension in Water of Cross-Bonded Waxy Corn. A = 1 cross-bond per 10,000 glucose units, B = 3 cross-bonds, C = 6 cross-bonds, and D = granule appearance. *Source:* Reprinted from L.H. Kruger and R. Murray, Starch Texture, in *Rheology and Texture in Food Quality*, J.M. deMan, P.W. Voisey, V.F. Raspar, and D.W. Stanley, eds., © 1976, Aspen Publishers, Inc.

Figure 8–37 Viscosity and Granule Appearance in the Viscoamylograph of Suspensions in Water of Corn and Tapioca Starch. A = 6% corn starch, B = 5% tapioca starch, C = corn granule appearance, and D = tapioca granule appearance. *Source:* Reprinted from L.H. Kruger and R. Murray, Starch Texture, in *Rheology and Texture in Food Quality*, J.M. deMan, P.W. Voisey, V.F. Raspar, and D.W. Stanley, eds., © 1976, Aspen Publishers, Inc.

molecular dispersions or solutions. Depending on the nature of the two phases, disperse systems can be classified into a number of types. A solid dispersed in a liquid is called a sol; for example, margarine, which has solid fat crystals dispersed in liquid oil, is a sol. Dispersions of liquid in liquid are emulsions; many examples of these are found among foods such as milk and mayonnaise. Dispersions of gas in liquid are foams (e.g., whipped cream). In many cases, these dis-

persions are more complex than one disperse phase. Many foods have several dispersed phases. For instance, in chocolate, solid cocoa particles as well as fat crystals are dispersed phases.

The production of disperse systems is often achieved by dispersion methods in which the disperse phase is subdivided into small particles by mechanical means. Liq-

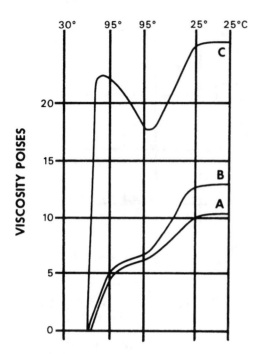

Figure 8–38 Viscoamylograph Viscosity Curves of Substituted Waxy Corn Starch. A = cross-bonded waxy corn, B = nonionic substituted cross-bonded waxy corn, and C = anionic substituted cross-bonded waxy corn. *Source*: Reprinted from L.H. Kruger and R. Murray, Starch Texture, in *Rheology and Texture in Food Quality*, J.M. deMan, P.W. Voisey, V.F. Raspar, and D.W. Stanley, eds., © 1976, Aspen Publishers, Inc.

uids are emulsified by stirring and homogenization; solids are subdivided by grinding, as, for instance, roller mills are used in chocolate making and colloid mills are used in other food preparations.

An important aspect of the subdivision of the disperse phase is the enormous increase in specific surface area. If a sphere with a radius $R = 1$ cm is dispersed into particles with radius $r = 10^{-6}$ cm, the area of the interface will increase by a factor of 10^6. The mechanical work dA needed to increase the interfacial area is proportional to the area increase, as follows:

$$dA = \sigma dO$$

where O = total interfacial area. The proportionality factor σ is the surface tension. In the production of emulsions, the surface tension is reduced by using surface active agents (see Chapter 2).

As particle size is reduced to colloidal dimensions, the particles are subject to Brownian movement. Brownian movement is the result of the random thermal movement of molecules, which impact on colloidal particles to give them a random movement as well (Figure 8–39). The size of dispersed particles has a profound effect on the properties of dispersions (Schubert 1987). Figure 8–40 shows the qualitative relationship of particle size and system properties. As particle size decreases, fracture resistance increases. The particles become increasingly uniform, which results in a grinding limit below which particles cannot be further reduced in size. The terminal settling rate, illustrated by a flour particle falling through the air, increases rapidly as a function of increasing particle size. According to Schubert (1987), a flour particle of 1 μm in size takes more than 6 hours to fall a distance of 1 meter in still air. Wetting becomes more difficult as size decreases. The specific surface area (the surface per unit volume) increases rapidly with decreasing particle size.

Colloidal systems, because of their large number of dispersed particles, show non-Newtonian flow behavior. For a highly dilute dispersion of spherical particles, the following equation has been proposed by Einstein:

$$\eta = \eta_o (1 + 2.5 \, \phi)$$

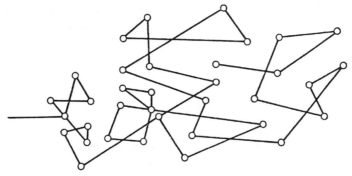

Figure 8–39 Flat Plane Projection of the Location of a Colloidal Particle Subject to Brownian Movement. *Source*: From H. Schubert, Food Particle Technology. Part 1: Properties of Particles and Particulate Food Systems, *J. Food Eng.*, Vol. 6, pp. 1–32, 1987, Elsevier Applied Science Publishers, Ltd.

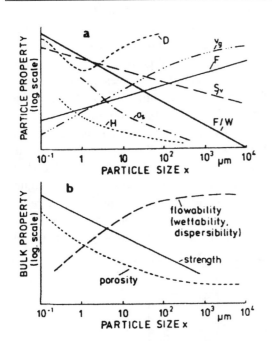

Figure 8–40 Relationship Between Particle Size and System Properties. D = particle deposition in fibrous fillers, F = adhesion force, H = homogeneity of a particle, S_v = surface area per unit volume, W = particle weight, Vg = terminal setting rate, σ_s = particle fracture resistance. *Source*: From H. Schubert, Food Particle Technology. Part 1: Properties of Particles and Particulate Food Systems, *J. Food Eng.*, Vol. 6, pp. 1–32, 1987, Elsevier Applied Science Publishers, Ltd.

where

η_o = viscosity of the continuous phase

ϕ = ratio of volumes of disperse and continuous phases

In this equation, viscosity is independent of the degree of dispersion. As soon as the ratio of disperse and continuous phases increases to the point where particles start to interact, the flow behavior becomes more complex. The effect of increasing the concentration of the disperse phase on the flow behavior of a disperse system is shown in Figure 8–41. The disperse phase, as well as the low solids dispersion (curves 1 and 2), shows Newtonian flow behavior. As the solids content increases, the flow behavior becomes non-Newtonian (curves 3 and 4). Especially with anisotropic particles, interaction between them will result in the formation of three-dimensional network structures. These network structures usually show non-Newtonian flow behavior and viscoelastic properties and often have a yield value. Network structure formation may occur in emulsions (Figure 8–42) as well as in particulate systems. The forces between particles that result in the formation of networks may be

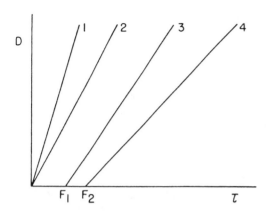

Figure 8–41 Effect of Increasing the Concentration of the Disperse Phase on the Flow Behavior of a Disperse System. 1—continuous phase, 2—low solids content, 3—medium solids content, 4—high solids content.

van der Waals forces, hydrophobic interactions, or covalent bonds. Network formation may result from heating or from chemical reactions that occur spontaneously either from components already present in the food or from added enzymes or coagulants. The formation of networks requires a minimal fraction of particles to be present, the critical fraction α_c, and the larger the number of sites f, used for bond formation, the sooner a network is formed. These two quantities are related as follows:

$$\alpha_c = 1/(f-1)$$

At particle concentrations below 10 percent, numerous contact points are required to form a network structure (Table 8–3). This means that only certain types of molecules or particles can form networks at this concentration. As the network is formed, the viscosity increases until, at a certain point, the product acquires plastic and/or viscoelastic properties. Network formation thus depends on particle concentration, reactive sites on the particles, and particle size and shape. Heertje et al. (1985) investigated structure formation in acid milk gels and found that the final texture of the products was influenced by many factors including heat, salt balance, pH, culture, and thickening agents. Structure formation in soy milk, induced by coagulants in the form of calcium or magnesium salts, results in a semisolid food called tofu, which has a fine internal protein network structure (Figure 8–43). Hermansson

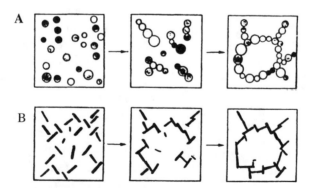

Figure 8–42 Structure Formation in Particulate Systems. (A) Flocculation of an emulsion. (B) Network formation in crystallized fat.

Table 8–3 Relationship Between Critical Particle Fraction (α_c) and Number of Bond Sites (f)

α_c	f
0.05	21
0.10	11
0.15	8
0.20	6
0.30	4
0.50	3

and Larsson (1986) reported on the structure of gluten gels and concluded that these gels consist of a continuous phase of densely packed protein units.

DeMan and Beers (1987) have reviewed the factors that influence the formation of three-dimensional fat crystal networks. The fat crystal networks in plastic fats (Figure 8–44) are highly thixotropic, and mechanical action on these products will result in a drastic reduction of hardness.

A variety of rheological tests can be used to evaluate the nature and properties of different network structures in foods. The strength of bonds in a fat crystal network can be evaluated by stress relaxation and by the decrease in elastic recovery in creep tests as a function of loading time (deMan et al. 1985). Van Kleef et al. (1978) have reported on the determination of the number of cross-links in a protein gel from its mechanical and swelling properties. Oakenfull (1984) used shear modulus measurements to estimate the size and thermodynamic stability of junction zones in noncovalently cross-linked gels.

Dynamic measurements of gels can provide information on the extent of cross-linking (Bell 1989). Systems with a relatively high storage modulus G' show a low value for G''/G', which indicates a highly cross-linked system such as an agar gel.

WATER ACTIVITY AND TEXTURE

Water activity (a_w) and water content have a profound influence on textural properties of foods. The three regions of the sorption isotherm can be used to classify foods on the basis of their textural properties (Figure 8–45). Region 3 is the high moisture area, which includes many soft foods. Foods in the intermediate moisture area (region 2) appear dry and firm. At lowest values of a_w (region 1), most products are hard and crisp (Bourne 1987).

Katz and Labuza (1981) examined the relationship between a_w and crispness in a study of the crispness of popcorn (Figure 8–46). They found a direct relationship between crispness and a_w.

Many foods contain biopolymers and low molecular weight carbohydrates. These can be present in a metastable amorphous state that is sensitive to temperature and the state

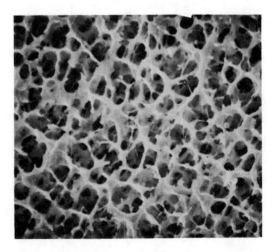

Figure 8–43 Microstructure of Soybean Curd (Tofu) as Seen in the Scanning Electron Microscope

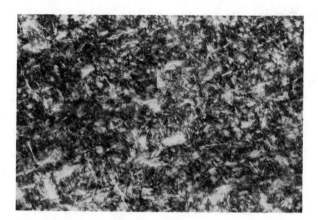

Figure 8–44 Fat Crystals in a Partially Crystallized Fat as Seen in the Polarizing Microscope

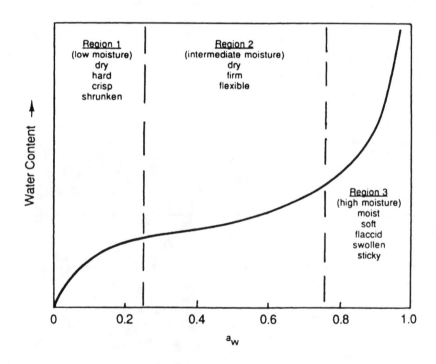

Figure 8–45 The Three Regions of the Sorption Isotherm Related to the Textural Properties of Food Systems. *Source*: Reprinted with permission from M.C. Bourne, Effects of Water Activity on Textural Properties of Food, in *Water Activity: Theory and Applications to Food*, L.B. Rockland and L.R. Beuchat, eds., p. 76, 1987, by courtesy of Marcel Dekker, Inc.

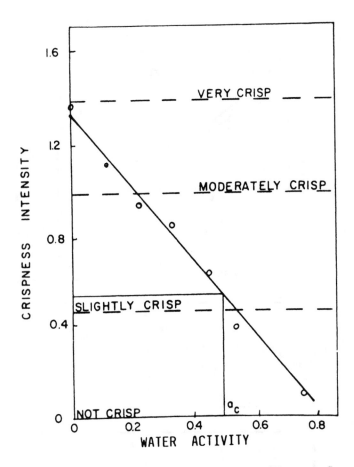

Figure 8–46 Relationship Between Water Activity and Crispness of Popcorn. *Source*: Reprinted with permission from E.E. Katz and T.P. Labuza, Effect of Water Activity on the Sensory Crispness and Mechanical Deformation of Snack Food Properties, *J. Food Sci.*, Vol. 46, p. 403, © 1981, Institute of Food Technologists.

of water. The amorphous state can be in the form of a rubbery structure or a very viscous glass, as shown in Figure 8–47 (Slade and Levine 1991; Levine and Slade 1992; Roos and Karel 1991). A more detailed analysis of the effect of temperature on textural properties expressed as modulus is presented in Figure 8–48. Below the melting temperature, the material enters a state of rubbery flow. As the temperature is lowered further, a leathery state is observed. In the leathery region the modulus increases sharply, until the glass transition temperature (T_g) is reached and the material changes to a glass.

Kapsalis et al. (1970) reported on a study of the textural properties of freeze-dried beef at different points of the moisture sorption isotherm over the complete range of water activity. Important changes in textural properties were observed at a_w values of 0.85 and at 0.15 to 0.30.

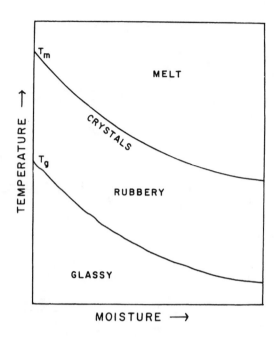

Figure 8–47 Rubbery and Glassy State of Moisture-Containing Foods as Affected by Temperature. T_m = melting point; T_g = glass transition temperature.

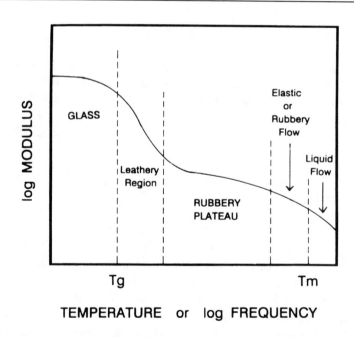

Figure 8–48 Effect of Temperature on the Texture as Expressed by Modulus. T_m = melting temperature, T_g = glass transition temperature.

REFERENCES

Bell, A.E. 1989. Gel structure and food biopolymers. In *Water and food quality*, ed. T.M. Hardman. New York: Elsevier Science.

Bjorksten, J., et al. 1967. A portable rotating knife tenderometer. *Food Technol.* 21: 84–86.

Bourne, M.C. 1966. Measure of shear and compression components of puncture tests. *J. Food Sci.* 31: 282–291.

Bourne, M.C. 1987. Effects of water activity on textural properties of food. In *Water activity: Theory and applications to food*, ed. L.B. Rockland and L.R. Beuchat. New York: Marcel Dekker.

Brandt, M.A., et al. 1963. Texture profile method. *J. Food Sci.* 28: 404–409.

Bratzler, L.J. 1932. Measuring the tenderness of meat by means of a mechanical shear. Master's thesis, Kansas State College.

Cairncross, S.E., and L.B. Sjöström. 1950. Flavor profiles—A new approach to flavor problems. *Food Technol.* 4: 308–311.

Chawla, P., and J.M. deMan. 1990. Measurement of the size distribution of fat crystals using a laser particle counter. *J. Am. Oil Chem. Soc.* 67: 329–332.

deMan, J.M. 1969. Food texture measurements with the penetration method. *J. Texture Studies* 1: 114–119.

deMan, J.M. 1997. Relationship between chemical, physical and textural properties of fats. Paper presented at International Conference on Physical Properties of Fats, Oils, Emulsifiers, September, Chicago.

deMan, J.M., and A.M. Beers. 1987. Fat crystal networks: Structure and rheological properties. *J. Texture Studies* 18: 303–318.

deMan, J.M., and F.W. Wood. 1959. Hardness of butter. II. Influence of setting. *J. Dairy Sci.* 42: 56–61.

deMan, J.M., et al. 1985. Viscoelastic properties of plastic fat products. *J. Am. Oil Chem. Soc.* 62: 1672–1675.

deMan, L., et al. 1991. Physical and textural characteristics of some North American shortenings. *J. Am. Oil Chem. Soc.* 68: 63–69.

deMan, L., and J.M. deMan. 1996. Textural measurements of some plastic fats: Oils, fats, lipids. *Proceedings of the International Society for Fat Research (ISF), the Hague 1995*, 3: 543–546. Bridgewater, UK: P.J. Barnes & Assoc.

Haighton, A.J. 1959. The measurement of hardness of margarine and fats with the penetrometers. *J. Am. Oil Chem. Soc.* 36: 345–348.

Heertje, I., et al. 1985. Structure formation in acid milk gels. *Food Microstructure* 4: 267–278.

Heertje, I., et al. 1987. Product morphology of fat products. *Food Microstructure* 6: 1–8.

Hermansson, A.M., and K. Larsson. 1986. The structure of gluten gels. *Food Microstructure* 5: 233–240.

Howard, R.D., and M.D. Judge. 1968. Comparison of sarcomere length to other predictors of beef tenderness. *J. Food Sci.* 33: 456–460.

Johansson, D., and B. Bergenstahl. 1995. Sintering of fat crystals in oil during post-crystallization processes. *J. Am. Oil Chem. Soc.* 72: 911–920.

Johnston, C.W., and C.H. Brower. 1966. Rheological and gelation measurements of plastisols. *Soc. Plastics Eng. J.* 22, no. 11: 45–52.

Kamel, B.S., and J.M. deMan. 1975. Evaluation of gelatin gel texture by penetration tests. *Lebensm. Wiss. Technol.* 8: 123–127.

Kapsalis, J.G., et al. 1970. A physico-chemical study of the mechanical properties of low and intermediate moisture foods. *J. Texture Studies* 1: 464–483.

Katz, E.E., and T.P. Labuza. 1981. Effect of water activity on the sensory crispness and mechanical deformation of snack food properties. *J. Food Sci.* 46: 403–409.

Kleinert, J. 1976. Rheology of chocolate. In *Rheology and texture in food quality*, ed. J.M deMan et al. Westport, CT: AVI Publishing Co.

Kokini, J.L. 1985. Fluid and semi-solid food texture and texture-taste interactions. *Food Technol.* 11: 86–94.

Kramer, A. 1961. The Shear-Press, a basic tool for the food technologist. *Food Scientist* 5: 7–16.

Kramer, A., and J.C. Cooler. 1962. An instrumental method for measuring quality of raw and canned sweet corn. *Proc. Am. Soc. Hort. Sci.* 81: 421–427.

Kruger, L.H., and R. Murray. 1976. Starch texture. In *Rheology and texture in food quality*, ed. J.M. deMan et al. Westport, CT: AVI Publishing Co.

Levine, H., and L. Slade. 1992. Glass transitions in foods. In *Physical chemistry of foods*, ed. H.G. Schwartzberg and R.W. Hartel. New York: Marcel Dekker.

Miyada, D.S., and A.L. Tappel. 1956. Meat tenderization. I. Two mechanical devices for measuring texture. *Food Technol.* 10: 142–145.

Mohsenin, N.N. 1970. *Physical properties of plant and animal materials*, Vol. 1, *Structure, physical characteristics and mechanical properties.* New York: Gordon and Breach Science Publisher.

Morrow, C.T., and N.N. Mohsenin. 1966. Consideration of selected agricultural products as viscoelastic materials. *J. Food Sci.* 31: 686–698.

Oakenfull, D. 1984. A method for using measurements of shear modulus to estimate the size and thermodynamic stability of junction zones in non-covalently cross linked gels. *J. Food Sci.* 49: 1103–1110.

Prentice, J.H. 1954. An instrument for estimating the spreadability of butter. *Lab. Practice* 3: 186–189.

Proctor, B.E., et al. 1956. A recording strain-gage denture tenderometer for foods. II. Studies on the masticatory force and motion and the force penetration relationship. *Food Technol.* 10: 327–331.

Reeve, R.M., and M.S. Brown. 1968a. Histological development of the bean pod as related to culinary texture. 1. Early stages of pod development. *J. Food Sci.* 33: 321–326.

Reeve, R.M., and M.S. Brown. 1968b. Histological development of the bean pod as related to culinary texture. 2. Structure and composition of edible maturity. *J. Food Sci.* 33: 326–331.

Roos, Y., and M. Karel. 1991. Plasticizing effect of water on thermal behavior and crystallization of amorphous food models. *J. Food Sci.* 56: 38–43.

Saravacos, G.D. 1968. Tube viscometry of fruit purees and juices. *Food Technol.* 22: 1585–1588.

Saravacos, G.D., and J.C. Moyer. 1967. Heating rates of fruit products in an agitated kettle. *Food Technol.* 21: 372–376.

Scherr, H.J., and L.P. Wittnauer. 1967. The application of a capillary extrusion rheometer to the determination of the flow characteristics of lard. *J. Am. Oil Chem. Soc.* 44: 275–280.

Schubert, H. 1987. Food particle technology. Part 1: Properties of particles and particulate food systems. *J. Food Eng.* 6: 1–32.

Sherman, P. 1969. A texture profile of foodstuffs based upon well-defined rheological properties. *J. Food Sci.* 34: 458–462.

Sherman, P. 1973. Structure and textural properties of foods. In *Texture measurement of foods*, ed. A. Kramer and A.S. Szczesniak. Dordrecht, The Netherlands: D. Reidel Publishing Co.

Shoemaker, C.F., et al. 1987. Instrumentation for rheological measurements of food. *Food Technol.* 41, no. 3: 80–84.

Slade, L., and H. Levine. 1991. A food polymer science approach to structure-property relationships in aqueous food systems. In *Water relationships in foods*, ed. H. Levine and L. Slade. New York: Plenum Press.

Sperring, D.D., et al. 1959. Tenderness in beef muscle as measured by pressure. *Food Technol.* 13: 155–158.

Steffe, J.F., and F.A. Osorio. 1987. Back extrusion of non-Newtonian fluids. *Food Technol.* 41, no. 3: 72–77.

Sterling, C. 1968. Effect of solutes and pH on the structure and firmness of cooked carrot. *J. Food Technol.* 3: 367–371.

Stoner, D.L., et al. 1974. A mechanical model for postmortem striated muscle. *J. Texture Studies* 4: 483–493.

Szczesniak, A.S. 1963. Classification of textural characteristics. *J. Food Sci.* 28: 385–389.

Szczesniak, A.S., and D.H. Kleyn. 1963. Consumer awareness of texture and other food attributes. *Food Technol.* 17: 74–77.

Szczesniak, A.S., et al. 1970. Behavior of different foods in the standard shear compression cell of the shear press and the effect of sample weight on peak area and maximum force. *J. Texture Studies* 1: 356–378.

Tanaka, M., et al. 1971. Measurement of textural properties of foods with a constant speed cone penetrometer. *J. Texture Studies* 2: 301–315.

Van Kleef, F.S.M., et al. 1978. Determination of the number of crosslinks in a protein gel from its mechanical and swelling properties. *Biopolymers* 17: 225–235.

Vasic, I., and J.M. deMan. 1967. Measurement of some rheological properties of plastic fat with an extrusion modification of the shear press. *J. Am. Oil Chem. Soc.* 44: 225–228.

Vasic, I., and J.M. deMan. 1968. Effect of mechanical treatment on some rheological properties of butter. In *Rheology and texture of foodstuffs*, 251–264. London: Society of Chemical Industry.

Voisey, P.W. 1970. Test cells for objective textural measurements. *Can. Inst. Food Sci. Technol. J.* 4: 91–103.

Voisey, P.W. 1971. The Ottawa texture measuring system. *Can. Inst. Food Sci. Technol. J.* 4: 91–103.

Voisey, P.W., and J.M. deMan. 1970. A recording food grinder. *Can. Inst. Food Sci. Technol. J.* 3: 14–18.

Voisey, P.W., and H. Hansen. 1967. A shear apparatus for meat tenderness evaluation. *Food Technol.* 21: 37A–42A.

Wehrli, H.P., and Y. Pomeranz. 1969. The role of chemical bonds in dough. *Bakers' Digest* 43: no. 6: 22–26.

CHAPTER 9

Vitamins

INTRODUCTION

Vitamins are minor components of foods that play an essential role in human nutrition. Many vitamins are unstable under certain conditions of processing and storage (Table 9–1), and their levels in processed foods, therefore, may be considerably reduced. Synthetic vitamins are used extensively to compensate for these losses and to restore vitamin levels in foods. The vitamins are usually divided into two main groups, the water-soluble and the fat-soluble vitamins. The occurrence of the vitamins in the various food groups is related to their water- or fat-solubility. The relative importance of certain types of foods in supplying some of the important vitamins is shown in Table 9–2. Some vitamins function as part of a coenzyme, without which the enzyme would be ineffective as a biocatalyst. Frequently, such coenzymes are phosphorylated forms of vitamins and play a role in the metabolism of fats, proteins, and carbohydrates. Some vitamins occur in foods as provitamins—compounds that are not vitamins but can be changed by the body into vitamins. Vitamers are members of the same vitamin family.

Lack of vitamins has long been recognized to result in serious deficiency diseases.

It is now also recognized that overdoses of certain vitamins, especially some of the fat-soluble ones, may result in serious toxic effects. For this reason, the addition of vitamins to foods should be carefully controlled.

The sources of vitamins in significant amounts by food groups have been listed by Combs (1992) as follows:

- Meats, poultry, fish, and beans provide thiamin, riboflavin, niacin, pyridoxine, pantothenic acid, biotin, and vitamin B_{12}.
- Milk and milk products provide vitamins A and D, riboflavin, pyridoxine, and vitamin B_{12}.
- Bread and cereals provide thiamin, riboflavin, niacin, pyridoxine, folate, pantothenic acid, and biotin.
- Fruits and vegetables provide vitamins A and K, ascorbic acid, riboflavin, and folate.
- Fats and oils provide vitamins A and E.

FAT-SOLUBLE VITAMINS

Vitamin A (Retinol)

The structural formula of vitamin A is shown in Figure 9–1. It is an alcohol that occurs in nature predominantly in the form

Table 9–1 Stability of Vitamins under Different Conditions

Vitamin	Vitamer	UV Light	Heat[a]	O_2	Acid	Base	Metals[b]	Most Stable
vitamin A	retinol	+		+	+		+	dark, seal
	retinal			+	+		+	seal
	retinoic acid							good stability
	dehydroret.			+				seal
	ret. esters							good stability
	β-carotene			+				seal
vitamin D	D_2	+	+	+	+		+	dark, cool, seal
	D_3	+	+	+	+		+	dark, cool, seal
vitamin E	tocopherols		+	+	+	+	+	cool, neutral pH
	tocopherol esters				+	+		good stability
vitamin K	K	+		+		+	+	avoid reductants[c]
	MK	+		+		+	+	avoid reductants[c]
	menadione	+				+	+	avoid reductants[c]
vitamin C	ascorbic acid			+[b]		+	+	seal, neutral pH
thiamine	disulfide form		+	+	+	+	+	neutral pH[c]
	hydrochloride[d]		+	+	+	+	+	seal, neutral pH[c]
riboflavin	riboflavin	+[e]	+			+	+	dark, pH 1.5–4[c]
niacin	nicotinic acid							good stability
	nicotinamide							good stability
vitamin B_6	pyridoxal	+	+					cool
	pyridoxol (HCl)							good stability
biotin	biotin			+		+		seal, neutral pH
pantothenic acid	free acid[f]	+		+		+		cool, neutral pH
	Ca salty[d]		+					seal, pH 6–7
folate	FH_4	+	+	+	+[g]		+	good stability[c]
vitamin B_{12}	CN-B_{12}	+			+[h]		+[i]	good stability[c]

[a]i.e., 100°C
[b]in solution with Fe^{+++} and Cu^{++}
[c]unstable to reducing agents
[d]slightly hygroscopic
[e]especially in alkaline solution
[f]very hygroscopic
[g]pH < 5
[h]pH < 3
[i]pH > 9

Source: Reprinted with permission from G.F. Combs, *The Vitamins: Fundamental Aspects in Nutrition and Health*, p. 449, © 1992, Academic Press.

Table 9–2 Contributions (%) of Various Food Groups to the Vitamin Intake of Americans

Foods	Vitamin A	Vitamin C	Thiamin	Riboflavin	Niacin	Vitamin B_6	Vitamin B_{12}
vegetables	39.4	51.8	11.7	6.9	12.0	22.2	–
legumes	–	–	5.4	–	8.2	5.4	–
fruits	8.0	39.0	4.4	2.2	2.5	8.2	
grain products	–	–	41.2	22.1	27.4	10.2	1.6
meats	22.5	2.0	27.1	22.2	45.0	40.0	69.2
milk products	13.2	3.7	8.1	39.1	1.4	11.6	20.7
eggs	5.8	–	2.0	4.9	–	2.1	8.5
fats and oils	8.2	–	–	–	–	–	–
other	2.7	3.4	–	–	3.3	–	–

Source: Reprinted with permission from G.F. Combs, *The Vitamins: Fundamental Aspects in Nutrition and Health*, p. 441, © 1992, Academic Press.

of fatty acid esters. Highest levels of vitamin A are found in certain fish liver oils, such as cod and tuna. Other important sources are mammalian liver, egg yolk, and milk and milk products. The levels of vitamin A and its provitamin carotene in some foods are listed in Table 9–3.

The structural formula of Figure 9–1 shows the unsaturated character of vitamin A. The all-*trans* form is the most active biologically. The 13-*cis* isomer is known as neovitamin A; its biological activity is only about 75 percent of that of the all-*trans* form. The amount of neo-vitamin A in natural vitamin A preparations is about one-third of the total. The amount is usually much less in synthetic vitamin A. The synthetic vitamin A is made as acetate or palmitate and marketed commercially in the form of oil solutions, stabilized powders, or aqueous emulsions. The compounds are insoluble in water but soluble in fats, oils, and fat solvents.

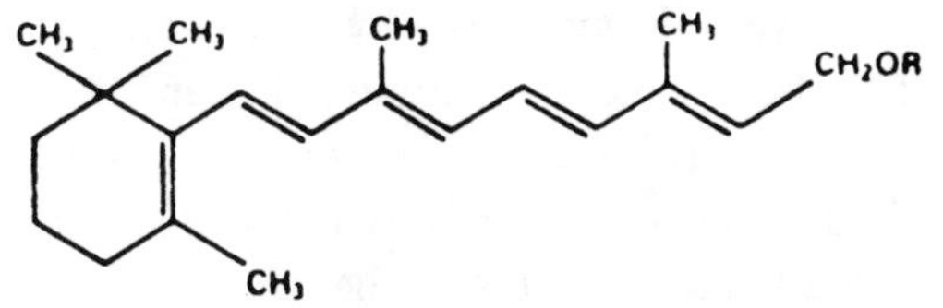

Table 9–3 Vitamin A and Carotene Content of Some Foods

Product	Vitamin A (IU/100 g)	Carotene (mg/100 g)
Beef (grilled sirloin)	37	0.04
Butter (May–November)	2363–3452	0.43–0.77
Cheddar cheese	553–1078	0.07–0.71
Eggs (boiled)	165–488	0.01–0.15
Herring (canned)	178	0.07
Milk	110–307	0.01–0.06
Tomato (canned)	0	0.5
Peach	0	0.34
Cabbage	0	0.3
Broccoli (boiled)	0	2.5
Spinach (boiled)	0	6.0

Figure 9–1 Structural Formula of Vitamin A. Acetate: $R = CO \cdot CH_3$. Palmitate: $R = CO \cdot (CH_2)_{14} \cdot CH_3$.

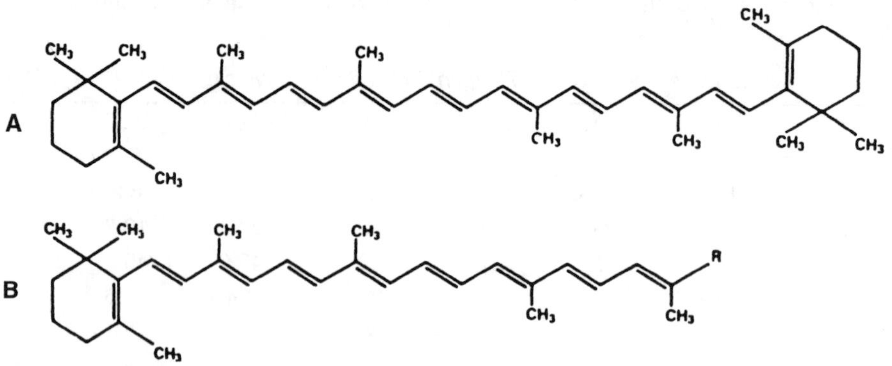

Figure 9–2 Structural Formulas of Some Provitamins A. (A) β-carotene, and (B) apocarotenal (R = CHO) and apocarotenoic acid ester (R = COOC₂H₅).

There are several provitamins A; these belong to the carotenoid pigments. The most important one is β-carotene, and some of the pigments that can be derived from it are of practical importance. These are β-apo-8′-carotenal and β-apo-8′-carotenoic acid ethyl ester (Figure 9–2). Other provitamins are α- and γ-carotene and cryptoxanthin.

Beta-carotene occurs widely in plant products and has a high vitamin A activity. In theory, one molecule of β-carotene could yield two molecules of vitamin A. The enzyme 15-15′-dioxygenase is able to cleave a β-carotene molecule symmetrically to produce two molecules of vitamin A (Figure 9–3). This enzyme occurs in intestinal mucosa, but the actual conversion is much less efficient. As shown in Figure 9–3, there are other reactions that may cause the yield of retinol to be less than 2. After cleavage of the β-carotene, the first reaction product is retinal, which is reduced to retinol (Rouseff and Nagy 1994). A general requirement for the conversion of a carotenoid to vitamin A is an unsubstituted β-ionone ring. Citrus fruits are a good source of provitamin A, which results mostly from

the presence of β-cryptoxanthin, β-carotene, and α-carotene. Gross (1987) reported a total of 16 carotenoids with provitamin A activity in citrus fruits.

Vitamin A levels are frequently expressed in International Units (IU), although this unit is officially no longer accepted. One IU equals 0.344 μg of crystalline vitamin A acetate, or 0.300 μg vitamin A alcohol, or 0.600 μg β-carotene. The recommended daily allowance (RDA) of vitamin A of the National Research Council Food and Nutrition Board is 5000 IU for an adult. Other sources quote the human requirement at about 1 μg/day. Conditions of rapid growth, pregnancy, or lactation increase the need for vitamin A.

Vitamin A, or retinol, is also known as vitamin A₁. Another form, vitamin A₂, is found in fish liver oils and is 3-dehydroretinol.

The Food and Agriculture Organization and the World Health Organization of the United Nations (FAO/WHO) and the National Academy of Sciences of the United States (1974a) have recommended that vitamin A activity be reported as the equivalent weight of retinol. To calculate total retinol equiva-

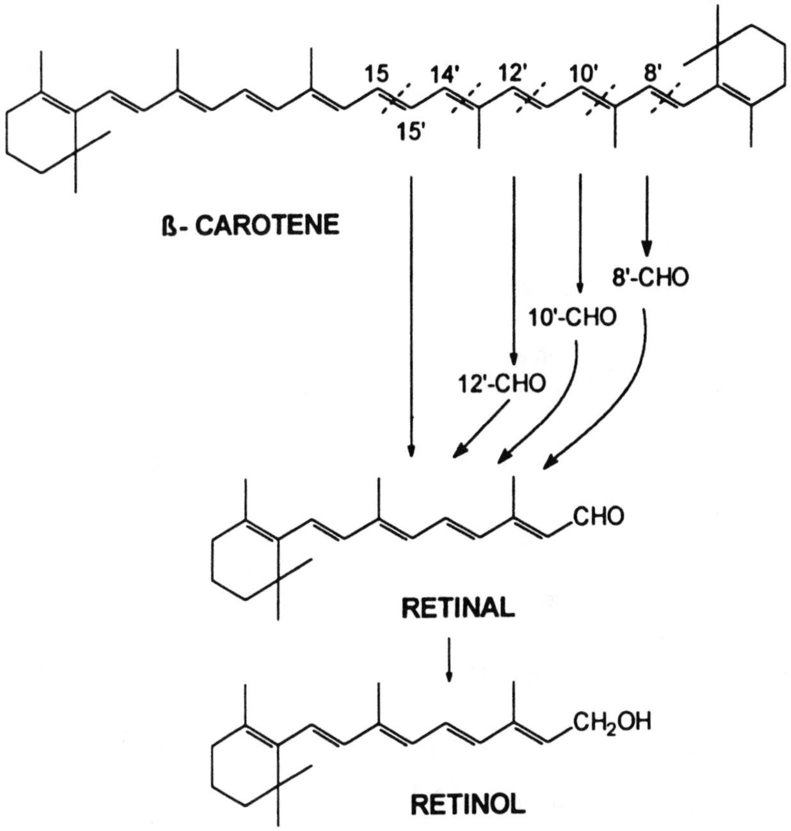

Figure 9–3 Conversion of Beta-Carotene to Vitamin A. *Source*: Reprinted with permission from R.R. Rouseff and S. Nagy, Health and Nutritional Benefits of Citrus Fruit Components, *Food Technology*, Vol. 48, No. 11, p. 125, © 1994, Institute of Food Technologists.

lents, it is proposed that food analyses list retinol, carotene, and other provitamin A carotenoids separately. It is also desirable to distinguish between the *cis-* and *trans-* forms of the provitamins in cooked vegetables. By definition, 1 retinol equivalent is equal to 1 μg of retinol, or 6 μg of β-carotene, or 12 μg of other provitamin A carotenoids. The National Academy of Sciences (1974a) states that 1 retinol equivalent is equal to 3.3 IU of retinol or 10 IU of β-carotene.

Vitamin A occurs only in animals and not in plants. The A_1 form occurs in all animals

and fish, the A_2 form in freshwater fish and not in land animals. The biological value of the A_2 form is only about 40 percent of that of A_1. Good sources of provitamin A in vegetable products are carrots, sweet potatoes, tomatoes, and broccoli. In milk and milk products, vitamin A and carotene levels are subject to seasonal variations. Hartman and Dryden (1965) report the levels of vitamin A in fluid whole milk in winter at 1,083 IU/L and in summer at 1,786 IU/L. Butter contains an average of 2.7 μg of carotene and 5.0 μg of vitamin A per g during winter and 6.1 μg

Table 9–4 Vitamin A and Carotene Stability in Foods

Product	Nutrient Content	Storage Conditions	Retention (%)
Vitamin A			
Butter	17,000–30,000 IU/lb	12 mo @ 5°C	66–98
		5 mo @ 28°C	64–68
Margarine	15,000 IU/lb	6 mo @ 5°C	89–100
		6 mo @ 23°C	83–100
Nonfat dry milk	10,000 IU/lb	3 mo @ 37°C	94–100
		12 mo @ 23°C	69–89
Fortified ready-to-eat cereal	4000 IU/oz	6 mo @ 23°C	83
Fortified potato chips	700 IU/100 g	2 mo @ 23°C	100
Carotene			
Margarine	3 mg/lb	6 mo @ 5°C	98
		6 mo @ 23°C	89
Lard	3.3 mg/lb	6 mo @ 5°C	100
		6 mo @ 23°C	100
Dried egg yolk	35.2 mg/100 g	3 mo @ 37°C	94
		12 mo @ 23°C	80
Carbonated beverage	7.6 mg/29 oz	2 mo @ 30°C	94
		2 mo @ 23°C	94
Canned juice drinks	0.6–1.3 mg/8 fl oz	12 mo @ 23°C	85–100

Source: From E. deRitter, Stability Characteristics of Vitamins in Processed Foods, *Food Technol.*, Vol. 30, pp. 48–51, 54, 1976.

---

of carotene and 7.6 µg of vitamin A per g during summer.

Vitamin A is used to fortify margarine and skim milk. It is added to margarine at a level of 3,525 IU per 100 g. Some of the carotenoids (provitamin A) are used as food colors.

Vitamin A is relatively stable to heat in the absence of oxygen (Table 9–4). Because of the highly unsaturated character of the molecule, it is quite susceptible to oxidation—especially under the influence of light, whether sunlight or artificial light. Vitamin A is unstable in the presence of mineral acids but stable in alkali. Vitamin A and the carotenoids have good stability during various food processing operations. Losses may occur at high temperatures in the presence of oxygen. These compounds are also susceptible to oxidation by lipid peroxides, and conditions favoring lipid oxidation also result in vitamin A breakdown. The prooxidant copper is especially harmful, as is iron to a lesser extent. Pasteurization of milk does not result in vitamin A loss, but exposure to light does. It is essential, therefore, that sterilized milk be packaged in light-impervious containers. Possible losses during storage of foods are more affected by duration of storage than by storage temperature. Blanching of fruits and vegetables helps prevent losses during frozen storage.

Vitamin A added to milk is more easily destroyed by light than the native vitamin A. This is not because natural and synthetic vitamin A are different, but because these two types of vitamin A are dispersed differently in the milk (deMan 1981). The form in which vitamin A is added to food products may influence its stability. Vitamin A in beadlet form is more stable than that added as a solution in oil. The beadlets are stabilized by a protective coating. If this coating is damaged by water, the stability of the vitamin is greatly reduced (de Man et al. 1986).

Vitamin D

This vitamin occurs in several forms; the two most important are vitamin D_2, or ergocalciferol, and vitamin D_3, or cholecalciferol. The structural formulas of these compounds are presented in Figure 9–4. Vitamin D does not occur in plant products. Vitamin D_2 occurs in small amounts in fish liver oils; vitamin D_3 is widely distributed in animal products, but large amounts occur only in fish liver oils. Smaller quantities of vitamin D_3 occur in eggs, milk, butter, and cheese (Table 9–5).

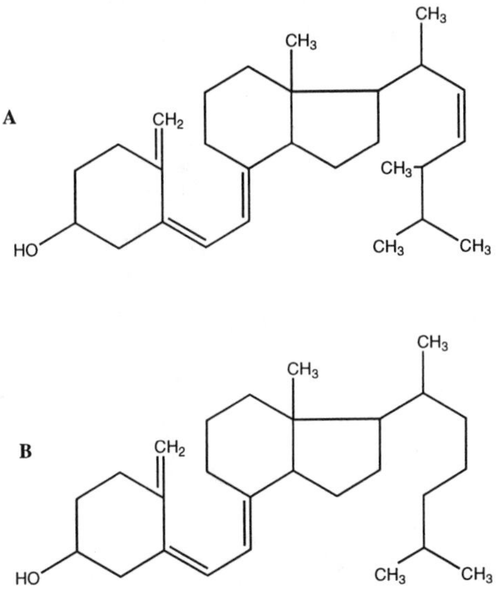

The precursors of vitamins D_2 and D_3 are ergosterol and 7-dehydrocholesterol, respectively. These precursors or provitamins can be converted into the respective D vitamins by irradiation with ultraviolet light. In addition to the two major provitamins, there are several other sterols that can acquire vitamin D activity when irradiated. The provitamins can be converted to vitamin D in the human skin by exposure to sunlight. Because very few foods are good sources of vitamin D, humans have a greater likelihood of vitamin D deficiency than of any other vitamin deficiency. Enrichment of some foods with vitamin D has significantly helped to eradicate rickets, which is a vitamin D deficiency disease. Margarine and milk are the foods commonly used as carrier for added vitamin D.

The unit of activity of vitamin D is the IU, which is equivalent to the activity of 1 mg of a standard preparation issued by the WHO. One IU is also equivalent to the activity of 0.025 μg of pure crystalline vitamin D_2 or D_3. The human requirement amounts to 400

Figure 9–4 Structural Formulas of (A) Vitamin D_2 and (B) Vitamin D_3

Table 9–5 Vitamin D Content of Some Foods

Product	Vitamin D (μg/1000 g Edible Portion)
Liver (beef, pork)	2–5
Eggs	44
Milk	0.9
Butter	2–40
Cheese	12–47
Herring oil	2,500

to 500 IU but increases to 1,000 IU during pregnancy and lactation. Adults who are regularly exposed to sunlight are likely to have a sufficient supply of vitamin D. Excessive intakes are toxic.

Vitamin D is extremely stable, and little or no loss is experienced in processing and storage. Vitamin D in milk is not affected by pasteurization, boiling, or sterilization (Hartman and Dryden 1965). Frozen storage of milk or butter also has little or no effect on vitamin D levels, and the same result is obtained during storage of dry milk.

The vitamin D potency of milk can be increased in several ways: by feeding cows substances that are high in vitamin D activity, such as irradiated yeast; by irradiating milk; and by adding vitamin D concentrates. The latter method is now the only commonly used procedure. The practice of irradiating milk to increase the vitamin D potency has been discontinued, undoubtedly because of the deteriorative action of the radiation on other milk components. Vitamin D is added to milk to provide a concentration of 400 IU

per quart. Addition of vitamin D to margarine is at a level of 550 IU per 100 g.

Tocopherols (Vitamin E)

The tocopherols are derivatives of tocol, and the occurrence of a number of related substances in animal and vegetable products has been demonstrated. Cottonseed oil was found to contain α-, β-, and γ-tocopherol, and a fourth, δ-tocopherol, was isolated from soybean oil. Several other tocopherols have been found in other products, and Morton (1967) suggests that there are four tocopherols and four tocotrienols. The tocotrienols have three unsaturated isoprenoid groups in the side chain. The structure of tocol is given in Figure 9–5 and the structures of the tocopherols and tocotrienols in Figure 9–6. The four tocopherols are characterized by a saturated side chain consisting of three isoprenoid units. The tocotrienols have three double bonds at the 3′, 7′, and 11′ carbons of the isoprenoid side chain (Figure

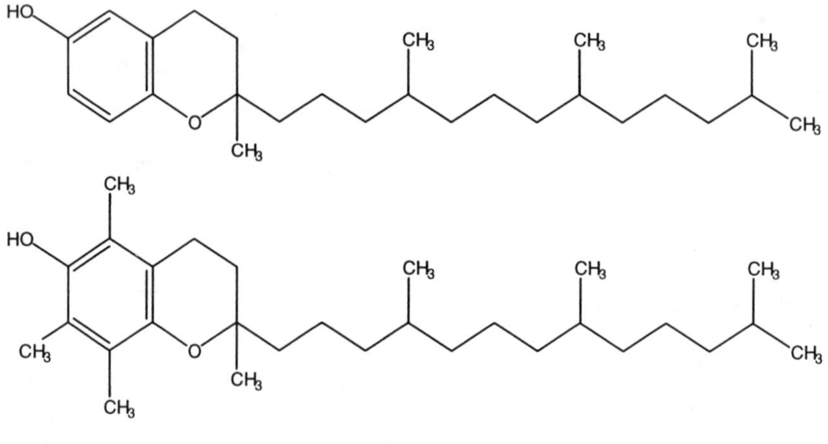

Figure 9–5 Structural Formula of (A) Tocol and (B) α-Tocopherol

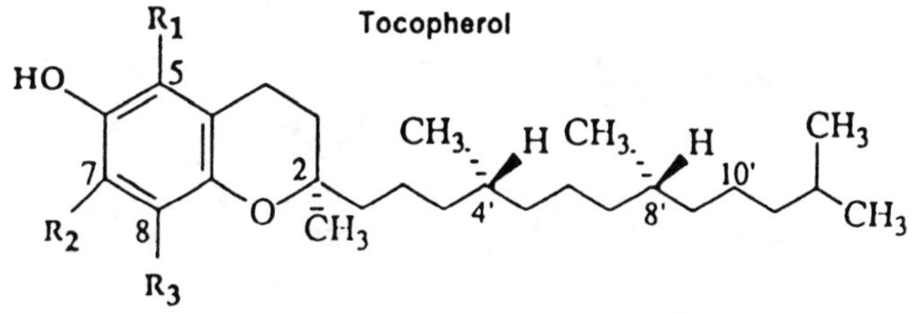

Tocopherol	Tocopherol	R_1	R_2	R_3
α	5,7,8 - Trimethyl	CH_3	CH_3	CH_3
β	5,8 - Dimethyl	CH_3	H	CH_3
γ	7,8 - Dimethyl	H	CH_3	CH_3
δ	8 - Methyl	H	H	CH_3

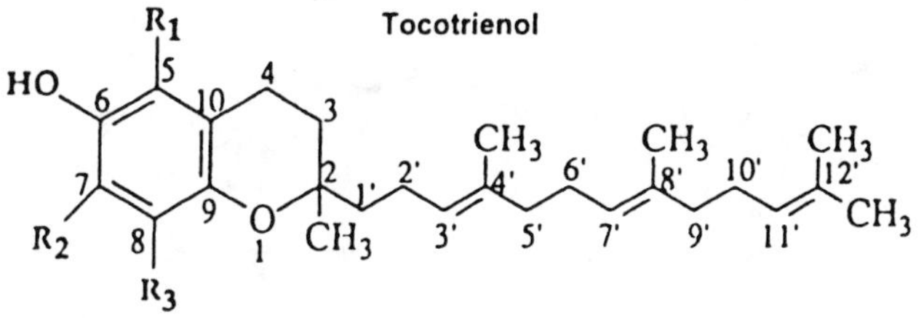

Tocotrienol	Tocotrienol	R_1	R_2	R_3
α	5,7,8 - Trimethyl	CH_3	CH_3	CH_3
β	5,8 - Dimethyl	CH_3	H	CH_3
γ	7,8 - Dimethyl	H	CH_3	CH_3
δ	8 - Methyl	H	H	CH_3

Figure 9–6 Chemical Structure of the Tocopherols and Tocotrienols

9–6). The carbons at locations 4′ and 8′ in the side chains of the tocopherols are asymmetric, as is the number 2 carbon in the chroman ring. The resulting possible isomers are described as having R or S rotation. The natural tocopherols and tocotrienols are predominantly RRR isomers. Morton (1967) has summarized the chemistry of the tocopherols as shown in Figure 9–7.

On oxidation, α-tocopherol can form a meta-stable epoxide that can be irreversibly converted to α-tocopherolquinone. Reduction of the quinone yields a quinol. Tocopherolquinones occur naturally. Oxidation with nitric acid yields the *o*-quinone or tocopherol red, which is not found in nature. Alpha-tocopheronic acid and α-tocopheronolactone are some of the products of metabolism of tocopherol. Much of the biological activity of the tocopherols is related to their antioxidant activity. Because α-tocopherol is the most abundant of the different tocopherols, and because it appears to have the greatest biological activity, the α-

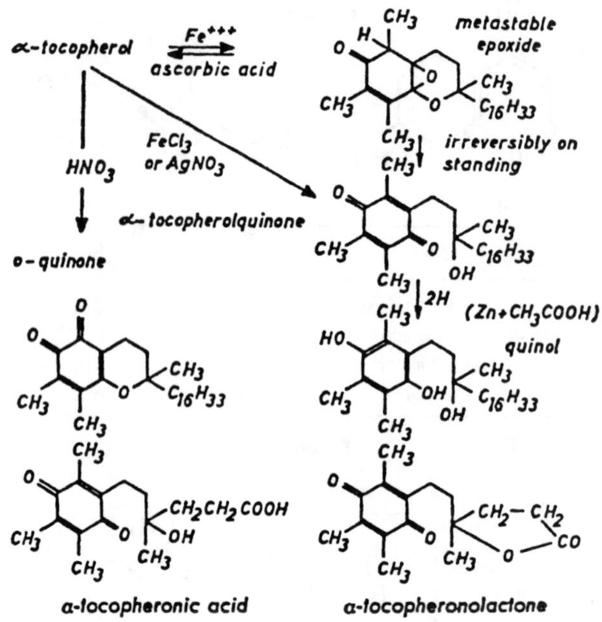

Figure 9–7 Chemistry of the Tocopherols. *Source:* From R.A. Morton, The Chemistry of Tocopherols, in *Tocopherole*, K. Lang, ed., 1967, Steinkopff Verlag, Darmstadt, Germany.

tocopherol content of foods is usually considered to be most important.

The biological activity of the tocopherols and tocotrienols varies with the number and position of the methyl groups on the chroman ring and by the configuration of the asymmetric carbons in the side chain. The R configuration at each chiral center has the highest biological activity. Because the different isomers have different activities, it is necessary to measure each homolog and convert these to RRR-α-tocopherol equivalents (α-TE). One α-TE is the activity of 1 mg of RRR-α-tocopherol (Eitenmiller 1997). The vitamin E activity of α-tocopherol isomers and synthetic tocopherols is listed in Table 9–6.

Tocopherols are important as antioxidants in foods, especially in vegetable oils. With few exceptions, animal and vegetable products contain from about 0.5 to 1.5 mg/100 g; vegetable oils from 10 to 60 mg/100 g; and cereal germ oils, which are a very good source, from 150 to 500 mg/100 g. Vegetable oils have the highest proportion of α-tocopherol, which amounts to about 60 percent of the total tocopherols. Refining of vegetable oils, carried out under normal precautions (such as excluding air), appears to result in little destruction of tocopherol. The tocopherol and tocotrienol content of selected fats and oils and their primary homologs are listed in Table 9–7. The seed oils contain only tocopherol. Tree oils, palm, palm kernel, coconut oil, and rice bran oil also contain major amounts of tocotrienols. The processing of vegetable oils by deodorization or physical refining removes a considerable

Table 9–6 Vitamin E Activity of α-Tocopherol Isomers and Synthetic Tocopherols

Name	IU/mg
d-α-tocopherol (2R4′R8′R) RRR-α-tocopherol	1.49
1-α-tocopherol (2S4′R8′R)	0.46
dl-α-tocopherol all-rac-α-tocopherol	1.10
2R4′R8′S-α-tocopherol	1.34
2S4′R8′S-α-tocopherol	0.55
2S4′S8′S-α-tocopherol	1.09
2S4′S8′R-α-tocopherol	0.31
2R4′S8′R-α-tocopherol	0.85
2S4′S8′S-α-tocopherol	1.10
d-α-tocopheryl acetate RRR-α-tocopheryl acetate	1.36
dl-α-tocopherol all-rac-α-tocopherol acetate	1.00

Source: Reprinted with permission from R.R. Eitenmiller, Vitamin E Content of Fats and Oils: Nutritional Implications, *Food Technol.*, Vol. 51, no. 5, p. 79, © 1997, Institute of Food Technologists.

portion of the tocopherols, and these steam-volatile compounds accumulate in the fatty acid distillate (Ong 1993). This product is an important source of natural vitamin E preparations. Baltes (1967) carried out tests in which two easily oxidizable fats, lard and partially hydrogenated whale oil, were stabilized with α-tocopherol and ascorbylpalmitate and citric acid as synergists. Without antioxidants, these fats cannot be used in the commercial food chain. Amounts of α-tocopherol ranging from 0.5 to 10 mg/100 g were effective in prolonging the storage life of some samples up to two years.

The tocopherol content of some animal and vegetable products as reported by Thaler (1967) is listed in Table 9–8. Cereals and cereal products are good sources of tocopherol (Table 9–9). The distribution of tocopherol throughout the kernels is not uniform, and flour of different degrees of extraction can have different tocopherol levels. This was shown by Menger (1957) in a study of wheat flour (Table 9–10).

Processing and storage of foods can result in substantial tocopherol losses. An example is given in Table 9–11, where the loss of tocopherol during frying of potato chips is reported. After only two weeks' storage of the chips at room temperature, nearly half of the tocopherol was lost. The losses were only slightly smaller during storage at freezer temperature. Boiling of vegetables in water for up to 30 minutes results in only minor losses of tocopherol. Baking of white bread results in a loss of about 5 percent of the tocopherol in the crumb.

The human daily requirement of vitamin E is estimated at 30 IU. Increased intake of polyunsaturated fatty acids increases the need for this vitamin.

Vitamin K

This vitamin occurs in a series of different forms, and these can be divided into two groups. The first is vitamin K_1 (Figure 9–8), characterized by one double bond in the side chain. The vitamins K_2 have a side chain consisting of a number of regular units of the type

$$R-[CH_2-CH=\overset{\displaystyle CH_3}{C}-CH_2]_n-H$$

where *n* can equal 4, 5, 6, 7, and so forth.

Vitamin K_1 is slowly decomposed by atmospheric oxygen but is readily destroyed by light. It is stable against heat, but unstable against alkali.

Table 9–7 Tocopherol (T) and Tocotrienol (T3) Content of Vegetable Oils and Their Primary Homologs

Fats and Oils	Total T+T3 (mg/100g)	α-TE/ 100g	%T	%T3	Primary Homologs
Sunflower	46–67	35–63	100	0	α-T, γ-T
Cottonseed	78	43	100	0	α-T, γ-T
Safflower	49–80	41–46	100	0	α-T, δ-T, γ-T, β-T
Safflower—high linolenic	41	41	100	0	α-T, β-T
Safflower—high oleic	32	31	100	0	α-T, β-T, γ-T
Palm	89–117	21–34	17–55	45–83	α-T, α-T3, δ-T3, α-T, δ-T3
Canola	65	25	100	0	γ-T, α-T, δ-T, α-T3(Tr), β-T(Tr)
Corn	78–109	20–34	95	5	γ-T, α-T, δ-T, γ-T3, δ-T3
Soybean	96–115	17–20	100	0	γ-T, δ-T, α-T
Rice bran	9–160	0.9–41	19–49	51–81	γ-T3, αT, α-T3, β-T, β-T3
Peanut	37	16	100	0	γ-T, α-T, δ-T
Olive	5.1	5.1	100	0	α-T
Cocoa butter	20	3.0	99	1	γ-T, δ-T, α-T, α-T3
Palm kernel	3.4	1.9	38	62	α-T3, α-T
Butter	1.1–2.3	1.1–2.3	100	0	α-T
Lard	0.6	0.6	100	0	α-T
Coconut	1.0–3.6	0.3–0.7	31	69	γ-T3, α-T3, δ-T, α-T, β-T3

Source: Reprinted with permission from R.R. Eitenmiller, Vitamin E Content of Fats and Oils: Nutritional Implications, *Food Technol.*, Vol. 51, no. 5, p. 80, © 1997, Institute of Food Technologists.

The human adult requirement is estimated at about 4 mg per day. Menadione (2-methyl 1,4-naphtoquinone) is a synthetic product and has about twice the activity of naturally occurring vitamin K.

Vitamin K occurs widely in foods and is also synthesized by the intestinal flora. Good sources of vitamin K are dark green vegetables such as spinach and cabbage leaves, and also cauliflower, peas, and cereals. Animal products contain little vitamin K_1, except for pork liver, which is a good source.

The Vitamin K levels in some foods, expressed in menadione units, are given in Table 9–12.

WATER-SOLUBLE VITAMINS

Vitamin C (L-Ascorbic Acid)

This vitamin occurs in all living tissues, where it influences oxidation-reduction reactions. The major source of L-ascorbic acid in foods is vegetables and fruits (Table 9–13).

Table 9–8 Tocopherol Content of Some Animal and Vegetable Food Products

Product	Total Tocopherol as α-Tocopherol (mg/100 g)
Beef liver	0.9–1.6
Veal, lean	0.9
Herring	1.8
Mackerel	1.6
Crab, frozen	5.9
Milk	0.02–0.15
Cheese	0.4
Egg	0.5–1.5
Egg yolk	3.0
Cabbage	2–3
Spinach	0.2–6.0
Beans	1–4
Lettuce	0.2–0.8 (0.06)
Peas	4–6
Tomato	0.9 (0.4)
Carrots	0.2 (0.11)
Onion	0.3 (0.22)
Potato	? (0.12)
Mushrooms	0.08

Source: From H. Thaler, Concentration and Stability of Tocopherols in Foods, in *Tocopherols*, K. Lang, ed., 1967, Steinkopff Verlag, Darmstadt, Germany.

Table 9–9 Tocopherol Content of Cereals and Cereal Products

Product	Total Tocopherol as α-Tocopherol (mg/100 g)
Wheat	7–10
Rye	2.2–5.7
Oats	1.8–4.9
Rice (with hulls)	2.9
Rice (polished)	0.4
Corn	9.5
Whole wheat meal	3.7
Wheat flour	2.3–5.4
Whole rye meal	2.0–4.5
Oat flakes	3.85
Corn grits	1.17
Corn flakes	0.43
White bread	2.15
Whole rye bread	1.3
Crisp bread	4.0

Source: From H. Thaler, Concentration and Stability of Tocopherols in Foods, in *Tocopherols*, K. Lang, ed., 1967, Steinkopff Verlag, Darmstadt, Germany.

L-ascorbic acid (Figure 9–9) is a lactone (internal ester of a hydroxycarboxylic acid) and is characterized by the enediol group, which makes it a strongly reducing compound. The D form has no biological activity. One of the isomers, D-isoascorbic acid, or erythorbic acid, is produced commercially for use as a food additive. L-ascorbic acid is readily and reversibly oxidized to dehydro-L-ascorbic acid (Figure 9–10), which retains vitamin C activity. This compound can be further oxidized to diketo-L-gulonic acid, in a nonreversible reaction. Diketo-L-gulonic acid has no biological activity, is unstable, and is further oxidized to several possible compounds, including 1-threonic acid. Dehydration and decarboxylation can lead to the formation of furfural, which can polymerize to form brown pigments or combine with amino acids in the Strecker degradation.

Humans and guinea pigs are the only primates unable to synthesize vitamin C. The human requirement of vitamin C is not well defined. Figures ranging from 45 to 75 mg/day have been listed as daily needs. Continued stress and drug therapy may increase the need for this vitamin.

Vitamin C is widely distributed in nature, mostly in plant products such as fruits (espe-

Table 9–10 Tocopherol Content of Wheat and Its Milling Products

Product	Ash (%)	Tocopherol mg/100 g (Dry Basis)
Whole wheat	2.05	5.04
Flour 1 (fine)	1.68	5.90
Flour 2	1.14	4.27
Flour 3	0.84	3.48
Flour 4	0.59	2.55
Flour 5	0.47	2.35
Flour 6 (coarse)	0.48	2.13
Germ	4.10	25.0

Source: From A. Menger, Investigation of the Stability of Vitamin E in Cereal Milling Products and Baked Goods, *Brot. Gebäck,* Vol. 11, pp. 167–173, 1957 (German).

Table 9–11 Tocopherol Losses During Processing and Storage of Potato Chips

	Tocopherol (mg/100 g)	Loss (%)
Oil before use	82	—
Oil after use	73	11
Oil from fresh chips	75	—
After two weeks at room temperature	39	48
After one month at room temperature	22	71
After two months at room temperature	17	77
After one month at −12°C	28	63
After two months at −12.°C	24	68

cially citrus fruits), green vegetables, tomatoes, potatoes, and berries. The only animal sources of this vitamin are milk and liver. Although widely distributed, very high levels of the vitamin occur only in a few products, such as rose hips and West Indian cherries. The concentration varies widely in different tissues of fruits; for example, in apples, the concentration of vitamin C is two to three times as great in the peel as in the pulp.

Vitamin C is the least stable of all vitamins and is easily destroyed during processing and storage. The rate of destruction is increased by the action of metals, especially copper and iron, and by the action of enzymes. Exposure to oxygen, prolonged heating in the presence of oxygen, and exposure to light are all harmful to the vitamin C content of foods. Enzymes containing copper or iron in their prosthetic groups are efficient catalysts of ascorbic acid decomposition. The most important enzymes of this group are ascorbic acid oxidase, phenolase, cytochrome oxidase, and peroxidase. Only ascorbic acid oxidase involves a direct reaction among enzyme, substrate, and molecular oxygen. The other enzymes oxidize the vitamin indirectly. Phenolase catalyzes the oxidation of mono-

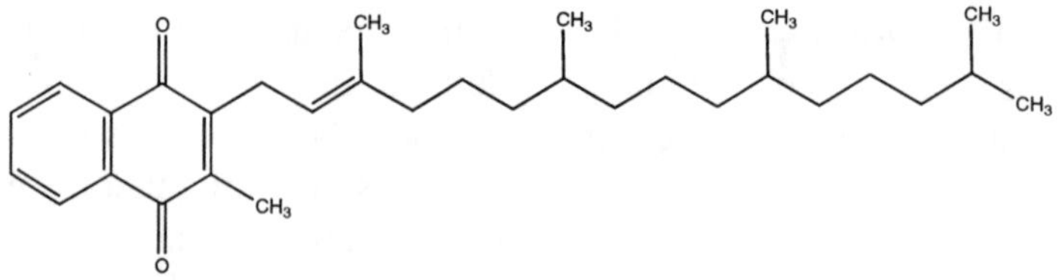

Figure 9–8 Structural Formula of Vitamin K$_1$

Table 9–12 Vitamin K in Some Foods (Expressed as Menadione Units per 100 g of Edible Portion)

Product	Units/100 g
Cabbage, white	70
Cabbage, red	18
Cauliflower	23
Carrots	5
Honey	25
Liver (chicken)	13
Liver (pork)	111
Milk	8
Peas	50
Potatoes	10
Spinach	161
Tomatoes (green)	24
Tomatoes (ripe)	12
Wheat	17
Wheat bran	36
Wheat germ	18

Table 9–13 Vitamin C Content of Some Foods

Product	Ascorbic Acid (mg/100 g)
Black currants	200
Brussels sprouts	100
Cauliflower	70
Cabbage	60
Spinach	60
Orange	50
Orange juice	40–50
Lemon	50
Peas	25
Tomato	20
Apple	5
Lettuce	15
Carrots	6
Milk	2.1–2.7
Potatoes	30

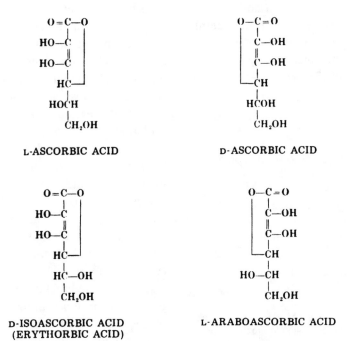

L-ASCORBIC ACID

D-ASCORBIC ACID

D-ISOASCORBIC ACID
(ERYTHORBIC ACID)

L-ARABOASCORBIC ACID

Figure 9–9 Structural Formulas of L-Ascorbic Acid and Its Stereoisomers

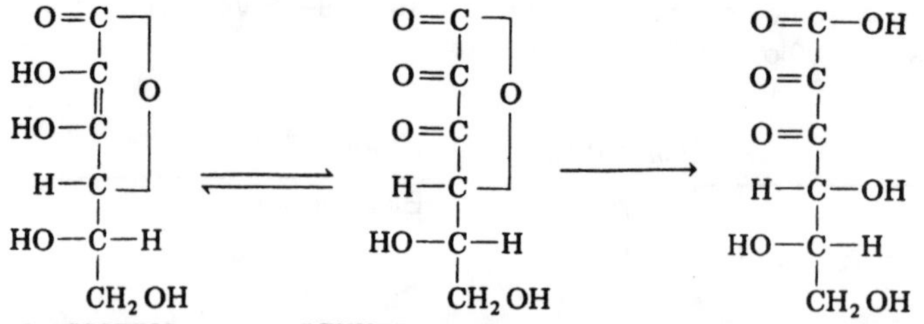

Figure 9–10 Oxidation of L-Ascorbic Acid

and dihydroxy phenols to quinones. The quinones react directly with the ascorbic acid. Cytochrome oxidase oxidizes cytochrome to the oxidized form and this reacts with L-ascorbic acid.

Peroxidase, in combination with phenolic compounds, utilizes hydrogen peroxide to bring about oxidation. The enzymes do not act in intact fruits because of the physical separation of enzyme and substrate. Mechanical damage, rot, or senescence lead to cellular disorganization and initiate decomposition. Inhibition of the enzymes in vegetables is achieved by blanching with steam or by electronic heating. Blanching is necessary before vegetables are dried or frozen. In fruit juices, the enzymes can be inhibited by pasteurization, deaeration, or holding at low temperature for a short period. The effect of blanching methods on the ascorbic acid content of broccoli was reported by Odland and Eheart (1975). Steam blanching was found to result in significantly smaller losses of ascorbic acid (Table 9–14). The retention of ascorbic acid in frozen spinach depends on storage temperature. At a very low temperature ($-29°C$), only 10 percent of the initially present ascorbic acid was lost after one year. At $-12°$, the loss after one year was much higher, 55 percent. The presence of metal chelating compounds stabilizes vitamin C. These compounds include anthocyanins and flavonols, polybasic or polyhydroxy acids such as malic and citric acids, and polyphosphates.

Ascorbic acid is oxidized in the presence of air under neutral and alkaline conditions. At acid pH (for example, in citrus juice), the vitamin is more stable. Because oxygen is required for the breakdown, removal of oxygen should have a stabilizing effect. For the production of fruit drinks, the water should be deaerated to minimize vitamin C loss. The type of container may also affect the extent

Table 9–14 Effect of Blanching Method on Ascorbic Acid Levels of Broccoli

Factor Effect	Ascorbic Acid (mg/100 g)		
	Reduced	Dehydro	Total
Raw	94.0	4.0	98.2
Water blanch	45.3	5.7	51.0
Steam blanch	48.8	7.4	56.2

Source: From D. Odland and M.S. Eheart, Ascorbic Acid, Mineral and Quality Retention in Frozen Broccoli Blanched in Water, Steam, and Ammonia-Steam, *J. Food Sci.*, Vol. 40, pp. 1004–1007, 1975.

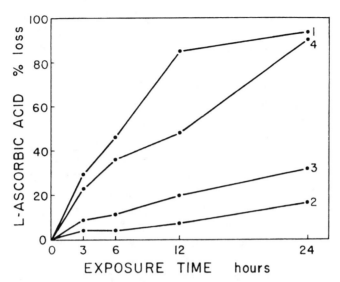

Figure 9–11 Effect of Exposure Time at Light Intensity of 200 Ft-C on the Loss of Ascorbic Acid in Milk. Packaging materials: (1) clear plastic pouch, (2) laminated nontransparent pouch, (3) carton, (4) plastic 3-quart jug. *Source:* From A. Sattar and J.M. deMan, Effect of Packaging Material on Light-Induced Quality Deterioration of Milk, *Can. Inst. Food Sci. Technol. J.*, Vol. 6, pp. 170–174, 1973.

of ascorbic acid destruction. Use of tin cans for fruit juices results in rapid depletion of oxygen by the electrochemical process of corrosion. In bottles, all of the residual oxygen is available for ascorbic acid oxidation. To account for processing and storage losses, it is common to allow for a loss of 7 to 14 mg of ascorbic acid per 100 mL of fruit juice. Light results in rapid destruction of ascorbic acid in milk. It has been shown (Sattar and deMan 1973) that transparent packaging materials permit rapid destruction of vitamin C (Figure 9–11). The extent of ascorbic acid destruction is closely parallel to the development of off-flavors. The destruction of ascorbic acid in milk by light occurs under the influence of riboflavin as a sensitizer. The reaction occurs in the presence of light and oxygen, and the riboflavin is converted to lumichrome.

Factors that affect vitamin C destruction during processing include heat treatment and leaching. The severity of processing conditions can often be judged by the percentage of ascorbic acid that has been lost. The extent of loss depends on the amount of water used. During blanching, vegetables that are covered with water may lose 80 percent; half covered, 40 percent; and quarter covered, 40 percent of the ascorbic acid. Particle size affects the size of the loss; for example, in blanching small pieces of carrots, losses may range from 32 to 50 percent, and in blanching large pieces, only 22 to 33 percent. Blanching of cabbage may result in a 20 percent loss of ascorbic acid, and subsequent dehydration may increase this to a total of 50 percent. In the processing of milk, losses may occur at various stages. From an initial level of about 22 mg/L in raw milk, the content in the product reaching the consumer may be well below 10 mg/L. Further losses may occur in the household during storage of the opened container.

The processing of milk into various dairy products may result in vitamin C losses. Ice cream contains no vitamin C, nor does cheese. The production of powdered milk involves a 20 to 30 percent loss, evaporated milk a 50 to 90 percent loss. Bullock et al. (1968) studied the stability of added vitamin C in evaporated milk and found that adding 266 mg of sodium ascorbate per kg was sufficient to ensure the presence of at least 140 mg/L of ascorbic acid during 12 months of storage at 21°C. Data on the stability of vitamin C in fortified foods have been assembled by deRitter (1976) (Table 9–15).

There are many technical uses of ascorbic acid in food processing. It is used to prevent browning and discoloration in vegetables and fruit products; as an antioxidant in fats, fish products, and dairy products; as a stabilizer of color in meat; as an improver of flour; as an oxygen acceptor in beer processing; as a reducing agent in wine, partially replacing sulfur dioxide; and as an added nutrient. The vitamin is protected by sulfur dioxide, presumably by inhibiting polyphenolase.

Vitamin B₁ (Thiamin)

This vitamin acts as a coenzyme in the metabolism of carbohydrates and is present in all living tissues. It acts in the form of thiamin diphosphate in the decarboxylation of α-keto acids and is referred to as cocarboxylase. Thiamin is available in the form of its chloride or nitrate, and its structural formula is shown in Figure 9–12. The molecule contains two basic nitrogen atoms; one is in the primary amino group, the other in the quater-

Table 9–15 Vitamin C Stability in Fortified Foods and Beverages after Storage at 23°C for 12 Months, Except as Noted

Product	No. of Samples	Retention Mean (%)	Range (%)
Ready-to-eat cereal	4	71	60–87
Dry fruit drink mix	3	94	91–97
Cocoa powder	3	97	80–100
Dry whole milk, air pack	2	75	65–84
Dry whole milk, gas pack	1	93	—
Dry soy powder	1	81	—
Potato flakes[1]	3	85	73–92
Frozen peaches	1	80	—
Frozen apricots[2]	1	80	—
Apple juice	5	68	58–76
Cranberry juice	2	81	78–83
Grapefruit juice	5	81	73–86
Pineapple juice	2	78	74–82
Tomato juice	4	80	64–93
Vegetable juice	2	68	66–69
Grape drink	3	76	65–94
Orange drink	5	80	75–83
Carbonated beverage	3	60	54–64
Evaporated milk	4	75	70–82

[1]Stored for 6 months at 23°C.
[2]Thawed after storage in freezer for 5 months.

Source: From E. deRitter, Stability Characteristics of Vitamins in Processed Foods, *Food Technol.*, Vol. 30, pp. 48–51, 54, 1976.

nary ammonium group. It forms salts with inorganic and organic acids. The vitamin contains a primary alcohol group, which is usually present in the naturally occurring vitamin in esterified form with ortho-, di-, or

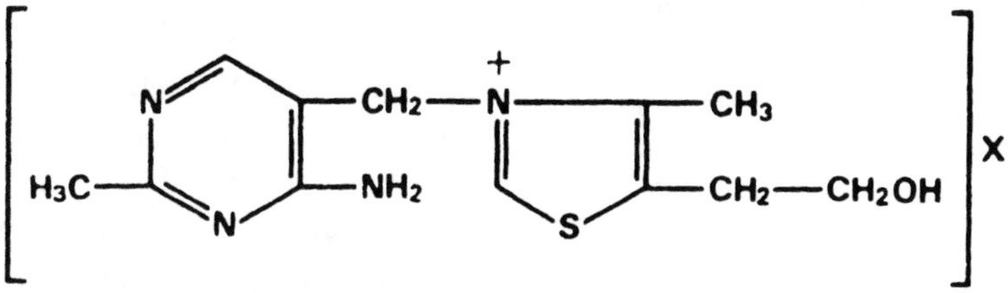

Figure 9–12 Structural Formula of Thiamin. Hydrochloride: $X = Cl^-$, HCl; Mononitrate: $X = NO_3^-$.

triphosphoric acid. In aqueous solution, the compound may occur in different forms, depending on pH. In acid solution, the equilibrium favors the formation of positive ions (Figure 9–13). The thiol- form is favored in alkaline medium. This form can react with compounds containing sulfhydryl groups to form disulfide bridges. It has been suggested that thiamin occurs in some foods linked to protein by disulfide bridges.

Small quantities of thiamin are present in almost all foods of plant and animal origin. Good sources are whole cereal grains; organ meats such as liver, heart, and kidney; lean pork; eggs; nuts; and potatoes (Table 9–16). Although thiamin content is usually mea-

sured in mg per 100 g of a food, another unit has been used occasionally, the IU corresponding to 3 µg of thiamin-hydrochloride. The human daily requirement is related to the carbohydrate level of the diet. A minimum intake of 1 mg per 2,000 kcal is considered essential. Increased metabolic activity, such as that which results from heavy work, pregnancy, or disease, requires higher intake.

Thiamin is one of the more unstable vitamins. Various food processing operations may considerably reduce thiamin levels. Heat, oxygen, sulfur dioxide, leaching, and neutral or alkaline pH may all result in destruction of thiamin. Light has no effect. The enzyme is stable under acid conditions;

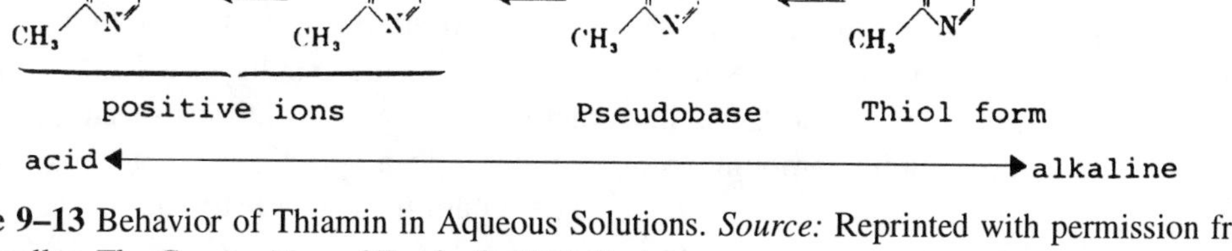

positive ions   Pseudobase   Thiol form

acid ◀──────────────────────────────────▶ alkaline

Figure 9–13 Behavior of Thiamin in Aqueous Solutions. *Source:* Reprinted with permission from J. Schormuller, *The Composition of Foods,* © 1965, Springer.

Table 9–16 Thiamin Content of Some Foods

Product	Thiamin (mg/100 g) Edible Portion
Almonds	0.24
Corn	0.37
Egg	0.11
Filberts	0.46
Beef heart	0.53
Beef liver	0.25
Macaroni (enriched)	0.88
Macaroni (not enriched)	0.09
Milk	0.03
Peas	0.28
Pork, lean	0.87
Potatoes	0.10
Wheat (hard red spring)	0.57
Wheat flour (enriched)	0.44
Wheat flour (not enriched)	0.08

at pH values of 3.5 or below, foods can be autoclaved at 120°C with little or no loss of thiamin. At neutral or alkaline pH, the vitamin is destroyed by boiling or even by storage at room temperature. Even the slight alkalinity of water used for processing may have an important effect. Bender (1971) reports that cooking rice in distilled water reduced thiamin content negligibly, whereas cooking in tap water caused an 8 to 10 percent loss, and cooking in well water caused a loss of up to 36 percent.

Some fish species contain an enzyme that can destroy thiamin. Sulfur dioxide rapidly destroys thiamin. For this reason, sulfur dioxide is not permitted as an additive in foods that contain appreciable amounts of thiamin.

Baking of white bread may result in thiamin loss of 20 percent. Thiamin loss in milk processing is as follows: pasteurization, 3 to 20 percent; sterilization, 30 to 50 percent; spray drying, 10 percent; and roller drying, 20 to 30 percent. Cooking of meat causes losses that are related to size of cut, fat content, and so on. Boiling loss is 15 to 40 percent; frying, 40 to 50 percent; roasting, 30 to 60 percent; and canning, 50 to 75 percent. Similar losses apply to fish. Because thiamin and other vitamins are located near the bran of cereal grains, there is a great loss during milling. White flour, therefore, has a greatly reduced content of B vitamins and vitamin E (Figure 9–14). Not only is thiamin content lowered by milling, but also storage of whole grain may result in losses. This depends on moisture content. At normal moisture level of 12 percent, five months' storage results in a 12 percent loss; at 17 percent moisture, a 30 percent loss; and at 6 percent moisture, no loss at all. Because of the losses that are likely to occur in cereal grain processing and in the processing of other foods, a program of fortification of flour is an important factor in preventing vitamin deficiencies. Table 9–17 lists the nutrients and recommended levels for grain products fortification (National Academy of Sciences 1974b). A summary of data relating processing treatment to thiamin stability has been given by deRitter (1976) (Table 9–18).

Vitamin B$_2$ (Riboflavin)

The molecule consists of a d-ribitol unit attached to an isoalloxazine ring (Figure 9–15). Anything more than a minor change in the molecule results in a loss of vitamin activity. Aqueous solutions of riboflavin are yellow with a yellowish-green fluorescence. The vitamin is a constituent of two coenzymes, flavin mononucleotide (FMN) and flavin adenine dinucleotide (FAD). FMN is

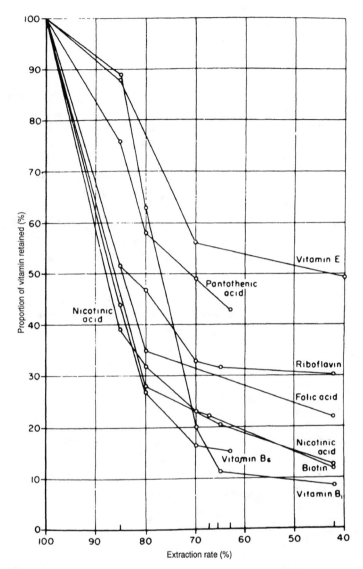

Figure 9–14 Relation Between Extraction Rate and Proportion of Total Vitamins of the Grains Retained in Flour. *Source:* Reprinted with permission from W.R. Aykroyd and J. Doughty, *Wheat in Human Nutrition,* © 1970, Food and Agriculture Organization of the United Nations.

riboflavin-5′-phosphate and forms part of several enzymes, including cytochrome c reductase. The flavoproteins serve as electron carriers and are involved in the oxidation of glucose, fatty acids, amino acids, and purines.

Very good sources of riboflavin are milk and milk products; other sources are beef muscle, liver, kidney, poultry, tomatoes, eggs, green vegetables, and yeast (Table 9–19).

Riboflavin is stable to oxygen and acid pH but is unstable in alkaline medium and is

Table 9–17 Nutrients and Levels Recommended for Inclusion in Fortification of Cereal-Grain Products[1]

Nutrient	Level (mg/lb)	(mg/100 g)
Vitamin A[2]	2.2	0.48
Thiamin	2.9	0.64
Riboflavin	1.8	0.40
Niacin	24.0	5.29
Vitamin B_6	2.0	0.44
Folic acid	0.3	0.07
Iron	40	8.81
Calcium	900	198.2
Magnesium	200	44.1
Zinc	10	2.2

[1]Wheat flour, corn grits, cornmeal, rice. Other cereal-grain products in proportion to their cereal-grain content.

[2]Retinol equivalent.

Source: Reprinted with permission from National Academy of Sciences, *Recommended Dietary Allowances*, 8th rev. ed., © 1974, National Academy of Sciences.

Table 9–18 Thiamin Stability in Foods

Product	Treatment	Retention (%)
Nine canned vegetables	Processing	31–89
Four canned vegetables	Storage, 2–3 mo @ room temperature	73–94
Cereals	Extrusion cooking	48–90
Fortified ready-to-eat cereal	Storage, 12 mo @ 23°C	100
Bread (white, whole wheat)	Commercial baking	74–79
Devil's food cake (pH 9)	Baking	0–7

Source: From E. deRitter, Stability Characteristics of Vitamins in Processed Foods, *Food Technol.*, Vol. 30, pp. 48–51, 54, 1976.

very sensitive to light. When exposed to light, the rate of destruction increases as pH and temperature increase. Heating under neutral or acidic conditions does not destroy the vitamin.

The human requirement for riboflavin varies with metabolic activity and body weight and ranges from 1 to 3 mg per day. Normal adult requirement is 1.1 to 1.6 mg per day. In most cases, the riboflavin of foods is present in the form of the dinucleotide, the phosphoric acid ester, or is bound to protein. Only in milk does riboflavin occur mostly in the free form.

Under the influence of light and alkaline pH, riboflavin is transformed into lumiflavin, an inactive compound with a yellowish-green fluorescence. Under acid conditions, riboflavin is transformed into another inactive derivative, lumichrome, and ribitol. This compound has a blue fluorescence. The trans-

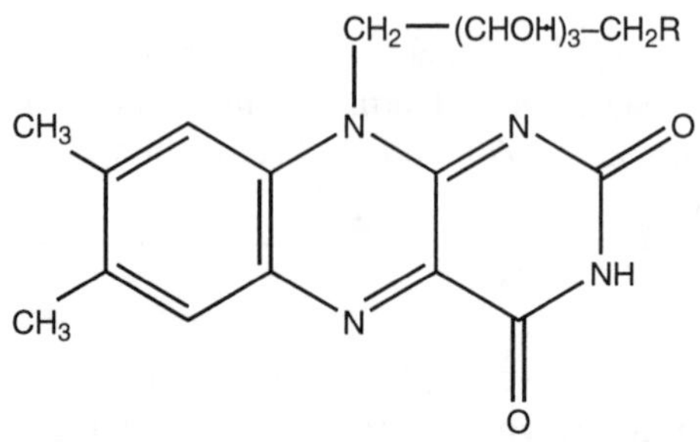

Figure 9–15 Structural Formula of Riboflavin. *Riboflavin: R = OH; Riboflavin phosphate: R = PO_3NaOH.*

Table 9–19 Riboflavin Content of Some Foods

Product	Riboflavin (mg/100 g) Edible Portion
Beef	0.16
Cabbage	0.05
Eggs	0.30
Chicken	0.19
Beef liver	3.26
Chicken liver	2.49
Beef kidney	2.55
Peas	0.29
Spinach	0.20
Tomato	0.04
Yeast (dry)	5.41
Milk	0.17
Nonfat dry milk	1.78

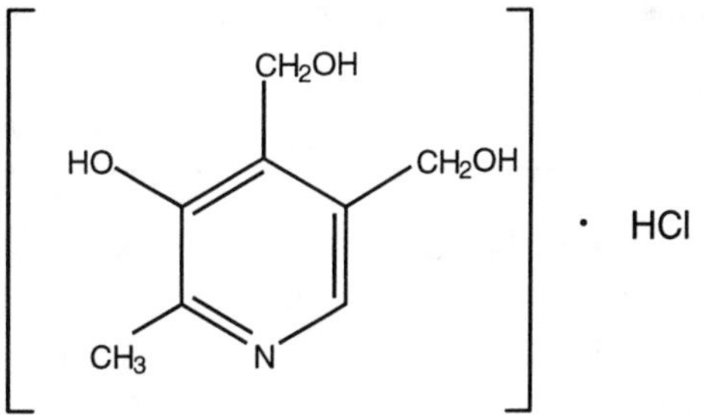

Figure 9–16 Structural Formula of Pyridoxine

formation into lumiflavin in milk results in the destruction of ascorbic acid.

The light sensitivity of riboflavin results in losses of up to 50 percent when milk is exposed to sunlight for two hours. The nature of the packaging material significantly affects the extent of riboflavin destruction. It appears that the wavelengths of light responsible for the riboflavin destruction are in the visible spectrum below 500 to 520 nm. Ultraviolet light has been reported to have no destructive effect on riboflavin (Hartman and Dryden 1965). Riboflavin is stable in dry milk for storage periods of up to 16 months. Pasteurization of milk causes only minor losses of riboflavin.

Vitamin B_6 (Pyridoxine)

There are three compounds with vitamin B_6 activity. The structural formula of pyri-

doxine is presented in Figure 9–16. The other two forms of this vitamin are different from pyridoxine—they have another substituent on carbon 4 of the benzene ring. Pyridoxal has a –CHO group in this position and pyridoxamine has a $-CH_2NH_2$ group. All three compounds can occur as salts. Vitamin B_6 plays an important role in the metabolism of amino acids, where it is active in the coenzyme form pyridoxal-5-phosphate. The three forms of vitamin B_6 are equally active in rats; although it can be expected that the same applies for humans, this has not been definitely established.

Vitamin B_6 is widely distributed in many foods (Table 9–20), and deficiencies of this vitamin are uncommon. The recommended allowance for adults has been established at 2 mg per day. The requirement appears to increase with the consumption of high-protein diets.

Vitamin B_6 occurs in animal tissues in the form of pyridoxal and pyridoxamine or as their phosphates. Pyridoxine occurs in plant products.

Pyridoxine is stable to heat and strong alkali or acid; it is sensitive to light, especially ultraviolet light and when present in alkaline solutions. Pyridoxal and pyridoxam-

Table 9–20 Vitamin B_6 Content of Some Foods

Product	Vitamin B_6 (µg/g)
Wheat	3.2–6.1
Whole wheat bread	4.2
White bread	1.0
Orange juice	0.52–0.60
Apple juice	0.35
Tomatoes	1.51
Beans, canned	0.42–0.81
Peas, canned	0.44–0.53
Beef muscle	0.8–4.0
Pork muscle	1.23–6.8
Milk, pasteurized	0.5–0.6
Yeast	50

ine are rapidly destroyed when exposed to air, heat, or light. Pyridoxamine is readily destroyed in food processing operations.

Because it is difficult to determine this vitamin in foods, there is a scarcity of information on its occurrence. Recent data establish the level in milk as 0.54 mg per liter. Other sources are meats, liver, vegetables, whole grain cereals, and egg yolk.

The effects of processing on pyridoxine levels in milk and milk products have been reviewed by Hartman and Dryden (1965). No significant losses have been reported to result from pasteurization, homogenization, or production of dried milk. Heat sterilization of milk, however, has been reported to result in losses ranging from 36 to 49 percent. Losses occur not only during the heat treatment but also during subsequent storage of milk. These storage losses have been attributed to a conversion of pyridoxal to pyridoxamine and then to a different form of the vitamin. Wendt and Bernhart (1960) have identified this compound as *bis*-4-pyridoxal disulfide (Figure 9–17). This compound is formed by reaction of pyridoxal and active sulfhydryl groups. The latter are formed during heat treatment of milk proteins. Exposure of milk to daylight in clear glass bottles for eight hours resulted in a vitamin B_6 loss of 21 percent.

Food canning results in losses of vitamin B_6 of 20 to 30 percent. Milling of wheat may result in losses of up to 80 to 90 percent. Baking of bread may result in losses of up to 17 percent.

A review of some stability data of vitamin B_6 as prepared by deRitter (1976) is given in Table 9–21.

Niacin

The term *niacin* is used in a generic sense for both nicotinic acid and nicotinamide

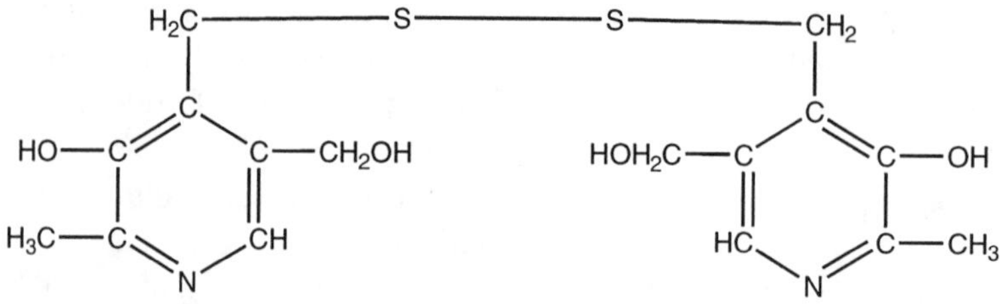

Figure 9–17 Structural Formula of *bis*-4-Pyridoxal Disulfide

Table 9–21 Vitamin B$_6$ Stability in Foods

Product	Treatment	Retention (%)		
Bread (added B$_6$)	Baking	100		
Enriched corn meal	12 mo @ 38°C + 50% relative humidity	90–95		
Enriched macaroni	12 mo @ 38°C + 50% relative humidity	100		
		Saccharomyces Carlsbergensis	Chick	Rat
Whole milk	Evaporation and sterilization	30	55	65
	Evaporation and sterilization + 6 mo @ room temperature	18	44	41
Infant formula, liquid	Processing and sterilization	33–50 (natural) 84 (added)		
Infant formula, dry	Spray drying	69–83		
Boned chicken	Canning	57		
	Irradiation (2.79 megarads)	68		

Source: From E. deRitter, Stability Characteristics of Vitamins in Processed Foods, *Food Technol.*, Vol. 30, pp. 48–51, 54, 1976.

(Figure 9–18). Nicotinamide acts as a component of two important enzymes, NAD and NADP, which are involved in glycolysis, fat synthesis, and tissue respiration. Niacin is also known as the pellagra preventive factor. The incidence of pellagra has declined but is still a serious problem in parts of the Near East, Africa, southeastern Europe, and in North American populations that subsist on corn diets. When corn is treated with alkali or lime, as for the tortilla preparation in Central America, the amount of available niacin can be greatly increased. Tryptophan can be converted by the body into niacin. Many diets causing pellagra are low in good-quality protein as well as in vitamins. Corn protein is low in tryptophan. The niacin of corn and other cereals may occur in a bound form,

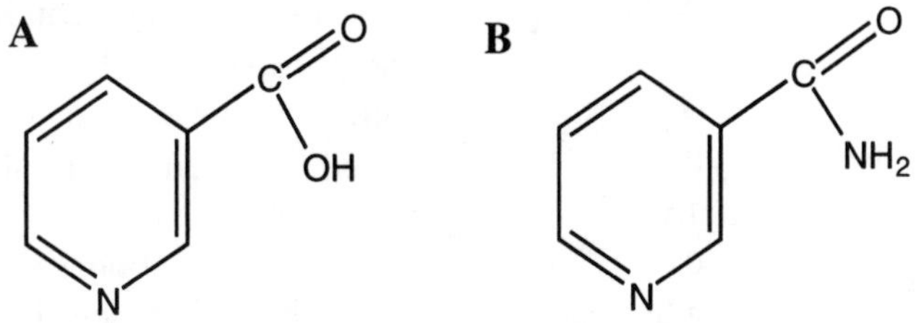

Figure 9–18 Structural Formulas of (A) Nicotinic Acid and (B) Nicotinamide

called niacytin, that can be converted into niacin by alkali treatment.

The human requirement of niacin is related to the intake of tryptophan. Animal proteins contain approximately 1.4 percent of tryptophan, vegetable proteins about 1 percent. A dietary intake of 60 mg of tryptophan is considered equivalent to 1 mg of niacin. When this is taken into account, average diets in the United States supply 500 to 1,000 mg tryptophan per day and 8 to 17 mg niacin for a total niacin equivalent of 16 to 33 mg. The RDA for adults, expressed as niacin, is 6.6 mg per 1,000 kcal, and not less than 13 mg when caloric intake is less than 2,000 kcal.

Table 9–22 Niacin Content of Some Foods

Product	Niacin (mg/100 g) Edible Portion
Barley (pearled)	3.1
Beans (green, snap)	0.5
Beans (white)	2.4
Beef (total edible)	4.4
Beef kidney	6.4
Beef liver	13.6
Chicken (dark meat)	5.2
Chicken (light meat)	10.7
Corn (field)	2.2
Haddock	3.0
Milk	0.1
Mushrooms	4.2
Peanuts	17.2
Peas	2.9
Potatoes	1.5
Spinach	0.6
Wheat	4.3
Yeast (dry)	36.7

Good dietary sources of this vitamin are liver, kidney, lean meat, chicken, fish, wheat, barley, rye, green peas, yeast, peanuts, and leafy vegetables. In animal tissues, the predominant form of niacin is the amide. Niacin content of some foods are listed in Table 9–22.

Niacin is probably the most stable of the B vitamins. It is unaffected by heat, light, oxygen, acid, or alkali. The main loss resulting from processing involves leaching into the process water. Blanching of vegetables may cause a loss of about 15 percent. Processes in which brines are used may cause losses of up to 30 percent. Processing of milk, such as pasteurization, sterilization, evaporation, and drying have little or no effect on nicotinic acid level. Virtually all the niacin in milk occurs in the form of nicotinamide. In many foods, application of heat, such as roasting or baking, increases the amount of available niacin. This results from the change of bound niacin to the free form.

Vitamin B$_{12}$ (Cyanocobalamine)

This vitamin possesses the most complex structure of any of the vitamins and is unique in that it has a metallic element, cobalt, in the molecule (Figure 9–19). The molecule is a coordination complex built around a central tervalent cobalt atom and consists of two major parts—a complex cyclic structure that closely resembles the porphyrins and a nucleotide-like portion, 5,6-dimethyl-1-(α-D-ribofuranosyl) benzimidazole-3′-phosphate. The phosphate of the nucleotide is esterified with 1-amino-2-propanol; this, in turn, is joined by means of an amide bond with the propionic acid side chain of the large cyclic structure. A second linkage with the large structure is through the coordinate bond between the cobalt atom and one of the nitro-

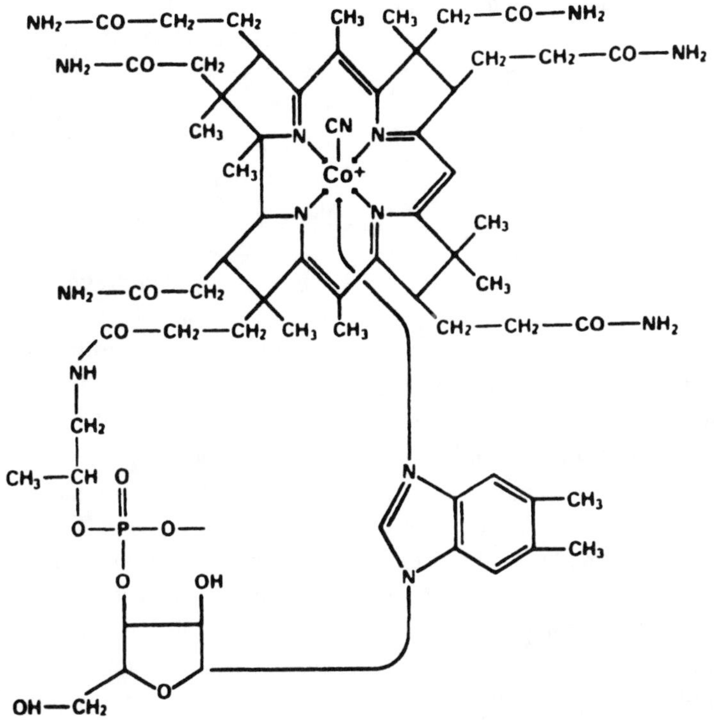

Figure 9–19 Structural Formula of Cyanocobalamine

gen atoms of the benzimidazole. The cyanide group can be split off relatively easily, for example, by daylight. This reaction can be reversed by removing the light source. The cyano group can also be replaced by other groups such as hydroxo, aquo, and nitroto. Treatment with cyanide will convert these groups back to the cyano form. The different forms all have biological activity.

Cyanocobalamine is a component of several coenzymes and has an effect on nucleic acid formation through its action in cycling 5-methyl-tetrahydrofolate back into the folate pool. The most important dietary sources of the vitamin are animal products. Vitamin B_{12} is also produced by many microorganisms. It is not surprising that vitamin B_{12} deficiency of dietary origin only occurs in vegetarians.

The average diet in the United States is considered to supply between 5 and 15 μg/day. In foods, the vitamin is bound to proteins via peptide linkages but can be readily absorbed in the intestinal tract. The RDA is 3 μg for adults and adolescents.

Few natural sources are rich in vitamin B_{12}. However, only very small amounts are required in the diet. Good sources are lean meat, liver, kidney, fish, shellfish, and milk (Table 9–23). In milk, the vitamin occurs as cobalamine bound to protein.

Vitamin B_{12} is not destroyed to a great extent by cooking, unless the food is boiled in alkaline solution. When liver is boiled in water for 5 minutes, only 8 percent of the vitamin B_{12} is lost. Broiling of meat may result in higher losses. Pasteurization causes

Table 9–23 Vitamin B$_{12}$ Content of Some Foods

Product	Vitamin B$_{12}$
Beef muscle	0.25–3.4 µg/100 g
Beef liver	14–152 µg/100 g
Milk	3.2–12.4 µg/L
Shellfish	600–970 µg/100 g (dry wt)
Egg yolk	0.28–1.556 µg/100 g

only a slight destruction of vitamin B$_{12}$ in milk; losses range from 7 to 10 percent depending on pasteurization method. More drastic heat treatment results in higher losses. Boiling milk for two to five minutes causes a 30 percent loss, evaporation about 50 percent, and sterilization up to 87 percent. The loss in drying of milk is smaller; in the production of dried skim milk, the vitamin B$_{12}$ loss is about 30 percent. Ultra-high-temperature sterilization of milk does not cause more vitamin B$_{12}$ destruction than does pasteurization.

Folic Acid (Folacin)

Folic acid is the main representative of a series of related compounds that contain three moieties: pterin, p-aminobenzoic acid, and glutamic acid (Figure 9–20). The commercially available form contains one glutamic acid residue and is named pteroylglutamic acid (PGA). The naturally occurring forms are either PGA or conjugates with varying numbers of glutamic acid residues, such as tri- and heptaglutamates. It has been suggested that folic acid deficiency is the most common vitamin deficiency in North America and Europe. Deficiency is especially likely to occur in pregnant women.

The vitamin occurs in a variety of foods, especially in liver, fruit, leafy vegetables, and yeast (Table 9–24) (Hurdle et al. 1968; Streiff 1971). The usual form of the vitamin in these products is a polyglutamate. The action of an enzyme (conjugase) is required to liberate the folic acid for metabolic activity; this takes place in the intestinal mucosa. The folacin of foods can be divided into two main groups on the basis of its availability to *L. casei*: (1) the so-called free folate, which is available to *L. casei* without conjugase treatment; and (2) the total folate, which also includes the conjugates that are not normally available to *L. casei*. About 25 percent of the dietary folacin occurs in free form. The folate in vegetables occurs mainly in the conjugated form; the folate in liver occurs in the free form.

The RDA for folacin is 400 µg for adults. There is an additional requirement of 400 µg/day during pregnancy and 200 µg/day during breastfeeding.

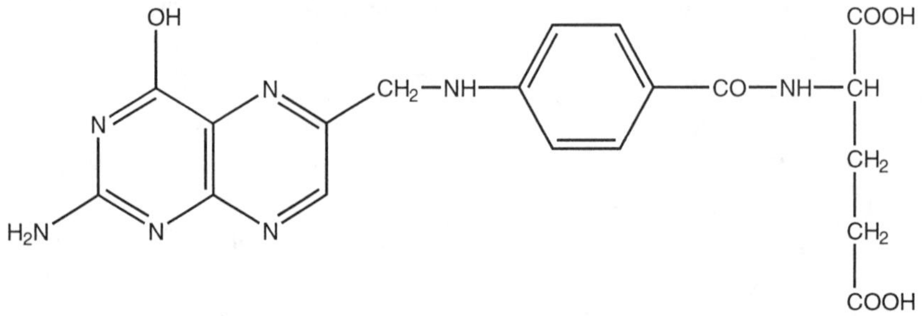

Figure 9–20 Structural Formula of Folic Acid

Table 9–24 Folate Content of Some Foods

Product	Folate (μg/g)
Beef, boiled	0.03
Chicken, roasted	0.07
Cod, fried	0.16
Eggs, boiled	0.30
Brussels sprouts, boiled	0.20
Cabbage, boiled	0.11
Lettuce	2.00
Potato, boiled	0.12
Spinach, boiled	0.29
Tomato	0.18
Orange	0.45
Milk	0.0028
Bread, white	0.17
Bread, brown	0.38
Orange juice, frozen reconstituted	0.50
Tomato juice, canned	0.10

Many of the naturally occurring folates are extremely labile and easily destroyed by cooking. Folic acid itself is stable to heat in an acid medium but is rapidly destroyed under neutral and alkaline conditions. In solution, the vitamin is easily destroyed by light. Folate may occur in a form more active than PGA; this is called folinic acid or citrovorum factor, which is N5-formyl-5, 6, 7, 8-tetrahydro PGA (Figure 9–21). The folate of milk consists of up to 20 percent of folinic acid. It has been reported that pasteurization and sterilization of milk involve only small losses or no loss. Hurdle et al. (1968) reported that boiling of milk causes no loss in folate; however, boiling of potato results in a 90 percent loss and boiling of cabbage a 98 percent loss. Reconstitution of dried milk followed by sterilization as can occur with baby formulas may lead to significant folacin losses. Fermentation of milk and milk products may result in greatly increased folate levels. Blanching of vegetables and cooking of meat do not appear to cause folic acid losses. Table 9–25 contains a summary of folate stability data prepared by deRitter (1976). Citrus fruit and juices are relatively good sources of folic acid, which is present mostly as the reduced 5-methyl tetrahydrofolate (monoglutamate form). There are also polyglutamate derivatives present (White et al. 1991).

Pantothenic Acid

The free acid (Figure 9–22) is very unstable and has the appearance of a hygroscopic

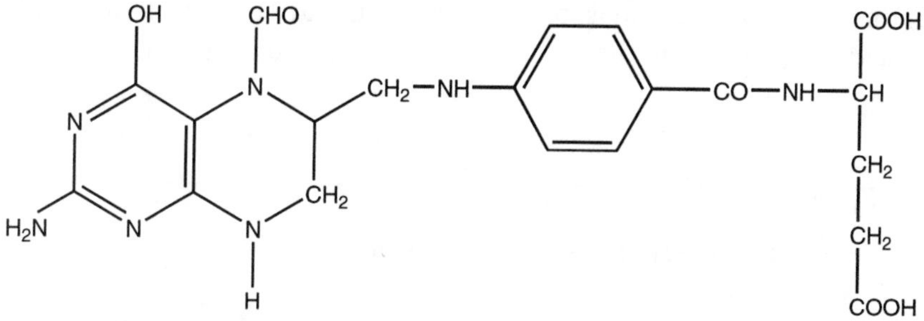

Figure 9–21 Structural Formula of Folinic Acid

Table 9–25 Folic Acid Stability in Foods

Product	Treatment	Retention of Folic Acid Free (%)	Retention of Folic Acid Total (%)
Cabbage	Boiled 5 min	32	54
Potatoes	Boiled 5 min	50	92
Rice	Boiled 15 min	—	10
Beef, pork, and chicken	Boiled 15 min	<50	<50
Various foods	Cooked	27	55

Source: From E. deRitter, Stability Characteristics of Vitamins in Processed Foods, *Food Technol.*, Vol. 30, pp. 48–51, 54, 1976.

Table 9–26 Pantothenic Acid Content of Some Foods

Product	Pantothenic Acid $(\mu g/g)$
Beef, lean	10
Wheat	11
Potatoes	6.5
Split peas	20–22
Tomatoes	1
Orange	0.7
Walnuts	8
Milk	1.3–4.2
Beef liver	25–60
Eggs	8–48
Broccoli	46

oil. The calcium and sodium salts are more stable. The alcohol (panthenol) has the same biological activity as the acid. Only the dextrorotatory or D form of these compounds has biological activity. Pantothenic acid plays an important role as a component of coenzyme A, and this is the form in which it occurs in most foods.

Pantothenic acid occurs in all living cells and tissues and is, therefore, found in most food products. Good dietary sources include meats, liver, kidney, fruits, vegetables, milk, egg yolk, yeast, whole cereal grains, and nuts (Table 9–26). In animal products, most of the pantothenic acid is present in the bound form, but in milk only about one-fourth of the vitamin is bound.

There is no recommended dietary allowance for this vitamin because of insufficient evidence to base one on. It is estimated that adult dietary intake in the United States ranges from 5 to 20 mg/day, and 5 to 10 mg/day probably represents an adequate intake.

The vitamin is stable to air, and labile to dry heat. It is stable in solution in the pH range of 5 to 7 and less stable outside this range. Pasteurization and sterilization of milk result in very little or no loss. The production and storage of dried milk involves little or no loss of pantothenic acid. Manu-

$$CH_2OH-C(CH_3)_2-CHOH-CO-NH-CH_2-CH_2-R$$

Figure 9–22 Structural Formula of Pantothenic Acid. Pantothenic acid: R = COOH; Panthenol: R = CH$_2$OH.

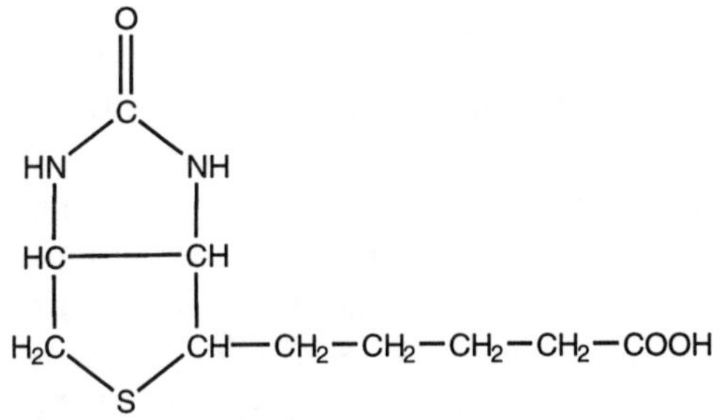

Figure 9–23 Structural Formula of Biotin

Table 9–27 Biotin Content of Some Foods

Product	Biotin (µg/100 g)
Milk	1.1–3.7
Tomatoes	1
Broad beans	3
Cheese	1.1–7.6
Wheat	5.2
Beef	2.6
Beef liver	96
Lettuce	3.1
Mushrooms	16
Potatoes	0.6
Spinach	6.9
Apples	0.9
Oranges	1.9
Peanuts	34

facture of cheese involves large losses during processing, but during ripening the pantothenic acid content increases, due to synthesis by microorganisms. Blanching of vegetables may involve losses of up to 30 percent. Boiling in water involves losses that depend on the amount of water used.

Biotin

The structural formula (Figure 9–23) contains three asymmetric carbon atoms, and eight different stereoisomers are possible. Only the dextrorotatory D-biotin occurs in nature and has biological activity. Biotin occurs in some products in free form (vegetables, milk, and fruits) and in other products is bound to protein (organ meats, seeds, and yeast). Good sources of the vitamin are meat, liver, kidney, milk, egg yolk, yeast, vegetables, and mushrooms (Table 9–27).

Biotin is important in a number of metabolic reactions, especially in fatty acid synthesis. The biotin supply of the human organism is only partly derived from the diet.

An important factor in biotin's availability is that some of the vitamin is derived from synthesis by intestinal microorganisms; this is demonstrated by the fact that three to six

times more biotin is excreted in the urine than is ingested with the food. The daily intake of biotin is between 100 and 300 µg. No recommended dietary allowance has been established. Biotin is deactivated by raw egg white. This is caused by the glycoprotein avidin. Heating of avidin will destroy the inactivator capacity for biotin.

Data on the stability of biotin are limited. The vitamin appears to be quite stable. Heat treatment results in relatively small losses. The vitamin is stable to air and is stable at neutral and acid pH. Pasteurization and sterilization of milk result in losses of less than 10 percent. In the production of evaporated and dried milk, losses do not exceed 15 percent.

VITAMINS AS FOOD INGREDIENTS

In addition to their role as essential micronutrients, vitamins may serve as food ingre-

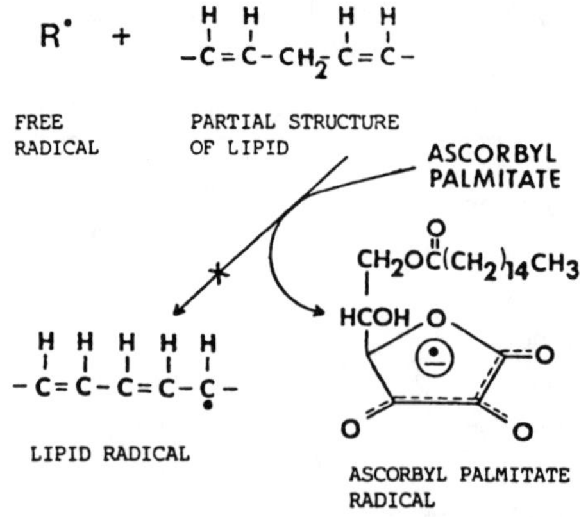

Figure 9–24 Prevention of Lipid Free Radical Formation by Ascorbyl Palmitate. *Source:* From M.L. Liao and P.A. Seib, Selected Reactions of L-Ascorbic Acid Related to Foods, *Food Technol.*, Vol. 41, no. 11, pp. 104–107, 1987.

dients for their varied functional properties (Institute of Food Technologists 1987). Vitamin C and vitamin E have found widespread use as antioxidants. In lipid systems, vitamin E may be used as an antioxidant in fats that have little or no natural tocopherol content. Ascorbic acid in the form of its palmitic acid ester, ascorbyl palmitate, is an effective antioxidant in lipid systems. Ascorbyl palmitate prevents the formation of lipid free radicals (Figure 9–24) and thereby delays the initiation of the chain reaction that leads to the deterioration of the fat (Liao and Seib 1987). Ascorbyl palmitate is used in vegetable oils because it acts synergistically with naturally occurring tocopherols. The tocopherols are fat-soluble antioxidants that are used in animal fats. Ascorbic acid reduces nitrous acid to nitric oxide and prevents the formation of N-nitrosamine. The reaction of nitrous acid and ascorbic acid is given in Figure 9–25 (Liao and Seib 1987). Ascorbic acid is also widely used to prevent enzymic browning in fruit products. Phenolic compounds are oxidized by polyphenoloxidase to quinones. The quinones rapidly polymerize to form brown pigments. This reaction is easily reversed by ascorbic acid (Figure 9–26).

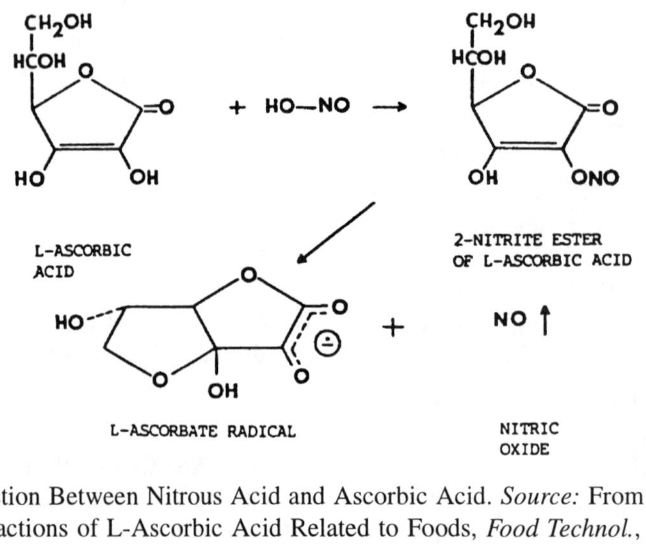

Figure 9–25 Reaction Between Nitrous Acid and Ascorbic Acid. *Source:* From M.L. Liao and P.A. Seib, Selected Reactions of L-Ascorbic Acid Related to Foods, *Food Technol.*, Vol. 41, no. 11, pp. 104–107, 1987.

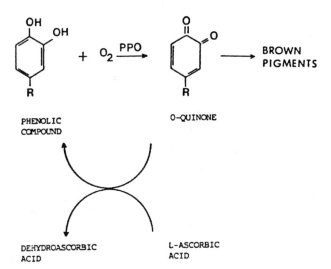

PHENOLIC COMPOUND

O-QUINONE

BROWN PIGMENTS

DEHYDROASCORBIC ACID

L-ASCORBIC ACID

Figure 9–26 Reduction of Ortho-Quinone by Ascorbic Acid During Enzymic Browning. *Source:* From M.L. Liao and P.A. Seib, Selected Reactions of L-Ascorbic Acid Related to Foods, *Food Technol.*, Vol. 41, no. 11, pp. 104–107, 1987.

The carotenoids β-carotene and β-apo-8-carotenal are used as colorants in fat-based as well as water-based foods.

Other functions of ascorbic acid are inhibition of can corrosion in canned soft drinks, protection of flavor and color of wine, prevention of black spot formation in shrimp, stabilization of cured meat color, and dough improvement in baked goods (Institute of Food Technologists 1987).

REFERENCES

Baltes, J. 1967. Tocopherols as fat stabilizers. In *Tocopherol*, K. Lang, ed. Darmstadt, Germany: Steinkopff Verlag.

Bender, A.E. 1971. The fate of vitamins in food processing operations. In *Vitamins*, ed. M. Stein. London: Churchill Livingstone.

Bullock, D.H., et al. 1968. Stability of vitamin C in enriched commercial evaporated milk. *J. Dairy Sci.* 51: 921–923.

Combs, G.F. 1992. *The vitamins: Fundamental aspects in nutrition and health.* San Diego: Academic Press.

deMan, J.M. 1981. Light-induced destruction of vitamin A in milk. *J. Dairy Sci.* 64: 2031–2032.

deMan, J.M., et al. 1986. Stability of vitamin A beadlets in nonfat dry milk. *Milchwissenschaft* 41: 468–469.

deRitter, E. 1976. Stability characteristics of vitamins in processed foods. *Food Technol.* 30: 48–51, 54.

Eitenmiller, R.R. 1997. Vitamin E content of fats and oils: Nutritional implications. *Food Technol.* 51, no. 5: 78–81.

Gross, J. 1987. *Pigments in fruits.* London: Harcourt Brace Jovanovich.

Hartman, A.M., and Dryden, L.P. 1965. Vitamins in milk and milk products. Am. Dairy Sci. Assoc. Champaign, IL.

Hurdle, A.D.F., et al. 1968. A method for measuring folate in food and its application to a hospital diet. *Am. J. Clin. Nutr.* 21: 1202–1207.

Institute of Food Technologists. 1987. Use of vitamins as additives in processed foods. *Food Technol.* 41, no. 9: 163–168.

Liao, M.-L., and P.A. Seib. 1987. Selected reactions of L-ascorbic acid related to foods. *Food Technol.* 41, no. 11: 104–107, 111.

Menger, A. 1957. Investigation of the stability of vitamin E in cereal milling products and baked goods. *Brot. Gebäck* 11: 167–173 (German).

Morton, R.A. 1967. The chemistry of tocopherols. In *Tocopherol,* ed. K. Lang. Darmstadt, Germany: Steinkopff Verlag.

National Academy of Sciences. 1974a. *Recommended dietary allowances,* 8th rev. ed. Washington, DC.

National Academy of Sciences. 1974b. *Proposed fortification policy for cereal-grain products.* Washington, DC.

Odland, D., and M.S. Eheart. 1975. Ascorbic acid, mineral and quality retention in frozen broccoli blanched in water, steam and ammonia-steam. *J. Food Sci.* 40: 1004–1007.

Ong, A.S.H. 1993. Natural sources of tocotrienols. In *Vitamin E in health and disease,* ed. L. Packer and J. Fuchs. New York: Marcel Dekker.

Rouseff, R.L., and S. Nagy. 1994. Health and nutritional benefits of citrus fruit components. *Food Technol.* 48, no. 11: 126–132.

Sattar, A., and J.M. deMan. 1973. Effect of packaging material on light-induced quality deterioration of milk. *Can. Inst. Food Sci. Technol. J.* 6: 170–174.

Streiff, R.R. 1971. Folate levels in citrus and other juices. *Am. J. Clin. Nutr.* 24: 1390–1392.

Thaler, H. 1967. Concentration and stability of tocopherols in foods. In *Tocopherol*, ed. K. Lang. Darmstadt, Germany: Steinkopff Verlag

Wendt, G., and E.W. Bernhart. 1960. The structure of a sulfur-containing compound with vitamin B_6 activity. *Arch. Biochem. Biophys.* 88: 270–272.

White, D.R., et al. 1991. Reverse phase HPLC/EC determination of folate in citrus juice by direct injection with column switching. *J. Agr. Food Chem.* 39: 714–717.

CHAPTER **10**

Enzymes

INTRODUCTION

Enzymes, although minor constituents of many foods, play a major and manifold role in foods. Enzymes that are naturally present in foods may change the composition of those foods; in some cases, such changes are desirable but in most instances are undesirable, so the enzymes must be deactivated. The blanching of vegetables is an example of an undesirable change that is deactivated. Some enzymes are used as indicators in analytical methods; phosphatase, for instance, is used in the phosphatase test of pasteurization of milk. Enzymes are also used as processing aids in food manufacturing. For example, rennin, contained in extract of calves' stomachs, is used as a coagulant for milk in the production of cheese.

Food science's emphasis in the study of enzymes differs from that in biochemistry. The former deals mostly with decomposition reactions, hydrolysis, and oxidation; the latter is more concerned with synthetic mechanisms. Whitaker (1972) has prepared an extensive listing of the uses of enzymes in food processing (Table 10–1) and this gives a good summary of the many and varied possible applications of enzymes.

NATURE AND FUNCTION

Enzymes are proteins with catalytic properties. The catalytic properties are quite specific, which makes enzymes useful in analytical studies. Some enzymes consist only of protein, but most enzymes contain additional nonprotein components such as carbohydrates, lipids, metals, phosphates, or some other organic moiety. The complete enzyme is called *holoenzyme*; the protein part, *apoenzyme*; and the nonprotein part, *cofactor*. The compound that is being converted in an enzymic reaction is called *substrate*. In an enzyme reaction, the substrate combines with the holoenzyme and is released in a modified form, as indicated in Figure 10–1. An enzyme reaction, therefore, involves the following equations:

$$\text{Enzyme + substrate} \underset{k_2}{\overset{k_1}{\rightleftharpoons}} \text{complex}$$

$$\overset{k_3}{\longrightarrow} \text{enzyme + products}$$

The equilibrium for the formation of the complex is given by

$$K_m = \frac{[E][S]}{[ES]}$$

389

Table 10–1 Uses and Suggested Uses of Enzymes in Food Processing

Enzyme	Food	Purpose or Action
Amylases	Baked goods	Increase sugar content for yeast fermentation
	Brewing	Conversion of starch to maltose for fermentation; removal of starch turbidities
	Cereals	Conversion of starch to dextrins, sugar; increase water absorption
	Chocolate-cocoa	Liquidification of starches for free flow
	Confectionery	Recovery of sugar from candy scraps
	Fruit juices	Remove starches to increase sparkling properties
	Jellies	Remove starches to increase sparkling properties
	Pectin	An aid in preparation of pectin from apple pomace
	Syrups and sugars	Conversion of starches to low molecular weight dextrins (corn syrup)
	Vegetables	Hydrolysis of starch as in tenderization of peas
Cellulase	Brewing	Hydrolysis of complex carbohydrate cell walls
	Coffee	Hydrolysis of cellulose during drying of beans
	Fruits	Removal of graininess of pears; peeling of apricots, tomatoes
Dextran-sucrase	Sugar syrups	Thickening of syrup
	Ice cream	Thickening agent, body
Invertase	Artificial honey	Conversion of sucrose to glucose and fructose
	Candy	Manufacture of chocolate-coated, soft, cream candies
Lactase	Ice Cream	Prevent crystallization of lactose, which results in grainy, sandy texture
	Feeds	Conversion of lactose to galactose and glucose
	Milk	Stabilization of milk proteins in frozen milk by removal of lactose
Tannase	Brewing	Removal of polyphenolic compounds
Pentosanase	Milling	Recovery of starch from wheat flour
Naringinase	Citrus	Debittering citrus pectin juice by hydrolysis of the glucoside, naringin
Pectic enzymes (useful)	Chocolate-cocoa	Hydrolytic activity during fermentation of cocoa
	Coffee	Hydrolysis of gelatinous coating during fermentation of beans
	Fruits	Softening
	Fruit juices	Improve yield of press juices, prevent cloudiness, improve concentration processes
	Olives	Extraction of oil
	Wines	Clarification

continues

Table 10–1 continued

Enzyme	Food	Purpose or Action
Pectic enzymes (deteriorative)	Citrus juice	Destruction and separation of pectic substances of juices
	Fruits	Excessive softening action
Proteases (useful)	Baked goods	Softening action in doughs; cut mixing time, increase extensibility of doughs; improvement in grain, texture, loaf volume; liberate β-amylase
	Brewing	Body, flavor and nutrients development during fermentation; aid in filtration and clarification, chill-proofing
	Cereals	Modify proteins to increase drying rate, improve product handling characteristics; manufacture of miso and tofu
	Cheese	Casein coagulation; characteristic flavors during aging
	Chocolate-cocoa	Action on beans during fermentation
	Eggs, egg products	Improve drying properties
	Feeds	Use in treatment of waste products for conversion to feeds
	Meats and fish	Tenderization; recovery of protein from bones, trash fish; liberation of oils
	Milk	In preparation of soybean milk
	Protein hydrolysates	Condiments such as soy sauce and tamar sauce; specific diets; bouillon, dehydrated soups, gravy powders, processed meats
	Wines	Clarification
Proteases (deteriorative)	Eggs	Shelf life of fresh and dried whole eggs
	Crab, lobster	Overtenderization if not inactivated rapidly
	Flour	Influence on loaf volume, texture if too active
Lipase (useful)	Cheese	Aging, ripening, and general flavor characteristics
	Oils	Conversion of lipids to glycerol and fatty acids
	Milk	Production of milk with slightly cured flavor for use in milk chocolate
Lipase (deteriorative)	Cereals	Overbrowning of oat cakes; brown discoloration of wheat bran
	Milk and dairy products	Hydrolytic rancidity
	Oils	Hydrolytic rancidity
Phosphatases	Baby foods	Increase available phosphate
	Brewing	Hydrolysis of phosphate compounds
	Milk	Detection of effectiveness of pasteurization
Nucleases	Flavor enhancers	Production of nucleotides and nucleosides
Peroxidases (useful)	Vegetables	Detection of effectiveness of blanching
	Glucose determinations	In combination with glucose oxidase

continues

Table 10–1 continued

Enzyme	Food	Purpose or Action
Peroxidases (deteriorative)	Vegetables	Off-flavors
	Fruits	Contribution to browning action
Catalase	Milk	Destruction of H_2O_2 in cold pasteurization
	Variety of products	To remove glucose and/or oxygen to prevent browning and/or oxidation; used in conjunction with glucose oxidase
Glucose oxidase	Variety of products	Removal of oxygen and/or glucose from products such as beer, cheese, carbonated beverages, dried eggs, fruit juices, meat and fish, milk powder, wine to prevent oxidation and/or browning; used in conjunction with catalase
	Glucose determination	Specific determination of glucose; used in conjunction with peroxidase
Polyphenol oxidase (useful)	Tea, coffee, tobacco	Development of browning during ripening, fermentation, and/or aging process
Polyphenol oxidase (deteriorative)	Fruits, vegetables	Browning, off-flavor development, loss of vitamins
Lipoxygenase	Vegetables	Destruction of essential fatty acids and vitamin A; development of off-flavors
Ascorbic acid oxidase	Vegetables, fruits	Destruction of vitamin C (ascorbic acid)
Thiaminase	Meats, fish	Destruction of thiamine

Source: Reprinted with permission from J.R. Whitaker, *Principles of Enzymology for the Food Sciences*, 1972, by courtesy of Marcel Dekker, Inc.

where

E, S, and *ES* are the enzyme, substrate, and complex, respectively

K_m is the equilibrium constant

This can be expressed in the form of the Michaelis-Menten equation, as follows:

$$v = V\frac{[S]}{[S] + K_m}$$

where

v is the initial short-time velocity of the reaction at substrate concentration $[S]$

V is the maximum velocity that can be attained at a high concentration of the substrate where all of the enzyme is in the form of the complex

This equation indicates that when v is equal to one-half of V, the equilibrium constant K_m is numerically equal to S. A plot of the reaction rates at different substrate concentrations can be used to determine K_m. Because it is not always possible to attain the maximum reaction rate at varying substrate concentrations, the Michaelis-Menten equation has been modified by using reciprocals and in this

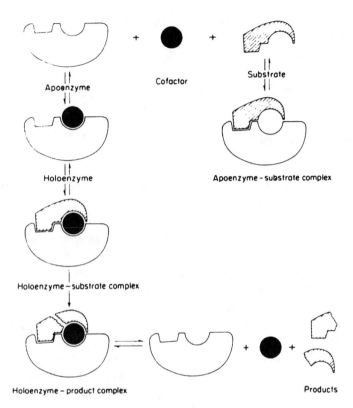

Figure 10–1 The Nature of Enzymes—Substrate Reactions

form is known as the Lineweaver-Burke equation,

$$\frac{1}{v} = \frac{1}{V} + \frac{K_m}{V[S]}$$

Plots of $1/v$ as a function of $1/[S]$ result in straight lines; the intercept on the Y-axis represents $1/V$; the slope equals K_m/V; and from the latter, K_m can be calculated.

Enzyme reactions follow either zero-order or first-order kinetics. When the substrate concentration is relatively high, the concentration of the enzyme-substrate complex will be maintained at a constant level and the amount of product formed is a linear function of the time interval. Zero-order reaction kinetics are characteristic of catalyzed reactions and can be described as follows:

$$\frac{d[S]}{dt} = k^\circ$$

where

S is substrate and k° is the zero-order reaction constant

First-order reaction kinetics are characterized by a graduated slowdown of the formation of product. This is because the rate of its formation is a function of the concentration of unreacted substrate, which decreases as the concentration of product increases. First-order reaction kinetics follow the equation,

$$\frac{d[S]}{dt} = k^1\,([S] - [P])$$

where

P is product and k^1 is the first-order reaction constant

For relatively short reaction times, the amount of substrate converted is proportional to the enzyme concentration.

Each enzyme has one—and some enzymes have more—optimum pH values. For most enzymes this is in the range of 4.5 to 8.0. Examples of pH optima are amylase, 4.8; invertase, 5.0; and pancreatic α-amylase, 6.9. The pH optimum is usually quite narrow, although some enzymes have a broader optimum range; for example, pectin methylesterase has a range of 6.5 to 8.0. Some enzymes have a pH optimum at very high or very low values, such as pepsin at 1.8 and arginase at 10.0.

Temperature has two counteracting effects on the activity of enzymes. At lower temperatures, there is a Q_{10} of about 2, but at temperatures over 40°C, the activity quickly decreases because of denaturation of the protein part of the enzymes. The result of these factors is a bell-shaped activity curve with a distinct temperature optimum.

Enzymes are proteins that are synthesized in the cells of plants, animals, or microorganisms. Most enzymes used in industrial applications are now obtained from microorganisms. Cofactors or coenzymes are small, heat-stable, organic molecules that may readily dissociate from the protein and can often be removed by dialysis. These coenzymes frequently contain one of the B vitamins; examples are tetrahydrofolic acid and thiamine pyrophosphate.

Specificity

The nature of the enzyme-substrate reaction as explained in Figure 10–1 requires that each enzyme reaction is highly specific. The shape and size of the active site of the enzyme, as well as the substrate, are important. But this complementarity may be even further expanded to cover amino acid residues in the vicinity of the active site, hydrophobic areas near the active site, or the presence of a positive electrical charge near the active site (Parkin 1993). Types of specificity may include group, bond, stereo, and absolute specificity, or some combination of these. An example of the specificity of enzymes is given in Figure 10–2, which illustrates the specificity of proline-specific peptidases (Habibi-Najafi and Lee 1996). The amino acid composition of casein is high in proline, and the location of this amino acid in the protein chain is inaccessible to common aminopeptidases and the di- and tripeptidases with broad specificity. Hydrolysis of the proline bonds requires proline-specific peptidases, including several exopeptidases and an endopeptidase. Figure 10–2 illustrates that this type of specificity is related to the type of amino acid in a protein as well as its location in the chain. Neighboring amino acids also determine the type of peptidase required to hydrolyze a particular peptide bond.

Classification

Enzymes are classified by the Commission on Enzymes of the International Union of Biochemistry. The basis for the classification is the division of enzymes into groups according to the type of reaction catalyzed. This, together with the name or names of substrate(s), is used to name individual enzymes. Each well-defined enzyme can be

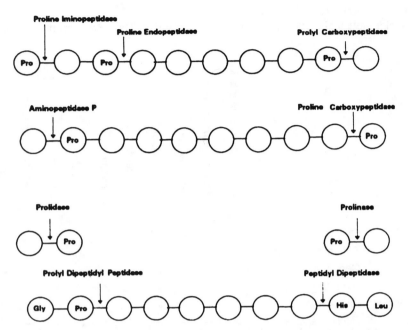

Figure 10–2 Mode of Action of Proline-Specific Peptidases. *Source*: Reprinted with permission from M.B. Habibi-Najafi and B.H. Lee, Bitterness in Cheese: A Review, *Crit. Rev. Food Sci. Nutr.*, Vol. 36, No. 5, p. 408. Copyright CRC Press, Boca Raton, Florida.

described in three ways—by a systematic name, by a trivial name, and by a number of the Enzyme Commission (EC). Thus, the enzyme α-amylase (trivial name) has the systematic name α-1,4-glucan-4-glucanohydrolase, and the number EC 3.2.1.1. The system of nomenclature has been described by Whitaker (1972; 1974) and Parkin (1993).

Enzyme Production

Some of the traditionally used industrial enzymes (e.g., rennet and papain) are prepared from animal and plant sources. Recent developments in industrial enzyme production have emphasized the microbial enzymes (Frost 1986). Microbial enzymes are very heat stable and have a broader pH optimum. Most of these enzymes are made by submerged cultivation of highly developed strains of microorganisms. Developments in

biotechnology will make it possible to transfer genes for the elaboration of specific enzymes to different organisms. The major industrial enzyme processes are listed in Table 10–2.

HYDROLASES

The hydrolases as a group include all enzymes that involve water in the formation of their products. For a substrate *AB*, the reaction can be represented as follows:

$$AB + HOH \rightarrow HA + BOH$$

The hydrolases are classified on the basis of the type of bond hydrolyzed. The most important are those that act on ester bonds, glycosyl bonds, peptide bonds, and C–N bonds other than peptides.

Table 10–2 Major Industrial Enzymes and the Process Used for Their Production

Enzyme	Source	Submerged Fermentation	Surface Fermentation	Intracellular	Extracellular	Concentration	Precipitation	Drying	Pelleting	Further Purification	Solid Product	Solution Product	Immobilized Product
Proteases													
Rennet	Calf stomach	—	—	—	✓	—	—	✓	—	—	✓	✓	—
Trypsin	Animal pancreas	—	—	—	✓	✓	✓	✓	—	✓	✓	—	—
Papain	Carica papaya fruit	—	—	—	✓	—	—	✓	—	—	✓	✓	—
Fungal	Aspergillus oryzae	✓	✓	—	✓	✓	✓	✓	—	—	✓	—	—
Fungal (rennins)	Mucor spp.	✓	✓	—	✓	✓	—	✓	—	—	✓	✓	—
Bacterial	Bacillus spp.	✓	—	—	✓	✓	✓	✓	✓	—	✓	✓	—
Glycosidases													
Bacterial α-amylase	Bacillus spp.	✓	—	—	✓	✓	—	✓	—	—	✓	✓	—
Fungal α-amylase	Aspergillus oryzae	✓	—	—	✓	✓	—	✓	—	—	✓	✓	—
β-amylase	Barley	—	—	—	✓	✓	✓	✓	—	—	✓	—	—
Amyloglucosidase	Aspergillus niger	✓	—	—	✓	✓	—	—	—	—	—	✓	—
Pectinase	Aspergillus niger	—	✓	—	✓	✓	—	✓	—	—	✓	✓	—
Cellulase	Molds	✓	—	—	✓	✓	✓	—	—	✓	✓	—	—
Yeast lactase	Kluyveromyces spp.	✓	—	✓	—	—	✓	✓	—	—	✓	✓	—
Mold lactase	Aspergillus spp.	✓	✓	—	✓	—	✓	✓	—	—	✓	—	✓
Others													
Glucose isomerase	Various microbial sources	✓	—	✓	—	✓	✓	✓	—	—	—	—	✓
Glucose oxidase	Aspergillus niger	✓	—	✓	—	—	✓	—	—	✓	✓	✓	—
Mold catalase	Aspergillus niger	✓	—	✓	—	—	✓	—	—	—	—	✓	—
Animal catalase	Liver	—	—	✓	—	✓	✓	✓	—	✓	✓	✓	—
Lipase	Molds	✓	—	—	✓	✓	✓	✓	—	—	✓	—	—

Source: From G.M. Frost, Commercial Production of Enzymes, in *Developments in Food Proteins*, B.J.F. Hudson, ed., 1986, Elsevier Applied Science Publishers Ltd.

Esterases

The esterases are involved in the hydrolysis of ester linkages of various types. The products formed are acid and alcohol. These enzymes may hydrolyze triglycerides and include several lipases; for instance, phospholipids are hydrolyzed by phospholipases, and cholesterol esters are hydrolyzed by cholesterol esterase. The carboxylesterases are enzymes that hydrolyze triglycerides such as tributyrin. They can be distinguished from lipases because they hydrolyze soluble substrates, whereas lipases only act at the water-lipid interfaces of emulsions. Therefore, any condition that results in increased surface area of the water-lipid interface will increase the activity of the enzyme. This is the reason that lipase activity is much greater in homogenized (not pasteurized) milk than in the non-homogenized product. Most of the lipolytic enzymes are specific for either the acid or the alcohol moiety of the substrate, and, in the case of esters of polyhydric alcohols, there may also be a positional specificity.

Lipases are produced by microorganisms such as bacteria and molds; are produced by plants; are present in animals, especially in the pancreas; and are present in milk. Lipases may cause spoilage of food because the free fatty acids formed cause rancidity. In other cases, the action of lipases is desirable and is produced intentionally. The boundary between flavor and off-flavor is often a very narrow range. For instance, hydrolysis of milk fat in milk leads to very unpleasant off-flavors at very low free fatty acid concentration. The hydrolysis of milk fat in cheese contributes to the desirable flavor. These differences are probably related to the background upon which these fatty acids are superimposed and to the specificity for particular groups of fatty acids of each enzyme.

In seeds, lipases may cause fat hydrolysis unless the enzymes are destroyed by heat. Palm oil produced by primitive methods in Africa used to consist of more than 10 percent of free fatty acids. Such spoilage problems are also encountered in grains and flour. The activity of lipase in wheat and other grains is highly dependent on water content. In wheat, for example, the activity of lipase is five times higher at 15.1 percent than at 8.8 percent moisture. The lipolytic activity of oats is higher than that of most other grains.

Lipases can be divided into those that have a positional specificity and those that do not. The former preferentially hydrolyze the ester bonds of the primary ester positions. This results in the formation of mono- and diglycerides, as represented by the following reaction:

During the progress of the reaction, the concentration of diglycerides and monoglycerides increases, as is shown in Figure 10–3.

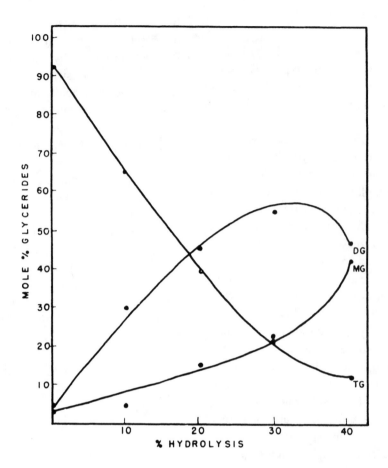

Figure 10–3 The Course of Pancreatic Lipase Hydrolysis of Tricaprylin. MG = monoglycerides, DG = diglycerides, TG = triglycerides. *Source*: From A. Boudreau and J.M. deMan, The Mode of Action of Pancreatic Lipase on Milkfat Glycerides, *Can. J. Biochem.*, Vol. 43, pp. 1799–1805, 1965.

The β-monoglycerides formed are resistant to further hydrolysis. This pattern is characteristic of pancreatic lipase and has been used to study the triglyceride structure of many fats and oils.

The hydrolysis of triglycerides in cheese is an example of a desirable flavor-producing process. The extent of free fatty acid formation is much higher in blue cheese than in Cheddar cheese, as is shown in Table 10–3. This is most likely the result of lipases elabo-

rated by organisms growing in the blue cheese, such as *P. roqueforti*, *P. camemberti*, and others. The extent of lipolysis increases with age, as is demonstrated by the increasing content of partial glycerides during the aging of cheese (Table 10–4). In many cases, lipolysis is induced by the addition of lipolytic enzymes. In the North American chocolate industry, it is customary to induce some lipolysis in chocolate by means of lipase. In the production of Italian cheeses, lipolysis is

Table 10–3 Free Fatty Acids in Some Dairy Products

Product	Free Fatty Acids (mg/kg)
Fresh milk	415
Moderately rancid cream	1,027
Butter	2,733
Cheddar cheese	1,793 (avg of 12 samples)
Blue cheese	23,500 to 66,700 (range 3 samples)

Source: From E.A. Day, Role of Milk Lipids in Flavors of Dairy Products, in *Flavor Chemistry*, R.F. Gould, ed., 1966, American Chemical Society.

induced by the use of pregastric esterases. These are lipolytic enzymes obtained from the oral glands located at the base of the tongue in calves, lambs, or kids.

Specificity for certain fatty acids by some lipolytic enzymes has been demonstrated. Pancreatic lipase and milk lipase are broad-spectrum enzymes and show no specificity for any of the fatty acids found in fats. Instead, the fatty acids that are released from

Table 10–4 Formation of Partial Glycerides in Cheddar Cheese

Product Type	Diglycerides (wt %)	Mono-glycerides (wt %)
Mild	7.4–7.6	1.0–2.0
Medium	7.6–9.7	0.5–1.4
Old	11.9–15.6	1.1–3.2

the glycerides occur in about the same ratio as they are present in the original fat. Specificity was shown by Nelson (1972) in calf esterase and in a mixed pancreatin-esterase preparation (Table 10–5). Pregastric esterases and lipase from *Aspergillus* species primarily hydrolyze shorter chain-length fatty acids (Arnold et al. 1975).

Specificity of lipases may be expressed in a number of different ways—substrate specific, regiospecific, nonspecific, fatty acyl specific, and stereospecific. Examples of these specificities have been presented by Villeneuve and Foglia (1997) (Table 10–6).

Substrate specificity is the ability to hydrolyze a particular glycerol ester, such as when

Table 10–5 Free Fatty Acids Released from Milkfat by Several Lipolytic Enzymes

Fatty Acid	Milk Lipase	Steapsin	Pancreatic Lipase	Calf Esterase	Esterase Pancreatin
4:0	13.9	10.7	14.4	35.00	15.85
6:0	2.1	2.9	2.1	2.5	3.6
8:0	1.8	1.5	1.4	1.3	3.0
10:0	3.0	3.7	3.3	3.1	5.5
12:0	2.7	4.0	3.8	5.1	4.4
14:0	7.7	10.7	10.1	13.2	8.5
16:0	21.6	21.6	24.0	15.9	19.3
18:1 and 18:2	29.2	24.3	25.5	14.2	21.1
18:0	10.5	13.4	9.7	3.2	10.1

Source: From J.H. Nelson, Enzymatically Produced Flavors for Fatty Systems, *J. Am. Oil Chem. Soc.*, Vol. 49, pp. 559–562, 1972.

Table 10–6 Examples of Lipase Specificities

Specificity	Lipase
Substrate specific	
Monoacylglyercols	Rat adipose tissue
Mono- and diacylglyc-erols	Penicillium camem-bertii
Triacylglycerols	Penicillium sp.
Regiospecific	
1,3-regioselective	Aspergilllus niger
	Rhizopus arrhizus
	Mucor miehei
sn-2-regioselective	Candida antarctica A
Nonspecific	Penicillium expan-sum
	Aspergillus sp.
	Pseudomonas cepacia
Fatty acylspecific	
Short-chain fatty acid (FA)	Penicillium roqueforti
	Premature infant gastric
cis-9 unsaturated FA	Geotrichum candi-dum
Long-chain unsatur-ated FA	Botrytis cinerea
Stereospecific	
sn-1 stereospecific	Humicola lanuginosa
	Pseudomonas aeruginosa
sn-3 stereospecific	Fusarium solani cutinase
	Rabbit gastric

Source: Reprinted with permission from P. Ville-neuve and T.A. Foglia, Lipase Specificities: Potential Application in Lipid Bioconversions, *J. Am. Oil Chem. Soc.*, Vol. 8, p. 641, © 1997, AOCS Press.

a lipase can rapidly hydrolyze a triacylglyc-erol, but acts on a monoacylglycerol only slowly. Regiospecificity involves a specific action on either the sn-1 and sn-3 positions or reaction with only the sn-2 position. The 1,3-specific enzymes have been researched extensively, because it is now recognized that lipases in addition to hydrolysis can catalyze the reverse reaction, esterification or transes-terification. This has opened up the possibil-ity of tailor-making triacylglycerols with a specific structure, and this is especially important for producing high-value fats such as cocoa butter equivalents. The catalytic activity of lipases is reversible and depends on the water content of the reaction mixture. At high water levels, the hydrolytic reaction prevails, whereas at low water levels the syn-thetic reaction is favored. A number of lipase catalyzed reactions are possible, and these have been summarized in Figure 10–4 (Ville-neuve and Foglia 1997). Most of the lipases used for industrial processes have been developed from microbes because these usu-ally exhibit high temperature tolerance. Lipases from *Mucor miehei* and *Candida antarctica* have been cloned and expressed in industry-friendly organisms. Lipases from genetically engineered strains will likely be of major industrial importance in the future (Godtfredsen 1993). Fatty acid–specific li-pases react with either short-chain fatty acids (*Penicillium roqueforti*) or some long-chain fatty acids such as *cis*-9-unsaturated fatty acids (*Geotrichum candidum*). Stereospecific lipases react with only fatty acids at the sn-1 or sn-3 position.

The applications of microbial lipases in the food industry involve the hydrolytic as well as the synthetic capabilities of these enzymes and have been summarized by Godtfredsen (1993) in Table 10–7.

The lipase-catalyzed interesterification process can be used for the production of tri-acylglycerols with specific physical proper-ties, and it also opens up possibilities for making so-called structured lipids. An exam-ple is a triacylglycerol that carries an essen-

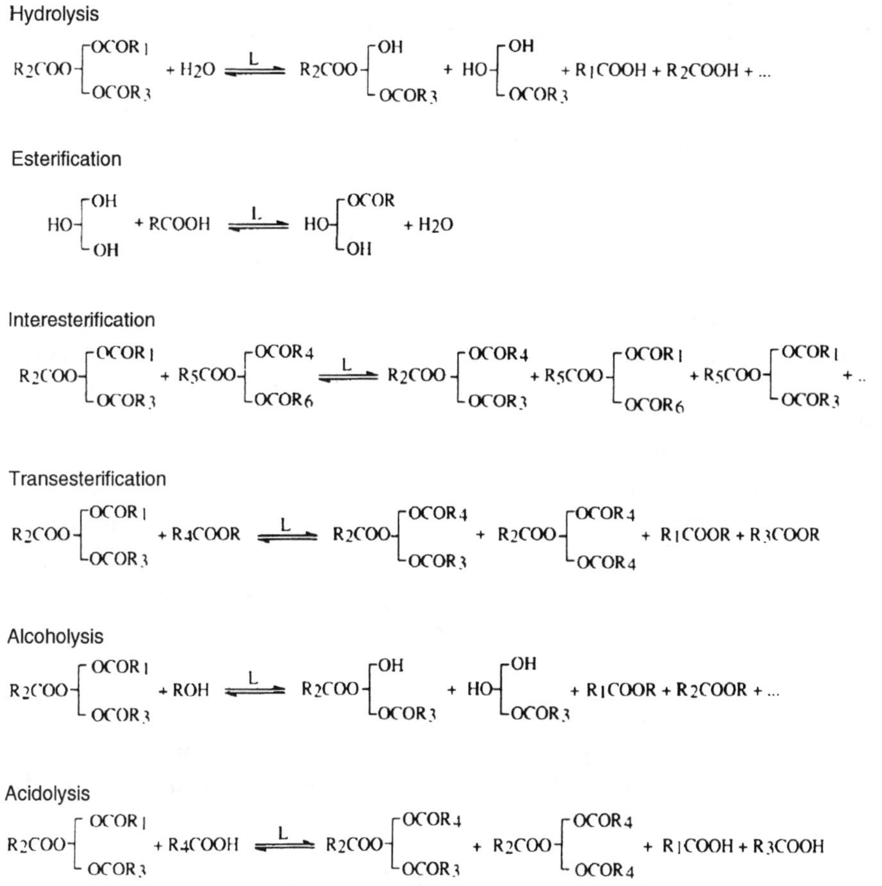

Hydrolysis

$$R_2COO{-}\begin{bmatrix}OCOR_1\\[4pt]OCOR_3\end{bmatrix} + H_2O \rightleftharpoons^{L} R_2COO{-}\begin{bmatrix}OH\\[4pt]OCOR_3\end{bmatrix} + HO{-}\begin{bmatrix}OH\\[4pt]OCOR_3\end{bmatrix} + R_1COOH + R_2COOH + \ldots$$

Esterification

$$HO{-}\begin{bmatrix}OH\\[4pt]OH\end{bmatrix} + RCOOH \rightleftharpoons^{L} HO{-}\begin{bmatrix}OCOR\\[4pt]OH\end{bmatrix} + H_2O$$

Interesterification

$$R_2COO{-}\begin{bmatrix}OCOR_1\\[4pt]OCOR_3\end{bmatrix} + R_5COO{-}\begin{bmatrix}OCOR_4\\[4pt]OCOR_6\end{bmatrix} \rightleftharpoons^{L} R_2COO{-}\begin{bmatrix}OCOR_4\\[4pt]OCOR_3\end{bmatrix} + R_5COO{-}\begin{bmatrix}OCOR_1\\[4pt]OCOR_6\end{bmatrix} + R_5COO{-}\begin{bmatrix}OCOR_1\\[4pt]OCOR_3\end{bmatrix} + \ldots$$

Transesterification

$$R_2COO{-}\begin{bmatrix}OCOR_1\\[4pt]OCOR_3\end{bmatrix} + R_4COOR \rightleftharpoons^{L} R_2COO{-}\begin{bmatrix}OCOR_4\\[4pt]OCOR_3\end{bmatrix} + R_2COO{-}\begin{bmatrix}OCOR_4\\[4pt]OCOR_4\end{bmatrix} + R_1COOR + R_3COOR$$

Alcoholysis

$$R_2COO{-}\begin{bmatrix}OCOR_1\\[4pt]OCOR_3\end{bmatrix} + ROH \rightleftharpoons^{L} R_2COO{-}\begin{bmatrix}OH\\[4pt]OCOR_3\end{bmatrix} + HO{-}\begin{bmatrix}OH\\[4pt]OCOR_3\end{bmatrix} + R_1COOR + R_2COOR + \ldots$$

Acidolysis

$$R_2COO{-}\begin{bmatrix}OCOR_1\\[4pt]OCOR_3\end{bmatrix} + R_4COOH \rightleftharpoons^{L} R_2COO{-}\begin{bmatrix}OCOR_4\\[4pt]OCOR_3\end{bmatrix} + R_2COO{-}\begin{bmatrix}OCOR_4\\[4pt]OCOR_4\end{bmatrix} + R_1COOH + R_3COOH$$

Figure 10–4 Lipase Catalyzed Reactions Used in Oil and Fat Modification. *Source:* Reprinted with permission from P. Villeneuve and T.A. Foglia, Lipase Specificities: Potential Application in Lipid Bioconversions, *J. Am. Oil Chem. Soc.*, Vol. 8, p. 642, © 1997, AOCS Press.

tial fatty acid (e.g., DHA-docosahexaenoic acid) in the sn-2 position and short-chain fatty acids in the sn-1 and sn-3 positions. Such a structural triacylglycerol would rapidly be hydrolyzed in the digestive tract and provide an easily absorbed monoacylglycerol that carries the essential fatty acid (Godtfredsen 1993).

The lipases that have received attention for their ability to synthesize ester bonds have been obtained from yeasts, bacteria, and fungi. Lipases can be classified into three groups according to their specificity (Macrae 1983). The first group contains nonspecific lipases. These show no specificity regarding the position of the ester bond in the glycerol molecule, or the nature of the fatty acid. Examples of enzymes in this group are lipases of *Candida cylindracae, Corynebacterium acnes*, and *Staphylococcus aureus*. The second group contains lipases with position specificity for the 1- and 3-positions of the glycerides. This is common among microbial lipases and is the result of the steri-

Table 10–7 Application of Microbial Lipases in the Food Industry

Industry	Effect	Product
Dairy	Hydrolysis of milk fat	Flavor agents
	Cheese ripening	Cheese
	Modification of butter fat	Butter
Bakery	Flavor improvement and shelf-life prolongation	Bakery products
Beverage	Improved aroma	Beverages
Food dressing	Quality improvement	Mayonnaise, dressing, and whipped toppings
Health food	Transesterification	Health foods
Meat and fish	Flavor development and fat removal	Meat and fish products
Fat and oil	Transesterification	Cocoa butter, margarine
	Hydrolysis	Fatty acids, glycerol, mono- and diglycerides

Source: Reprinted with permission from S.E. Godtfredsen, Lipases, Enzymes in Food Processing, T. Nagodawithana and G. Reed, eds., p. 210, © 1993, Academic Press.

cally hindered ester bond of the 2-position's inability to enter the active site of the enzyme. Lipases in this group are obtained from *Aspergillus niger, Mucor javanicus,* and *Rhizopus arrhizus.* The third group of lipases show specificity for particular fatty acids. An example is the lipase from *Geotrichum candidum,* which has a marked specificity for long-chain fatty acids that contain a *cis* double bond in the 2-position. The knowledge of the synthetic ability of lipases has opened a whole new area of study in the modification of fats. The possibility of modifying fats and oils by immobilized lipase technology may result in the production of food fats that have a higher essential fatty acid content and lower *trans* levels than is possible with current methods of hydrogenation.

Amylases

The amylases are the most important enzymes of the group of glycoside hydro-lases. These starch-degrading enzymes can be divided into two groups, the so-called debranching enzymes that specifically hydrolyze the 1,6-linkages between chains, and the enzymes that split the 1,4-linkages between glucose units of the straight chains. The latter group consists of endoenzymes that cleave the bonds at random points along the chains and exoenzymes that cleave at specific points near the chain ends. This behavior has been represented by Marshall (1975) as a diagram of the structure of amylopectin (Figure 10–5). In this molecule, the 1,4-α-glucan chains are interlinked by 1,6-α-glucosidic linkages resulting in a highly branched molecule. The molecule is composed of three types of chains; the *A* chains carry no substituent, the *B* chains carry other chains linked to a primary hydroxyl group, and the molecule contains only one *C* chain with a free reducing glucose unit. The chains are 25 to 30 units in length in starch and only 10 units in glycogen.

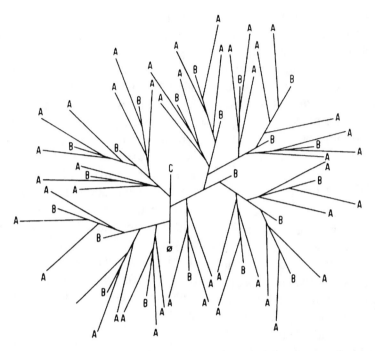

Figure 10–5 Diagrammatic Representation of Amylopectin Structure. Lines represent α-D-glucan chains linked by 1,4-bonds. The branch points are 1,6-α glucosidic bonds. *Source*: From J.J. Marshall, Starch Degrading Enzymes, Old and New, *Starke*, Vol. 27, pp. 377–383, 1975.

Alpha-amylase (α-1,4-Glucan 4-Glucanohydrolase)

This enzyme is distributed widely in the animal and plant kingdoms. The enzyme contains 1 gram-atom of calcium per mole. Alpha-amylase (α-1,4-glucan-4-glucanohydrolase) is an endoenzyme that hydrolyzes the α-1,4-glucosidic bonds in a random fashion along the chain. It hydrolyzes amylopectin to oligosaccharides that contain two to six glucose units. This action, therefore, leads to a rapid decrease in viscosity, but little monosaccharide formation. A mixture of amylose and amylopectin will be hydrolyzed into a mixture of dextrins, maltose, glucose, and oligosaccharides. Amylose is completely hydrolyzed to maltose, although there usually is some maltotriose formed, which hydrolyzes only slowly.

Beta-amylase (α-1,4-Glucan Maltohydrolase)

This is an exoenzyme and removes successive maltose units from the nonreducing end of the glucosidic chains. The action is stopped at the branch point where the α-1,6 glucosidic linkage cannot be broken by α-amylase. The resulting compound is named *limit dextrin*. Beta-amylase is found only in higher plants. Barley malt, wheat, sweet potatoes, and soybeans are good sources. Beta-amylase is technologically important in the baking, brewing, and distilling industries, where starch is converted into the fermentable sugar maltose. Yeast ferments maltose, sucrose, invert sugar, and glucose but does not ferment dextrins or oligosaccharides containing more than two hexose units.

Glucoamylase (α-1,4-Glucan Glucohydrolase)

This is an exoenzyme that removes glucose units in a consecutive manner from the non-reducing end of the substrate chain. The product formed is glucose only, and this differentiates this enzyme from α- and β-amylase. In addition to hydrolyzing the α-1,4 linkages, this enzyme can also attack the α-1,6 linkages at the branch point, albeit at a slower rate. This means that starch can be completely degraded to glucose. The enzyme is present in bacteria and molds and is used industrially in the production of corn syrup and glucose.

A problem in the enzymic conversion of corn starch to glucose is the presence of transglucosidase enzyme in preparations of α-amylase and glucoamylase. The transglucosidase catalyzes the formation of oligosaccharides from glucose, thus reducing the yield of glucose.

Nondamaged grains such as wheat and barley contain very little α-amylase but relatively high levels of β-amylase. When these grains germinate, the β-amylase level hardly changes, but the α-amylase content may increase by a factor of 1,000. The combined action of α- and β-amylase in the germinated grain greatly increases the production of fermentable sugars. The development of α-amylase activity during malting of barley is shown in Table 10–8. In wheat flour, high α-

amylase activity is undesirable, because too much carbon dioxide is formed during baking.

Raw, nondamaged, and ungelatinized starch is not susceptible to β-amylase activity. In contrast, α-amylase can slowly attack intact starch granules. This differs with the type of starch; for example, waxy corn starch is more easily attacked than potato starch. In general, extensive hydrolysis of starch requires gelatinization. Damaged starch granules are more easily attacked by amylases, which is important in bread making Alpha-amylase can be obtained from malt, from fungi (*Aspergillus oryzae*), or from bacteria (*B. subtilis*). The bacterial amylases have a higher temperature tolerance than the malt amylases.

Beta-galactosidase (β-D-Galactoside Galactohydrolase)

This enzyme catalyzes the hydrolysis of β-D-galactosides and α-L-arabinosides. It is best known for its action in hydrolyzing lactose and is, therefore, also known as lactase. The enzyme is widely distributed and occurs in higher animals, bacteria, yeasts, and

Table 10–8 Development of α-Amylase During Malting of Barley at 20°C

Days of Steeping and Germination	α-Amylase (20° Dextrose Units)
0	0
3	55
5	110
7	130
8	135

Source: From S.R. Green, New Use of Enzymes in the Brewing Industry, *MBAA Tech. Quar.*, Vol. 6, pp. 33–39, 1969.

plants. Beta-galactosidase or lactase is found in humans in the cells of the intestinal mucous membrane. A condition that is widespread in non-Caucasian adults is characterized by an absence of lactase. Such individuals are said to have lactose intolerance, which is an inability to digest milk properly.

The presence of galactose inhibits lactose hydrolysis by lactase. Glucose does not have this effect.

Pectic Enzymes

The pectic enzymes are capable of degrading pectic substances and occur in higher plants and in microorganisms. They are not found in higher animals, with the exception of the snail. These enzymes are commercially important for the treatment of fruit juices and beverages to aid in filtration and clarification and increasing yields. The enzymes can also be used for the production of low methoxyl pectins and galacturonic acids. The presence of pectic enzymes in fruits and vegetables can result in excessive softening. In tomato and fruit juices, pectic enzymes may cause "cloud" separation.

There are several groups of pectic enzymes, including pectinesterase, the enzyme that hydrolyzes methoxyl groups, and the depolymerizing enzymes polygalacturonase and pectate lyase.

Pectinesterase (Pectin Pectyl-Hydrolase)

This enzyme removes methoxyl groups from pectin. The enzyme is referred to by several other names, including pectase, pectin methoxylase, pectin methyl esterase, and pectin demethylase. Pectinesterases are found in bacteria, fungi, and higher plants, with very large amounts occurring in citrus fruits and tomatoes. The enzyme is specific for galacturonide esters and will not attack non-galacturonide methyl esters to any large extent. The reaction catalyzed by pectin esterase is presented in Figure 10–6. It has been suggested that the distribution of methoxyl groups along the chain affects the reaction velocity of the enzyme (MacMillan and Sheiman 1974). Apparently, pectinesterase requires a free carboxyl group next to an esterified group on the galacturonide chain to act, with the pectinesterase moving down the chain linearly until an obstruction is reached.

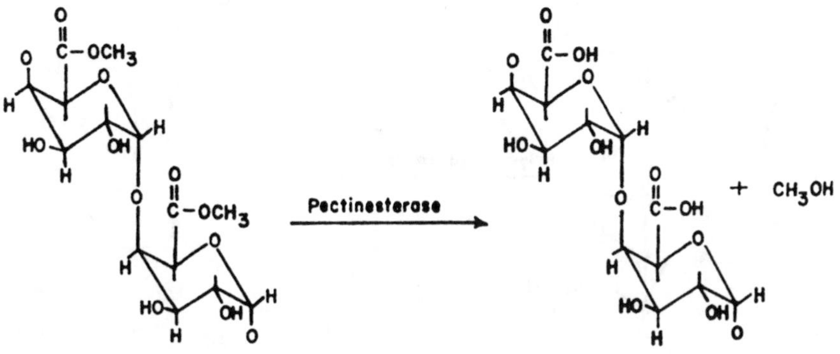

Figure 10–6 Reaction Catalyzed by Pectinesterase

To maintain cloud stability in fruit juices, high-temperature–short-time (HTST) pasteurization is used to deactivate pectolytic enzymes. Pectin is a protective colloid that helps to keep insoluble particles in suspension. Cloudiness is required in commercial products to provide a desirable appearance. The destruction of the high levels of pectinesterase during the production of tomato juice and puree is of vital importance. The pectinesterase will act quite rapidly once the tomato is broken. In the so-called hot-break method, the tomatoes are broken up at high temperature so that the pectic enzymes are destroyed instantaneously.

Polygalacturonase (Poly-α-1,4-Galacturonide Glycanohydrolase)

This enzyme is also known as pectinase, and it hydrolyzes the glycosidic linkages in pectic substances according to the reaction pattern shown in Figure 10–7. The polygalacturonases can be divided into endoenzymes that act within the molecule on α-1,4 linkages and exoenzymes that catalyze the stepwise hydrolysis of galacturonic acid molecules from the nonreducing end of the chain. A further division can be made by the fact that some polygalacturonases act principally on methylated substrates (pectins), whereas others act on substrates with free carboxylic acid groups (pectic acids). These enzymes are named polymethyl galacturonases and polygalacturonases, respectively. The preferential mode of hydrolysis and the preferred substrates are listed in Table 10–9. Endopolygalacturonases occur in fruits and in filamentous fungi, but not in yeast or bacteria. Exopolygalacturonases occur in plants (for example, in carrots and peaches), fungi, and bacteria.

Pectate Lyase (Poly-α-1,4-D-Galacturonide Lyase)

This enzyme is also known as trans-eliminase; it splits the glycosidic bonds of a glucuronide chain by trans elimination of hydrogen from the 4- and 5-positions of the glucuronide moiety. The reaction pattern is presented in Figure 10–8. The glycosidic bonds in pectin are highly susceptible to this reaction. The pectin lyases are of the endotype and are obtained exclusively from fila-

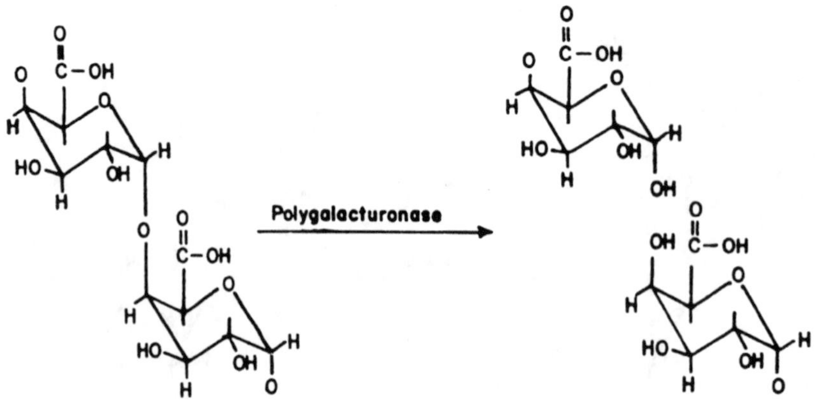

Figure 10–7 Reaction Catalyzed by Polygalacturonase

Table 10–9 Action of Polygalacturonases

Type of Attack	Enzyme	Preferred Substrate
Random	Endo-polymethylgalacturonase	Pectin
Random	Endo-polygalacturonase	Pectic acid
Terminal	Exo-polymethylgalacturonase	Pectin
Terminal	Exo-polygalacturonase	Pectic acid

mentous fungi, such as *Aspergillus niger.* The purified enzyme has an optimum pH of 5.1 to 5.2 and isoelectric point between 3 and 4 (Albersheim and Kilias 1962).

Commercial Use

Pectic enzymes are used commercially in the clarification of fruit juices and wines and for aiding the disintegration of fruit pulps. By reducing the large pectin molecules into smaller units and eventually into galacturonic acid, the compounds become water soluble and lose their suspending power; also, their viscosity is reduced and the insoluble pulp particles rapidly settle out.

Most microorganisms produce at least one but usually several pectic enzymes. Almost all fungi and many bacteria produce these enzymes, which readily degrade the pectin layers holding plant cells together. This leads to separation and degradation of the cells, and the plant tissue becomes soft. Bacterial degradation of pectin in plant tissues is responsible for the spoilage known as "soft rot" in fruits and vegetables. Commercial food grade pectic enzyme preparations may contain several different pectic enzymes. Usually, one type predominates; this depends on the intended use of the enzyme preparation.

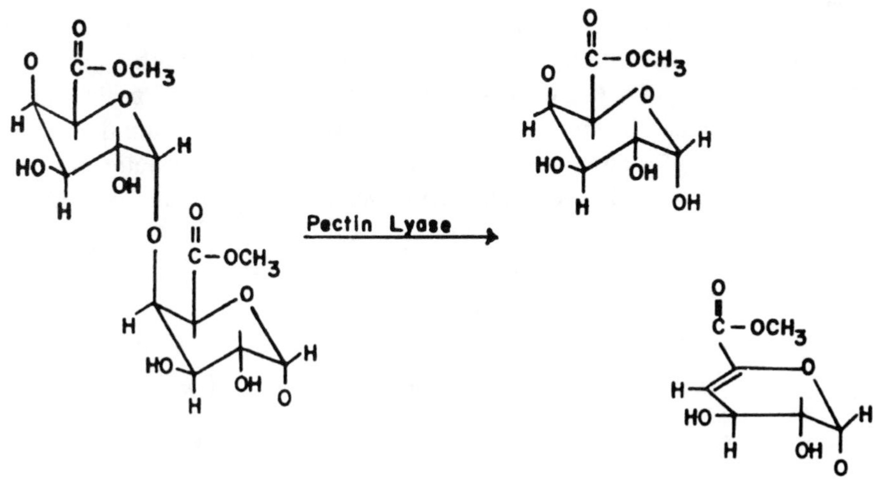

Figure 10–8 Reaction Catalyzed by Pectin Lyase

Proteases

Proteolytic enzymes are important in many industrial food processing procedures. The reaction catalyzed by proteolytic enzymes is the hydrolysis of peptide bonds of proteins; this reaction is shown in Figure 10–9. Whitaker (1972) has listed the specificity requirements for the hydrolysis of peptide bonds by proteolytic enzymes. These include the nature of R_1 and R_2 groups, configuration of the amino acid, size of substrate molecule, and the nature of the X and Y groups. A major distinguishing factor of proteolytic enzymes is the effect of R_1 and R_2 groups. The enzyme α-chymotrypsin hydrolyzes peptide bonds readily only when R_1 is part of a tyrosyl, phenylalanyl, or tryptophanyl residue. Trypsin requires R_1 to belong to an arginyl or lysyl residue. Specific requirement for the R_2 groups is exhibited by pepsin and the carboxypeptidases; both require R_2 to belong to a phenylalanyl residue. The enzymes require the amino acids of proteins to be in the L-configuration but frequently do not have a strict requirement for molecular size. The nature of X and Y permits the division of proteases into endopeptidases and exopeptidases. The former split peptide bonds in a random way in the interior of the substrate molecule and show maximum activity when X and Y are derived. The carboxypeptidases require that Y be a hydroxyl group, the aminopeptidases require that X be a hydrogen, and the dipeptidases require that X and Y both be underived.

Proteolytic enzymes can be divided into the following four groups: the acid proteases, the serine proteases, the sulfhydryl proteases, and the metal-containing proteases.

Acid Proteases

This is a group of enzymes with pH optima at low values. Included in this group are pepsin, rennin (chymosin), and a large number of microbial and fungal proteases. Rennin, the pure enzyme contained in rennet, is an extract of calves' stomachs that has been used for thousands of years as a coagulating agent in cheese making. Because of the scarcity of calves' stomachs, rennet substitutes are now widely used, and the coagulants used in cheese making usually contain mixtures of rennin and pepsin and/or microbial proteases. Some of the microbial proteases have been used for centuries in the Far East in the production of fermented foods such as soy sauce.

Rennin is present in the fourth stomach of the suckling calf. It is secreted in an inactive form, a zymogen, named prorennin. The crude extract obtained from the dried stomachs (vells) contains both rennin and prorennin. The conversion of prorennin to rennin can be speeded up by addition of acid. This conversion involves an autocatalytic process, in which a limited proteolysis of the proren-

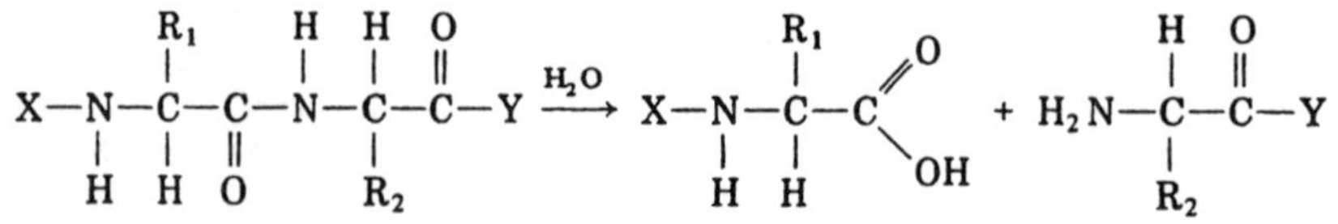

Figure 10–9 Reaction Catalyzed by Proteases

nin occurs, thus reducing the molecular weight about 14 percent. The conversion can also be catalyzed by pepsin. The process involves the release of peptides from the N-terminal end of prorennin, which reduces the molecular weight from about 36,000 to about 31,000. The molecule of prorennin consists of a single peptide chain joined internally by three disulfide bridges. After conversion to rennin, the disulfide bridges remain intact. As the calves grow older and start to eat other feeds as well as milk, the stomach starts to produce pepsin instead of rennin. The optimum activity of rennin is at pH 3.5, but it is most stable at pH 5; the clotting of cheese milk is carried out at pH values of 5.5 to 6.5.

The coagulation or clotting of milk by rennin occurs in two stages. In the first, the enzymic stage, the enzyme acts on κ-casein so that it can no longer stabilize the casein micelle. The second, or nonenzymic stage, involves the clotting of the modified casein micelles by calcium ions. The enzymic stage involves a limited and specific action on the κ-casein, resulting in the formation of insoluble *para*-κ-casein and a soluble macropeptide. The latter has a molecular weight of 6,000 to 8,000, is extremely hydrophilic, and contains about 30 percent carbohydrate. The glycomacropeptide contains galactosamine, galactose, and N-acetyl neuraminic acid (sialic acid). The splitting of the glycomacropeptide from κ-casein involves the breaking of a phenylalanine-methionine bond in the peptide chain. Other clotting enzymes—including pepsin, chymotrypsin, and microbial proteases—break the same bond and produce the same glycomacropeptide.

Pepsin is elaborated in the mucosa of the stomach lining in the form of pepsinogen. The high acidity of the stomach aids in the autocatalytic conversion into pepsin. This conversion involves splitting several peptide fragments from the N-terminal end of pepsinogen. The fragments consist of one large peptide and several small ones. The large peptide remains associated with pepsinogen by noncovalent bonds and acts as an inhibitor. The inhibitor dissociates from pepsin at a pH of 1 to 2. In the initial stages of the conversion of pepsinogen to pepsin, six peptide bonds are broken, and continued action on the large peptide (Figure 10–10) results in three more bonds being hydrolyzed. In this process, the molecular weight changes from 43,000 to 35,000 and the isoelectric point changes from 3.7 to less than 1. The pepsin molecule consists of a single polypeptide chain that contains 321 amino acids. The tertiary structure is stabilized by three disulfide bridges and a phosphate linkage. The phosphate group is attached to a seryl residue and is not essential for enzyme activity. The pH optimum of pepsin is pH 2 and the enzyme is stable from pH 2 to 5. At higher pH values, the enzyme is rapidly denatured and loses its activity. The primary specificity of pepsin is toward the R_2 group (see the equation shown in Figure 10–9), and it prefers this to be a phenylalanyl, tyrosyl, or tryptophanyl group.

The use of other acid proteases as substitutes for rennin in cheese making is determined by whether bitter peptides are formed during ripening of the cheese and by whether initial rapid hydrolysis causes excessive protein losses in the whey. Some of the acid proteases used in cheese making include preparations obtained from the organisms *Endothia parasitica*, *Mucor miehei*, and *Mucor pusillus*. Rennin contains the enzyme chymosin, and the scarcity of this natural enzyme preparation for cheese making resulted in the use of pepsin for this purpose. Pepsin and chymosin have primary structures that have about 50 percent homology

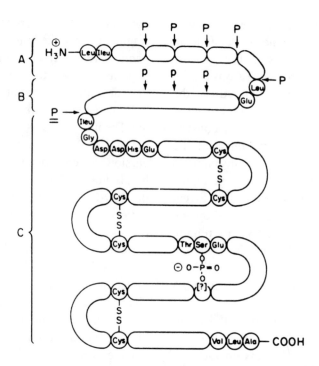

Figure 10–10 Structure of Pepsinogen and Its Conversion to Pepsin. *Source*: From F.A. Bovey and S.S. Yanari, Pepsin, in *The Enzymes*, Vol. 4, P.D. Boyer et al., eds., 1960, Academic Press.

and quite similar tertiary structures. The molecular mass of the two enzymes is similar, 35 kDa, but chymosin has a higher pI. Much of the chymosin used in cheese making is now obtained by genetic engineering processes. In the production of soy sauce and other eastern food products, such as miso (an oriental fermented food) and ketjap (Indonesian type soy sauce), the acid proteases of *Aspergillus oryzae* are used. Other products involve the use of the fungus *Rhizopus oligosporus*. Acid proteases also play a role in the ripening process of a variety of soft cheeses. This includes the *Penicillia* used in the blue cheeses, such as Roquefort, Stilton, and Danish blue, and in Camembert and Brie. The molds producing the acid proteases may grow either on the surface of the cheese or throughout the body of the cheese.

Serine Proteases

This group includes the chymotrypsins, trypsin, elastase, thrombin, and subtilisin. The name of this group of enzymes refers to the seryl residue that is involved in the active site. As a consequence, all of these enzymes are inhibited by diisopropylphosphorofluoridate, which reacts with the hydroxyl group of the seryl residue. They also have an imidazole group as part of the active site and they are all endopeptides. The chymotrypsins, trypsin and elastase, are pancreatic enzymes that carry out their function in the intestinal tract. They are produced as inactive zymogens and are converted into the active form by limited proteolysis.

Sulfhydryl Proteases

These enzymes obtain their name from the fact that a sulfhydryl group in the molecule is essential for their activity. Most of these enzymes are of plant origin and have found widespread use in the food industry. The only sulfhydryl proteases of animal origin are two of the cathepsins, which are present in the tissues as intracellular enzymes. The most important enzymes of this group are papain, ficin, and bromelain. Papain is an enzyme present in the fruit, leaves, and trunk of the papaya tree (*Carica papaya*). The commercial enzyme is obtained by purification of the exudate of full-grown but unripe papaya fruits. The purification involves use of affinity chromatography on a column containing an inhibitor (Liener 1974). This process leads to the full activation of the enzyme, which then contains 1 mole of sulfhydryl per mole of protein. The crude papain is not fully active and contains only 0.5 mole of sulfhydryl per mole of protein. Bromelain is obtained from the fruit or stems of the pineapple plant (*Ananas comosus*). The stems are pressed and the enzyme precipitated from the juice by acetone. Ficin is obtained from the latex of tropical fig trees (*Ficus glabrata*). The enzyme is not homogeneous and contains at least three different proteolytic components.

The active sites of these plant enzymes contain a cysteine and a histidine group that are essential for enzyme activity. The pH optimum is fairly broad and ranges from 6 to 7.5. The enzymes are heat stable up to temperatures in the range of 60 to 80°C. The papain molecule consists of a single polypeptide chain of 212 amino acids. The molecular weight is 23,900. Ficin and bromelain contain carbohydrate in the molecule; papain does not. The molecular weights of the enzymes are quite similar; that of ficin is 25,500 and that of bromelain, 20,000 to 33,200. These enzymes catalyze the hydrolysis of many different compounds, including peptide, ester, and amide bonds. The variety of peptide bonds split by papain appears to indicate a low specificity. This has been attributed (Liener 1974) to the fact that papain has an active site consisting of seven subsites that can accommodate a variety of amino acid sequences in the substrate. The specificity in this case is not determined by the nature of the side chain of the amino acid involved in the susceptible bond but rather by the nature of the adjacent amino acids.

Commercial use of the sulfhydryl proteases includes stabilizing and chill proofing of beer. Relatively large protein fragments remaining after the malting of barley may cause haze in beer when the product is stored at low temperatures. Controlled proteolysis sufficiently decreases the molecular weight of these compounds so that they will remain in solution. Another important use is in the tenderizing of meat. This can be achieved by injecting an enzyme solution into the carcass or by applying the enzyme to smaller cuts of meat. The former method suffers from the difficulty of uneven proteolysis in different parts of the carcass with the risk of overtenderizing some parts of the carcass.

Metal-Containing Proteases

These enzymes require a metal for activity and are inhibited by metal-chelating compounds. They are exopeptidases and include carboxypeptidase A (peptidyl-L-amino-acid hydrolase) and B (peptidyl-L-lysine hydrolase), which remove amino acids from the end of peptide chains that carry a free α-carboxyl group. The aminopeptidases remove amino acids from the free α-amino end of the peptide chain. The metalloexopeptidases require a divalent metal as a cofactor; the carboxypeptidases contain zinc. These enzymes are quite specific in the action; for example, carboxypeptidase B requires the C-terminal amino acid to be either arginine or lysine; the requirement for carboxypeptidase A is phenylalanine, tryptophan, or isoleucine. These specificities are compared with those of some other proteolytic enzymes in Figure 10–11. The carboxypeptidases are relatively small molecules; molecular weight of carboxypeptidase A is 34,600. The amino peptidases have molecular weights around 300,000. Although many of the aminopeptidases are found in animal tissues, several are present in microorganisms (Riordan 1974).

Protein Hydrolysates

Protein hydrolysates is the name given to a family of protein breakdown products obtained by the action of enzymes. It is also possible to hydrolyze proteins by chemical means, acids, or alkali, but the enzymatic method is preferred. Many food products such as cheese and soy sauce are obtained by enzymatic hydrolysis. The purpose of the production of protein hydrolysates is to improve nutritional value, cost, taste, antigenicity, solubility, and functionality. The proteins most commonly selected for producing hydrolysates are casein, whey protein, and soy protein (Lahl and Braun 1994). Proteins can be hydrolyzed in steps to yield a series of proteoses, peptones, peptides, and finally amino acids (Table 10–10). These products should not be confused with hydrolyzed vegetable proteins, which are intended as flavoring substances.

The extent of hydrolysis of protein hydrolysates is measured by the ratio of the amount of amino nitrogen to the total amount of nitrogen present in the raw material (AN/TN ratio). Highly hydrolyzed materials have AN/TN ratios of 0.50 to 0.60. To obtain the desired level of hydrolysis in a protein, a combination of proteases is selected. Serine protease prepared from *Bacillus licheniformis* has broad specificity and some preference for terminal hydrophobic amino acids. Peptides containing terminal hydrophobic amino acids cause bitterness. Usually a mixture of different proteases is employed. The hydrolysis reaction is terminated by adjust-

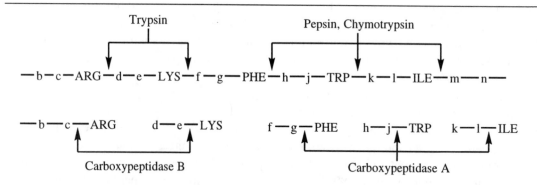

Figure 10–11 Specificity of Some Proteolytic Enzymes

Table 10–10 Protein Hydrolysate Products Produced from Casein and Whey Protein Concentrate (WPC)

Hydrolysate[a]	Protein Source	Average Molecular Weight[b]	AN/TN[c]
Intact protein	Casein	28,500	0.07
	WPC	25,000	0.06
Proteose	Casein	6,000	0.13
	WPC	6,800	0.11
Peptone	Casein	2,000	0.24
	WPC	1,400	0.24
Peptides	Casein	400	0.48
	WPC	375	0.43
Peptides and free amino acids	Casein	260	0.55
	WPC	275	0.58

[a] Commercial hydrolysates produced by Deltown Specialties, Fraser, NY.

[b] Determined by reverse-phase HPLC.

[c] Ratio of amino nitrogen present in the hydrolysate to the total amount of nitrogen present in the substrate.

Source: Reprinted with permission from W.J. Lahl and S.D. Braun, Enzymatic Production of Protein Hydrolysates for Food Use, *Food Technology*, Vol. 48, No. 10, p. 69, © 1994, Institute of Food Technologists.

ing the pH and increasing the temperature to inactivate the enzymes. The process for producing hydrolysates is shown in Figure 10–12 (Lahl and Braun 1994). Protein hydrolysates can be used as food ingredients with specific functional properties or for physiological or medical reasons. For example, hydrolyzed proteins may lose allergenic properties by suitably arranged patterns of hydrolysis (Cordle 1994).

OXIDOREDUCTASES

Phenolases

The enzymes involved in enzymic browning are known by the name polyphenoloxidase and are also called polyphenolase or phenolase. It is generally agreed (Mathew and Parpia 1971) that these terms include all enzymes that have the capacity to oxidize phenolic compounds to *o*-quinones. This can be represented by the conversion of *o*-dihydroxyphenol to *o*-quinone,

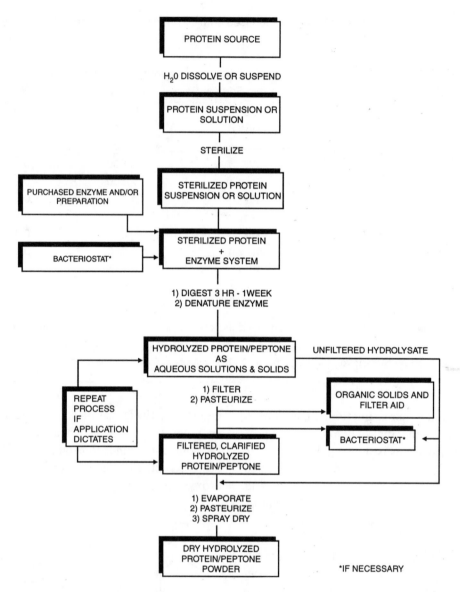

Figure 10–12 Process for the Production of Protein Hydrolysates. *Source*: Reprinted with permission from W.J. Lahl and S.D. Braun, Enzymatic Production of Protein Hydrolysates for Food Use, *Food Technology*, Vol. 48, No. 10, p. 70, © 1994, Institute of Food Technologists.

The action of polyphenolases is detrimental when it leads to browning in bruised and broken plant tissue but is beneficial in the processing of tea and coffee. The enzyme occurs in almost all plants, but relatively high levels are found in potatoes, mushrooms, apples, peaches, bananas, avocados, tea leaves, and coffee beans.

In addition to changing *o*-diphenols into *o*-quinones, the enzymes also catalyze the conversion of monophenols into *o*-diphenols, as follows:

OH

+ O$_2$ + BH$_2$ ⟶

CH$_3$

p-cresol

OH

OH

+ B + H$_2$O

CH$_3$

4-methyl-catechol

where BH$_2$ stands for an *o*-diphenolic compound.

To distinguish this type of activity from the one mentioned earlier, it is described as cresolase activity, whereas the other is referred to as catecholase activity. For both types of activity, the involvement of copper is essential. Copper has been found as a component of all polyphenolases. The activity of cresolase involves three steps, which can be represented by the following overall equation (Mason 1956):

$$\text{Protein-Cu}_2^+\text{-O}_2 + \text{monophenol} \rightarrow$$
$$\text{Protein-Cu}_2^+ + o\text{-quinone} + \text{H}_2\text{O}$$

The protein copper-oxygen complex is formed by combining one molecule of oxygen with the protein to which two adjacent cuprous atoms are attached.

Catecholase activity involves oxidizing two molecules of *o*-diphenols to two molecules of *o*-quinones, resulting in the reduction of one molecule of oxygen to two molecules of water. The action sequence as presented in Figure 10–13 has been proposed by Mason (1957). The enzyme-oxygen complex serves as the hydroxylating or dehydroxylating intermediate, and (Cu)*n* represents the actual charge designation of the copper at the active site. In preparations high in cresolase activity, $n = 2$, and in preparations high in catecholase activity, $n = 1$. The overall reaction involves the use of one molecule of oxygen, one atom of which goes into the formation of the diphenol, and the other, which is reduced to water. This can be expressed in the following equation given by Mathew and Parpia (1971):

$$\text{Monophenol} + \text{O}_2 + o\text{-diphenol} \xrightarrow{\text{enzyme}}$$
$$o\text{-diphenol} + \text{quinone} + \text{H}_2\text{O}$$

The substrates of the polyphenol oxidase enzymes are phenolic compounds present in plant tissues, mainly flavonoids. These include catechins, anthocyanidins, leucoanthocyanidins, flavonols, and cinnamic acid derivatives. Polyphenol oxidases from different sources show distinct differences in their activity for different substrates. Some specific examples of polyphenolase substrates are chlorogenic acid, caffeic acid, dicatechol, protocatechuic acid, tyrosine, catechol, dihydroxyphenylalanine, pyrogallol, and catechins.

To prevent or minimize enzymic browning of damaged plant tissue, several approaches are possible. The first and obvious one, although rarely practical, involves the exclusion of molecular oxygen. Another approach is the addition of reducing agents that can prevent the accumulation of *o*-quinones.

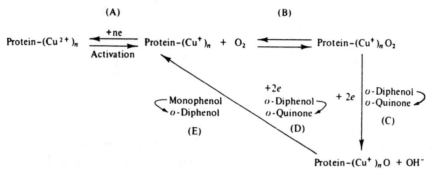

Figure 10–13 Phenolase Catalyzed Reactions. (A) activation of phenolase. (B,C,D) Two-step four-electron reduction of oxycuprophenolase, and the associated hydroxylation of monophenols. *Source*: From H.S. Mason, Mechanisms of Oxygen Metabolism, *Adv. Enzymol.*, Vol. 19, pp. 79–233, © 1957.

Heat treatment is effective in deactivating the enzymes. Metal complexing agents may deactivate the enzyme by making the copper unavailable.

One of the most useful methods involves the use of L-ascorbic acid as a reducing agent. This is practiced extensively in the commercial production of fruit juices and purees. The ascorbic acid reacts with the o-quinones and changes them back into o-diphenols (Figure 10–14).

Glucose Oxidase (β-D-Glucose: Oxygen Oxidoreductase)

This enzyme catalyzes the oxidation of D-glucose to δ-D-gluconolactone and hydrogen peroxide in the presence of molecular oxygen, as follows:

$$C_6H_{12}O_6 \xrightarrow{\text{enzyme}} C_6H_{10}O_6 + H_2O_2$$

The enzyme is present in many fungi and is highly specific for β-D-glucopyranose. It has been established that the enzyme does not oxidize glucose by direct combination with molecular oxygen. The mechanism as described by Whitaker (1972) involves the oxidized form of the enzyme, flavin adenine dinucleotide (FAD), which serves as a dehydrogenase. Two hydrogen atoms are removed from the glucose to form the reduced state of the enzyme, $FADH_2$, and δ-D-gluconolactone. The enzyme is then reoxidized by molecular oxygen. The gluconolactone is hydrolyzed in the presence of water to form D-gluconic acid.

In food processing, glucose oxidase is used to remove residual oxygen in the head space of bottled or canned products or to remove glucose. Light has a deteriorative effect on citrus beverages. Through the catalytic action of light, peroxides are formed that lead to oxidation of other components, resulting in very unpleasant off-flavors. Removing the oxygen by the use of a mixture of glucose oxidase and catalase will prevent these peroxides from forming. The glucose oxidase promotes the formation of gluconic acid with uptake of one molecule of oxygen. The cata-

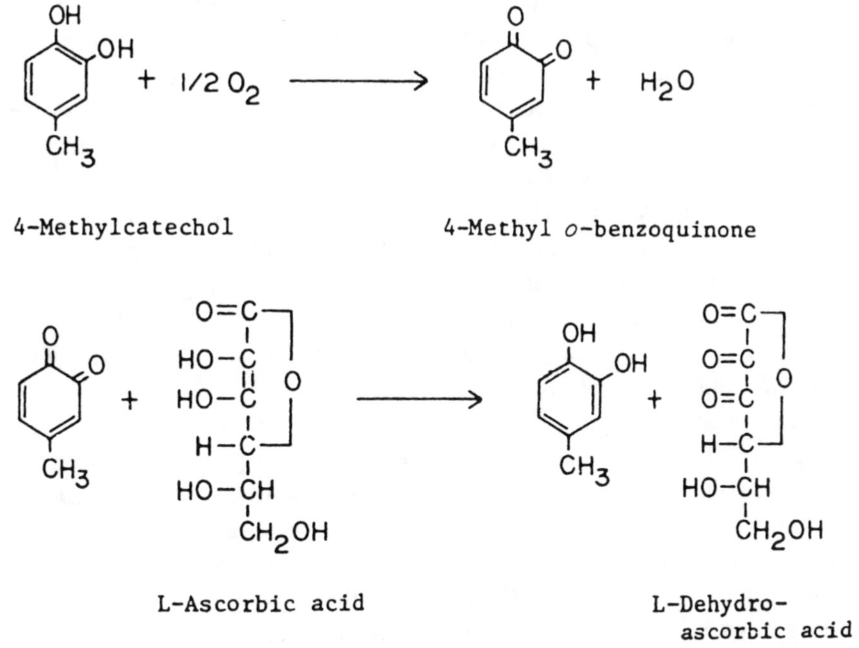

Figure 10–14 Reaction of L-Ascorbic Acid with *o*-Quinone in the Prevention of Enzymic Browning

lase decomposes the hydrogen peroxide formed into water and one half-molecule of oxygen. The net result is the uptake of one half-molecule of oxygen. The overall reaction can be written as follows:

$$\text{glucose} + \tfrac{1}{2}\,O_2 \xrightarrow[\text{catalase}]{\text{glucose oxidase}} \text{gluconic acid}$$

Recent information suggests that this application is not effective because of reversible inhibition of the glucose oxidase by the dyes used in soft drinks below pH 3 (Hammer 1993).

This enzyme mixture can also remove glucose from eggs before drying to prevent Maillard type browning reactions in the dried product.

Catalase (Hydrogen Peroxide: Hydrogen Peroxide Oxidoreductase)

Catalase catalyzes the conversion of two molecules of hydrogen peroxide into water and molecular oxygen as follows:

$$2\,H_2O_2 \xrightarrow{\text{catalase}} 2\,H_2O + O_2$$

This enzyme occurs in plants, animals, and microorganisms. The molecule has four subunits; each of these contains a protohemin group, which forms part of four independent active sites. The molecular weight is 240,000. Catalase is less stable to heat than is peroxidase. At neutral pH, catalase will rapidly lose activity at 35°C. In addition to catalyzing the reaction shown above (catalatic activity), catalase can also have peroxidatic activity. This occurs at low concentrations of hydrogen peroxide and in the presence of hydrogen donors (e.g., alcohols).

In plants, catalase appears to have two functions. First is the ability to dispose of the excess H_2O_2 produced in oxidative metabolism, and second is the ability to use H_2O_2 in the oxidation of phenols, alcohols, and other hydrogen donors. The difference in heat stability of catalase and peroxidase was demonstrated by Lopez et al. (1959). They found that blanching of southern peas for one minute in boiling water destroys 70 to 90 percent of the peroxidase activity and 80 to 100 percent of the catalase activity.

The combination of glucose oxidase and catalase is used in a number of food processing applications, including the removal of trace glucose or oxygen from foods and in the production of gluconic acid from glucose. Greenfield and Lawrence (1975) have studied the use of these enzymes in their immobilized form on an inorganic support.

Peroxidase (Donor: Hydrogen Peroxide Oxidoreductase)

The reaction type catalyzed by peroxidase involves hydrogen peroxide as an acceptor, and a compound AH_2 as a donor of hydrogen atoms, as shown:

$$H_2O_2 + AH_2 \xrightarrow{\text{peroxidase}} 2 H_2O + A$$

In contrast to the action of catalase, no molecular oxygen is formed.

The peroxidases can be classified into the two groups, iron-containing peroxidases and flavoprotein peroxidases. The former can be further subdivided into ferriprotoporphyrin peroxidases and verdoperoxidases. The first group contains ferriprotoporphyrin III (protohemin) as the prosthetic group (Figure 10–15). The common plant peroxidases (horseradish, fig, and turnip) are in this group and the enzymes are brown when highly purified. The second group includes the peroxidases of animal tissue and milk (lactoperoxidase). In these enzymes, the prosthetic group is an iron porphyrin nucleus but not protohemin. When highly purified, these enzymes are green in color. Flavoprotein peroxidases occur in microorganisms and animal tissues. The prosthetic group is FAD.

The linkage between the iron-containing prosthetic group and the protein can be stabilized by bisulfite (Embs and Markakis 1969). It is suggested that the bisulfite forms a complex with the peroxidase iron, which stabilizes the enzyme.

Because of the widespread occurrence of peroxidase in plant tissues, Nagle and Haard (1975) have suggested that it plays an important role in the development and senescence of plant tissues. It plays a role in biogenesis of ethylene; in regulating ripening and senescence; in the degradation of chlorophyll; and in the oxidation of indole-3-acetic acid.

The enzyme can occur in a variety of multiple molecular forms, named isoenzymes or isozymes. Such isoenzymes have the same enzymatic activity but can be separated by electrophoresis. Nagle and Haard (1975) separated the isoperoxidases of bananas into six anionic and one cationic component by gel electrophoresis. By using other methods of separation, an even greater number of isoenzymes was demonstrated.

Peroxidase has been implicated in the formation of the "grit cells" or "stone cells" of pears (Ranadive and Haard 1972). Bound peroxidase but not total peroxidase activity was higher in the fruit that contained excessive stone cells. The stone cells or sclereids are lignocellulosic in nature. The presence

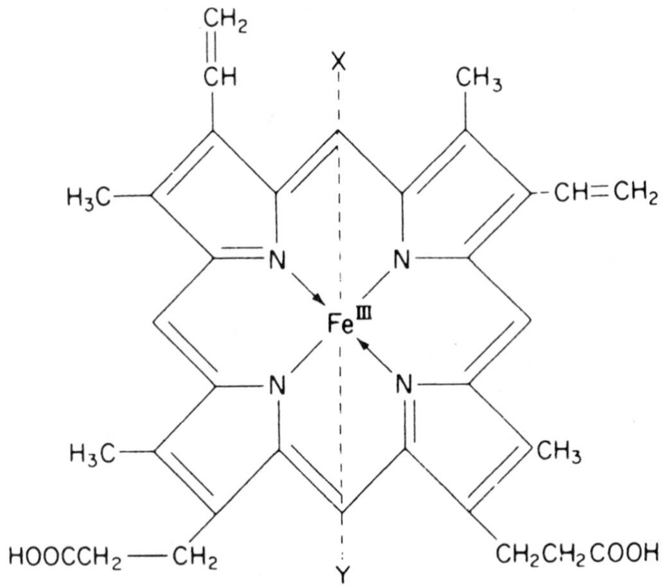

Figure 10–15 Structural Formula of Ferriprotoporphyrin III (Protohemin). *Source*: From J.R. Whitaker, *Food Related Enzymes*, 1974, American Chemical Society.

of calcium ions causes the release of wall bound peroxidase and a consequent decrease in the deposition of lignin.

The peroxidase test is used as an indicator of satisfactory blanching of fruits and vegetables. However, it has been found that the enzymes causing off-flavors during frozen storage can, under some conditions, be regenerated. Regeneration of enzymes is a relatively common phenomenon and is more likely to occur the faster the temperature is raised to a given point in the blanching process. The deactivation and reactivation of peroxidase by heat was studied by Lu and Whitaker (1974). The rate of reactivation was at a maximum at pH 9 and the extent of reactivation was increased by addition of hematin.

The deactivation of peroxidase is a function of heating time and temperature. Lactoperoxidase is completely deactivated by heating at 85°C for 13 seconds. The effect of heating time at 76°C on the deactivation of

lactoperoxidase is represented in Figure 10–16, and the effect of heating temperature on the deactivation constant is shown in Figure 10–17. Lactoperoxidase can be regenerated under conditions of high temperature short time (HTST) pasteurization. Figure 10–18 shows the regeneration of lactoperoxidase activity in milk that is pasteurized for 10 seconds at 85°C. The regeneration effect depends greatly on storage temperature; the lower the storage temperature, the smaller the regeneration effect.

Lactoperoxidase is associated with the serum proteins of milk. It has an optimum pH of 6.8 and a molecular weight of 82,000.

Lipoxygenase (Linoleate: Oxygen Oxidoreductase)

This enzyme, formerly named lipoxidase, is present in plants and catalyzes the oxidation of unsaturated fats. The major source of lipoxygenase is legumes, soybeans, and other beans and peas. Smaller amounts are present in peanuts, wheat, potatoes, and radishes. Lipoxygenase is a metallo-protein with an iron atom in its active center. In plants two types of lipoxygenase exist: type I lipoxygenase peroxidizes only free fatty acids with a high stereo- and regioselectivity; type II lipoxygenase is less specific for free linoleic acid and acts as a general catalyst for autoxidation. Type I reacts with fats in a food only after free fatty acids have been formed by lipase action; type II acts directly on triacylglycerols.

Lipoxygenase is highly specific and attacks the *cis-cis*-1,4-pentadiene group contained in the fatty acids linoleic, linolenic, and arachidonic, as follows:

$$-CH=CH-CH_2-CH=CH-$$

The specificity of this enzyme requires that both double bonds are in the *cis* configuration; in addition, there is a requirement that the central methylene group of the 1,4-pentadiene group occupies the ω-8 position on the fatty acid chain and also that the hydrogen to be removed from the central methylene group be in the L-position. Although the exact mechanism of the reaction is still in some doubt, there is agreement that the essential steps are as represented in Figure 10–19. Initially, a hydrogen atom is abstracted from the ω-8 methylene group to produce a free radical. The free radical isomerizes, causing conjugation of the double bond and isomerization to the *trans* configuration. The free radical then reacts to form the ω-6 hydroperoxide.

Lipoxygenase is reported to have a pH optimum of about 9. However, these values are determined with linoleic acid as substrate, and in natural systems the substrate is usually present in the form of triglycerides. The enzyme has a molecular weight of

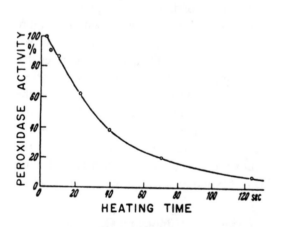

Figure 10–16 Deactivation of Lactoperoxidase as a Function of Heating Time. *Source:* From F. Kiermeier and C. Kayser, Heat Inactivation of Lactoperoxidase (in German), *Z. Lebensm. Untersuch. Forsch.*, Vol. 113, 1960.

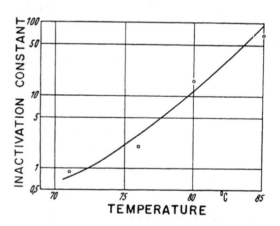

Figure 10–17 Dependence of the Rate Constant of Heat Deactivation of Lactoperoxidase on the Heating Temperature. *Source:* From F. Kiermeier and C. Kayser, Heat Inactivation of Lactoperoxidase (in German), *Z. Lebensm. Untersuch. Forsch.*, Vol. 113, 1960.

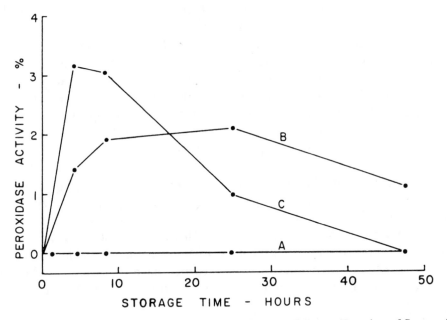

Figure 10–18 Regeneration in Ultra High Temperature Treated Milk as a Function of Storage Temperature. (A) 2°C; (B) 20°C; (C) 37°C. *Source*: From F. Kiermeier and C. Kayser, Heat Inactivation of Lactoperoxidase (in German), *Z. Lebensm. Untersuch. Forsch.*, Vol. 113, 1960.

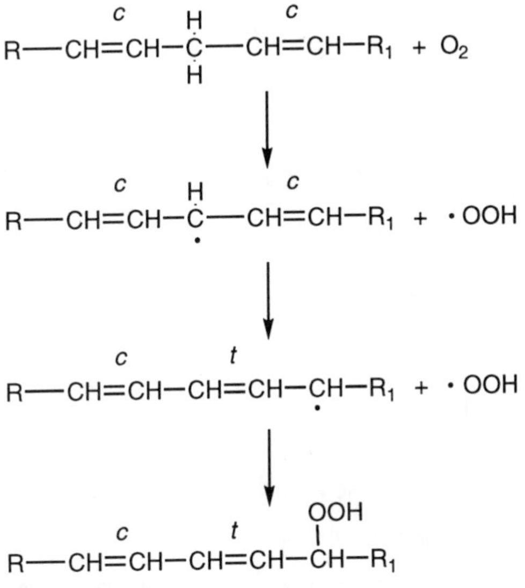

Figure 10–19 Essential Steps in the Mechanism of the Lipoxygenase-Catalyzed Oxidation of the 1,4-Pentadiene Group

102,000 and an isoelectric point of 5.4. The peroxide formation by lipoxygenase is inhibited by the common lipid antioxidants. The antioxidants are thought to react with the free radicals and interrupt the oxidation mechanism.

Most manifestations of lipoxygenase in foods are undesirable. However, it is used in baking to bring about desirable changes. Addition of soybean flour to wheat flour dough results in a bleaching effect, because of oxidation of the xanthophyll pigments. In addition, there is an effect on the rheological and baking properties of the dough. It has been suggested that lipoxygenase acts indirectly in the oxidation of sulfhydryl groups in the gluten proteins to produce disulfide bonds. When raw soybeans are ground with water to produce soy milk, a strong and unpleasant flavor develops that is called painty, green, or beany. Carrying out the grinding in boiling water instantly deactivates the enzyme, and no off-flavor is formed. Blanching of peas and beans is essential in preventing the lipoxygenase-catalyzed development of off-flavor. In addition to the development of off-flavors, the enzyme may be responsible for destruction of carotene and vitamin A, chlorophyll, bixin, and other pigments.

In some cases the action of lipoxygenase leads to development of a characteristic aroma. Galliard et al. (1976) found that the main aroma compounds of cucumber, 2-*trans* hexenol and 2-*trans*, 6-*cis*-nonadienal, are produced by reaction of linolenic acid and lipoxygenase to form hydroperoxide

(Figure 10–20); these are changed into *cis* unsaturated aldehydes by hydroperoxide lyase. The *cis* unsaturated aldehydes are transformed by isomerase into the corresponding *trans* isomers. These same substances in another matrix would be experienced as off-flavors. The use of lipoxygenase as a versatile biocatalyst has been described by Gardner (1996).

Xanthine Oxidase (Xanthine: Oxygen Oxidoreductase)

This enzyme catalyzes the conversion of xanthine and hypoxanthine to uric acid. The reaction equation is given in Figure 10–21; heavy arrows indicate the reactions catalyzed

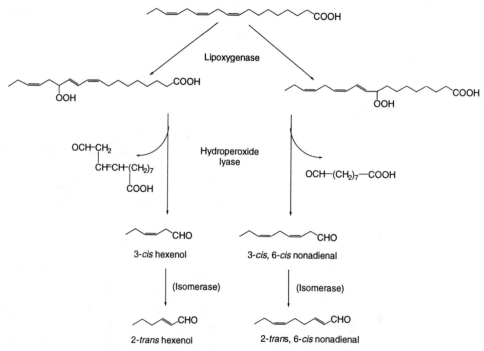

Figure 10–20 Lipoxygenase Catalyzed Formation of Aroma Compounds in Cucumber. *Source:* Reprinted from *Biochim. Biophys. Acta.*, Vol. 441, T. Galliard, D.R. Phillips, and J. Reynolds, The Formation of *cis*-3-nonenal, *trans*-2-nonenal and Hexanol from Linoleic Acid Hydroperoxide Isomers by a Hydroperoxide Cleavage Enzyme System in Cucumber (Cucumis Sativus) Fruits, p. 184, Copyright 1976, with permission from Elsevier Science.

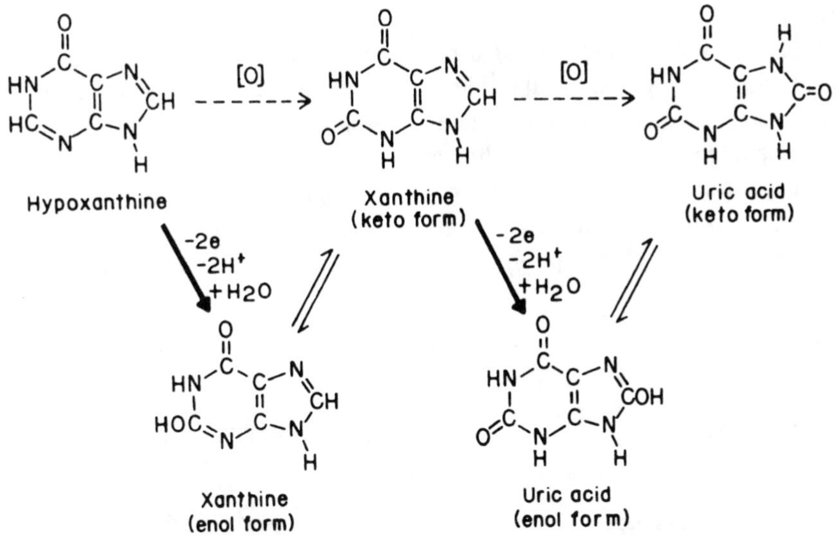

Figure 10–21 Oxidation of Hypoxanthine and Xanthine to Uric Acid by Xanthine Oxidase. *Source*: From J.R. Whitaker, *Principles of Enzymology for the Food Sciences*, 1972, Marcel Dekker, Inc.

by the enzyme and the dashed arrows represent the net result of the catalytic process (Whitaker 1972). Although xanthine oxidase is a nonspecific enzyme and many substances can serve as substrate, the rate of oxidation of xanthine and hypoxanthine is many times greater than that of other substrates.

Xanthine oxidase has been isolated from milk and obtained in the crystalline state. The molecular weight is 275,000. One mole of the protein contains 2 moles of FAD, 2 gram-atoms of molybdenum, 8 gram-atoms of nonheme iron, and 8 labile sulfide groups. The 8 labile sulfide groups are liberated in the form of H_2S upon acidification or boiling at pH 7. The optimum pH for activity is 8.3. The xanthine oxidase in milk is associated with the fat globules and, therefore, follows the fat into the cream when milk is separated. It seems to be located in small particles (microsomes) that are attached to the fat globules. The microsomes also contain the enzyme alkaline phosphatase. The microsomes can be dislodged from the fat globules by mechanical treatment such as pumping and agitation and by heating and cooling. The enzyme is moderately stable to heat but no less so than peroxidase.

IMMOBILIZED ENZYMES

One of the most important recent developments in the use of enzymes for industrial food processing is the fixing of enzymes on water-insoluble inert supports. The fixed enzymes retain their activity and can be easily added to or removed from the reaction mixture. The use of immobilized enzymes permits continuous processing and greatly increased use of the enzyme. Various possible methods of immobilizing enzymes have been listed by Weetall (1975) and Hultin (1983). A schematic representation of the available methods is given in Figure 10–22. The immobilizing methods include adsorption on organic polymers, glass, metal oxides, and siliceous materials such as bentonite and silica; entrapment in natural or synthetic polymers, usually polyacrylamide; microencapsulation in polymer membranes; ion exchange; cross-linking; adsorption and cross-linking combined; copolymerization; and covalent attachment to organic polymers. The chemistry of immobilizing enzymes has been covered in detail by Stanley and Olson (1974).

A summary of immobilization methods has been provided by Adlercreutz (1993) and is presented in Exhibit 10–1. In membrane

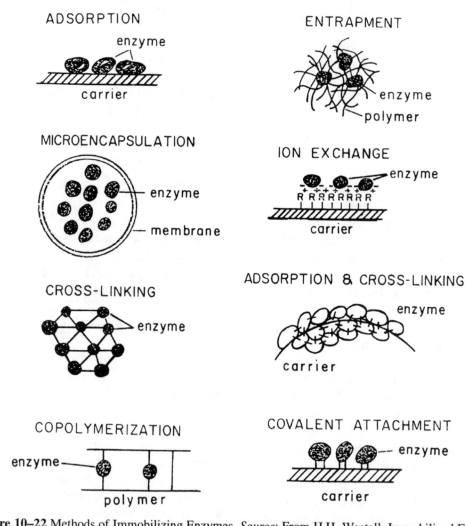

Figure 10–22 Methods of Immobilizing Enzymes. *Source*: From H.H. Weetall, Immobilized Enzymes and Their Application in the Food and Beverage Industry, *Process Biochem.*, Vol. 10, pp. 3–6, 1975.

Exhibit 10–1 Summary of Enzyme
Immobilization Methods

1. chemical methods
   • covalent binding
   • cross-linking
2. physical methods
   • adsorption
   • physical deposition
   • entrapment
     –in polymer gels
     –in microcapsules
   • membranes
3. two-phase systems
   • organic-aqueous
   • aqueous-aqueous

Immobilizing enzymes is likely to change their stability, and the method of attachment to the carrier also affects the degree of stability. When a high molecular weight substrate is used, the immobilizing should not be done by entrapment, microencapsulation, or copolymerization, because enzyme and substrate cannot easily get in contact. One of the promising methods appears to be covalent coupling of enzymes to inorganic carriers such as porous silica glass particles. Not all of the immobilized enzyme is active, due to either inactivation or steric hindrance. Usually, only about 30 to 50 percent of the bound enzyme is active.

Immobilized enzymes can be used in one of two basic types of reactor systems. The first is the stirred tank reactor where the immobilized enzyme is stirred with the substrate solution. This is a batch system and, after the reaction is complete, the immobilized enzyme is separated from the product. The other system employs continuous flow columns in which the substrate flows through the immobilized enzyme contained in a column or similar device. A simplified flow diagram of such a system is given in Figure 10–23.

reactors, the reaction product is separated from the reaction mixture by a semipermeable membrane. In two-phase systems, a hydrophobic reaction product can be separated from the aqueous reaction mixture by transfer to the organic solvent phase. In aqueous-aqueous systems, two incompatible polymers in aqueous solution form a two-phase system.

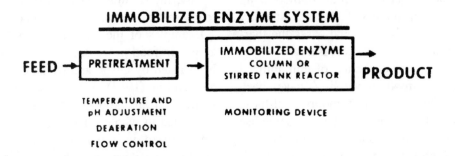

Figure 10–23 Flow Diagram of an Immobilized Enzyme System (Column Operation of Lactase Immobilized on Phenol-Formaldehyde Resin with Glutaraldehyde). *Source*: From W.L. Stanley and A.C. Olson, The Chemistry of Immobilizing Enzymes, *J. Food Sci.*, Vol. 39, pp. 660–666, 1974.

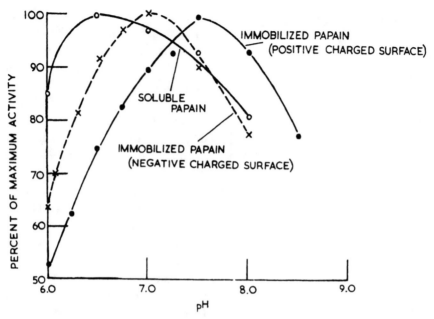

Figure 10–24 Effect of Immobilizing on the pH Optimum of Papain. *Source*: From H.H. Weetall, Immobilized Enzymes and Their Application in the Food and Beverage Industry, *Process Biochem.*, Vol. 10, pp. 3–6, 1975.

The characteristics of immobilized enzymes are likely to be somewhat different than those of the original enzyme. The pH optimum can be shifted; this depends on the surface charge of the carrier (Figure 10–24). Another property that can be changed is the Michaelis constant, K_m. This value can become either larger or smaller. Immobilizing may result in increased thermal stability (Figure 10–25), but in some cases the thermal stability is actually decreased.

Many examples of the use of immobilized enzymes in food processing have been reported. One of the most important of these is the use of immobilized glucose isomerase obtained from *Streptomyces* for the production of high-fructose corn syrup (Mermelstein 1975). In this process, the enzyme is bound to an insoluble carrier such as diethyl amino ethyl cellulose or a slurry of the fixed enzyme coated onto a pressure-leaf filter. The filter then serves as the continuous reactor through which the corn syrup flows. The product obtained by this process is a syrup with 71 percent solids that contains about 42 percent fructose and 50 percent glucose; it has high sweetening power, high fermentability, high humectancy, reduced tendency to crystallize, low viscosity, and good flavor.

Examples of the use of immobilized enzymes in food processing and analysis have been listed by Olson and Richardson (1974) and Hultin (1983). L-aspartic acid and L-malic acid are produced by using enzymes contained in whole microorganisms that are immobilized in a polyacrylamide gel. The enzyme aspartase from *Escherichia coli* is used for the production of aspartic acid. Fumarase from *Brevibacterium ammoniagenes* is used for L-malic acid production.

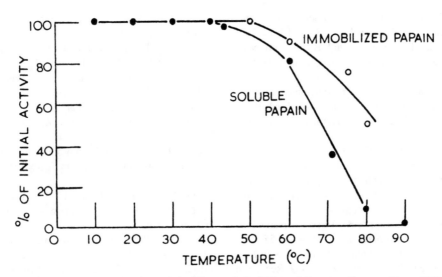

Figure 10–25 Effect of Immobilizing of the Thermal Stability of Papain. *Source*: From H.H. Weetall, Immobilized Enzymes and Their Application in the Food and Beverage Industry, *Process Biochem.*, Vol. 10, pp. 3–6, 1975.

The most widely used immobilized enzyme process involves the use of the enzyme glucose isomerase for the conversion of glucose to fructose in corn syrup (Carasik and Carroll 1983). The organism *Bacillus coagulans* has been selected for the production of glucose isomerase. The development of the immobilized cell slurry has not proceeded to the point where half-lives of the enzyme are more than 75 days. A half-life is defined as the time taken for a 50 percent decrease in activity. Such immobilized enzyme columns can be operated for periods of over three half-lives.

The second important application of immobilized enzymes is the hydrolysis of lactose to glucose and galactose in milk and milk products by lactase (Sprössler and Plainer 1983). Several lactase sources are available; from yeast, *Saccharomyces lactis* and *S. fragilis*, or from fungi, *Aspergillus oryzae* or *A. niger*. The enzymes vary in their optimum pH and optimum temperature, as well as other conditions.

It is to be expected that the use of immobilized enzymes in food processing will continue to grow rapidly in the near future.

REFERENCES

Adlercreutz, P. 1993. Immobilized enzymes. In *Enzymes in food processing*, ed. T. Nagodaurithana and G. Reed. San Diego, CA: Academic Press, Inc.

Albersheim, P., and U. Kilias. 1962. Studies relating to the purification and properties of pectin transeliminase. *Arch. Biochem. Biophys.* 97: 107–115.

Arnold, R.G., et al. 1975. Application of lipolytic enzymes to flavor development in dairy products. *J. Dairy Sci.* 58: 1127–1143.

Carasik, W., and J.O. Carroll. 1983. Development of immobilized enzymes for production of high-fructose corn syrup. *Food Technol.* 37, no. 10: 85–91.

Cordle, C.T. 1994. Control of food allergies using protein hydrolysates. *Food Technol.* 48, no. 10: 72–76.

Embs, R.J., and P. Markakis. 1969. Bisulfite effect on the chemistry and activity of horseradish peroxidase. *J. Food Sci.* 34: 500–501.

Frost, G.M. 1986. Commercial production of enzymes. In *Developments in food proteins*, Vol. 4, ed. B.J.F. Hudson. New York: Elsevier Applied Science Publishers.

Galliard, T., et al. 1976. The formation of *cis*-3-nonenal, *trans*-2-nonenal and hexanal from linoleic acid hydroperoxide isomers by a hydroperoxide cleavage enzyme system in cucumber (*Cucumis sativus*) fruits. *Biochim. Biophys. Acta* 441: 181–192.

Gardner, H.W. 1996. Lipoxygenase as a versatile biocatalyst. *J. Am. Oil Chem. Soc.* 73: 1347–1357.

Godtfredsen, S.E. 1993. Lipases. In *Enzymes in food processing*, ed. T. Nagodawithana and G. Reed. San Diego, CA: Academic Press, Inc.

Greenfield, P.F., and R.L. Lawrence. 1975. Characterization of glucose oxidase and catalase on inorganic supports. *J. Food Sci.* 40: 906–910.

Habibi-Najafi, M.B., and B.H. Lee. 1996. Bitterness in cheese: A review. *Crit. Rev. Food Sci. Nutr.* 36: 397–411.

Hammer, F.E. 1993. Oxidoreductases. In *Enzymes in food processing*, ed. T. Nagodawithana and G. Reed. San Diego, CA: Academic Press, Inc.

Hultin, H.O. 1983. Current and potential uses of immobilized enzymes. *Food Technol.* 37, no. 10: 66–82, 176.

Lahl, W.J., and S.D. Braun. 1994. Enzymatic production of protein hydrolysates for food use. *Food Technol.* 48, no. 10: 68–71.

Liener, I.E. 1974. The sulfhydryl proteases. In *Food related enzymes*, ed. J.R. Whitaker. Advances in Chemistry Series 136. Washington, DC: American Chemical Society.

Lopez, A., et al. 1959. Catalase and peroxidase activity in raw and blanched southern peas, *Vigna senensis*. *Food Res.* 24: 548–551.

Lu, A.T., and J.R. Whitaker. 1974. Some factors affecting rates of heat inactivation and reactivation of horseradish peroxidase. *J. Food Sci.* 39: 1173–1178.

MacMillan, J.D., and M.I. Sheiman. 1974. Pectic enzymes. In *Food related enzymes*, ed. J.R. Whitaker. Advances in Chemistry Series 136. Washington, DC: American Chemical Society.

Macrae, A.R. 1983. Lipase catalyzed interesterification of oils and fats. *J. Am. Oil Chem. Soc.* 60: 291–294.

Marshall, J.J. 1975. Starch degrading enzymes, old and new. *Starke* 27: 377–383.

Mason, H.S. 1956. Structures and functions of the phenolase complex. *Nature* 177: 79–81.

Mason, H.S. 1957. Mechanisms of oxygen metabolism. *Adv. Enzymol.* 19: 79–233.

Mathew, A.G., and H.A.B. Parpia. 1971. Food browning as a polyphenol reaction. *Adv. Food Res.*, 19: 75–145.

Mermelstein, N.H. 1975. Immobilized enzymes produce high-fructose corn syrup. *Food Technol.* 29, no. 6: 20–26.

Nagle, N.E., and N.F. Haard. 1975. Fractionation and characterization of peroxidase from ripe banana fruit. *J. Food Sci.* 40: 576–579.

Nelson, J.H. 1972. Enzymatically produced flavors for fatty systems. *J. Am. Oil Chem. Soc.* 49: 559–562.

Olson, N.F., and T. Richardson. 1974. Immobilized enzymes in food processing and analysis. *J. Food Sci.* 39: 653–659.

Parkin, K.L. 1993. General characteristics of enzymes. In *Enzymes in food processing*, ed. T. Nagodawithana and G. Reed. San Diego, CA: Academic Press, Inc.

Ranadive, A.S., and N.F. Haard. 1972. Peroxidase localization and lignin formation in developing pear fruit. *J. Food Sci.* 37: 381–383.

Riordan, J.F. 1974. Metal-containing exopeptidases. In *Food related enzymes*, ed. J.R. Whitaker. Advances in Chemistry Series 136. Washington, DC: American Chemical Society.

Sprössler, B., and H. Plainer. 1983. Immobilized lactase for processing whey. *Food Technol.* 37, no. 10: 93–95.

Stanley, W.L., and A.C. Olson. 1974. The chemistry of immobilizing enzymes. *J. Food Sci.* 39: 660–666.

Villeneuve, P., and T.A. Foglia. 1997. Lipase specificities: Potential application in lipid bioconversions. *J. Am. Oil Chem. Soc.* 8: 640–650.

Weetall, H.H. 1975. Immobilized enzymes and their application in the food and beverage industry. *Process Biochem.* 10: 3–6.

Whitaker, J.R. 1972. *Principles of enzymology for the food sciences*. New York: Marcel Dekker.

Whitaker, J.R. 1974. *Food related enzymes*. Advances in Chemistry Series 136. Washington, DC: American Chemical Society.

Chapter 11

Additives and Contaminants

INTRODUCTION

The possibility of harmful or toxic substances becoming part of the food supply concerns the public, the food industry, and regulatory agencies. Toxic chemicals may be introduced into foods unintentionally through direct contamination, through environmental pollution, and as a result of processing. Many naturally occurring food compounds may be toxic. A summary of the various toxic chemicals in foods (Exhibit 11–1) was presented in a scientific status summary of the Institute of Food Technologists (1975). Many toxic substances present below certain levels pose no hazard to health. Some substances are toxic and at the same time essential for good health (such as vitamin A and selenium). An understanding of the properties of additives and contaminants and how these materials are regulated by governmental agencies is important to the food scientist. Regulatory controls are dealt with in Chapter 12.

Food additives can be divided into two major groups, intentional additives and incidental additives. Intentional additives are chemical substances that are added to food for specific purposes. Although we have little control over unintentional or incidental additives, intentional additives are regulated

by strict governmental controls. The U.S. law governing additives in foods is the Food Additives Amendment to the Federal Food, Drug and Cosmetic Act of 1958. According to this act, a food additive is defined as follows:

The term food additive means any substance the intended use of which results, or may reasonably be expected to result, directly or indirectly in its becoming a component or otherwise affecting the characteristics of any food (including any substance intended for use in producing, manufacturing, packing, processing, preparing, treating, packaging, transporting, or holding food; and including any source of radiation intended for any such use), if such a substance is not generally recognized, among experts qualified by scientific training and experience to evaluate its safety, as having been adequately shown through scientific procedures (or, in the case of a substance used in food prior to January 1, 1958, through either scientific procedures or experience based on common use in food) to be safe under the condition of its intended use; except that such a term does not include pesticides, color

Exhibit 11–1 Toxic Chemicals in Foods

NATURAL

- normal components of natural food products
- natural contaminants of natural food products
  - microbiological origin: toxins
  - nonmicrobiological origin: toxicants (e.g., Hg, Se) consumed in feeds by animals used as food sources

MAN-MADE

- agricultural chemicals (e.g., pesticides, fertilizers)

- food additives
- chemicals derived from food packaging materials
- chemicals produced in processing of foods (e.g., by heat, ionizing radiation, smoking)
- inadvertent or accidental contaminants
  - food preparation accidents or mistakes
  - contamination from food utensils
  - environmental pollution
  - contamination during storage or transport

---

additives and substances for which prior sanction or approval was granted.

The law of 1958 thus recognizes the following three classes of intentional additives:

1. additives generally recognized as safe (GRAS)
2. additives with prior approval
3. food additives

Coloring materials and pesticides on raw agricultural products are covered by other laws. The GRAS list contains several hundred compounds, and the concept of such a list has been the subject of controversy (Hall 1975).

Before the enactment of the 1958 law, U.S. laws regarding food additives required that a food additive be nondeceptive and that an added substance be either safe and therefore permitted, or poisonous and deleterious and

therefore prohibited. This type of legislation suffered from two main shortcomings: (1) it equated poisonous with harmful, and (2) the onus was on the government to demonstrate that any chemical used by the food industry was poisonous. The 1958 act distinguishes between toxicity and hazard. *Toxicity* is the capacity of a substance to produce injury. *Hazard* is the probability that injury will result from the intended use of the substance. It is now well recognized that many components of our foods, whether natural or added, are toxic at certain levels but harmless or even nutritionally essential at lower levels. The ratio between effective dose and toxic dose of many compounds, including such common nutrients as amino acids and salts, is of the order of 1 to 100. It is now mandatory that any user of an additive must petition the government for permission to use the material and must supply evidence that the compound is safe.

An important aspect of the act is the so-called Delaney clause, which specifies that no additive shall be deemed safe if it is found to induce cancer in man or animal. Such special consideration in the case of cancer-producing compounds is not incorporated in the food laws of many other countries.

INTENTIONAL ADDITIVES

Chemicals that are intentionally introduced into foods to aid in processing, to act as preservatives, or to improve the quality of the food are called intentional additives. Their use is strictly regulated by national and international laws. The National Academy of Sciences (1973) has listed the purposes of food additives as follows:

- to improve or maintain nutritional value
- to enhance quality
- to reduce wastage
- to enhance consumer acceptability
- to improve keeping quality
- to make the food more readily available
- to facilitate preparation of the food

The use of food additives is in effect a food processing method, because both have the same objective—to preserve the food and/or make it more attractive. In many food processing techniques, the use of additives is an integral part of the method, as is smoking, heating, and fermenting. The National Academy of Sciences (1973) has listed the following situations in which additives should *not* be used:

- to disguise faulty or inferior processes
- to conceal damage, spoilage, or other inferiority
- to deceive the consumer

- if use entails substantial reduction in important nutrients
- if the desired effect can be obtained by economical, good manufacturing practices
- in amounts greater than the minimum necessary to achieve the desired effects

There are several ways of classifying intentional food additives. One such method lists the following three main types of additives:

1. complex substances such as proteins or starches that are extracted from other foods (for example, the use of caseinate in sausages and prepared meats)
2. naturally occurring, well-defined chemical compounds such as salt, phosphates, acetic acid, and ascorbic acid
3. substances produced by synthesis, which may or may not occur in nature, such as coal tar dyes, synthetic β-carotene, antioxidants, preservatives, and emulsifiers

Some of the more important groups of intentional food additives are described in the following sections.

Preservatives

Preservatives or antimicrobial agents play an important role in today's supply of safe and stable foods. Increasing demand for convenience foods and reasonably long shelf life of processed foods make the use of chemical food preservatives imperative. Some of the commonly used preservatives—such as sulfites, nitrate, and salt—have been used for centuries in processed meats and wine. The choice of an antimicrobial agent has to be based on a knowledge of the antimicrobial spectrum of the preservative, the chemical and physical properties of both food and preservative, the conditions of storage and handling, and the assurance of a high initial quality of the food to be preserved (Davidson and Juneja 1990).

Benzoic Acid

Benzoic acid occurs naturally in many types of berries, plums, prunes, and some spices. As an additive, it is used as benzoic acid or as benzoate. The latter is used more often because benzoic acid is sparsely soluble in water (0.27 percent at 18°C) and sodium benzoate is more soluble (66.0 g/100 mL at 20°C). The undissociated form of benzoic acid is the most effective antimicrobial agent. With a pK_a of 4.2, the optimum pH range is from 2.5 to 4.0. This makes it an effective antimicrobial agent in high-acid foods, fruit drinks, cider, carbonated beverages, and pickles. It is also used in margarines, salad dressings, soy sauce, and jams.

Parabens

Parabens are alkyl esters of p-hydroxybenzoic acid. The alkyl groups may be one of the following: methyl, ethyl, propyl, butyl, or heptyl. Parabens are colorless, tasteless, and odorless (except the methyl paraben). They are nonvolatile and nonhygroscopic. Their solubility in water depends on the nature of the alkyl group; the longer the alkyl chain length, the lower the solubility. They differ from benzoic acid in that they have antimicrobial activity in both acid and alkaline pH regions.

The antimicrobial activity of parabens is proportional to the chain length of the alkyl group. Parabens are more active against molds and yeasts than against bacteria, and more active against gram-positive than gram-negative bacteria. They are used in fruitcakes, pastries, and fruit fillings. Methyl and propyl parabens can be used in soft drinks. Combinations of several parabens are often used in applications such as fish products, flavor extracts, and salad dressings.

Sorbic Acid

Sorbic acid is a straight-chain, trans-trans unsaturated fatty acid, 2,4-hexadienoic acid. As an acid, it has low solubility (0.15 g/100 mL) in water at room temperature. The salts, sodium, or potassium are more soluble in water. Sorbates are stable in the dry form; they are unstable in aqueous solutions because they decompose through oxidation. The rate of oxidation is increased at low pH, by increased temperature, and by light exposure.

Sorbic acid and sorbates are effective against yeasts and molds. Sorbates inhibit yeast growth in a variety of foods including wine, fruit juice, dried fruit, cottage cheese, meat, and fish products. Sorbates are most effective in products of low pH including salad dressings, tomato products, carbonated beverages, and a variety of other foods.

The effective level of sorbates in foods is in the range of 0.5 to 0.30 percent. Some of the common applications are shown in Table 11-1. Sorbates are generally used in sweetened wines or wines that contain residual sugars to prevent refermentation. At the levels generally used, sorbates do not affect food flavor. However, when used at higher levels, they may be detected by some people as an unpleasant flavor. Sorbate can be degraded by certain microorganisms to produce off-flavors. Molds can metabolize sorbate to produce 1,3 pentadiene, a volatile compound with an odor like kerosene. High levels of microorganisms can result in the

Table 11–1 Applications of Sorbates as Antimicrobial Agents

Products	Levels (%)
Dairy products: aged cheeses, processed cheeses, cottage cheese, cheese spreads, cheese dips, sour cream, yogurt	0.05–0.30
Bakery products: cakes, cake mixes, pies, fillings, mixes, icings, fudges, toppings, doughnuts	0.03–0.30
Vegetable products: fermented vegetables, pickles, olives, relishes, fresh salads	0.02–0.20
Fruit products: dried fruit, jams, jellies, juices, fruit salads, syrups, purees, concentrates	0.02–0.25
Beverages: still wines, carbonated and noncarbonated beverages, fruit drinks, low-calorie drinks	0.02–0.10
Food emulsions: mayonnaise, margarine, salad dressings	0.05–0.10
Meat and fish products: smoked and salted fish, dry sausages	0.05–0.30
Miscellaneous: dry sausage casings, semimoist pet foods, confectionery	0.05–0.30

Source: Reprinted with permission from J.N. Sofos and F.F. Busta, Sorbic Acid and Sorbates, in *Antimicrobials in Foods*, P.M. Davidson and A.L. Branen, eds., p. 62, 1993, by courtesy of Marcel Dekker, Inc.

degradation of sorbate in wine and result in the off-flavor known as geranium off-odor (Edinger and Splittstoesser 1986). The compounds responsible for the flavor defect are ethyl sorbate, 4-hexenoic acid, 1-ethoxy-hexa-2,4-diene, and 2-ethoxyhexa-3,5-diene. The same problem may occur in fermented vegetables treated with sorbate.

Sulfites

Sulfur dioxide and sulfites have long been used as preservatives, serving both as antimicrobial substance and as antioxidant. Their use as preservatives in wine dates back to Roman times. Sulfur dioxide is a gas that can be used in compressed form in cylinders. It is liquid under pressure of 3.4 atm and can be injected directly in liquids. It can also be used to prepare solutions in ice cold water. It dissolves to form sulfurous acid. Instead of sulfur dioxide solutions, a number of sulfites can be used (Table 11–2) because, when dissolved in water, they all yield active SO_2.

The most widely used of these sulfites is potassium metabisulfite. In practice, a value of 50 percent of active SO_2 is used. When sulfur dioxide is dissolved in water, the following ions are formed:

$$SO_2 \text{ (gas)} \rightarrow SO_2 \text{ (aq)}$$
$$SO_2 \text{ (aq)}^+ \rightarrow H_2O \rightarrow H_2SO_3$$
$$H_2SO_3 \rightarrow H^+ + HSO_3^- \quad (K_1 = 1.7 \times 10^{-2})$$
$$HSO_3^- \rightarrow H^+ + SO_3^{2-} \quad (K_2 = 5 \times 10^{-6})$$
$$2HSO_3^- \rightarrow S_2O_5^{2-} + H_2O$$

All of these forms of sulfur are known as free sulfur dioxide. The bisulfite ion (HSO_3^-) can react with aldehydes, dextrins, pectic substances, proteins, ketones, and certain sugars to form addition compounds.

Table 11–2 Sources of SO_2 and Their Content of Active SO_2

Chemical	Formula	Content of Active SO_2
Sulfur dioxide	SO_2	100.00%
Sodium sulfite, anhydrous	Na_2SO_3	50.82%
Sodium sulfite, heptahydrate	$Na_2SO_3 \cdot 7\ H_2O$	25.41%
Sodium hydrogen sulfite	$NaHSO_3$	61.56%
Sodium metabisulfite	$Na_2S_2O_5$	67.39%
Potassium metabisulfite	$K_2S_2O_5$	57.63%
Calcium sulfite	$CaSO_3$	64.00%

The addition compounds are known as bound sulfur dioxide. Sulfur dioxide is used extensively in wine making, and in wine acetaldehyde reacts preferentially with bisulfite. Excess bisulfite reacts with sugars. It is possible to classify bound SO_2 into three forms: aldehyde sulfurous acid, glucose sulfurous acid, and rest sulfurous acid. The latter holds the SO_2 in a less tightly bound form. Sulfites in wines serve a dual purpose: (1) antiseptic or bacteriostatic and (2) antioxidant. These activities are dependent on the form of SO_2 present. The various forms of SO_2 in wine are represented schematically in Figure 11–1. The free SO_2 includes the water-soluble SO_2 and the undissociated H_2SO_3 and constitutes about 2.8 percent of the total. The bisulfite form constitutes 96.3 percent and the sulfite form 0.9 percent (all at pH 3.3 and 20°C). The bound SO_2 is mostly (80 percent) present as acetaldehyde SO_2, 1 percent as glucose SO_2, and 10 to 20 percent as rest SO_2. The various forms of sulfite have different activities. The two free forms are the only ones with antiseptic activity. The antioxidant activity is limited to the SO_3^{2-} ion (Figure 11–1). The antiseptic activity of SO_2 is highly dependent on the pH, as indicated in Table 11–3. The lower the pH the greater the

antiseptic action of SO_2. The effect of pH on the various forms of sulfur dioxide is shown in Figure 11–2.

Sulfurous acid inhibits molds and bacteria and to a lesser extent yeasts. For this reason, SO_2 can be used to control undesirable bacteria and wild yeast in fermentations without affecting the SO_2-tolerant cultured yeasts. According to Chichester and Tanner (1968), the undissociated acid is 1,000 times more active than HSO_3^- for *Escherichia coli*, 100 to 500 times for *Saccharomyces cerevisiae*, and 100 times for *Aspergillus niger*.

The amount of SO_2 added to foods is self-limiting because at levels from 200 to 500 ppm the product may develop an unpleasant off-flavor. The acceptable daily intake (ADI) is set at 1.5 mg/kg body weight. Because large intakes can result from consumption of wine, there have been many studies on reducing the use of SO_2 in wine making. Although some other compounds (such as sorbic acid and ascorbic acid) may partially replace SO_2, there is no satisfactory replacement for SO_2 in wine making.

The use of SO_2 is not permitted in foods that contain significant quantities of thiamine, because this vitamin is destroyed by SO_2. In the United States, the maximum per-

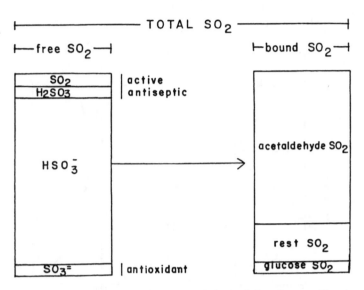

Figure 11–1 The Various Forms of SO$_2$ in Wine and Their Activity. *Source:* Reprinted with permission from J.M. deMan, 500 Years of Sulfite Use in Winemaking, *Am. Wine Soc. J.*, Vol. 20, pp. 44–46, © 1988, American Wine Society.

mitted level of SO$_2$ in wine is 350 ppm. Modern practices have resulted in much lower levels of SO$_2$. In some countries SO$_2$ is used in meat products; such use is not permitted in North America on the grounds that this would result in consumer deception. SO$_2$ is also widely used in dried fruits, where levels may be up to 2,000 ppm. Other applications are in dried vegetables and dried potato products. Because SO$_2$ is volatile and easily lost to the atmosphere, the residual levels may be much lower than the amounts originally applied.

Table 11–3 Effect of pH on the Proportion of Active Antiseptic SO$_2$ of Wine Containing 100 mg/L Free SO$_2$

pH	Active SO$_2$ (mg/L)
2.2	37.0
2.8	8.0
3.0	5.0
3.3	3.0
3.5	1.8
3.7	1.2
4.0	0.8

Nitrates and Nitrites

Curing salts, which produce the characteristic color and flavor of products such as bacon and ham, have been used throughout history. Curing salts have traditionally contained nitrate and nitrite; the discovery that nitrite was the active compound was made in about 1890. Currently, nitrate is not considered to be an essential component in curing mixtures; it is sometimes suggested that nitrate may be transformed into nitrite, thus forming a reservoir for the production of nitrite. Both nitrates and nitrites are thought to have antimicrobial action. Nitrate is used in the production of Gouda cheese to prevent gas formation by butyric acid–forming bacteria. The action of nitrite in meat curing is

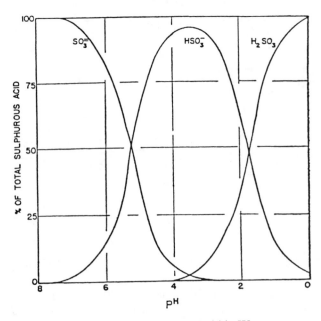

Figure 11–2 Effect of pH on the Ionization of Sulfurous Acid in Water

considered to involve inhibition of toxin formation by *Clostridium botulinum*, an important factor in establishing safety of cured meat products. Major concern about the use of nitrite was generated by the realization that secondary amines in foods may react to form nitrosamines, as follows:

$$\begin{matrix} R_1 \\ \diagdown \\ \quad NH + HNO_2 \longrightarrow \\ \diagup \\ R \end{matrix}$$

$$\begin{matrix} R_1 \\ \diagdown \\ \quad N-N{=}O + H_2O \\ \diagup \\ R \end{matrix}$$

The nitrosamines are powerful carcinogens, and they may be mutagenic and teratogenic as well. It appears that very small amounts of nitrosamines can be formed in certain cured meat products. These levels are in the ppm or the ppb range and, because analytical procedures are difficult, there is as yet no clear picture of the occurrence of nitrosamines. The nitrosamines may be either volatile or nonvolatile, and only the latter are usually included in analysis of foods. Nitrosamines, especially dimethyl-nitrosamine, have been found in a number of cases when cured meats were surveyed at concentrations of a few μg/kg (ppb). Nitrosamines are usually present in foods as the result of processing methods that promote their formation (Havery and Fazio 1985). An example is the spray drying of milk. Suitable modifications of these process conditions can drastically reduce the nitrosamine levels. Considerable further research is necessary to establish why nitrosamines are present only in some samples and what the toxicological importance of nitrosamines is at these levels. There appears to be no suitable replacement for nitrite in the production of cured meats such as ham and bacon. The ADI of nitrite has been set at 60 mg per person per day. It is estimated that the daily intake per person in Canada is about 10 mg.

Cassens (1997) has reported a dramatic decline in the residual nitrite levels in cured meat products in the United States. The current residual nitrite content of cured meat products is about 10 ppm. In 1975 an average residual nitrite content in cured meats was reported as 52.5 ppm. This reduction of nitrite levels by about 80 percent has been attributed to lower ingoing nitrite, increased use of ascorbates, improved process control, and altered formulations.

The nitrate-nitrite intake from natural sources is much higher than that from processed foods. Fassett (1977) estimated that the nitrate intake from 100 g of processed meat might be 50 mg and from 100 g of high-nitrate spinach, 200 mg. Wagner and Tannenbaum (1985) reported that nitrate in cured meats is insignificant compared to nitrite produced endogenously. Nitrate is produced in the body and recirculated to the oral cavity, where it is reduced to nitrite by bacterial action.

Hydrogen Peroxide

Hydrogen peroxide is a strong oxidizing agent and is also useful as a bleaching agent. It is used for the bleaching of crude soya lecithin. The antimicrobial action of hydrogen peroxide is used for the preservation of cheese milk. Hydrogen peroxide decomposes slowly into water and oxygen; this process is accelerated by increased temperature and the presence of catalysts such as catalase, lacto-peroxidase and heavy metals. Its antimicrobial action increases with temperature. When hydrogen peroxide is used for cheese making, the milk is treated with 0.02 percent hydrogen peroxide followed by catalase to remove the hydrogen peroxide. Hydrogen peroxide can be used for sterilizing food processing equipment and for sterilizing packaging material used in aseptic food packaging systems.

Sodium Chloride

Sodium chloride has been used for centuries to prevent spoilage of foods. Fish, meats, and vegetables have been preserved with salt. Today, salt is used mainly in combination with other processing methods. The antimicrobial activity of salt is related to its ability to reduce the water activity (a_w), thereby influencing microbial growth. Salt has the following characteristics: it produces an osmotic effect, it limits oxygen solubility, it changes pH, sodium and chloride ions are toxic, and salt contributes to loss of magnesium ions (Banwart 1979). The use of sodium chloride is self-limiting because of its effect on taste.

Bacteriocins

Nisin is an antibacterial polypeptide produced by some strains of *Lactococcus lactis*. Nisin-like substances are widely produced by lactic acid bacteria. These inhibitory substances are known as bacteriocins. Nisin has been called an antibiotic, but this term is avoided because nisin is not used for therapeutic purposes in humans or animals. Nisin-producing organisms occur naturally in milk. Nisin can be used as a processing aid against gram-positive organisms. Because its effectiveness decreases as the bacterial load increases, it is unlikely to be used to cover up unhygienic practices.

Nisin is a polypeptide with a molecular weight of 3,500, which is present as a dimer of molecular weight 7,000. It contains some unusual sulfur amino acids, lanthionine and β-methyl lanthionine. It contains no aromatic amino acids and is stable to heat.

The use of nisin as a food preservative has been approved in many countries. It has been used effectively in preservation of processed cheese. It is also used in the heat treatment of nonacid foods and in extending the shelf life of sterilized milk.

A related antibacterial substance is natamycin, identical to pimaricin. Natamycin is effective in controlling the growth of fungi but has no effect on bacteria or viruses. In fermentation industries, natamycin can be used to control mold or yeast growth. It has a low solubility and therefore can be used as a surface treatment on foods. Natamycin is used in the production of many varieties of cheese.

Acids

Acids as food additives serve a dual purpose, as acidulants and as preservatives. Phosphoric acid is used in cola soft drinks to reduce the pH. Acetic acid is used to provide tartness in mayonnaise and salad dressings. A similar function in a variety of other foods is served by organic acids such as citric, tartaric, malic, lactic, succinic, adipic, and fumaric acid. The properties of some of the common food acids are listed in Table 11–4 (Peterson and Johnson 1978). Members of the straight-chain carboxylic acids, propionic and sorbic acids, are used for their antimicrobial properties. Propionic acid is mainly used for its antifungal properties. Propionic acid applied as a 10 percent solution to the surface of cheese and butter retards the growth of molds. The fungistatic effect is higher at pH 4 than at pH 5. A 5 percent solution of calcium propionate acidified with lactic acid

to pH 5.5 is as effective as a 10 percent unacidified solution of propionic acid. The sodium salts of propionic acid also have antimicrobial properties.

Antioxidants

Food antioxidants in the broadest sense are all of the substances that have some effect on preventing or retarding oxidative deterioration in foods. They can be classified into a number of groups (Kochhar and Rossell 1990).

Primary antioxidants terminate free radical chains and function as electron donors. They include the phenolic antioxidants, butylated hydroxyanisole (BHA), butylated hydroxytoluene (BHT), tertiary butyl hydroquinone (TBHQ), alkylgalates, usually propylgallate (PG), and natural and synthetic tocopherols and tocotrienols.

Oxygen scavengers can remove oxygen in a closed system. The most widely used compounds are vitamin C and related substances, ascorbyl palmitate, and erythorbic acid (the D-isomer of ascorbic acid).

Chelating agents or sequestrants remove metallic ions, especially copper and iron, that are powerful prooxidants. Citric acid is widely used for this purpose. Amino acids and ethylene diamine tetraacetic acid (EDTA) are other examples of chelating agents.

Enzymic antioxidants can remove dissolved or head space oxygen, such as glucose oxidase. Superoxide dismutase can be used to remove highly oxidative compounds from food systems.

Natural antioxidants are present in many spices and herbs (Lacroix et al. 1997; Six 1994). Rosemary and sage are the most potent antioxidant spices (Schuler 1990). The active principles in rosemary are carnosic acid and carnosol (Figure 11–3). Anti-

Table 11–4 Properties of Some Common Food Acids

Property	Acetic Acid	Adipic Acid	Citric Acid	Fumaric Acid	Glucono-Delta-Lactone	Lactic Acid	Malic Acid	Phosphoric Acid	Tartaric Acid
Structure	CH_3COOH	COOH–CH_2–CH_2–CH_2–CH_2–COOH	COOH–CH_2–C(OH)(COOH)–CH_2–COOH	HOOCCH=HCCOOH	O=C–HCOH–HOCH–HCOH–HC–CH_2OH (lactone ring)	CH_3–C(H)(OH)–COOH	COOH–C(H)(OH)–CH_2–COOH	H_3PO_4	COOH–C(H)(OH)–C(OH)(H)–COOH
Empirical formula	$C_2H_4O_2$	$C_6H_{10}O_4$	$C_6H_8O_7$	$C_4H_4O_4$	$C_6H_{10}O_6$	$C_3H_6O_3$	$C_4H_6O_5$	H_3PO_4	$C_4H_6O_6$
Physical form	Oily Liquid	Crystalline	Crystalline	Crystalline	Crystalline	85% Water Solution	Crystalline	85% Water Solution	Crystalline
Molecular weight	60.05	146.14	192.12	116.07	178.14	90.08	134.09	82.00	150.09
Equivalent weight	60.05	73.07	64.04	58.04	178.14	90.08	67.05	27.33	75.05
Sol. in water (g/100 mL solv.)	∞	1.4	181.00	0.63	59.0	∞	144.0	∞	147.0
Ionization constants									
K_1	8×10^{-5}	3.7×10^{-5}	8.2×10^{-4}	1×10^{-3}	2.5×10^{-4} (gluconic acid)	1.37×10^{-4}	4×10^{-4}	7.52×10^{-3}	1.04×10^{-3}
K_2		2.4×10^{-6}	1.77×10^{-5}	3×10^{-5}			9×10^{-6}	6.23×10^{-8}	5.55×10^{-5}
K_3			3.9×10^{-6}					3×10^{-13}	

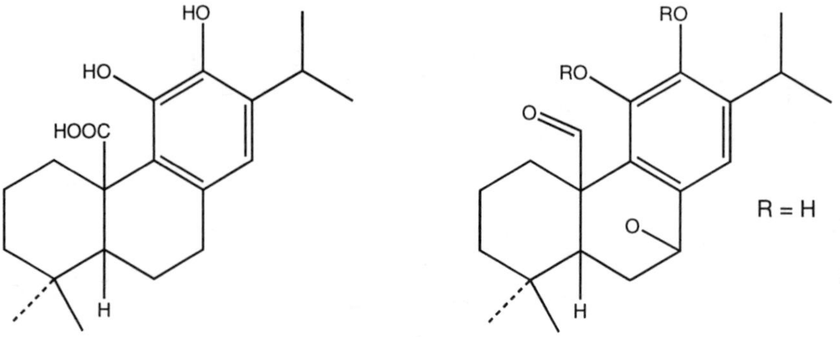

carnosic acid   carnosol

Figure 11–3 Chemical Structure of the Active Antioxidant Principles in Rosemary

oxidants from spices can be obtained as extracts or in powdered form by a process described by Bracco et al. (1981).

The level of phenolic antioxidants permitted for use in foods is limited. U.S. regulations allow maximum levels of 0.02 percent based on the fat content of the food.

Sometimes the antioxidants are incorporated in the packaging materials rather than in the food itself. In this case, a larger number of antioxidants is permitted, provided that no more than 50 ppm of the antioxidants become a component of the food.

Emulsifiers

With the exception of lecithin, all emulsifiers used in foods are synthetic. They are characterized as ionic or nonionic and by their hydrophile/lipophile balance (HLB). All of the synthetic emulsifiers are derivatives of fatty acids.

Lecithin is the commercial name of a mixture of phospholipids obtained as a byproduct of the refining of soybean oil. Phosphatidylcholine is also known as lecithin, but the commercial product of that name contains several phospholipids including phos-

phatidylcholine. Crude soybean lecithin is dark in color and can be bleached with hydrogen peroxide or benzoyl peroxide. Lecithin can be hydroxylated by treatment with hydrogen peroxide and lactic or acetic acid. Hydroxylated lecithin is more hydrophilic, and this makes for a better oil-in-water emulsifier. The phospholipids contained in lecithin are insoluble in acetone.

Monoglycerides are produced by transesterification of glycerol with triglycerides. The reaction proceeds at high temperature, under vacuum and in the presence of an alkaline catalyst. The reaction mixture, after removal of excess glycerol, is known as commercial monoglyceride, a mixture of about 40 percent monoglyceride and di- and triglycerides. The di- and triglycerides have no emulsifying properties. Molecular distillation can increase the monoglyceride content to well over 90 percent. The emulsifying properties, especially HLB, are determined by the chain length and unsaturation of the fatty acid chain.

Hydroxycarboxylic and fatty acid esters are produced by esterifying organic acids to monoglycerides. This increases their hydrophilic properties. Organic acids used are acetic, citric, fumaric, lactic, succinic, or tartaric acid. Succinylated monoglycerides are synthesized from distilled monoglycerides and succinic anhydride. They are used as dough conditioners and crumb softeners (Krog 1981). Acetic acid esters can be produced from mono- and diglycerides by reaction with acetic anhydride or by transesterification. They are used to improve aeration in foods high in fat content and to control fat crystallization. Other esters may be prepared: citric, diacetyl tartaric, and lactic acid. A product containing two molecules of lactic acid per emulsifier molecule, known as stearoyl-2-lactylate, is available as the sodium or calcium salt. It is used in bakery products.

Polyglycerol esters of fatty acids are produced by reacting polymerized glycerol with edible fats. The degree of polymerization of the glycerol and the nature of the fat provide a wide range of emulsifiers with different HLB values.

Polyethylene or propylene glycol esters of fatty acids are more hydrophilic than monoglycerides. They can be produced in a range of compositions.

Sorbitan fatty acid esters are produced by polymerization of ethylene oxide to sorbitan fatty acid esters. The resulting polyoxyethylene sorbitan esters are nonionic hydrophilic emulsifiers. They are used in bakery products as antistaling agents. They are known as polysorbates with a number as indication of the type of fatty acid used (e.g., lauric, stearic, or oleic acid).

Sucrose fatty acid esters can be produced by esterification of fatty acids with sucrose, usually in a solvent system. The HLB varies, depending on the number of fatty acids esterified to a sucrose molecule. Monoesters have an HLB value greater than 16, triesters less than 1. When the level of esterification increases to over five molecules of fatty acid,

the emulsifying property is lost. At high levels of esterification the material can be used as a fat replacer because it is not absorbed or digested and therefore yields no calories.

Bread Improvers

To speed up the aging process of wheat flour, bleaching and maturing agents are used. Benzoyl peroxide is a bleaching agent that is frequently used; other compounds—including the oxides of nitrogen, chlorine dioxide, nitrosyl chloride, and chlorine—are both bleaching and improving (or maturing) agents. Improvers used to ensure that dough will ferment uniformly and vigorously include oxidizing agents such as potassium bromate, potassium iodate, and calcium peroxide. In addition to these agents, there may be small amounts of other inorganic compounds in bread improvers, including ammonium chloride, ammonium sulfate, calcium sulfate, and ammonium and calcium phosphates. Most of these bread improvers can only be used in small quantities, because excessive amounts reduce quality. Several compounds used as bread improvers are actually emulsifiers and are covered under that heading.

Flavors

Included in this group is a wide variety of spices, oleoresins, essential oils, and natural extractives. A variety of synthetic flavors contain mostly the same chemicals as those found in the natural flavors, although the natural flavors are usually more complex in composition. For legislative purposes, three categories of flavor compounds have been proposed.

1. *Natural flavors and flavoring substances* are preparations or single substances obtained exclusively by phys-

ical processes from raw materials in their natural state or processed for human consumption.

2. *Nature-identical flavors* are produced by chemical synthesis or from aromatic raw materials; they are chemically identical to natural products used for human consumption.

3. *Artificial flavors* are substances that are not present in natural products.

The first two categories require considerably less regulatory control than the latter one (Vodoz 1977). The use of food flavors covers soft drinks, beverages, baked goods, confectionery products, ice cream, desserts, and so on. The amounts of flavor compounds used in foods are usually small and generally do not exceed 300 ppm. Spices and oleoresins are used extensively in sausages and prepared meats. In recent years, because of public perception, the proportion of natural flavors has greatly increased at the expense of synthetics (Sinki and Schlegel 1990). Numerous flavoring substances are on the generally recognized as safe (GRAS) list. Smith et al. (1996) have described some of the recent developments in the safety evaluation of flavors. They mention a significant recent development in the flavor industry—the production of flavor ingredients using biotechnology—and describe their safety assessment.

Flavor Enhancers

Flavor enhancers are substances that carry the property of umami (see Chapter 7) and comprise glutamates and nucleotides. Glutamic acid is a component amino acid of proteins but also occurs in many protein-containing foods as free glutamic acid. In spite of their low protein content, many vegetables have high levels of free glutamate, including mushrooms, peas, and tomatoes. Sugita (1990) has listed the level of bound and free glutamate in a variety of foods. Glutamate is an element of the natural ripening process that results in fullness of taste, and it has been suggested as the reason for the popularity of foods such as tomatoes, cheese, and mushrooms (Sugita 1990).

The nucleotides include disodium 5'-inosinate (IMP), adenosine monophosphate (AMP), disodium 5'-guanylate (GMP), and disodium xanthylate (XMP). IMP is found predominantly in meat, poultry, and fish; AMP is found in vegetables, crustaceans, and mollusks; GMP is found in mushrooms, especially shiitake mushrooms.

Monosodium glutamate (MSG) is the sodium salt of glutamic acid. The flavor-enhancing property is not limited to MSG. Similar taste properties are found in the L-forms of α-amino dicarboxylates with four to seven carbon atoms. The intensity of flavor is related to the chemical structure of these compounds. Other amino acids that have similar taste properties are the salts of ibotenic acid, tricholomic acid, and L-theanine.

The chemical structure of the nucleotides is shown in Figure 7–21. They are purine ribonucleotides with a hydroxyl group on carbon 6 of the purine ring and a phosphate ester group on the 5'-carbon of the ribose. Nucleotides with the ester group at the 2' or 3' position are tasteless. When the ester group is removed by the action of phosphomonoesterases, the taste activity is lost. It is important to inactivate such enzymes in foods before adding 5'-nucleotide flavor enhancers.

The taste intensity of MSG and its concentration are directly related. The detection threshold for MSG is 0.012 g/100 mL; for sodium chloride it is 0.0037 g/100 mL; and for sucrose it is 0.086 g/100 mL. There is a strong synergistic effect between MSG and IMP. The mixture of the two has a taste intensity that is 16 times stronger than the same amount of MSG. MSG contains 12.3 percent sodium; common table salt contains three times as much sodium. By using flavor enhancers in a food, it is possible to reduce the salt level without affecting the palatability or food acceptance. The mode of action of flavor enhancers has been described by Nagodawithana (1994).

Sweeteners

Sweeteners can be divided into two groups, nonnutritive and nutritive sweeteners. The nonnutritive sweeteners include saccharin, cyclamate, aspartame, acesulfame K, and sucralose. There are also others, mainly plant extracts, which are of limited importance. The nutritive sweeteners are sucrose; glucose; fructose; invert sugar; and a variety of polyols including sorbitol, mannitol, maltitol, lactitol, xylitol, and hydrogenated glucose syrups.

The chemical structure of the most important nonnutritive sweeteners is shown in Figure 11–4. Saccharin is available as the sodium or calcium salt of orthobenzosulfimide. The cyclamates are the sodium or calcium salts of cyclohexane sulfamic acid or the acid itself. Cyclamate is 30 to 40 times sweeter than sucrose, and about 300 times sweeter than saccharin. Organoleptic comparison of sweetness indicates that the medium in which the sweetener is tasted may affect the results. There is also a concentration effect. At higher concentrations, the sweetness intensity of the synthetic sweeteners increases at a lower rate than that which occurs with sugars. This has been ascribed to the bitter-

ness and strong aftertaste that appears at these relatively high concentrations.

Cyclamates were first synthesized in 1939 and were approved for use in foods in the United States in 1950. Continued tests on the safety of these compounds resulted in the 1967 finding that cyclamate can be converted by intestinal flora into cyclohexylamine, which is a carcinogen. Apparently, only certain individuals have the ability to convert cyclamate to cyclohexylamine (Collings 1971). In a given population, a portion are nonconverters, some convert only small amounts, and others convert large amounts.

Aspartame is a dipeptide derivative, L-aspartyl-L-phenylalanine methyl ester, which was approved in the United States in 1981 for use as a tabletop sweetener, in dry beverage mixes, and in foods that are not heat processed. This substance is metabolized in the body to phenylalanine, aspartic acid, and methanol. Only people with phenylketonuria cannot break down phenylalanine. Another compound, diketopiperazine, may also be formed. However, no harmful effects from this compound have been demonstrated. The main limiting factor in the use of aspartame is its lack of heat stability (Homler 1984).

A new sweetener, approved in 1988, is acesulfame K. This is the potassium salt of 6-methyl-1,2,3-oxathiozine-4(3H)-one-2, 2-dioxide (Figure 11–4). It is a crystalline powder that is about 200 times sweeter than sugar. The sweetening power depends to a certain degree on the acidity of the food it is used in. Acesulfame K is reportedly more stable than other sweeteners. The sweet taste is clean and does not linger. Sucralose is a trichloroderivative of the C-4 epimer galactosucrose. It is about 600 times sweeter than sucrose and has a similar taste profile. One of its main advantages is heat stability, so it can be used in baking.

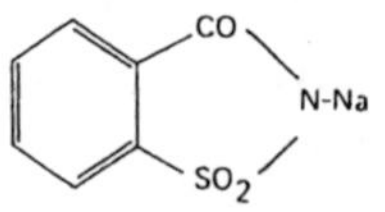

Na-Saccharin

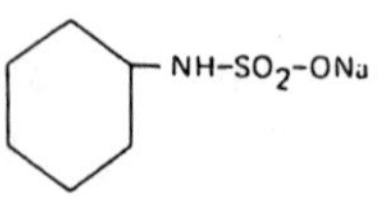

Na-cyclamate

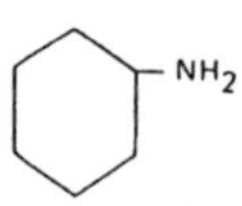

cyclohexylamine

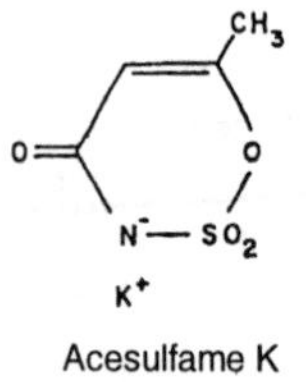

Acesulfame K

Figure 11–4 Chemical Structure of Sodium Saccharin, Sodium Cyclamate, Cyclohexylamine, and Acesulfame K

Blending of nonnutritive sweeteners may lead to improved taste, longer shelf life, lower production cost, and reduced consumer exposure to any single sweetener (Verdi and Hood 1993). The dihydrochalcone sweeteners are obtained from phenolic glycosides present in citrus peel. Such compounds can be obtained from naringin of grapefruit or from the flavonoid neohesperidin. The compound neohesperidin dihydrochalcone is rated 1,000 times sweeter than sucrose (Inglett 1971). Horowitz and Gentili (1971) investigated the relationship between chemical structure and sweetness, bitterness, and tastelessness. Several other natural compounds having intense sweetness have been described by Inglett (1971); these include glycyrrhizin (from licorice root) and a taste-modifying glycoprotein named miraculin that is obtained from a tropical fruit known as miracle berry. Stevioside is an extract from the leaves of a South American plant that is 300 times sweeter than sugar. Thaumatin, a protein mixture from a West African fruit, is 2,000 times sweeter than sugar, but its licorice-like aftertaste limits its usefulness.

It has been suggested that sugars from the L series could be used as low-calorie sweeteners. These sugars cannot be metabolized in the normal way, as D sugars would, and therefore pass through the digestive system unaltered. Their effect on the body has not been sufficiently explored.

Possible new sweeteners have been described by Gelardi (1987).

Phosphates

These compounds are widely used as food additives, in the form of phosphoric acid as acidulant, and as monophosphates and polyphosphates in a large number of foods and for a variety of purposes. Phosphates serve as buffering agents in dairy, meat, and fish products; anticaking agents in salts; firming agents in fruits and vegetables; yeast food in bakery products and alcoholic beverages;

and melting salts in cheese processing. Phosphorus oxychloride is used as a starch-modifying agent.

The largest group of phosphates and the most important in the food industry is the orthophosphates (Figure 11–5). The phosphate group has three replaceable hydrogens, giving three possible sodium orthophosphates—monosodium, disodium, and trisodium phosphate. The phosphates can be divided into othophosphates, polyphosphates, and metaphosphates, the latter having little practical importance. Polyphosphates have two or more phosphorus atoms joined by an oxygen bridge in a chain structure. The first members of this series are the pyrophosphates, which have one P-O-P linkage. The condensed phosphate with two linkages is tripolyphosphate. Alkali metal phosphates with chain lengths greater than three are usually mixtures of polyphosphates with varied chain lengths. The best known is sodium hexametaphosphate. The longer chain length salts are glasses. Hexametaphosphate is not a real metaphosphate, since these are ring structures and hexametaphosphate is a straight-chain polyphosphate. Sodium hexametaphosphate has an average chain length of 10 to 15 phosphate units.

Phosphates are important because they affect the absorption of calcium and other elements. The absorption of inorganic phosphorus depends on the amount of calcium, iron, strontium, and aluminum present in the diet. Chapman and Pugsley (1971) have suggested that a diet containing more phosphorus than calcium is as detrimental as a simple calcium deficiency. The ratio of calcium to phosphorus in bone is 2 to 1. It has been recommended that in early infancy, the ratio should be 1.5 to 1; in older infants, 1.2 to 1; and for adults, 1 to 1. The estimated annual per capita intake in the United States is 1 g Ca and 2.9 g P, thus giving a ratio of 0.35. The danger in raising phosphorus levels is that calcium may become unavailable.

Coloring Agents

In the United States two classes of color additives are recognized: colorants exempt from certification and colorants subject to certification. The former are obtained from vegetable, animal, or mineral sources or are synthetic forms of naturally occurring compounds. The latter group of synthetic dyes and pigments is covered by the Color Additives Amendment of the U.S. Food, Drug and Cosmetic Act. In the United States these color compounds are not known by their common names but as FD&C colors (Food, Drug and Cosmetic colors) with a color and a number (Noonan 1968). As an example,

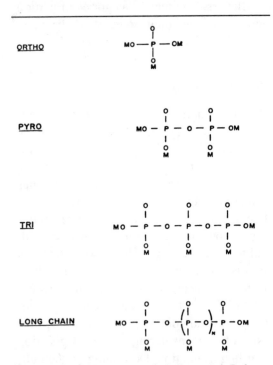

Figure 11–5 Structure of Ortho- and Polyphosphate Salts

FD&C red dye no. 2 is known as amaranth outside the United States. Over the years the originally permitted fat-soluble dyes have been removed from the list of approved dyes, and only water-soluble colors remain on the approved list.

According to Newsome (1990) only nine synthetic colors are currently approved for food use and 21 nature-identical colors are exempt from certification. The approved FD&C colors are listed in Exhibit 11–2. Citrus red no. 2 is only permitted for external use on oranges, with a maximum level of 2 ppm on the weight of the whole orange. Its use is not permitted on oranges destined for processing.

Lakes are insoluble forms of the dyes and are obtained by combining the color with aluminum or calcium hydroxide. The dyes provide color in solution, and the lakes serve as insoluble pigments.

Exhibit 11–2 Color Additives Permitted for Food Use in the United States and Their Common Names

- FD&C red no. 3 (erythrosine)
- FD&C red no. 40 (allura red)
- FD&C orange B
- FD&C yellow no. 6 (sunset yellow)
- FD&C yellow no. 5 (tartrazine)
- FD&C green no. 3 (fast green)
- FD&C blue no. 1 (brillian blue)
- FD&C blue no. 2 (indigotine)
- Citrus red no. 2

Source: Reprinted with permission from R.L. Newsome, Natural and Synthetic Coloring Agents, in *Food Additives*, A.L. Branen, P.M. Davidson, and S. Salminen, eds., p. 344, 1990, by courtesy of Marcel Dekker, Inc.

The average per capita consumption of food colors is about 50 mg per day. Food colors have been suspect as additives for many years, resulting in many deletions from the approved list. An example is the removal of FD&C red no. 2 or amaranth in 1976. In the United States, it was replaced by FD&C red no. 40. The removal from the approved list was based on the observation of reproductive problems in test animals that consumed amaranth at levels close to the ADI. As a consequence, the Food and Agriculture Organization (FAO)/World Health Organization (WHO) reduced the ADI to 0.75 mg/kg body weight from 1.5 mg/kg. Other countries, including Canada, have not delisted amaranth.

The natural or nature-identical colors are less stable than the synthetic ones, more variable, and more likely to introduce undesirable flavors. The major categories of natural food colors and their sources are listed in Table 11–5.

Food Irradiation

Food irradiation is the treatment of foods by ionizing radiation in the form of beta, gamma, or X-rays. The purpose of food irradiation is to preserve food and to prolong shelf life, as other processing techniques such as heating or drying have done. For regulatory purposes irradiation is considered a process, but in many countries it is considered to be an additive. This inconsistency in the interpretation of food irradiation results in great obstacles to the use of this process and has slowed down its application considerably. Several countries are now in the process of reconsidering their legislation regarding irradiation. Depending on the radiation dose, several applications can be distinguished. The unit of radiation is the Gray

Table 11–5 Major Categories of Natural Food Colors and Their Sources

Colorant	Sources
Anthocyanins	Grape skins, elderberries
Betalains	Red beets, chard, cactus fruits, pokeberries, bougainvillea, amaranthus
Caramel	Modified sugar
Carotenoids	
Annatto (bixin)	Seeds of *Bixa orellana*
Canthaxanthin	Mushrooms, crustaceans, fish, seaweed
β-apocarotenal	Oranges, green vegetables
Chlorophylls	Green vegetables
Riboflavin	Milk
Others	
Carmine (cochineal extract)	*Coccus cati* insect
Turmeric (curcuma)	*Curcuma longa*
Crocetin, crocin	Saffron

Source: Reprinted with permission from R.L. Newsome, Natural and Synthetic Coloring Agents, in *Food Additives*, A.L. Branen, P.M. Davidson, and S. Salminen, eds., p. 333, 1990, by courtesy of Marcel Dekker, Inc.

(Gy), which is a measure of the energy absorbed by the food. It replaced the older unit rad (1 Gy = 100 rad).

Radiation sterilization produces foods that are stable at room temperature and requires a dose of 20 to 70 kGy. At lower doses, longer shelf life may be obtained, especially with perishable foods such as fruits, fish, and shellfish. The destruction of *Salmonella* in poultry is an application for radiation treatment. This requires doses of 1 to 10 kGy. Radiation disinfestation of spices and cereals may replace chemical fumigants, which have come under increasing scrutiny in recent years. Dose levels of 8 to 30 kGy would be required. Other possible applications of irradiation processing are inhibition of sprouting in potatoes and onions and delaying of the ripening of tropical fruits.

Nutrition Supplements

There are two fundamental reasons for the addition of nutrients to foods consumed by the public: (1) to correct a recognized deficiency of one or more nutrients in the diets of a significant number of people when the deficit actually or potentially adversely affects health; and (2) to maintain the nutritional quality of the food supply at a level deemed by modern nutrition science to be appropriate to ensure good nutritional health, assuming only that a reasonable variety of foods are consumed (Augustin and Scarbrough 1990).

A variety of compounds are added to foods to improve the nutritional value of a product, to replace nutrients lost during processing, or to prevent deficiency diseases. Most of the additives in this category are vitamins or minerals. Enrichment of flour and related products is now a well-recognized practice. The U.S. Food and Drug Administration (FDA) has established definitions and standards of identity for the enrichment of wheat flour, farina, corn meal, corn grits, macaroni, pasta products, and rice. These standards define minimum and maximum levels of addition of thiamin, riboflavin, niacin, and iron. In some cases, optional addition of calcium and vitamin D is allowed. Margarine contains added vitamins A and D, and vitamin D is added to fluid and evaporated milk. The addition of the fat-soluble vitamins is strictly controlled, because of the possible toxicity of overdoses of these vitamins. The vitamin D enrichment of foods has been an important measure in the elimination of rickets. Another example of the beneficial effect of enrichment programs is the addition of iodine to table salt. This measure has virtually eliminated goiter.

One of the main potential deficiencies in the diet is calcium. Lack of calcium is associated with osteoporosis and possibly several other diseases. The recommended daily allowance for adolescents/young adults and the elderly has increased from the previous recommendation of 800 to 1,200 mg/day to 1,500 mg/day. This level is difficult to achieve, and the use of calcium citrate in fortified foods has been recommended by Labin-Goldscher and Edelstein (1996). Sloan and Stiedemann (1996) highlighted the relationship between consumer demand for fortified products and complex regulatory issues.

Migration from Packaging Materials

When food packaging materials were mostly glass or metal cans, the transfer of packaging components to the food consisted predominantly of metal (iron, tin, and lead) uptake. With the advent of extensive use of plastics, new problems of transfer of toxicants and flavor and odor substances became apparent. In addition to polymers, plastics may contain a variety of other chemicals, catalysts, antioxidants, plasticizers, colorants, and light absorbers. Depending on the nature of the food, especially its fat content, any or all of these compounds may be extracted to some degree into the food (Bieber et al. 1985).

Awareness of the problem developed in the mid 1970s when it was found that mineral waters sold in polyvinyl chloride (PVC) bottles contained measurable amounts of vinyl chloride monomer. Vinyl chloride is a known carcinogen. The Codex Alimentarius Committee on Food Additives and Contaminants has set a guideline of 1 ppm for vinyl chloride monomer in PVC packaging and 0.01 ppm of the monomer in food (Institute of Food Technologists 1988). Another additive found in some PVC plastics is octyl tin mercaptoacetate or octyl tin maleate. Specific regulations for these chemicals exist in the Canadian Food and Drugs Act.

The use of plastic netting to hold and shape meat during curing resulted in the finding of N-nitrosodiethylamine and N-nitrosodibutylamine in hams up to levels of 19 ppb (parts per billion) (Sen et al. 1987). Later research established that the levels of nitrosamines present were not close to violative levels (Marsden and Pesselman 1993).

Plasticizers, antioxidants, and colorants are all potential contaminants of foods that are contained in plastics made with these chemicals. Control of potential migration of plastic components requires testing the containers with food simulants selected to yield information relevant to the intended type of food to be packaged (DeKruyf et al. 1983; Bieber et al. 1984).

Other Additives

In addition to the aforementioned major groups of additives, there are many others including clarifying agents, humectants, glazes, polishes, anticaking agents, firming agents, propellants, melting agents, and enzymes. These intentional additives present considerable scientific and technological problems as well as legal, health, and public relations challenges. Future introduction of new additives will probably become increasingly difficult, and some existing additives may be disallowed as further toxicological studies are carried out and the safety requirements become more stringent.

INCIDENTAL ADDITIVES OR CONTAMINANTS

Radionuclides

Natural radionuclides contaminate air, food, and water. The annual per capita intake of natural radionuclides has been estimated to range from 2 Becquerels (Bq) for ^{232}Th to about 130 Bq for ^{40}K (Sinclair 1988). The Bq is the International System of Units (SI) unit of radioactivity; 1 Bq = 1 radioactive disintegration per second. The previously used unit of radioactivity is the Curie (Ci); 1 Ci = 3.7×10^{10} disintegrations per second, and 1 Bq = 27×10^{-12} Ci. The quantity of radiation or energy absorbed is expressed in Sievert (Sv), which is the SI unit of dose equivalent. The absorbed dose (in Gy) is multiplied by a quality factor for the particular type of radiation. Rem is the previously used unit for dose equivalent; 100 rem = 1 Sv.

The effective dose of Th and K radionuclides is about 400 μSv per capita per year, with half of it resulting from ^{40}K. The total exposure of the U.S. population to natural radiation has been estimated at about 3 mSv. In addition, 0.6 mSv is caused by man-made radiation (Sinclair 1988).

Radioactive Fallout

Major concern about rapidly increasing levels of radioactive fallout in the environment and in foods developed as a result of the extensive testing of nuclear weapons by the United States and the Soviet Union in the 1950s. Nuclear fission generates more than 200 radioisotopes of some 60 different elements. Many of these radioisotopes are harmful to humans because they may be incorporated into body tissues. Several of these radioactive isotopes are absorbed efficiently by the organism because they are related chemically to important nutrients; for example, strontium-90 is related to calcium and cesium-137 to potassium. These radioactive elements are produced by the following nuclear reactions, in which the half-life is given in parentheses:

$$^{90}\text{Kr (33 sec)} \xrightarrow{\beta^-} {}^{90}\text{Rb (2.7 min)} \xrightarrow{\beta^-} {}^{90}\text{Sr (28 y)}$$

$$^{137}\text{I (22 sec)} \xrightarrow{\beta^-} {}^{137}\text{Xe (3.8 min)} \xrightarrow{\beta^-} {}^{137}\text{Cs (29 y)}$$

The long half-life of the two end products makes them especially dangerous. In an atmospheric nuclear explosion, the tertiary fission products are formed in the stratosphere and gradually come down to earth. Every spring about one-half to two-thirds of the fission products in the stratosphere come down and are eventually deposited by precipitation. Figure 11–6 gives a schematic outline of the pathways through which the fallout may reach us.

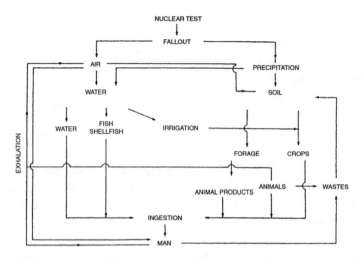

Figure 11–6 Pathways of the Transfer of Nuclear Fallout to Humans

Among the radioisotopes that can be taken up in the food chain, the most significant as internal radiation hazards are barium-140, cesium-137, iodine-131, iodine-133, strontium-89, and strontium-90.

[131]I is chemically similar to ordinary iodine and, therefore, accumulates in the thyroid gland. It has a half-life of eight days and is a beta-gamma emitter. Because milk is produced year round and is consumed within one half-life, the presence of this isotope in milk was a major concern during the atmospheric testing period in the early 1960s.

[137]Cs has a half-life of 29 years and, because of its chemical similarity to potassium, accumulates in muscle tissue. [137]Cs may cause several types of cell damage, including genetic damage. It is not retained in the body for a long time. [137]Cs is a gamma emitter.

[90]Sr has a half-life of 28 years and is a beta emitter. This isotope collects in the bones because of its chemical similarity to calcium. It can result in bone cancer and leukemia. Children are very sensitive to this isotope because they require large amounts of cal-

cium for bone formation and as a result deposit relatively more [90]Sr. They also face a longer life span, which is important because radiation effects are cumulative.

In 1964, the rate of fallout of ^{90}Sr was about 40 pc/day/m^2. Total intake of ^{90}Sr during that period was about 40 pc/day/person in some Western countries. Because about 3,000 m^2 of arable land are required to produce food for one person, the total amount of ^{90}Sr deposited on that surface was estimated to be 120 nc per day. This means a reduction of about 3000-fold, indicating a highly effective barrier mechanism. The amount of radioactivity gradually diminished after the United States and the Soviet Union ceased their atmospheric test programs. Emergency measures for decontaminating essential food items such as milk have been developed. Such procedures use ion-exchange methods to remove radioisotopes (Glascock 1965).

The distribution of radioactive fallout in the environment and therefore in foods is nonuniform. The distribution is influenced by latitude; most of the fallout comes down between 30° and 60° latitude (Miettinen

1967). Because the fallout comes down with precipitation, precipitation is a major factor. In addition, uptake by plants is influenced by soil type. Wiechen (1972) found that the ^{137}Cs content of milk from a small herd of cows averaged 26 pc per kg when the animals grazed on an area of sandy soil but increased to 244 pc/kg when they were transferred to moorland. The primary contamination level of the two soil types was 280 and 262 pc per kg, respectively. The higher transfer rate of ^{137}Cs in the moorland soil-grass-milk chain was the result of the low potassium content of this soil (180 mg/kg versus 720 mg/kg in sandy soil) and, to a lesser extent, different mobilities of Cs and K in the various soils. Lindell and Magi (1965) found a general similarity between precipitation distribution and distribution of ^{137}Cs levels in milk. In Sweden, the lowest levels of the isotope ^{90}Sr were found for the island of Gotland, which has low precipitation and a soil rich in calcium.

Johnson and Nayfield (1970) reported a case of true selective concentration of ^{137}Cs. They found that high levels of ^{137}Cs in game animals from the southeastern United States resulted from their feeding on mushrooms in wooded areas. Common gill mushrooms (*Agaricaceae*) from these areas had ^{137}Cs levels as high as 29,000 pc/kg wet weight, with a mean of 15,741 pc/kg. These elevated levels occurred without similar concentration of potassium-40. White-tailed deer in these regions had ^{137}Cs levels ranging from 250 to 152,940 pc/kg body weight.

The 1986 nuclear reactor accident at Chernobyl in the Soviet Union distributed radioactive fallout over most of Western Europe and the rest of the world. In addition to short-term problems with radionuclides of short half-life, there are ongoing concerns in countries far removed from the source of the contamination. In the United Kingdom there are concerns over the contamination of sheep, whose levels still exceed the interim limit of 1,000 Bq/kg; bentonite is being used experimentally to reduce Cs uptake from grazing.

In addition to the Chernobyl accident, there have been nuclear reactor accidents at Windscale in the United Kingdom in 1957 and at Three Mile Island in the United States in 1979. Low-level emissions from nuclear reactor plants apparently are not uncommon.

Pesticides

Contamination of food with residues of pesticides may result from the application of these chemicals in agricultural, industrial, or household use. Nearly 300 organic pesticides are in use, including insecticides, miticides, nematocides, rodenticides, fungicides, and herbicides. The most likely compounds to appear as food contaminants are insecticides, of which there are two main classes—chlorinated hydrocarbon insecticides and organophosphorous insecticides.

The chlorinated hydrocarbon insecticides can be divided into three classes—oxygenated compounds, benzenoid nonoxygenated compounds, and nonoxygenated nonbenzenoid compounds (Exhibit 11–3) (Mitchell 1966). In addition to the pesticide compounds, there may be residues of their metabolites, which may be equally toxic. Two important properties of the chlorinated hydrocarbons are their stability, which leads to persistence in the environment, and their solubility in fat, which results in their deposition and accumulation in fatty tissues. The structure of some of the chlorinated hydrocarbon insecticides is given in Figure 11–7. Aldrin is a technical compound containing about 95 percent of the compound

Exhibit 11–3 Classes of Chlorinated Hydrocarbon Insecticides

Class I—Oxygenated Compounds

- Chlorobenzilate
- Dicofol
- Dieldrin
- Endosulfan
- Endrin
- Kepone
- Methoxychlor
- Neotran
- Ovex
- Sulfenone
- Tetradifon

Class II—Benzenoid, Nonoxygenated Compounds

- BHC
- Chlorobenside
- DDT
- Lindane
- Perthane
- TDE
- Zectran

Class III—Nonoxygenated, Nonbenzenoid Compounds

- Aldrin
- Chlordan
- Heptachlor
- Mirex
- Strobane
- Toxaphene

Source: From L.E. Mitchell, Pesticides: Properties and Prognosis, in *Organic Pesticides in the Environment*, R.F. Gould, ed., 1966, American Chemical Society.

1,2,3,4,10,10-hexachloro-1,4,4a,5,8,8a-hexahydro-exo-1,4-*endo*-exo-5,8-dimethanonaphthalene. It has a molecular weight of 365, formula $C_{12}H_8Cl_6$, and contains 58 percent chlorine. Residues of this compound in animal and plant tissues are converted into dieldrin by epoxidation. The epoxide is the stable form and, thus, it is usual to consider these compounds together.

Dieldrin contains about 85 percent of the compound 1, 2, 3, 4, 10, 10-hexachloro-6, 7-epoxy-1, 4, 4a, 5, 6, 7, 8, 8a-octahydro-exo-1, 4-endo-exo-5, 8-dimethano-naphthalene (HEOD). It has a molecular weight of 381, formula $C_{12}H_8Cl_6O$, and contains 56 percent chlorine. DDT is a technical compound that contains about 70 percent of the active ingredient pp'-DDT. In addition, there are other isomers, including op'-DDT, as well as related compounds such as TDE or rhothane. The insecticide pp'-DDT is 1,1,1-trichloro-2,2-di-(4-chlorophenyl) ethane, formula $C_{14}H_9Cl_5$. It has a molecular weight of 334.5 and contains 50 percent chlorine. Residues of DDT in animal tissue are slowly dehydrochlorinated to pp'-DDE, which may occur at levels of up to 70 percent of the original DDT. It is usual to combine DDT, DDE, and TDE in one figure as "total DDT equivalent."

Heptachlor contains about 75 percent of 1,4,5,6,7,10,10-heptachloro-4,7,8,9-tetrahyro-4,7-methyleneindene, formula $C_{10}H_5Cl_7$. It has a molecular weight of 373.5 and contains 67 percent chlorine. In animal and plant tissues, it epoxidizes to heptachlor epoxide, which is analogous in structure to HEOD (dieldrin).

Although relatively stable, the organochlorine pesticides undergo a variety of reactions that may result in metabolites that are as toxic or more toxic to mammals than the original compound. An example is the effect of ultraviolet light on DDT (Van Middelem 1966). Under the influence of ultraviolet light and air, 4,4'-dichlorobenzophenone is formed. Without air, 2,3-dichloro-1,1,4,4-tetrakis-(p-chlorophenyl)-2-butene is formed. The latter may be oxidized to 4,4'-dichlorobenzophenone (Figure 11–8). In mammalian tissue, 2,2-bis (p-chlorophenyl) acetic acid (DDA) is formed by initial dehydrochlorination of DDT to DDE, followed by hydrolysis to DDA (Figure 11–9).

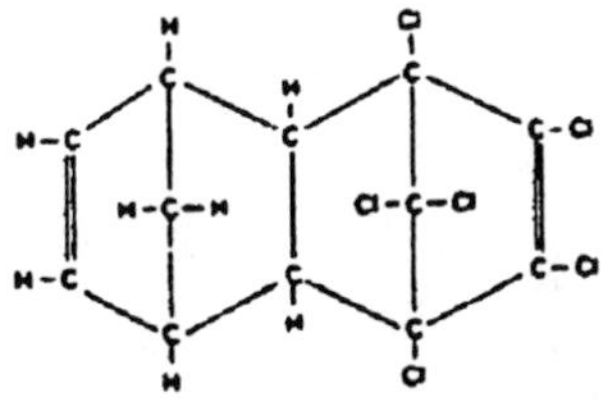

ALDRIN

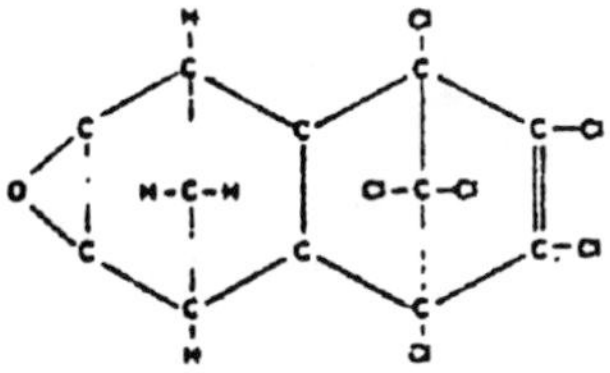

DIELDRIN

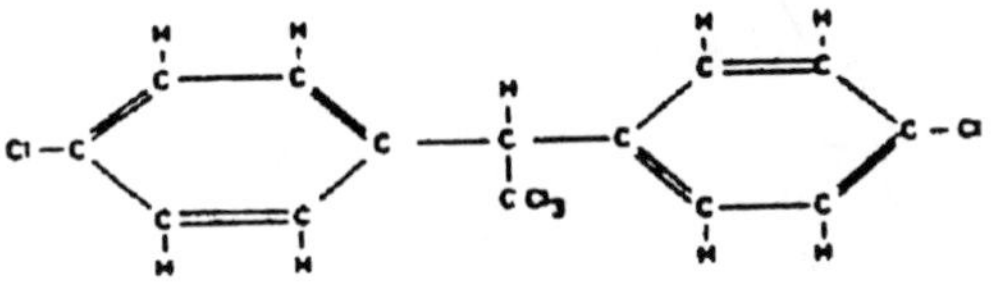

pp' – DDT

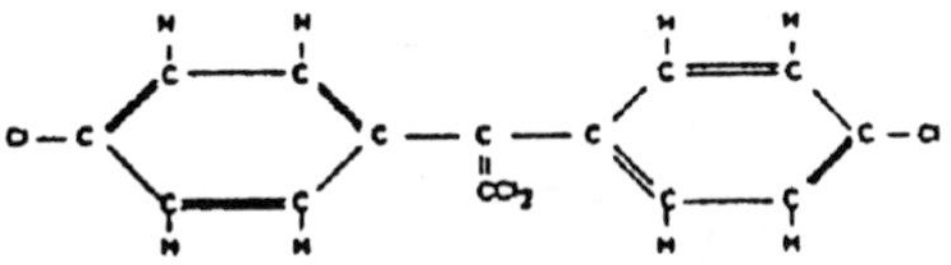

pp' – DDE

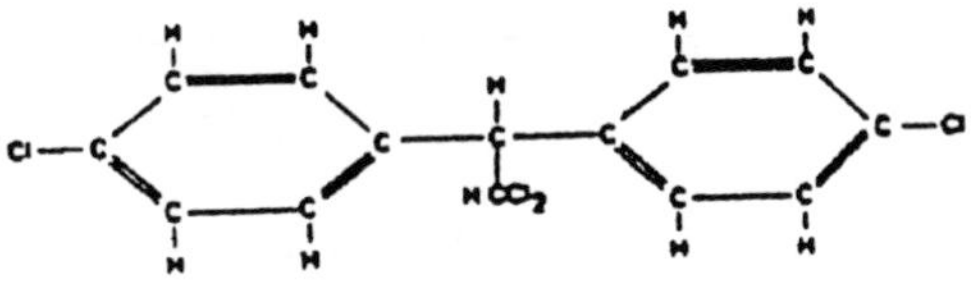

TDE (RHOTHANE)

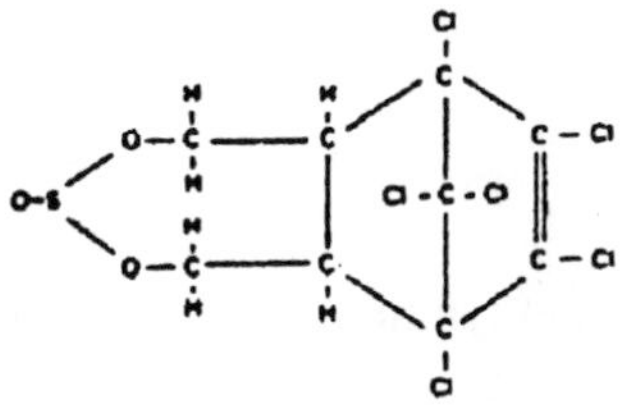

ENDOSULFAN

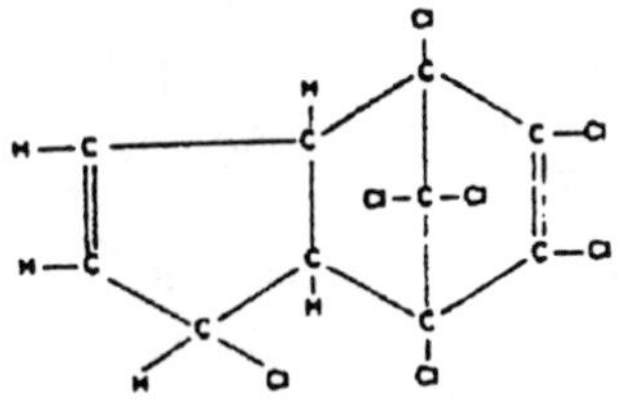

HEPTACHLOR

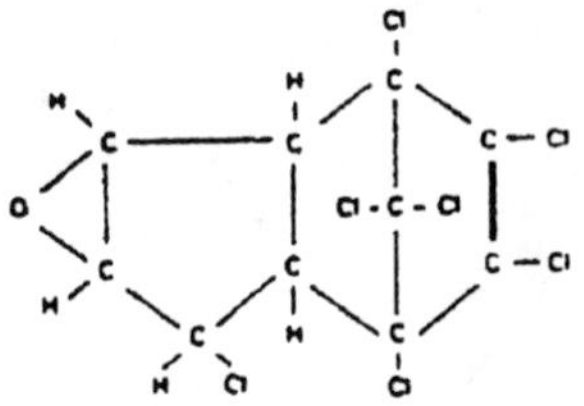

HEPTACHLOR EPOXIDE

Figure 11–7 Structure of Some Chlorinated Hydrocarbon Pesticides

The organophosphorous insecticides are inhibitors of cholinesterase and, because of their water solubility and volatility, create less of a problem as food contaminants than the chlorinated hydrocarbons. A large number of organophosphorous insecticides are in use; these can act by themselves or after oxidative conversions in plants and animals (Exhibit 11–4). The water solubility of these compounds varies widely, as is indicated by Table 11–6. The organophosphorous insecticides may be subject to oxidation, hydroly-

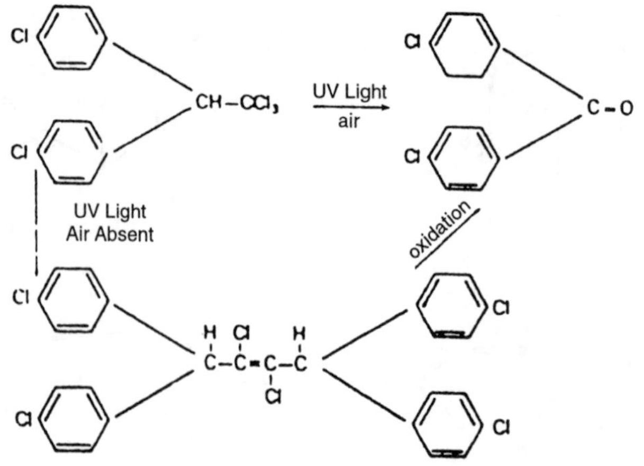

Figure 11–8 Decomposition of DDT by Ultraviolet Light and Air. *Source:* From C.H. Van Middelem, Fate and Persistence of Organic Pesticides in the Environment, in *Organic Pesticides in the Environment*, R.F. Gould, ed., 1966, American Chemical Society.

sis, and demethylation (Figure 11–10). Thiophosphates may be changed to sulfoxides and sulfones in animals and plants.

In animal products, chlorinated hydrocarbon residues are predominantly present in the lipid portion, organophosphates in both lipid and aqueous parts. In plant materials, the residue of chlorinated hydrocarbons are mostly surface bound or absorbed by waxy materials, but some can be translocated to inner parts. Extensive research has demon-

strated that processing methods such as washing, blanching, heating, and canning may remove large proportions of pesticide residues (Liska and Stadelman 1969; Farrow et al. 1969). An illustration of the removal of DDT and carbaryl from vegetables by washing, blanching, and canning is given in Figure 11–11.

It has been reported (Farrow et al. 1969) that 48 percent of DDT residues on spinach and 91 percent on tomatoes are removed by

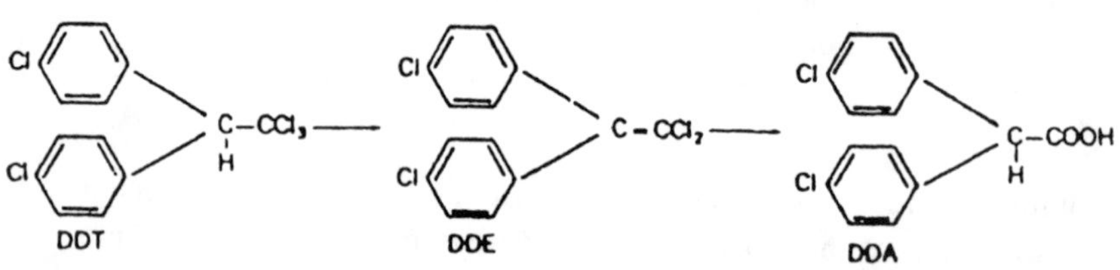

Figure 11–9 Dehydrochlorination and Hydrolysis of DDT to DDE and DDA in Mammals. *Source:* From C.H. Van Middelem, Fate and Persistence of Organic Pesticides in the Environment, in *Organic Pesticides in the Environment*, R.F. Gould, ed., 1966, American Chemical Society.

Exhibit 11–4 Classification of Organophospho-rous Insecticides

---

Aliphatic Derivatives

- Butonate
- Demeton
- Dichlorvos
- Dimefox
- Dimethoate
- Dithiodemeton
- Ethion
- Malathion
- Methyl demeton

- Mevinphos
- Mipefox
- Naled
- Phorate
- Phosphamidon
- Schradan
- Sulfotepp
- Tepp
- Trichlorofon

Aromatic (Cyclic) Derivatives

- Azinphosmethyl
- Carbophenothion
- Diazinon
- Dicapthon
- Endothion

- EPN
- Fenthion
- Methyl parathion
- Parathion
- Ronnel

Source: From L.E. Mitchell, Pesticides: Prop-erties and Prognosis, in *Organic Pesticides in the Environment*, R.F. Gould, ed., 1966, Ameri-can Chemical Society.

Table 11–6 Water Solubilities of Some Organophosphorus Insecticides

Insecticide	(ppm)
Carbophenothion	2
Parathion	24
Azinphosmethyl	33
Diazinon	40
Methyl parathion	50
Phorate	85
Malathion	145
Dichlorvos	1000
Dimethoate	7000
Mevinphos	∞

Source: From L.E. Mitchell, Pesticides: Properties and Prognosis, in *Organic Pesticides in the Environment*, R.F. Gould, ed., 1966, American Chemical Society.

---

washing. Elkins (1989) reported that wash-ing and blanching reduced carbaryl residues on spinach and broccoli by 97 and 98 per-cent, respectively. Washing, blanching, and canning reduced carbaryl pesticides on toma-toes and spinach by 99 percent. Although this pattern of removal generally holds true, Peterson and colleagues (1996) have pointed out that there are exceptions. Pesticides may accumulate in one part of an agricultural product. Friar and Reynolds (1991) reported that baking does not result in a decline in thi-abendazole residues in potatoes, and Elkins

et al. (1972) found that thermal processing does not result in a reduction of methoxy-chlor residues on apricots. Sometimes pro-cessing may cause a chemical to degrade, producing a compound that is more toxic than the original one.

The dietary intake of pesticide chemicals from foods is well below the acceptable daily intake (ADI) levels set by the FAO/WHO (Table 11–7). In recent years, severe restric-tions on the use of many chlorinated hydro-carbon pesticides have been instituted in many areas. As a result, the intake of these chemicals should further decrease in future years.

Dioxin

The term *dioxin* is used to represent two related groups of chlorinated organic com-pounds, polychlorinated dibenzo-*p*-dioxins (PCDD) and polychlorinated dibenzofurans

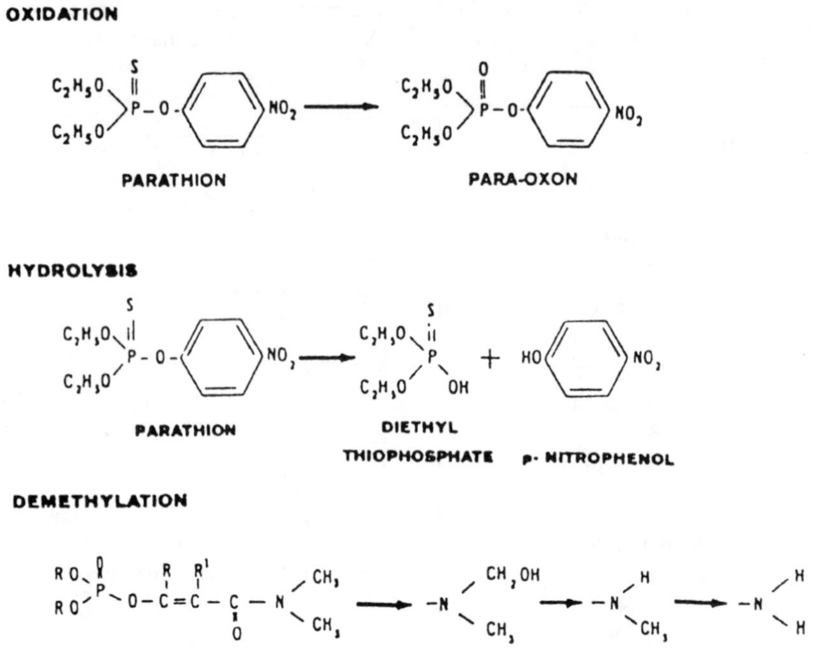

OXIDATION

HYDROLYSIS

DEMETHYLATION

Figure 11–10 Oxidation, Hydrolysis, and Demethylation Reactions of Organophosphorous Insecticides. *Source:* From L.E. Mitchell, Pesticides: Properties and Prognosis, in *Organic Pesticides in the Environment*, R.F. Gould, ed., 1966, American Chemical Society.

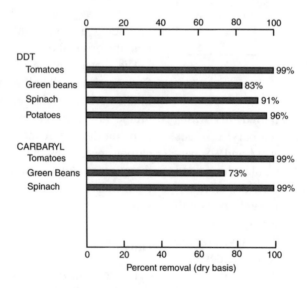

Figure 11–11 Removal of Pesticides DDT and Carbaryl by Washing, Blanching, and Canning. *Source:* From R.P. Farrow et al., Canning Operations That Reduce Insecticide Levels in Prepared Foods and in Solid Food Wastes, *Residue Rev.*, Vol. 29, pp. 73–78, 1969.

Table 11–7 Dietary Intake of Pesticide Chemicals

	Milligrams/Kilogram Body Weight/Day		
Pesticide Chemical	*WHO-FAO Acceptable Daily Intake*	*Average 1965–1969*	*Range*
Aldrin-dieldrin	0.0001	0.00008	(0.00006–0.00013)
Carbaryl	0.02	0.0005	(None–0.0021)
DDT, DDE, TDE	0.01 (0.005)[1]	0.0008	(0.0005–0.0010)
Lindane	0.012	0.00005	(0.00002–0.00007)
Heptachlor-heptachlor epoxide	0.0005	0.00003	(0.00002–0.00005)
Malathion	0.02	0.0001	(0.0001–0.0004)
Parathion	0.005	0.00001	(0.000001–0.00001)
Diazinon	0.002	0.00001	(0.000001–0.00002)
All chlorinated organics		0.001	(0.0008–0.0016)
All organophosphates		0.0002	(0.00007–0.00025)
All herbicides		0.0001	(0.00005–0.0001)

[1]Current value accepted 1969 Meeting

Source: From J.R. Wessel, Pesticide Residues in Foods, in *Environmental Contaminants in Foods*, Special Report No. 9, 1972, Cornell University.

(PCDF) (Figure 11–12). A total of eight carbon atoms in each molecule can carry chlorine substitution, which produces 75 possible isomers for PCDD and 135 for PCDF. These compounds are lipophilic, have low volatility, and are extremely stable. They are also very toxic, although the toxicity of each isomer may vary widely. These compounds may exhibit acute toxicity, carcinogenicity, and teratogenicity (birth defects). They are ubiquitous environmental contaminants and are present in human tissues.

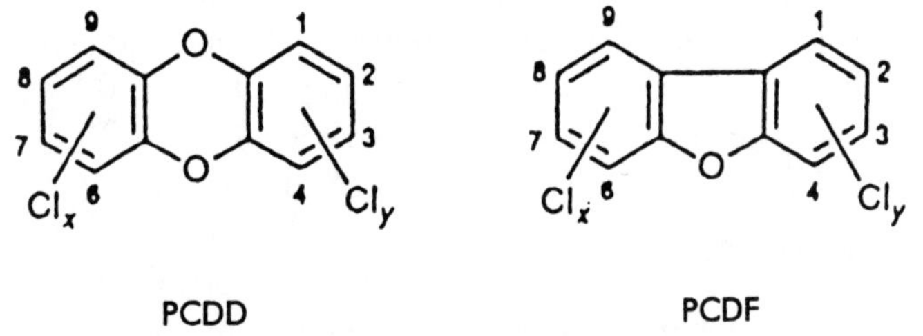

PCDD

PCDF

Figure 11–12 Chemical Structure of Polychlorinated Dibenzo-*p*-dioxins (PCDD) and Polychlorinated Dibenzofurans (PCDF)

The dioxins are produced as contaminants in the synthesis of certain herbicides and other chlorinated compounds, as a result of combustion and incineration, in the chlorine bleaching of wood pulp for paper making, and in some metallurgical processes (Startin 1991). Dioxins first attracted attention as a contaminant of the herbicide 2,4,5-trichloro-phenoxyacetic acid (2,4,5-T). The particular compound identified was 2,3,7,8-TCDD, which was for some time associated with the name *dioxin*. This compound was present in substantial concentration in the defoliant "Agent Orange" used by U.S. forces during the war in Vietnam.

The various isomers, also known as congeners, vary in toxicity with the 2,3,7,8-substituted ones being the most toxic. Humans appear to be less sensitive than other species.

Dioxins can be generated from chlorine bleaching of wood pulp in the paper- and cardboard-making process. This can not only lead to environmental contamination but also to incorporation of the dioxins in the paper used for making coffee filters, tea bags, milk cartons, and so forth. Dioxins can migrate into milk from cartons, even if the cartons have a polyethylene plastic coating. Unbleached coffee filters and cardboard containers have been produced to overcome this problem, and there have also been improvements in the production of wood pulp using alternative bleaching agents. The FDA guideline for dioxin in fish is 25 parts per trillion (Cordle 1981). Dioxin is considered a very potent toxin, but information on harmful effects on humans is controversial.

Polychlorinated Biphenyls (PCBs)

The PCBs are environmental contaminants that are widely distributed and have been found as residues in foods. PCBs are pre-pared by chlorination of biphenyl, which results in a mixture of isomers that have different chlorine contents. In North America, the industrial compounds are known as Aroclor; these are used industrially as dielectric fluids in transformers, as plasticizers, as heat transfer and hydraulic fluids, and so forth. The widespread industrial use of these compounds results in contamination of the environment through leakages and spills and seepage from garbage dumps. The PCBs may show up on chromatograms at the same time as chlorinated hydrocarbon pesticides. The numbering system used in PCBs and the prevalent substitution pattern are presented in Figure 11–13. Table 11–8 presents information on commercial Aroclor compounds. In the years prior to 1977 production of PCBs in North America amounted to about 50 million pounds per year. PCBs were first discovered in fish and wildlife in Sweden in 1966, and they can now be found in higher concentrations in fish than organochlorine pesticides (Zitko 1971).

PCBs decompose very slowly. It is estimated that between 1929 and 1977, about 550 million kg of PCBs were produced in the United States. Production was stopped voluntarily after a serious poisoning occurred in Japan in 1968. Large amounts are still present in, for example, transformers and could enter the environment for many years. Federal regulations specify the following limits in foods: 1.5 ppm in milk fat, 1.5 ppm in fat portion of manufactured dairy products, 3 ppm in poultry, and 0.3 ppm in eggs. The tolerance level for PCB in fish was reduced from 5 to 2 ppm in 1984. Although there has been a good deal of concern about the possible toxicity of PCBs, there is now evidence that PCBs are much less toxic than initially assumed (American Council on Science and Health 1985).

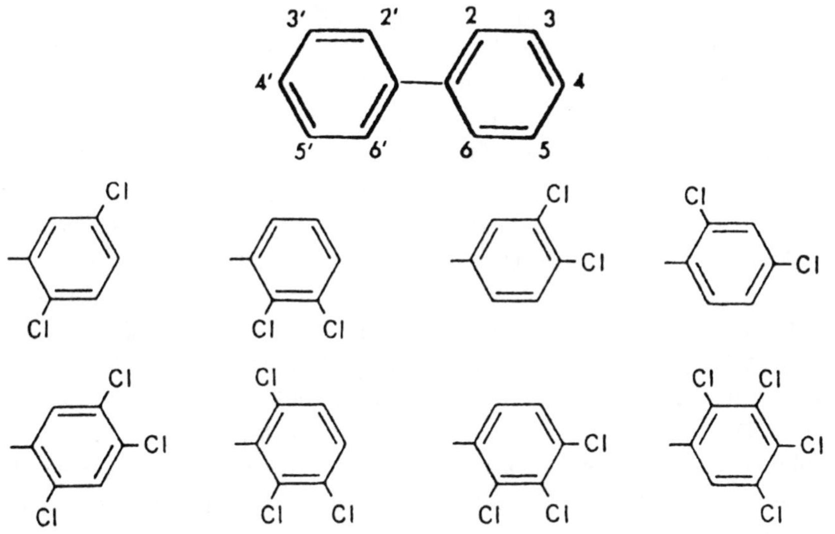

Figure 11–13 The Numbering System Used in PCBs and the Prevalent Substitution Pattern of Chlorine

Zabik and Zabik (1996) have reviewed the effect of processing on the removal of PCBs from several foods. In the processing of vegetable oil the PCB present in the crude oil was completely removed; some was removed by the hydrogenation catalyst, but most was lost by deodorization. The PCB was recovered in the deodorizer distillate.

Asbestos

Asbestos is widely distributed in the environment as a result of industrial pollution. Many water supplies contain asbestos fibers, which may become components of foods (especially beverages). An additional source of asbestos fibers may be asbestos filtration

Table 11–8 Information on Aroclor Preparations

Aroclor	% Cl	Average Number of Cl per Molecule	Average Molecular Weight
Aroclor 1221	21	1.15	192
Aroclor 1232	32	2.04	221
Aroclor 1242	42	3.10	261
Aroclor 1248	48	3.90	288
Aroclor 1254	54	4.96	327
Aroclor 1260	60	6.30	372
Aroclor 1262	62	6.80	389
Aroclor 1268	68	8.70	453

pads; such contamination has been suggested to occur in the filtration of beer (Pontefract 1974). The most common form of this contaminant is chrysotile asbestos, which occurs as minute fibers of about 24 nm in size. The amount of asbestos in water supplies is extremely low, in the nanogram range, but this may represent a large number of fibers. Many Canadian water supplies have been found to contain upward of 1 million asbestos fibers per liter. Chrysotile asbestos is a hydrated magnesium silicate of the composition $Mg_3Si_2O_5(OH)_4$. Inhalation of asbestos has been related to cancer among asbestos workers. It appears that asbestos fibers ingested with water or food may pass through the intestinal wall and enter into the bloodstream. Cunningham and Pontefract (1971) have reported the occurrence of asbestos fibers in a variety of beverages and in tap water. Some of their results are presented in Table 11–9.

Antibiotics

Growth-retarding or antimicrobial substances may be present in foods naturally,

may be produced in a food during processing, or may occur incidentally through the treatment of diseased animals. The latter problem has created the greatest concern. The use of antibiotics in therapy, prophylaxis, and growth promotion of animals may result in residues in foods. These residue levels rarely exceed the range of 1 to 0.1 ppm (where toxicological interest ceases). However, levels well below those of toxicological interest may be important in food processing, for example, in cheese making, by preventing starter development. The low levels may also be important in causing allergies and development of resistant organisms. Highly sensitized persons may experience allergic reactions from milk that contains extremely low amounts of penicillin. The various antibiotics used in agriculture, including some used in food processing, are listed in Exhibit 11–5. The tetracyclines, CTC and OTC, are broad-spectrum antibiotics and act against both gram-positive and gram-negative bacte-

Table 11–9 Asbestos Fibers in Beverages and Water

Sample	No. of Fibers/L × 10^6
Beer	4.3
Sherry	4.1
Soft drink	12.2
Tap water, Ottawa	2.0
Tap water, Toronto	4.4

Source: From H.M. Cunningham and R.D. Pontefract, Asbestos Fibers in Beverages and Drinking Water, *Nature*, Vol. 232, pp. 332–333, 1971.

Exhibit 11–5 Antibiotics Used in Agriculture and Food Processing

- Benzylpenicillin, cloxacillin, ampicillin, phenethicillin potassium
- Chlortetracycline and oxytetracycline
- Streptomycin and dihydrostreptomycin
- Neomycin, oleandomycin, spiramycin
- Chloramphenicol
- Framycetin, bacitracin, and polymyxins
- Tylosin[1] and nisin[1]
- Nystatin

[1]Not used in human therapeutics

ria. The action is bacteriostatic and not bactericidal. The tetracyclines have been used to delay spoilage in poultry and fish. Their effectiveness seems to decrease quite rapidly, because the contaminating flora quickly become resistant. Nisin, which is one of the few antibiotics not used in human therapeutics, has been found to be effective as an aid in heat sterilization of foods. It is a polypeptide with a molecular weight of about 7,000 and contains 18 amino acid residues. It is active against certain gram-positive organisms only, and all spores are sensitive to it.

Trace Metals

A variety of trace metals (such as mercury and lead) may become components of foods through industrial contamination of the environment. Some trace metals (such as tin and lead) may be introduced into foods through pickup from equipment and containers (especially tin cans).

Mercury

Large amounts of mercury are released into the environment by several industries.

Major mercury users are the chloralkali industry, where mercury is used in electrolytic cells; the pulp and paper industry, where mercury compounds are used as slimicides; and agriculture, where uses include seed dressings and sprays. Mercury is now known to be converted in sediments on river and lake bottoms into highly toxic methyl mercury compounds. This conversion scheme is shown in Figure 11–14. Formation of the more volatile dimethyl mercury is favored at alkaline pH. The less volatile monomethyl form is favored at acid pH. Because much of the mercury pollution ends up in rivers and lakes where it is converted into methyl mercury, contamination of fish with mercury has been a great concern. In many animal tissues, methyl mercury may comprise as much as 99 percent of the total mercury present. The present interest in mercury and its effect on humans and wildlife originated with the discovery of mercury as the causative agent in the Minamata disease in Japan. Near the town of Minamata, a chemical industry used mercury compounds as catalysts for the conversion of acetylene into acetaldehyde and vinyl chloride. Organic mercury compounds

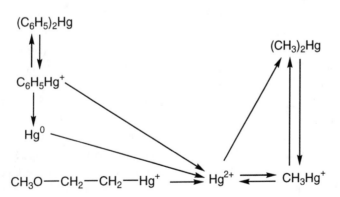

Figure 11–14 Conversion of Inorganic Mercury and Some Mercury-Containing Compounds to Methyl Mercury. *Source*: From N. Nelson, Hazards of Mercury, *Environmental Res.*, Vol. 4, pp. 41–50, 1971, Academic Press.

were released into the waters of Minamata Bay and contaminated fish and shellfish. Many cases of mercury poisoning occurred, resulting in the death of close to 50 patients. This event triggered research into mercury contamination in many areas of the world. Table 11–10 summarizes the mercury levels found in foods in various countries. The high mercury level in Japanese rice should be noted. High levels of mercury have been observed in fish in many lakes and streams as well as in the oceans. Table 11–11 presents the results of mercury analyses in Atlantic coast fish. The high level of mercury in tuna and swordfish is probably the result of the predatory lifestyle and long life of these fish. Generally, 0.5 ppm of mercury in fish is con-

Table 11–11 Mercury Levels in Atlantic Coast Fish

Species	Hg Level Range (ppm)
Clam	0.02–0.11
Cod	0.02–0.23
Crab	0.06–0.15
Flounder	0.07–0.17
Haddock	0.07–0.10
Herring	0.02–0.09
Lobster	0.08–0.20
Oyster	0.02–0.14
Swordfish	0.82–1.00
Tuna	0.33–0.86

Source: From E.G. Bligh, Mercury in Canadian Fish, *Can. Inst. Food Sci. Technol. J.*, Vol. 5, pp. A6–A14, 1972.

Table 11–10 Mercury Levels Found in Foods in Various Countries

Food	Country	Hg, Range (ng/g)
Haddock	United States	17–23
Herring	Baltic States	26–41
Apples	United Kingdom	20–120
Apples	New Zealand	11–135
Pears	Australia	40–260
Tomatoes	United Kingdom	12–110
Potatoes	United Kingdom	5–32
Wheat	Sweden	8–12
Rice	Japan	227–1000
Rice	United Kingdom (imports)	5–15
Carrots	United States	20
White bread	United States	4–8
Whole milk	United States	3–10
Beer	United States	4

Source: From N. Nelson, Hazards of Mercury, *Environmental Res.*, Vol. 4, pp. 41–50, 1971, Academic Press.

sidered the maximum permitted level. Fish in Canada exceeding this level cannot be sold.

Lead and Tin

The presence of lead in foods may be the result of environmental contamination, pickup of the metal from equipment, or the solder of tin cans. It has been estimated that nearly 90 percent of the ingested lead is derived from food (Somers and Smith 1971). However, only 5 percent of this is absorbed. In the early 1970s, the average North American car was reported to emit 2.5 kg of lead per year (Somers and Smith 1971), and Zuber and colleagues (1970) reported that crops grown near busy highways had a high lead content (in some cases, exceeding 100 ppm of lead in the dry matter). The removal of lead from gasoline has eliminated this source of contamination. Lead can also be picked up by acid foods such as fruit juices that are kept in glazed pottery made with lead-containing glazes. Both lead and tin may be taken up by foods from the tin of cans and from the solder used in their manufacture. The amounts of lead and tin taken up depend on the type of tin plate and solder used and on the composition and properties of the canned foods. In a study on the detinning of cans by spinach, Lambeth et al. (1969) found that detinning was significantly related to the oxalic acid content and pH of the product. Detinning in excess of 60 percent was observed during nine months' storage of high-oxalate spinach.

The present levels of lead that humans ingest cause concern because ADI calculations range from 0.1 to 0.8 mg lead per day. The average daily intake is in the vicinity of 0.4 mg lead per day. This means that lead is one of the few toxic food components for which the acceptable daily intake is approached or exceeded by the general population (Clarkson 1971).

Cadmium

As are lead and mercury, cadmium is a nonessential trace metal with high toxicity. Crustaceans have the ability to accumulate cadmium as well as other trace metals, such as zinc. Cadmium levels in oysters may reach 3 to 4 ppm, whereas in other foods, levels are only one-tenth or one-hundredth of these (Underwood 1973).

Polycyclic Aromatic Hydrocarbons (PAHs)

These compounds form a large group of materials that are now known to occur in the environment. The structural formulas of the major members of this group are presented in Figure 11–15. Several of these, especially

benzo(a)pyrene (3,4-benzopyrene), have been found to be carcinogenic. Usually, the polycyclic hydrocarbons occur together in foods, especially in smoked foods, because the aromatic hydrocarbons are constituents of wood smoke. Trace quantities of PAHs have been found in a variety of foods, and this may be the result of environmental contamination.

The PAHs may be carcinogenic and mutagenic. The level of carcinogenicity may vary widely between different members of this group. Minor constituents of PAH mixtures may make large contributions to the carcinogenic activity of the mixture. Certain methylchrysenes, particularly the 5-isomer, which is one of the most carcinogenic compounds known, may dominate the carcinogenic activity of a mixture (Bartle 1991).

Rhee and Bratzler (1968) analyzed hydrocarbons in smoke, and the amounts found in smoke and in the vapor phase (smoke filtered to remove particles) are listed in Table 11–12. Small amounts of these smoke constituents may be transferred to foods during smoking. Howard and Fazio (1969) reported the levels of aromatic polycyclic hydrocarbons in foods, and results for smoked foods are listed in Table 11–13. These compounds have also been found in unsmoked foods, as is shown in Table 11–14. Higher levels than those found in smoked food may occur as a result of barbecuing or charcoal broiling. Roasting of coffee and nuts results in formation of PAHs. The levels present in roasted coffee increase with more intense roasting; this is shown in Table 11–15, which is based on results obtained by Fritz (1968). PAHs occur in vegetables (Grimmer and Hildebrand 1965), and the levels are thought to be related to the leaf area and the relative level of atmospheric pollution.

Surprisingly, the largest proportion of the total human intake of PAHs does not come

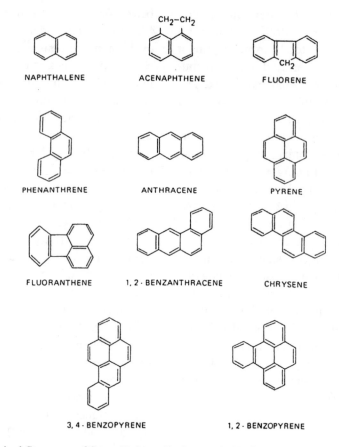

Figure 11–15 Chemical Structure of Some Polycyclic Aromatic Hydrocarbons

from smoked or roasted foods, but from other common products. Bartle (1991) has stated that cereals are likely to be a greater hazard, especially in the form of flour, than smoked or barbecued foods. Although cereal has a much lower PAH content than smoked or roasted foods do, cereal is consumed in much greater amounts.

Bacterial and Fungal Toxins

Microbial toxins are some of the most potent toxins known to humans. They may be the result of microbial growth in foods or, as in the case of fungal toxins, growth of molds during the production of many agricultural crops.

Bacterial toxins are produced mainly from species of the genera *Staphylococcus* and *Clostridium*. The staphylococcus poison usually results from improperly handled food in food service establishments and in the home, but rarely from food processing plants. Although the toxin seldom causes human death, it is highly toxic. In contrast, botulinum toxin has a high fatality rate. The neuro-

Table 11–12 Aromatic Polycyclic Hydrocarbons in Wood Smoke and in Wood Smoke Vapor Phase

	Amount, µg/45 kg Sawdust	
Hydrocarbon	Whole Smoke	Vapor Phase
Phenanthrene	51.5	28.4
Anthracene	3.8	1.9
Pyrene	5.5	4.1
Fluoranthene	5.7	4.2
1,2-Benzan-thracene	7.0	4.3
Chrysene	2.6	0.3
3,4-Benzopyrene	1.2	0.4
1,2-Benzopyrene	0.9	Trace

Source: From K.S. Rhee and L.J. Bratzler, Polycyclic Hydrocarbon Composition of Wood Smoke, *J. Food Sci.*, Vol. 33, pp. 626–632, 1968.

toxin is produced by six types of *Clostridium botulinum*, which differ in immunological properties. These organisms are gram-positive, spore-forming anaerobic bacteria. The prevention of the outgrowth of the spores is a major responsibility for the heat-sterilized processed-food industry.

Fungal toxins, also called mycotoxins, are produced by fungi or molds. Most of the interest in fungal toxins is concerned with the so-called storage fungi, molds that grow on relatively dry cereals and oilseeds. These belong to two common genera, *Aspergillus* and *Penicillium*. The most common of the fungal toxins are the aflatoxins formed by members of the *Aspergillus flavus* group. The aflatoxins were discovered as a result of widespread poisoning of turkeys in the early 1960s in England through feeding of toxic peanut meal. The aflatoxins belong to the most powerful toxins known and are highly

Table 11–13 Polycyclic Aromatic Hydrocarbons Found in Smoked Food Products (ppb)

Food Product	Benzo (a)-anthracene	Benzo (a)-pyrene	Benzo (e)-pyrene	Benzo (g,h,i,)-perylene	Fluoran-thene	Pyrene	4-Methyl-pyrene
Beef, chipped	0.4				0.6	0.5	
Cheese, Gouda					2.8	2.6	
Fish							
Herring					3.0	2.2	
Herring (dried)	1.7	1.0	1.2	1.0	1.8	1.8	
Salmon	0.5		0.4		3.2	2.0	
Sturgeon		0.8			2.4	4.4	
White					4.6	4.0	
Ham	2.8	3.2	1.2	1.4	14.0	11.2	2.0
Frankfurters					6.4	3.8	
Pork roll					3.1	2.5	

Source: From J.W. Howard and T. Fazio, A Review of Polycyclic Aromatic Hydrocarbons in Foods, *Agr. Food Chem.*, Vol. 17, pp. 527–531, 1969, American Chemical Society.

Table 11-14 Polycyclic Aromatic Hydrocarbons in Unsmoked Food Products

Food Product	Fluoran-thene (ppb)	Pyrene (ppb)
Cheese, cheddar	0.8	0.7
Fish, haddock	1.6	0.8
Fish, herring (salted)	0.8	1.0
Fish, salmon (canned)	1.8	1.4

Source: From J.W. Howard and T. Fazio, A Review of Polycyclic Aromatic Hydrocarbons in Foods, *Agr. Food Chem.*, Vol. 17, pp. 527–531, 1969, American Chemical Society.

Table 11-15 Polycyclic Aromatic Hydrocarbons in Coffee (µg/kg)

Compound	Heavy Roasting	Normal Roasting
Anthracene	6.2	1.5
Phenanthrene	74.0	28.0
Pyrene	28.0	3.5
Fluoranthene	34.0	3.9
1,2-Benzanthracene	14.2	1.5
Chrysene	14.8	—
3,4-Benzopyrene	5.8	0.3
1,2-Benzopyrene	7.0	0.7
Perylene	0.6	—
11,12-Benzfluoranthene	1.8	—
Anthanthrene	0.9	—
1,12-Benzperylene	2.2	—
3,4-Benzfluoranthene	1.2	—
Coronene	0.9	—
Indenopyrene	0.7	—

Source: From W. Fritz, Formation of Carcinogenic Hydrocarbons During Thermal Treatment of Foods, *Nahrung*, Vol. 12, pp. 799–804, 1968.

carcinogenic. A dose of 1 mg given to rats for short or long periods can result in liver cancer, and a diet containing 0.1 ppm of aflatoxin produces liver tumors in 50 percent of male rats (Spensley 1970). There are at least eight aflatoxins, of which the more important are designated B_1, B_2, G_1, and G_2. The names result from the blue and green fluorescence of these compounds when viewed under ultraviolet light. Aflatoxin B_1 is a very powerful liver carcinogen; a level of 15 ppb in the diet of rats resulted in tumors in 100 percent of cases after 68 weeks (Scott 1969). Ducklings are used as test animals because they are especially sensitive to aflatoxins. The aflatoxins, for which the formulas are shown in Figure 11-16, can occur in many foods but are particularly common in peanuts. Roasting of peanuts reduces the level of aflatoxin; for example, roasting for a half hour at 150° may reduce aflatoxin B_1 content by as much as 80 percent (Scott 1969). However, aflatoxin may still be carried over into peanut butter. In addition, aflatoxins have been found in cottonseed meal, rice, sweet potatoes, beans, nuts, and wheat. Through ingestion of moldy feed by animals, aflatoxins may end up as contaminants in milk and meat. Aflatoxins found in milk may be M1 or M2, where M stands for metabolic; these are also toxic. The development of aflatoxins depends very much on temperature and moisture conditions. With peanuts, contamination occurs mostly during the drying period. Improper drying and storage are responsible for most of the contamination. This has been found to apply for rice. Optimum conditions for growth of *Aspergillus flavus* are 25° to 40°C with a relative humidity greater than 85 percent.

Control measures for prevention of aflatoxin production focus on reduction of water

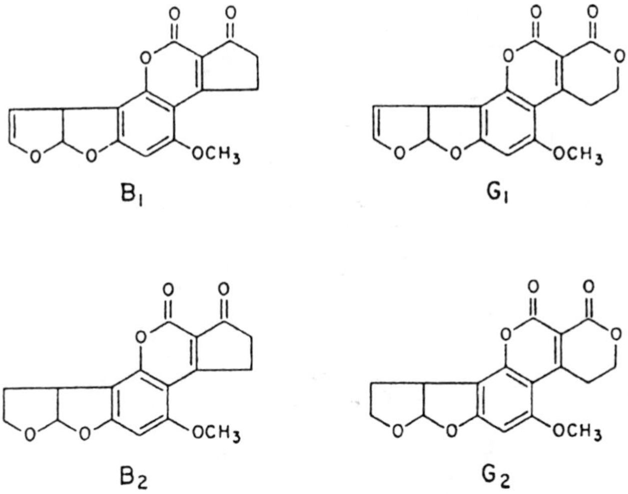

Figure 11–16 Chemical Structure of Aflatoxins B$_1$, B$_2$, G$_1$, and G$_2$. *Source:* From P.M. Scott, The Analysis of Foods for Aflatoxins and Other Fungal Toxins: A Review, *Can. Inst. Food Technol. J.*, Vol. 2, pp. 173–177, 1969.

activity to a point where the fungus is unable to grow and maintenance of low water activity during storage. A moisture content of 18.0 to 19.5 percent in cereal grains is required for growth and toxin production by *A. flavus*. Aflatoxin-contaminated commodities can be detoxified by treatment with ammonia, calcium hydroxide, or a combination of formaldehyde and calcium hydroxide (Palmgren and Hayes 1987).

Sterigmatocystin, which is a carcinogenic metabolite of *Aspergillus ochraceus*, has been found to be a natural contaminant of foods, especially corn. Molds of the species *Fusarium* produce several mycotoxins in countries with moderate climates (Andrews et al. 1981). Two of these are zearalenone and deoxynivalenol (Figure 11–17). Zearalenone, of F-2 toxin, is produced by *Fusarium* molds that grow on corn (Marasas et al. 1979) that is immature or high in moisture at harvest. Deoxynivalenol, also known as vomitoxin, has been found in wheat and barley (Trenholm et al. 1981; Scott et al. 1983).

During the wet summer of 1980, wheat grown in Ontario showed sprouting of kernels and pink discoloration. Experiments on milling showed that the vomitoxin was distributed throughout the milled products and was not destroyed by the bread-making process. Patulin is another *Aspergillus* metabolite and has been indicated as a food contaminant, especially in fruits, as a result of storage rot. It has been found as a constituent of apple juice (Harwig et al. 1973).

Natural Toxicants

In spite of the prevalent perception that "natural" is harmless or nontoxic, natural food products contain an abundance of toxic chemicals (Committee on Food Protection 1973; National Academy of Sciences 1973). The foods that are consumed by humans (with the exception of mother's milk consumed by infants) were not designed by nature for human use; rather, they were adapted by humans over the centuries, by

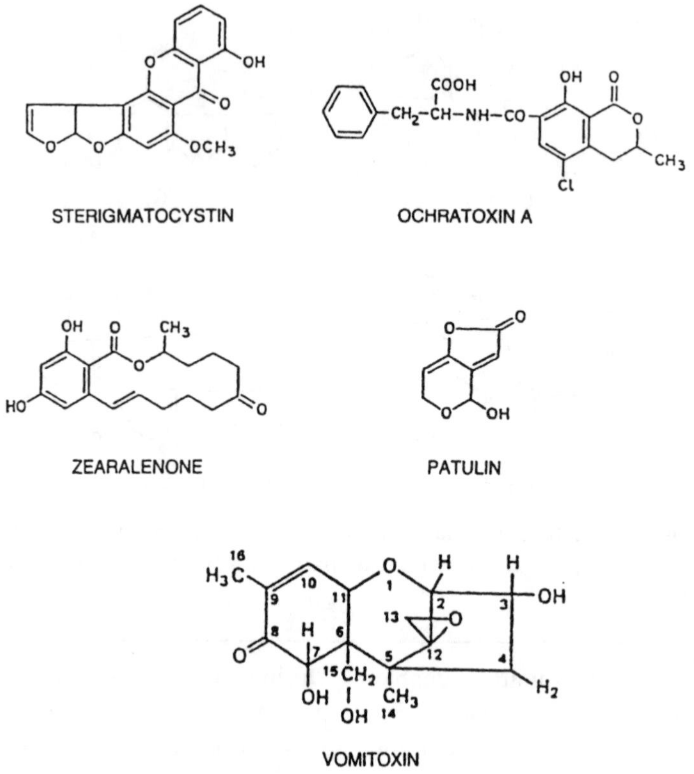

Figure 11–17 Chemical Structures of Sterigmatocystin, Ochratoxin A, Zearalenone, Deoxynivalenol (Vomitoxin), and Patulin. *Source:* From P.M. Scott, The Analysis of Foods for Aflatoxins and Other Fungal Toxins: A Review, *Can. Inst. Food Technol. J.*, Vol. 2, pp. 173–177, 1969; P.M. Scott et al., Effects of Experimental Flour Milling and Breadbaking on Retention of Deoxynivalenol (Vomitoxin) in Hard Red Spring Wheat, *Cereal Chem.*, Vol. 60, pp. 421–424, 1983.

trial and error. This process is effective in eliminating foods that cause acute symptoms of toxicity but is less effective in dealing with the long-term effects. Coon (1973) has stated that past experience has provided more knowledge on the safety margins of natural foods than animal experimentation. Many natural food components, such as caffeine, goitrogens, and cyanogenic glycosides, would not be approved for human consumption if examined with the techniques now required for intentional addi-

tives. Because our foods contain so many potentially toxic substances, the best defense is to consume a varied diet.

Some natural toxins such as seafood toxins or fungal toxins may occur at abnormally high levels only in unusual circumstances, causing disease in the normal individual who eats a normal amount of food. Other natural toxins are normal constituents of food and cause disease only in humans consuming abnormal amounts of that food. In addition, normal food components may cause harm in abnormal individuals (Coon 1973). The latter case is probably the most difficult to deal with in regulatory aspects. Banning of foods that are considered safe for most people is unthinkable, and protection of diseased or allergic individuals becomes a problem. Some examples of natural toxins in foods are given below.

Sulfur Compounds

Many cruciferous plants contain goitrogens, which are known as glucosinolates. They are harmful if ingested in excessive amounts. Plants of the genus *Allium*, including onions, chives, and garlic, contain precursors of sulfur-containing compounds that can be liberated by enzymic action.

Salunkhe and Wu (1977) have described the enzymic breakdown of the glucosinolate to isothiocyanate and goitrin (Figure 11–18). In addition to being goitrogens, the isothiocyanates produced from glucosinolates in rapeseed (canola) oil have been found to have a poisoning effect on the nickel catalysts employed in the hydrogenation of the oil (Abraham and deMan 1987).

Sulfur compounds of the *Allium* species result from enzymic action on precursors and are responsible for the odor and flavor as well as the lachrymatory properties of these vegetables. These precursors are S-substituted cysteine sulfoxides, or combinations with peptides. The enzyme alliinase liberates labile thiosulfinates, which slowly decompose into thiosulfonates and disulfides. The latter are volatile and are responsible for the characteristic flavor. Thiosulfinates have antibiotic activity.

Seafood Toxicants

Many of the seafood toxins are found in shellfish and usually result from the growth of certain marine algae. The toxins end up in the shellfish through the food chain. The sporadic occurrence of these events makes the problem more difficult to control. In 1987, mussels cultivated in Eastern Canada were found to be poisonous, and the cause was established as the toxin domoic acid. Paralytic shellfish poisoning has been observed in many of the world's fishing areas. Although some of these toxins have been identified, many remain uncharacterized (Shantz 1973). The poison saxitoxin, from California mussels and Alaska butter clams, is a dibasic salt and is highly soluble in water.

Caffeine

Caffeine is a naturally occurring chemical, 1,3,7-trimethylxanthine (Figure 11–19), which is found in the leaves, seeds, and fruits of more than 63 species of plants growing all over the world. It occurs as a constituent of coffee, tea, cocoa, and chocolate, and is an additive in soft drinks and other foods. Because humans have used it for thousands of years, caffeine has GRAS status in the United States. Roberts and Barone (1983) have estimated that daily caffeine consumption in the United States is 206 mg per person. Caffeine shows a number of physiological effects (Von Borstel 1983) and, as a result, its regulatory status has been under review (Miles 1983). Complicating the matter is the fact that caffeine is both a naturally occurring chemical as well as a food additive. Caffeine stimulates the central nervous system, can help people stay awake, and can relieve headaches. Adverse effects may include sleep disturbance, depression, and stomach upsets. Large overdoses of caffeine may be fatal (Leviton 1983). Caffeine is a good example of a widely occurring natural toxicant that has been part of our food supply for centuries.

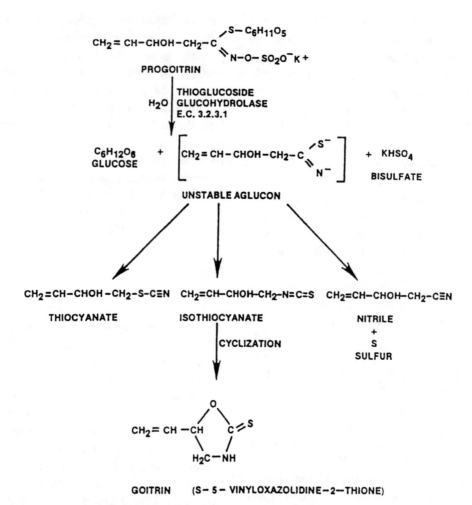

Figure 11–18 Formation of Goitrin and Isothiocyanates from Glucosinolates in Cruciferous Products. *Source:* From D.K. Salunkhe and M.T. Wu, Toxicants in Plants and Plant Products, *Food Sci. Nutr.*, Vol. 9, pp. 265–324, 1977.

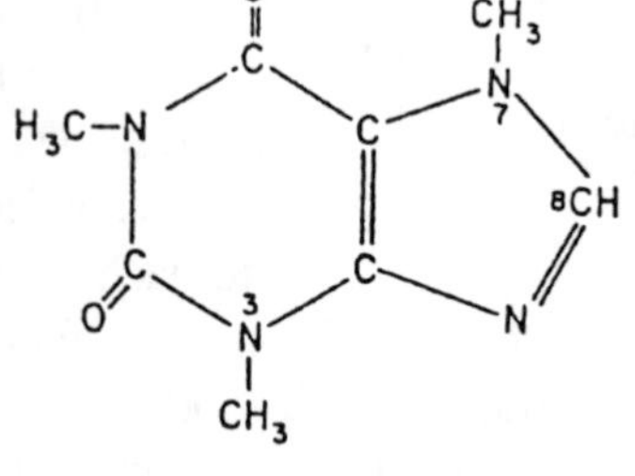

Figure 11–19 The Structure of Caffeine 1,3,7- Trimethylxanthine

REFERENCES

Abraham, V., and J.M. deMan. 1987. Effect of some isothiocyanates on the hydrogenation of canola oil. *J. Am. Oil Chem. Soc.* 64: 855–858.

American Council on Science and Health. 1985. *PCBs: Is the cure worth the cost?* New York.

Andrews, R.I., et al. 1981. A national survey of mycotoxins in Canada. *J. Am. Oil Chem. Soc.* 58: 989A–991A.

Augustin, J., and F.E. Scarbrough. 1990. Nutritional additives. In *Food additives*, ed. A.L. Branen et al. New York: Marcel Dekker.

Banwart, G.J. 1979. *Basic Food Microbiology.* Westport, CT: AVI Publishing Co.

Bartle, K.D. 1991. Analysis and occurrence of polycyclic aromatic hydrocarbons in food. In *Food contaminants: Sources and surveillance*, ed. L.S. Creases and R. Purchase. London: Royal Society of Chemistry.

Bieber, W.D., et al. 1984. Transfer of additives from plastics materials into foodstuffs and into food simulants: A comparison. *Food Chem. Toxic.* 22: 737–742.

Bieber, W.D., et al. 1985. Interaction between plastics packaging materials and foodstuffs with different fat content and fat release properties. *Food Additives Contaminants* 2: 113–124.

Bracco, U., et al. 1981. Production and use of natural antioxidants. *J. Am. Oil Chem. Soc.* 58: 686–690.

Cassens, R.G. 1997. Residual nitrate in cured meat. *Food Technol.* 51, no. 2: 53–55.

Chapman, D.G., and L.I. Pugsley. 1971. The public health aspects of the use of phosphates in foods. In *Symposium: Phosphates in food processing*, ed. J.M. deMan and P. Melnychyn. Westport, CT: AVI Publishing Co.

Chichester, D.F., and F.W. Tanner. 1968. Antimicrobial food additives. In *Handbook of food additives*, ed. T.E. Furia. Cleveland, OH: Chemical Rubber Co.

Clarkson, T.W. 1971. Epidemiological and experimental aspects of lead and mercury contamination of food. *Food Cosmet. Toxicol.* 9: 229–243.

Collings, A.J. 1971. The metabolism of sodium cyclamate. In *Sweetness and sweeteners*, ed. G.G. Birch et al. London: Applied Science Publishers Ltd.

Committee on Food Protection. 1973. *Toxicants naturally occurring in foods.* Washington, DC: National Academy of Sciences.

Coon, J.M. 1973. Toxicology of natural food chemicals: A perspective. In *Toxicants naturally occurring in foods.* Washington, DC: National Academy of Sciences.

Cordle, F. 1981. The use of epidemiology in the regulation of dioxins in the food supply. *Reg. Toxicol. Pharmacol.* 1: 379–387.

Cunningham, H.M., and R.D. Pontefract. 1971. Asbestos fibers in beverages and drinking water. *Nature* 232: 332–333.

Davidson, P.M., and V.K. Juneja. 1990. Antimicrobial agents. In *Food additives*, ed. A.L. Braneu et al. New York: Marcel Dekker.

DeKruyf, N., et al. 1983. Selection and application of a new volatile solvent as a fatty food simulant for determining the global migration of constituents of plastics materials. *Food Chem. Toxic.* 21: 187–191.

Edinger, W.D., and D.F. Splittstoesser. 1986. Production by lactic acid bacteria of sorbic alcohol, the precursor of the geranium odor compound. *Am. J. Enol. Vitic.* 37: 34.

Elkins, E.R. 1989. Effect of commercial processing on pesticide residues in selected fruits and vegetables. *J. Assoc. Off. Anal. Chem.* 72: 533–535.

Elkins, E.R., et al. 1972. The effect of heat processing and storage on pesticide residues in spinach and apricots. *Agr. Food Chem.* 20: 286–291.

Farrow, R.P., et al. 1969. Canning operations that reduce insecticide levels in prepared foods and in solid food wastes. *Residue Rev.* 29: 73–78.

Fassett, D.W. 1977. Nitrates and nitrites. In *Toxicants naturally occurring in foods.* Washington, DC: National Academy of Sciences.

Friar, P.M.K., and S.L. Reynolds. 1991. The effects of microwave-baking and oven-baking on thiobendazole residues in potatoes. *Food Additives Contaminants* 8: 617–626.

Fritz, W. 1968. Formation of carcinogenic hydrocarbons during thermal treatment of foods. *Nahrung* 12: 799–804.

Gelardi, R.C. 1987. The multiple sweetener approach and new sweeteners on the horizon. *Food Technol.* 41, no. 1: 123–124.

Glascock, R.F. 1965. A pilot plant for the removal of radioactive strontium from milk: An interim report. *J. Soc. Dairy Technol.* 18: 211–217.

Grimmer, G., and A. Hildebrand. 1965. Content of polycyclic hydrocarbons in different types of vegetables and lettuce. *Dtsch. Lebensm. Rundschau.* 61: 237–239.

Hall, R.L. 1975. GRAS: Concept and application. *Food Technol.* 29: 48–53.

Harwig, J., et al. 1973. Occurrence of patulin and patulin-producing strains of *Penicillium expansum* in natural rots of apple in Canada. *Can. Inst. Food Sci. Technol. J.* 6: 22–25.

Havery, D.C., and T. Fazio. 1985. Human exposure to nitrosamines from foods. *Food Technol.* 39, no. 1: 80–83.

Homler, B.E. 1984. Properties and stability of aspartame. *Food Technol.* 38, no. 7: 50–55.

Horowitz, R.M., and B. Gentili. 1971. Dihydrochalcone sweeteners. In *Sweetness and sweeteners*, ed. G.G. Birch et al. London: Applied Science Publishers Ltd.

Howard, J.W., and T. Fazio. 1969. A review of polycyclic aromatic hydrocarbons in foods. *Agr. Food Chem.* 17: 527–531.

Inglett, G.E. 1971. Intense sweetness of natural origin. In *Sweetness and sweeteners*, ed. G.G. Birch et al. London: Applied Science Publishers Ltd.

Institute of Food Technologists. 1975. Naturally occurring toxicants in foods: A scientific status summary. *J. Food Sci.* 40: 215–222.

Institute of Food Technologists. 1988. Migration of toxicants, flavors, and odor-active substances from flexible packaging materials to food. *Food Technol.* 42, no. 7: 95–102.

Johnson, W., and C.L. Nayfield. 1970. Elevated levels of cesium-137 in common mushrooms (*Agaricaceae*) with possible relationship to high levels of cesium-137 in whitetail deer, 1968–1969. *Radiological Health Data Reports* 11: 527–531.

Kochhar, S.P., and J.B. Rossell. 1990. Detection, estimation and evaluation of antioxidants in food systems. In *Food antioxidants*, ed. B.J.F. Hudson. London: Elsevier Applied Science.

Krog, N. 1981. Theoretical aspects of surfactants in relation to their use in breadmaking. *Cereal Chem.* 58: 158–164.

Labin-Goldscher, R., and S. Edelstein. 1996. Calcium citrate: A revised look at calcium fortification. *Food Technol.* 50, no. 6: 96–98.

Lacroix, M., et al. 1997. Prevention of lipid radiolysis by natural antioxidants from rosemary (Rosmarinus officinalis L) and thyme (Thymus vulgaris L). *Food Res. Intern.* 30: 457–462.

Lambeth, V.N., et al. 1969. Detinning by canned spinach as related to oxalic acid, nitrates and mineral composition. *Food Technol.* 23: 840–842.

Leviton, A. 1983. Biological effects of caffeine. Behavioral effects. *Food Technol.* 37, no. 9: 44–47.

Lindell, B., and A. Magi. 1965. The occurrence of ^{137}Cs in Swedish food, especially in dairy milk, and in the human body after the nuclear test explosions in 1961 and 1962. *Arkiv Fysik* 29: 69–96.

Liska, B.J., and W.J. Stadelman. 1969. Effects of processing on pesticides in foods. *Residue Rev.* 29: 61–72.

Marasas, W.F.O., et al. 1979. Incidence of *Fusarium* species and the mycotoxins, deoxynivalenol and zearalenone, in corn produced in esophageal cancer areas in Transkei. *J. Agric. Food Chem.* 27: 1108–1112.

Marsden, J., and R. Pesselman. 1993. Nitrosamines in food contact netting: Regulatory and analytical challenges. *Food Technol.* 47, no. 3: 131–134.

Miettinen, J.K. 1967. Radioactive food chains in subartic regions. *Nutrition Dieta* 9: 43–58.

Miles, C.I. 1983. Biological effects of caffeine. FDA status. *Food Technol.* 37, no. 9: 48–50.

Mitchell, L.E. 1966. Pesticides: Properties and prognosis. In *Organic pesticides in the environment*, ed. R.F. Gould. Advances in Chemistry Series 60. Washington, DC: American Chemical Society.

Nagodawithana, T. 1994. Flavor enhancers: Their probable mode of action. *Food Technol.* 46, no. 4: 79–85.

National Academy of Sciences. 1973. *The use of chemicals in food production, processing, storage, and distribution.* Washington, DC: National Academy of Sciences.

Newsome, R.L. 1990. Natural and synthetic coloring agents. In *Food additives*, ed. A.L. Branen et al. New York: Marcel Dekker, Inc.

Noonan, J. 1968. Color additives in food. In *Handbook of food additives*, ed. T.E. Furia. Cleveland, OH: Chemical Rubber Co.

Palmgren, M.S., and A.W. Hayes. 1987. Aflatoxins in food. In *Mycotoxins in food*, ed. P. Krogh. New York: Academic Press Ltd.

Peterson, B., et al. 1996. Pesticide degradation: Exceptions to the rule. *Food Technol.* 50, no. 5: 221–223.

Peterson, M.S., and A.H. Johnson. 1978. *Encyclopedia of food science.* Westport, CT: AVI Publishing Co. Inc.

Pontefract, R.D. 1974. Ingestion of asbestos. *Can. Res. Dev.* 7, no. 6: 21.

Rhee, K.S., and L.J. Bratzler. 1968. Polycyclic hydrocarbon composition of wood smoke. *J. Food Sci.* 33: 626–632.

Roberts, H.R., and J.J. Barone. 1983. Biological effects of caffeine. History and use. *Food Technol.* 37, no. 9: 32–39.

Salunkhe, D.K., and M.T. Wu. 1977. Toxicants in plants and plant products. *Crit. Rev. Food Sci. Nutr.* 9: 265–324.

Schuler, P. 1990. Natural antioxidants exploited commercially. In *Food antioxidants*, ed. B.J.F. Hudson. London: Elsevier Applied Science.

Scott, P.M. 1969. The analysis of foods for aflatoxins and other fungal toxins: A review. *Can. Inst. Food Technol. J.* 2: 173–177.

Scott, P.M., et al. 1983. Effects of experimental flour milling and breadbaking on retention of deoxynivalenol (vomitoxin) in hard red spring wheat. *Cereal Chem.* 60: 421–424.

Sen, N.B., et al. 1987. Volatile nitrosamines in cured meat packaged in elastic rubber nettings. *J. Agric. Food Chem.* 35: 346–350.

Shantz, E.J. 1973. Seafood toxicants. In *Toxicants naturally occurring in foods*. Washington, DC: National Academy of Sciences.

Sinclair, W.K. 1988. Radionuclides in the food chain. In *Radionuclides in the food chain*, ed. J.H. Harley et al. Berlin: Springer Verlag.

Sinki, G.S., and W.A.F. Schlegel. 1990. Flavoring agents. In *Food Additives*, ed. A.L. Branen et al. New York: Marcel Dekker, Inc.

Six, P. 1994. Current research in natural food antioxidants. *Inform* 5: 679–687.

Sloan, A.E., and M.K. Stiedemann. 1996. Food fortification: From public health solution to contemporary demand. *Food Technol.* 50, no. 6: 100–108.

Smith, R.L., et al. and the FEMA Expert Panel. 1996. GRAS flavoring substances 17. *Food Technol.* 50, no. 10: 72–81.

Somers, E., and D.M. Smith. 1971. Source and occurrence of environmental contaminants. *Food Cosmet. Toxicol.* 9: 185–193.

Spensley, P.C. 1970. Mycotoxins. *Royal Soc. Health J.* 90: 248–254.

Startin, J.R. 1991. Polychlorinated dibenzo-p-dioxins, polychlorinated dibenzo furans, and the food chain. In *Food contaminants: Sources and surveillance*, ed. L.S. Creaser and R. Purchase. London: Royal Society of Chemistry.

Sugita, Y. 1990. Flavor enhancers. In *Food additives*, ed. A.L. Braner et al. New York: Marcel Dekker.

Trenholm, H.L., et al. 1981. Survey of vomitoxin contamination of the 1980 white winter wheat crop in Ontario, Canada. *J. Am. Oil Chem. Soc.* 58: 992A–994A.

Underwood, E.J. 1973. Trace elements. In *Toxicants occurring naturally in foods*. Washington, DC: National Academy of Sciences.

Van Middelem, C.H. 1966. Fate and persistence of organic pesticides in the environment. In *Organic pesticides in the environment*, ed. R.F. Gould. Advances in Chemistry Series 60. Washington, DC: American Chemical Society.

Verdi, R.J., and L.L. Hood. 1993. Advantages of alternative sweetener blends. *Food Technol.* 47, no. 6: 94–101.

Vodoz, C.A. 1977. Flavour legislation: World trends. *Food Technol. Australia* 10: 393–399.

Von Borstel, R.W. 1983. Biological effects of caffeine. Metabolism. *Food Technol.* 37, no. 9: 40–47.

Wagner, D.A., and S.R. Tannenbaum. 1985. In vivo formation of n-nitroso compounds. *Food Technol.* 39, no. 1: 89–90.

Wiechen, A. 1972. Cause of the high Cs-137 content of milk from moorland. *Milchwissenschaft* 27: 82–84.

Zabik, M.E., and M.J. Zabik. 1996. Influence of processing on environmental contaminants in foods. *Food Technol.* 50, no. 5: 225–229.

Zitco, V. 1971. Polychlorinated biphenyls and organochlorine pesticides in some freshwater and marine fishes. *Bull. Environmental Contamination Toxicol.* 6: 464–470.

Zuber, R., et al. 1970. Lead as atmospheric pollutant and its accumulation on plants along heavily travelled roads. *Rech. Agron. Suisse* 9: 83–96 (French).

CHAPTER 12

Regulatory Control of Food Composition, Quality, and Safety

HISTORICAL OVERVIEW

Attempts at regulating the composition of foods go back to the Middle Ages. Primarily restricted to certain food items such as bread or beer, these ancient regulations were intended to protect the consumer from fraudulent practices. The original Bavarian beer purity law dating from the Middle Ages is still quoted today to indicate that nothing but water, malt, yeast, and hops have been used in the production of beer. The foundations for many of our modern food laws were laid in the last quarter of the 19th century. Increasing urbanization and industrialization meant that many people had less control over the food that had to be brought into the urban centers. Foodstuffs were deliberately contaminated to increase bulk or improve appearance. Chalk was mixed with flour, and various metal salts were added to improve color (Reilly 1991). Some of these added substances were highly toxic. One practice leading to disastrous results was the distillation of rum in stills constructed of lead.

The first food laws in the United Kingdom were enacted in 1860 and 1875, and the first Canadian food law was passed in 1875. In the United Sates the first comprehensive federal food law came into effect in 1906. This law prohibited the use of certain harmful chemicals in foods and the interstate commerce of misbranded or adulterated foods. Public-concern about adulteration and false health claims during the 1930s led to the federal Food, Drug and Cosmetic Act (FDCA) in 1938. A major weakness of this law was that the burden of proof of the toxicity of a chemical was entirely upon the government. Any substance could be used until such time when it was proven in a court of law that the substance was harmful to health. A select committee of the U.S. House of Representatives, the Delaney committee, studied the law and recommended its revision. The revised law, which went into effect in 1958, is known as the Food Additives Amendment of the federal Food, Drug and Cosmetic Act. Under this act, no chemical can be used in food until the manufacturer can demonstrate its safety. The U.S. Food and Drug Administration (FDA) is responsible only for evaluating the safety evidence submitted by the applicant. The principle of establishing the safety of chemicals before they can be used is now becoming widely accepted in U.S. and international food laws.

A peculiar aspect of the federal act of 1958 is the so-called Delaney clause, which stipulates that any substance that is found to cause

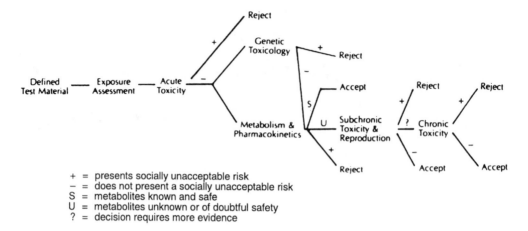

+ = presents socially unacceptable risk
− = does not present a socially unacceptable risk
S = metabolites known and safe
U = metabolites unknown or of doubtful safety
? = decision requires more evidence

Figure 12–1 Proposed System for Food Safety Assessment. From Food Safety Council, 1982.

cancer in humans or animals is banned from use in food at any level. This controversial clause has been the subject of much discussion over the years. Suspected carcinogens can be dealt with in other food law systems under the general provisions of safety.

The establishment of the safety of a chemical has become more and more difficult over the years. There are several reasons for this. First, analytical instrumentation can detect ever smaller levels of a substance. Where it was once common to have levels of detection of parts per million, now levels of detection can be as low as parts per billion or parts per trillion. At these levels, chemicals become toxicologically insignificant. Second, the requirements for safety have become more complex. Initially, the safety of a chemical was determined by its acute toxicity measured on animals and expressed as LD_{50}, the dose level that results in a 50 percent mortality in a given test population. As the science of toxicology has matured, safety requirements have increased; safety testing now follows a standard pattern as exemplified by the proposed system for food safety assessment shown in Figure 12–1. Third, new process-

ing techniques and novel foods have been developed. Many years of research were required to demonstrate the safety of radiation pasteurization of foods, and even now only limited use is made of radiation treatment of food and food ingredients. The issue of the safety of novel foods has gained new importance since the introduction of genetically modified crops. In addition to the requirements of the safety decision tree of Figure 12–1, the issue of allergenicity has arisen. Toxicity is assumed to affect everyone in a similar way, but allergic reactions affect only certain individuals. Allergic reactions can be of different degrees of severity. A major allergic reaction can result in anaphylactic shock and even death. Regulations are now being developed in several countries related to placing warning labels on foods containing certain allergens. One example of possible transfer of allergenicity to another food occurred when a company explored the genetic modification of soybeans to improve protein content. A Brazil nut storage protein gene was selected for transfer into the soybean genetic makeup. When it was found that people who were allergic to nuts also became allergic to the genetically altered soybean, the commercial development of this type of genetically modified soybean was abandoned. A fourth difficulty in regulatory control of food composition and quality is the often overlapping authority of different agencies. In many countries, the basic food law is the responsibility of the health department. However, control of meat products, animal health, and veterinary drug residues may reside in agriculture departments. Some countries such as Canada have a separate department dealing with fish and fisheries. Environmental issues sometimes come under the jurisdiction of industry departments. In addition, countries may have a federal structure where individual states or provinces exercise complete or partial control. Before the enactment of the FDCA in the United States, it was argued that food safety should be under the control of individual states. Canada is a federation, but the Canadian Food and Drugs Act is federal legislation that applies to all provinces and territories. In contrast, the situation in Australia, also a federation, makes each state responsible for its own food laws. Recent efforts there have tried to harmonize state food laws by introduction in each state of a "model food act" (Norris and Black 1989).

Usually, food laws are relatively short and simple documents that set out the general principles of food control. They are accompanied by regulations that provide specific details of how the principles set out in the food law should be achieved. In the United States the law deals with food, drugs, and cosmetics; in Canada the regulations deal with food and drugs. The tendency today is to provide laws that specifically deal with food. The separation of food laws and regulations makes sense because the regulations can be constantly updated without going through the difficult process of changing the law.

Food and drugs have traditionally been considered separate categories in the legislative process. Until relatively recently, health claims on foods were prohibited in many countries. However, in recent years consumers have been deluged with health information relating to their foods. Some of this information has been negative, such as information about the effect of fat on the incidence of heart disease; other information has been positive as for instance the beneficial effect of dietary fiber.

There is increasing interest in a group of substances known as nutraceuticals or functional foods and food supplements. A nutraceutical can be defined as any food or food ingredient that provides medical or health benefits, including the prevention and treatment of disease. These materials cover a gray area between foods and drugs and present difficulties in developing proper regulatory controls. It has been stated (Camire 1996) that dietary supplements in the United States of America enjoy a favored status. They do not require proof of either efficacy or safety. Dietary supplements include a large variety of substances such as vitamins, minerals, phytochemicals, and herbal or botanical extracts (Pszczola 1998).

SAFETY

The safety of foods—including food additives, food contaminants, and even some of the major natural components of foods—is becoming an increasingly complex issue. Prior to the enactment of the Food Additives Amendment to the FDCA, food additive control required that a food additive be non-deceptive and that an added substance be either safe and therefore permitted, or poisonous and deleterious and therefore prohibited. This type of legislation suffered from two main shortcomings: (1) it equated poisonous with harmful and (2) the onus was on the government to demonstrate that any chemical used by the food industry was poisonous. The 1958 act distinguishes between toxicity and hazard: *Toxicity* is the capacity of a substance to produce injury, and *hazard* is the probability that injury will result from the intended use of a substance. It is now well recognized that many components of our foods, whether natural or added, are toxic at certain levels but harmless or even nutritionally essential at lower levels. Some of the fat-soluble vitamins are in this category. The ratio between effective dose and toxic dose of many compounds, including such common nutrients as amino acids and salts, is of the order of 1 to 100. Today any user of an additive must petition the government for permission to use the material and supply evidence that the compound is safe.

The public demand for absolute safety is incompatible with modern scientific understanding of the issues. Safety is not absolute but rather a point on a continuum; the exact position involves judgments based on scientific evidence and other important factors including societal, political, legal, and economic issues. Modern legislation moves away as much as possible from the nonscience factors. Several recent issues have demonstrated how difficult this can be. In some cases scientific knowledge is unavailable, and decision making is difficult. In addition, we now know that food safety relates to all parts of the food chain, not merely the industrial processing of foods. What happens on the farm in terms of use of particular animal feeds or use of agricultural chemicals up to the handling of foods in food service establishments are all part of the food safety problem.

Scheuplein and Flamm (1989) stated that the assurance of safety by the FDA has moved away from a comfortable assurance of absolute safety to an assurance of some very small yet distinctly uncomfortable level of risk. It appears that the public is less inclined to accept even a very low level of risk related to the food supply than the often much greater risks of many of our daily activities.

In the United States, safety is often expressed as the principle of "reasonable certainty of no harm." This principle has replaced the earlier idea of "zero tolerance" for toxic substances. The idea of zero tolerance is incorporated in the Delaney clause of the Food Additives Amendment.

As the science of toxicology developed, the requirements for establishing safety became more demanding. At one time the LD_{50} was sufficient to establish safety. The effect of dose level is very important in toxicology. The effects, which vary from no effect dose (NED) levels to fatal effect, have been summarized in Figure 12–2 (Concon 1988). Two types of substances exist; type I shows no beneficial effects and type II shows nutritional and/or therapeutic beneficial effects.

LD_{50} is a measure of acute toxicity. Over time, many other test requirements have been added to establish safety as shown in the safety decision tree developed by the Food Safety Council (1982). In this system an organized sequence of tests is prescribed (see Figure 12–1). Other tests in this system involve genetic toxicity, metabolism, pharmacokinetics (the pathways of chemicals in the system and their possible accumulation in organs), subchronic toxicity, teratogenicity (birth defects), and chronic toxicity. To all this are added tests for carcinogenicity and

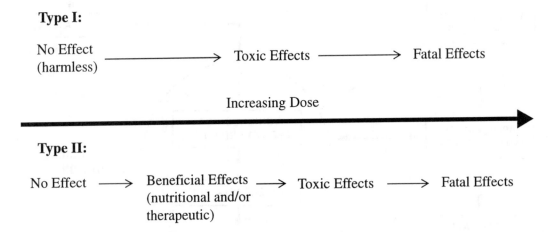

Type I:

No Effect
(harmless) ————————————→ Toxic Effects ————————→ Fatal Effects

Increasing Dose

Type II:

No Effect ——→ Beneficial Effects ——→ Toxic Effects ——→ Fatal Effects
(nutritional and/or
therapeutic)

Figure 12–2 Relationship Between Dose Level and Toxic Effects. *Source*: Reprinted with permission from J.M. Concon, *Food Toxicology. Part A—Principles and Concepts. Part B—Contaminants and Additives*, p. 16, 1988, by courtesy of Marcel Dekker, Inc.

allergenicity. Most of these tests are performed on animals. The no-effect level ascertained with animals is then divided by a safety factor of 100 to arrive at a safe level for humans. The idea of establishing a safety margin for chronic toxicity was accepted by the FDA in 1949.

The sequence of events leading from toxicological investigations to the formulation of regulations is shown in Figure 12–3 (Vettorazi 1989). The important part of this procedure is the interpretation. This is carried out by qualified experts who develop recommendations based on the scientific data produced. It is sometimes possible for different groups of experts (such as groups in different countries) to come up with differing recommendations based essentially on the same data.

U.S. FOOD LAWS

The basic U.S. law dealing with food safety and consumer protection is the Food, Drug and Cosmetic Act (FDCA) of 1938 as amended by the Food Additives Amendment of 1958. The FDCA applies to all foods distributed in the United States, including foods imported from other countries. A number of other acts are important for the production and handling of foods. Some of the more important ones include the following:

- *The Meat Inspection Act of 1906.* The responsibility for the safety and wholesomeness of meat and meat products falling under the provisions of this act is delegated to the U.S. Department of Agriculture (USDA). The USDA's responsibilities include inspection of meat-processing facilities and animals before and after slaughter, inspection of meat products and meat-processing laboratories, and premarket clearance of meat product labels. When a food product contains less than 3 percent meat, the product comes under the jurisdiction of the FDA. Similar laws are the Poultry Products Inspection Act and the Egg

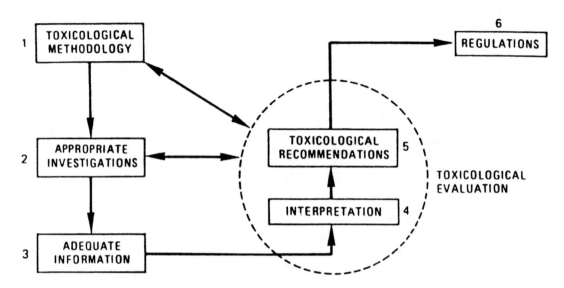

Figure 12–3 Critical Points and Objectives of Toxicological Evaluation of Food Additives. *Source*: Reprinted with permission from G. Vettorazi, Role of International Scientific Bodies, in *International Food Regulation Handbook*, R.D. Middlekauff and P. Shubik, eds., p. 489, 1989, by courtesy of Marcel Dekker, Inc.

Product Inspection Act. Both of these are the responsibility of USDA.

- *The Safe Drinking Water Act.* Passed in 1974, this law gives the FDA authority to regulate bottled drinking water and the Environmental Protection Agency authority to set standards for drinking water supplies.
- *The Nutrition Labeling and Education Act of 1990 (NLEA).* This is an extension of the FDCA and requires that all foods intended for retail sales are provided with nutrition labeling. Mandatory nutrition labeling is not required in most other countries unless a health claim is made.
- Alcoholic beverages come under the authority of the Bureau of Alcohol, Tobacco and Firearms (BATF), an organization unique to the United States. It is noteworthy that some of the labeling requirements for other foods do not apply to alcoholic beverages.

The various U.S. agencies involved in food control and their responsibilities are summarized in Table 12–1. The FDA is the agency primarily responsible for the control of food, and its authority derives from the U.S. Department of Health and Human Services. The USDA is responsible for meat, poultry, and egg products. These activities are carried out by a number of organizations within USDA. The Food Safety and Inspection Service (FSIS), the Food and Nutrition Service (FNS), and the Agricultural Marketing Service (AMS) are all part of this activity.

The Food Additives Amendment to the FDCA (see Chapter 11) recognizes the following three classes of intentional additives:

1. those generally recognized as safe (GRAS)
2. those with prior approval
3. food additives

Table 12–1 Food Safety Responsibilities of 12 U.S. Agencies

Agency	Responsibilities
Food and Drug Administration (FDA)	Ensures safety of all foods except meat, poultry, and egg products. Also, ensures safety of animal drugs and feeds.
Food Safety and Inspection Service (FSIS)	Ensures safety of meat, poultry, and egg products.
Animal and Plant Health Inspection Service (APHIS)	Protects animals and plants from disease and pests or when human health may be affected.
Grain Inspection, Packers and Stockyard Administration (GIPSA)[a]	Inspects grain, rice, and related products for quality and aflatoxin contamination.
Agricultural Marketing Service (AMS)	Grades quality of egg, dairy, fruit, vegetable, meat, and poultry products.
Agricultural Research Service (ARS)	Performs food safety research.
National Marine Fisheries Service (NMFS)	Conducts voluntary seafood inspection program.
Environmental Protection Agency (EPA)	Establishes pesticide tolerance levels.
Centers for Disease Control and Prevention (CDC)	Investigates foodborne disease problems.
Federal Trade Commission (FTC)	Regulates advertising of food products.
U.S. Customs Service (Customs)	Examines/collects food import samples.
Bureau of Alcohol, Tobacco and Firearms (ATF)	Regulates alcoholic beverages.

[a] GIPSA replaced USDA's Grain Inspection Service.

Coloring materials and pesticides on raw agricultural products are covered by other laws. The GRAS list contains several hundred compounds, and the concept of such a list has been the subject of a good deal of controversy (Hall 1975). The concept of a GRAS list is unique to the U.S. regulatory system; there is no equivalent in the legislation of other countries.

An important aspect of U.S. food laws is mandatory nutritional labeling. Nutritional labeling in Canada and Europe is voluntary and only becomes mandatory if a health claim is made.

Another trend in food legislation is the change from prescriptive regulations to the requirement of total quality assurance systems. This means that food industries will be required to adopt HACCP systems (hazard analysis critical control points).

CANADIAN FOOD LAWS

In May 1997 a completely reorganized system of food control in Canada went into effect with the creation of the Canadian Food Inspection Agency (CFIA). The CFIA combines into a single organization food control functions of at least four federal departments. This major change was intended to simplify a complex and fragmented system.

Prior to the formation of CFIA, food control responsibilities were shared by the following federal departments: Health Canada (HC), Agriculture and Agri-food Canada (AAFC), Fisheries and Oceans Canada (FOC), and Industry Canada (IC).

The major law relating to food safety is the Food and Drugs Act and regulations. Until May 1997 HC was responsible for food, health, safety, and nutrition as well as for administering the Food and Drugs Act and regulations (Smith and Jukes 1997). Food labeling regulations are part of Food and Drugs Act and regulations, but enforcement was shared with AAFC. AAFC administered the Meat Inspection Act and the Canadian Agricultural Products Act. FOC administered the Fish Inspection Act. The Consumer Packaging and Labeling Act standardizes the form and manner of essential information on the label of all prepackaged consumer products including foods. The required information includes the common name of the product, the net quantity, and name and address of the company or person responsible for the product. Canadian regulations require this information to be provided in both official languages, English and French.

Because the Food and Drugs Act is criminal law, it applies to all foods sold in Canada. The laws administered by AAFC and FOC are not criminal law and, therefore, do not apply to foods produced and sold within the same province. This is similar to the situation in the United States.

Provinces and municipalities have a certain level of involvement with food control. Provincial regulations are mainly concerned with health issues and the control of certain commodities such as dairy products.

The establishment of the CFIA in 1997 significantly changed the system. CFIA is responsible for the enforcement and/or administration of 11 statutes regulating food,

animal and plant health, and related products. This involves a consolidation of the inspection and animal and plant health services of HC, AAFC, and FOC. A single body, the CFIA, is now responsible for the federal control of all food products.

The establishment of the CFIA is only the first step in a complete overhaul of the Canadian food control system. One of the immediate goals is the development of a Canadian Food Act, and the harmonization of federal and provincial acts. Approximately 77 different federal, provincial, and territorial acts regulate food in Canada. Through the Canadian Food Inspection System (CFIS), a common regulatory base will be developed, as depicted in Figure 12–4. An important aspect of future food regulations will be the reliance on HACCP for safety assurance.

EUROPEAN UNION (EU) FOOD LAWS

The EU at this time involves 15 independent states, and one of the aims of the union is to facilitate trade among member states. To achieve the harmonization of food laws, a program was instituted to develop a common set of food laws. The EU food laws apply in all of the 15 member nations, but the enforcement remains with the individual member states. The EU is governed by three bodies, the European Council (the Council), which consists of ministers from the member countries; the European Parliament, which is formed from members elected in the member countries; and the European Commission (the Commission). The Commission is the working organization that develops laws. The Council approves the laws, and the Parliament has an advisory function. The EU laws, adopted by the Council, may take the following forms:

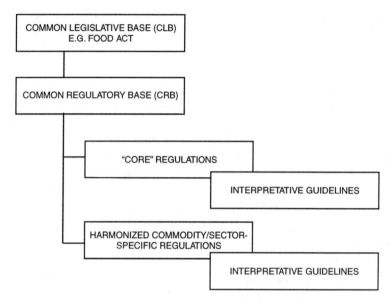

Figure 12–4 Common Regulatory Base Suggested for the Canadian Food System

- *Regulations.* These are directly applied without the need for national measures to implement them.
- *Directives.* These bind member states as to the objectives to be achieved while leaving the national authorities the power to choose the form and means to be used.
- *Decisions.* These are binding in all their aspects upon those to whom they are addressed. A decision may be addressed to any or all member states, to undertakings, or to individuals.
- *Recommendations and opinions.* These are not binding.

The Commission began preparing a comprehensive directive on food additives in 1988. The comprehensive directive on food additives will have two major parts: (1) a list of all the additives and their conditions of use, and (2) the purity criteria of these addi-

tives, together with other specifications such as sampling methods and methods of analysis.

An interesting development in EU food laws is the decision of the Commission to discontinue issuing vertical directives (vertical relates to commodity-specific issues) and to concern itself with horizontal regulations (horizontal relates to general issues across commodities).

An important recent issue concerns the Novel Food Regulation, which is a system of formal, mandatory, premarket evaluation and approval for most innovative foods and food production processes (Huggett and Conzelmann 1997). Novel foods are all foods and food ingredients that have not hitherto been used for human consumption to a significant degree in the EU. The Novel Food Regulation requires additional specific labeling of any characteristic, food property (such as composition, nutritional value, or nutritional effects), or intended use that renders the food no longer equivalent to its conventional counterpart. This regulation, therefore, requires specific labeling for foods produced through genetic engineering. U.S. regulations do not require labeling to describe the use of genetic engineering in developing a new variety of food.

A food safety crisis developed in Europe beginning in the late 1980s and early 1990s. The disease in cattle known as bovine spongiform encephalopathy (BSE), popularly know as mad cow disease, assumed epidemic proportions in England, and more than a million head of cattle had to be destroyed. The problem with BSE is twofold: the pathogenic agent(s) has not been identified, and the transmission to humans is suspected but not proven. There is a human spongiform encephalopathy, Creutzfeldt-Jakob disease (CJD), which is rare and usually affects older people; a new variant (vCJD) affects younger persons (Digiulio et al. 1997). Many unanswered questions about the disease and its possible effect on humans as well as incompetent handling of the issue by politicians created a great deal of unease by the public in Europe. The possibility of transfer of the pathogenic agent via rendered meat and bone meal (MBM) has been suggested.

The BSE scare reinforced the importance of involving consumers and other groups in the consultative process in the development of EU legislation (Figure 12–5).

The EU passed a directive in 1993 requiring all food companies in the EU to implement an effective HACCP system by December 1995. The directive covers not only large and medium-sized businesses but also small companies and even small bakery shops and catering establishments. This directive makes the food manufacturer liable for damages suffered as a result of product defects.

INTERNATIONAL FOOD LAW: CODEX ALIMENTARIUS

The Codex Alimentarius Commission is a joint effort by two organizations of the United Nations—the Food and Agriculture Organization (FAO), headquartered in Rome, and the World Health Organization (WHO), headquartered in Geneva. The Codex Alimentarius Commission is responsible for developing a set of rules known as the Codex Alimentarius (CA). The CA has no legal status, and its adoption is voluntary. Its purpose is to serve as a reference for food safety and standardization on a worldwide basis and to serve as a model for adoption by nations that do not have the resources to develop their own standards. Working under the commission are worldwide general subject committees, a series of worldwide commodity committees, and regional coordinating committees (Figure 12–6).

The fact that CA is a joint effort of FAO and WHO is fortunate and meaningful. Even today in the United States, the FDA is constantly searching to serve both the consuming public and the food industry without creating an impression of being partial to one side or the other.

Since its inception, the CA Commission has produced a large volume of standards, codes of practice, and guidelines. It has developed more than 220 commodity standards, more than 40 codes of practice, a model food law, a code of ethics, and limits for more than 500 food additives. In addition, the commission scrutinized 2,000 pesticides and established limits on 200 of them (Mendez 1993). The work on pesticide residues has resulted in establishing maximum residue limits (MRLs) for a wide range of pesticides in many food commodities. The

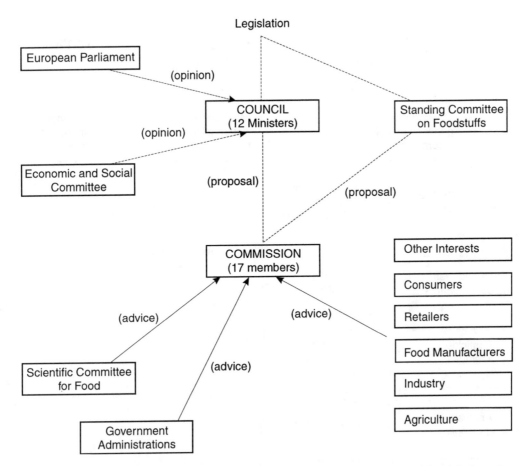

Figure 12–5 The Consultative Process Used in the Development of EU Food Legislation. *Source*: Reprinted with permission from R. Haigh and P. Deboyser, Food Additives and the European Economic Community, in *International Food Regulation Handbook*, R.D. Middlekauff and P. Shubik, eds., 1989, by courtesy of Marcel Dekker, Inc.

commission has studied the safety of a large variety of food additives, considering both toxicology and efficacy. The commission has also been active in the area of the safe use of veterinary drugs and has set maximum residue levels for these compounds. The codes of hygienic/technological practice have been developed for a wide range of food commodities.

An important recent development in the work of the CA is its change in emphasis. It is gradually moving away from the vertical approach to laws (that is, laws relating to a single commodity) to horizontal laws (more broadly based laws that apply across all foods and food commodities). The CA procedure for the elaboration of standards is a complex process involving eight steps. Recently, the CA Commission decided to discontinue work on a standard for mayonnaise. This trend of moving away from vertical standards is not confined to CA. It is also taking place in EU legislation and in many national systems.

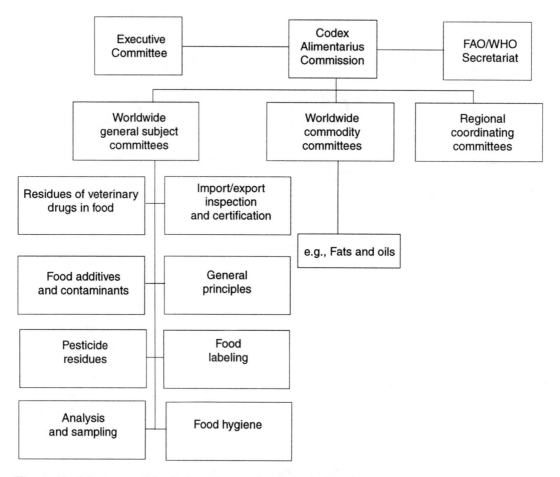

Figure 12–6 Structure of the Codex Alimentarius Commission

The importance of CA standards for international trade increased significantly as a result of the formation in 1995 of the World Trade Organization (WTO), headquartered in Geneva. The WTO is the successor to the General Agreement on Tariffs and Trade (GATT), and most trading nations of the world are members of WTO. One of the main purposes of WTO is to promote trade through the elimination of nontariff trade barriers. In the area of food trade, "health requirements" often were used as a trade barrier. To improve the rules that were in effect during the GATT period, WTO established the Agreement on Sanitary and Phytosanitary Measures, known as the S&P Agreement. This agreement deals with trade in agricultural products of animal and plant origin. Under this agreement, member states of WTO agree to settle trade disputes on the basis of scientific facts and use of the CA standards. A recent case that was brought before the WTO panel involved the refusal by the EU to allow importation of beef originating in the United States that is produced using growth hormones. The United States argued on the basis of scientific evidence that this practice did not result in any detectable residue of the hormones in the beef. The WTO panel has ruled in favor of the U.S. position.

The labeling of food causing severe allergic reaction in some people has resulted in the draft list of foods in May 1996. Severe allergic reactions may cause anaphylaxis and possible death in sensitive persons. The list includes the following foods:

- cereals containing gluten (wheat, rye, barley, oats, spelt, or their hybridized strains and products of these)
- crustacea and products of these
- eggs and egg products

- fish and fish products
- peanuts, soybeans, and products of these
- milk and milk products (lactose included)
- tree nuts and nut products
- sulfite in concentrations of 10 mg/kg or more

This CA proposal is likely to be adopted for inclusion in the food laws of many countries.

The possibility of transfer of allergenicity from an existing food to a new genetically engineered variety is one of the major concerns relating to novel foods produced by genetic engineering. Assessment of the allergenic potential is a critical component of the safety assessment of crops developed by using plant biotechnology (Fuchs and Astwood 1996).

Table 12–2 Comparison of Flour Enrichment Requirements in Canada and the United States

Nutrient	Canada (Flour, White Flour, Enriched Flour, or Enriched White Flour)		United States (Enriched Flour)
	Minimum per 100 g	Maximum per 100 g	Amount per 100 g
Mandatory			
Thiamine	0.44 mg	0.77 mg	0.64 mg
Riboflavin	0.27 mg	0.48 mg	0.40 mg
Niacin	3.5 mg	6.4 mg	5.29 mg
Folic acid	—	—	0.15 mg
Iron	2.9 mg	4.3 mg	4.40 mg
Optional			
Vitamin B_6	0.25 mg	0.31 mg	—
Folic acid	0.04 mg	0.05 mg	—
Pantothenic acid	1.0 mg	1.3 mg	—
Magnesium	150 mg	190 mg	—
Calcium	110 mg	140 mg	211 mg

Source: Reprinted from Health Canada, Health Protection Branch consultative document on draft proposals-subjects: (1) fortification of flour and pasta with folic acid, (2) harmonization of flour enrichment with the United States of America, (3) optional enrichment of flour.

HARMONIZATION

Harmonization of food laws between nations and trading blocks is important for the promotion of international trade. Harmonization does not necessarily mean that food laws have to become identical in different jurisdictions. It may rather be a case of establishing the principle of equivalency. It can be assumed that if the basic principles of the different laws are essentially the same (the assurance of a safe and wholesome food supply) and their enforcement is satisfactory, then products produced in one country can be accepted as complying with the law in another country. Harmonization of food laws between trading partners in free trade groups is important in promoting free trade. The best example is the efforts of harmonizing food laws in the countries of the EU. The establishment of the WTO has increased the importance of the CA and will have an effect in establishing CA as the worldwide reference for settling disputes about nontariff trade barriers.

Other efforts at harmonizing food laws occur between partners of the North American Free Trade Agreement (NAFTA) involving the United States, Canada, and Mexico, and between Australia and New Zealand. A suggested approach to legislative harmonization is depicted in Figure 12–4. The existence of a model food act developed by CA should be an incentive in bringing about harmonization within and between countries.

Food laws developed in various countries reflect the way governments are organized and the state of development of the food industry. Different interpretations of scientific and nutritional information can result in establishment of different standards. This is demonstrated by the comparison of Canadian and U.S. rules on flour enrichment (Table 12–2).

The increasing efforts of harmonization of food laws around the world will continue as international trade in food products continues to grow.

REFERENCES

Camire, M.E. 1996. Blurring the distinction between dietary supplements and foods. *Food Technol.* 50, no. 6: 160.

Concon, J.M. 1988. *Food toxicology. Part A: Principles and concepts. Part B: Contaminants and additives.* New York: Marcel Dekker.

Digulio, K., et al. 1997. International symposium on spongiform encephalopathies: Generating rational policy in the face of public fears. *Trends Food Sci. Technol.* 8: 204–206.

Food Safety Council. 1982. *A proposed food safety evaluation process: Final report of board of trustees.* Washington, DC.

Fuchs, R.L., and J.D. Astwood. 1996. Allergenicity assessment of foods derived from genetically modified plants. *Food Technol.* 50, no. 2: 83–88.

Hall, R.L. 1975. GRAS: Concept and application. *Food Technol.* 29: 48–53.

Huggett, A.C., and C. Conzelmann. 1997. EU regulation on novel foods: Consequences for the food industry. *Trends Food Sci. Technol.* 8: 133–139.

Mendez, G.R., Jr. 1993. Codex Alimentarius promotes international co-operation. *Food Technol.* 47, no. 6: 14.

Norris, B., and A.L. Black. 1989. Food administration in Australia. In *International food regulation handbook,* ed. R.D. Middlekauff and P. Shubik. New York: Marcel Dekker.

Pszczola, D.E. 1998. The ABC's of nutraceutical ingredients. *Food Technol.* 52, no. 3: 30–37.

Reilly, C. 1991. *Metal contamination of food.* 2nd ed. London: Elsevier Applied Science.

Scheuplein, R.J., and W.G. Flamm. 1989. A historical perspective on FDA's use of risk assessment. In *International food regulation handbook*. ed. R.D. Middlekauff and P. Shubik. New York: Marcel Dekker.

Smith, T.M., and D.J. Jukes. 1997. Food control systems in Canada. *Crit. Rev. Food Sci. Nutr.* 37: 299–251.

Vettorazi, G. 1989. Role of international scientific bodies. In *International food regulation handbook*. ed. R.D. Middlekauff and P. Shubik. New York: Marcel Dekker.

Units and Conversion Factors

The International System (SI) of the Units rests upon seven base units and two supplementary units as shown in Table A–1. From the base units, derived units can be obtained to express various quantities such as area, power, force, etc. Some of these have special names as listed in Table A–2. Multiples and submultiples are obtained by using prefixes as shown in Table A–3.

Older units in the metric system and the avoirdupois system are still widely used in the literature, and the information supplied in this appendix is given for convenience in converting these units, Table A–4.

Table A–1 Base Units and Supplementary Units

Quantity	Unit	Symbol
Base Units		
Length	meter	m
Mass	kilogram	kg
Time	second	s
Electric current	ampere	A
Temperature	kelvin	K
Luminous intensity	candela	cd
Amount of substance	mole	mol
Supplementary Units		
Plane angle	radian	rad
Solid angle	steradian	sr

TEMPERATURE

$0\ °C = 273\ °K$
Celsius was formerly called Centigrade
$100\ °C = (100 \times 1.8) + 32\ °F = 212\ °F$
$0\ °C = 32\ °F$
$°F = (°C \times 1.8) + 32$
$°C = (°F - 32) \div 1.8$

Table A–2 Derived Units with Special Names

Quality	Unit	Symbol	Formula
Force	newton	N	$kg.m/s^2$
Energy	joule	J	N.m
Power	watt	W	J/s
Pressure	pascal	Pa	N/m^2
Electrical potential	volt	V	W/A
Electrical resistance	ohm	Ω	V/A
Electrica conductance	siemens	S	$1/\Omega$
Electrical charge	coulomb	C	A.s
Electrical capacitance	farad	F	C/V
Magnetic flux	weber	Wb	V.s
Magnetic flux density	tesla	T	Wb/m^2
Inductance	henry	H	Wb/A
Frequency	hertz	Hz	$2\pi/s$
Illumination	lux	lx	$cd.sr/m^2$
Luminous flux	lumen	lm	cd.sr

Table A–3 Multiples and Submultiples

Multiplier	Exponent Form	Prefix	SI Symbol
1 000 000 000 000	10^{12}	tera	T
1 000 000 000	10^9	giga	G
1 000 000	10^6	mega	M
1 000	10^3	kilo	k
1 00	10^2	hecto	h
10	10^{-1}	deca	da
0.1	10^{-1}	deci	d
0.01	10^{-2}	centi	c
0.001	10^{-3}	milli	m
0.000 001	10^{-6}	micro	μ
0.000 000 001	10^{-9}	nano	n
0.000 000 000 001	10^{-12}	pico	p

Table A–4 Conversion Factors

	To Convert ⟶	Into	Multiply By
0.30480	meters (m)	feet (ft) (= 12 in)	3.28084
0.09290	m^2	ft^2	10.76391
0.02832	m^3	cu ft (ft^3)	35.31467
28.31685	dm^3 - liters (L)	ft^3	0.03531
3.78541	liter (= 1000 cc)	US gal (= 128 US fl. oz)	0.26417
4.54609	liter (= 1000 mL)	Imp gal (= 160 l fl. oz)	0.21997
35.2383	liter (L)	US bushel	0.02838
0.06	m^3/h	L/min	16.66667
1.69901	m^3/h	cu ft/min	0.58858
0.22712	m^3/h	USGPm	4.40287
0.27277	m^3/h	IGPM	3.66615
0.10197	kg (= 1000 g)	Newton (N)	9.80665
0.45359	kg	lb (av) (= 16 oz)	2.20462
0.90718	Metric ton (MT)	Short ton (= 2000 lbs)	1.10231
1.01605	M ton (= 1000 kg)	Long ton (= 2240 lbs)	0.96421
0.01602	kg/dm^3	lb/ft^3	62.42789
0.06895	bar (= 10 N/cm^2)	psi	14.50377
0.001	bar	mbar (= 100 Pascals)	1.0×10^3
0.09807	bar	mH_2O	10.19716
1.33331	mbar	mmHg (torr)	0.75001
33.77125	mbar	inHg (60°F)	0.02961
4.1868	kJ (kiloJoule)	kcal	0.23885
1.05504	kJ	BTU	0.94783
3.6×10^3	kJ	kWh	0.27778×10^{-3}
0.85985×10^3	kcal	kWh	1.16300×10^{-3}
1.16300×10^{-3}	kW	kcal/h	0.85985×10^3
0.29307×10^{-3}	kW (kJ/sec)	BTU/h	3.41219×10^3
0.25199	kcal/h	BTU/h	3.96838
0.746	kW	HP (electr.)	1.34048
0.73550	kW	Metric hp	1.35962
0.0935	foot-candle (ft-c)	lux	10.76
1	centipoise (cp)	mPa.s	1
1	centisokes (cSt)	mm^2/s	1
Multiply By	*Into* ⟵	*To Convert*	

Greek Alphabet

Greek Character	Greek Name	Roman Equivalent
A α	alpha	A a
B β	beta	B b
Γ γ	gamma	G g
Δ δ	delta	D d
E ε, ϵ	epsilon	Ĕ ĕ
Z ζ	zeta	Z z
H η	eta	Ē ē
Θ φ, θ	theta	Th th
I ι	iota	I i
K ℵ, κ	kappa	K k
Λ λ	lambda	L l
M μ	mu	M m
N ν	nu	N n
Ξ ξ	xi	X x
O o	omicron	Ŏ ŏ
Π π, ϖ	pi	P p
P ρ	rho	R r
Σ σ, ς	sigma	S s
T τ	tau	T t
Υ υ	upsilon	Y y
Φ φ, ϕ	phi	Ph ph
X χ	chi	Ch ch
Ψ ψ	psi	Ps ps
Ω ω	omega	Ō ō

Index